TOPICS IN
NUMBER THEORY
VOLUMES I AND II

WILLIAM JUDSON LeVEQUE

DOVER PUBLICATIONS, INC.
Mineola, New York

Published in the United Kingdom by David & Charles, Brunel House,
Forde Close, Newton Abbot, Devon TQ12 4PU.

Bibliographical Note

This Dover edition, first published in 2002, is an unabridged republi-
cation of the work published in two volumes by the Addison-Wesley
Publishing Company, Inc., Reading, Massachusetts, 1956. The Errata
List was prepared especially for this edition by the author.

The two volumes contained in this book are paginated separately (and
have separate tables of contents). Volume II begins following page 202 of
Volume I.

Library of Congress Cataloging-in-Publication Data

LeVeque, William Judson.
    Topics in number theory / by William Judson LeVeque.—Dover ed.
       p. cm.
    Originally published: Reading, Mass., Addison-Wesley Pub., [1956].
    Includes bibliographical references and index.
    ISBN 0-486-42539-8 (pbk. : set)
      1. Number theory. I. Title.

QA241 .L58 2002
512'.7—dc21

2002067433

Manufactured in the United States of America
Dover Publications, Inc., 31 East 2nd Street, Mineola, N.Y. 11501

TOPICS IN
NUMBER THEORY

VOLUME I

To A. J. Kempner

PREFACE

The theory of numbers, one of the oldest branches of mathematics, has engaged the attention of many gifted mathematicians during the past 2300 years. The Greeks, Indians, and Chinese made significant contributions prior to 1000 A.D., and in more modern times the subject has developed steadily since Fermat, one of the fathers of Western mathematics. It is therefore rather surprising that there has never been a strong tradition in number theory in America, although a few men of the stature of L. E. Dickson have emerged to keep the flame alive. But in most American universities the theory of numbers is regarded as a slightly peripheral subject, which has an unusual flavor and unquestioned historical importance, but probably merits no more than a one-term course on the senior or first-year graduate level. It seems to me that this is an inappropriate attitude to maintain toward a subject which is flourishing in European hands, and which has contributed so much to the mathematics of the past and which promises exciting developments in the future. Changing its status is complicated, however, by the paucity of advanced works suitable for use as textbooks in American institutions. There are several excellent elementary texts available, and an ever-increasing number of monographs, mostly European, but to the best of my knowledge no general book designed for a second course in the theory of numbers has appeared since Dickson ceased writing. In Volume II of the present work I have attempted partially to fill this gap.

When I began to write Volume II, the number of introductory texts was very small, and no one of them contained all the information I found occasion to refer to. Since I had already written lecture notes for a first course, there seemed to be some advantage in expanding them into a more complete exposition of the standard elementary topics. Volume I is the result; it is designed to serve either as a self-contained textbook for a one-term course in number theory, or as a preliminary to the second volume. The two volumes together are intended to provide an introduction to some of the important techniques and results of classical and modern number theory; I hope they will prove useful as a first step in the training of students who are or might become seriously interested in the subject.

In view of the diversity of problems and methods grouped together under the name of number theory, it is clearly impossible to write even an introductory treatment which in any sense covers the field completely. My choice of topics was made partly on the basis of my own taste and knowledge, of course, but also more objectively on the grounds of the technical importance of the methods developed or of the results obtained. It was this consideration which led me, for example, to give a standard function-theoretic proof of the Prime Number Theorem in the second volume: the analytic method has proved to be extremely powerful and is applicable to a large variety of problems, so that it must be considered an essential tool in the subject, while the elementary Erdös-Selberg method has found only limited applications, and so for the time being must be regarded as an isolated device, of great interest to the specialist but of secondary importance to the beginner.

In a similar vein, I have on several occasions given proofs which are neither the shortest nor most elegant known, but which seem to me to be the most natural, or to lead to the deepest understanding of the phenomena under consideration. For example, the proof given in Chapter 8, Volume I, of Hurwitz' theorem on the approximation of an irrational number by rationals is perhaps not as elegant as some others known, but of those which make no use of continued fractions, it is the only one I am familiar with which does not require prior knowledge of the special role played by the number $\sqrt{5}$. To my mind, these other proofs are inferior pedagogically, in that they give no hint as to how the student might attack a similar problem.

Most of the material in the first volume is regularly included in various elementary courses, although it would probably be impossible to cover the entire volume in one semester. This allows the instructor to choose topics to suit his taste and, what is even more important for my general purpose, it presents the student with an opportunity for further reading in the subject.

I consider the first volume suitable for presentation to advanced undergraduate and beginning graduate students, insofar as the difficulty of the subject matter is concerned. No technical knowledge is assumed except in Section 3–5 and in Chapter 6, where calculus is used. On the other hand, elementary number theory is by no means easy, and that vaguely defined quality called mathematical maturity is of great value in developing a sound feeling for the subject. I doubt, though, that it should be considered a prerequisite, even if it could be measured; studying number theory is perhaps as good a way as any of acquiring it.

Rather few of the problems occurring at the ends of sections are of the routine computational type; I assume that the student can devise such problems as well as I. It has been my experience that many of those included offer some difficulty to most students. For this reason I have appended hints in profusion, and have indicated by asterisks a few problems that remain more difficult than the average.

The development of continued fractions in the final chapter may be sufficiently novel to warrant a word of explanation. I have chosen to regard as the basic problem that of finding the "good" rational approximations to a real number, and have derived the regular continued fraction as the solution. This procedure seems to me to be pedagogically better than the classical treatment, in which one simply defines a continued fraction and verifies that the convergents have the requisite property. Moreover, this same approach looks promising for the corresponding problem of approximating complex numbers by the elements of a fixed quadratic field, while earlier attempts to define a useful complex continued fraction algorithm have been conspicuously unsuccessful. The idea of associating an interval with each Farey point is derived from work by K. Mahler, who, with J. W. S. Cassels and W. Ledermann, investigated the much more complicated Gaussian case [*Philosophical Transactions of the Royal Society*, A (London) **243**, 585–628 (1951)].

I am grateful to Professors T. Apostol, A. Brauer, B. W. Jones, and K. Mahler for their many constructive criticisms, to Mrs. Edith Fisher for her help in typing the manuscript, and to Mr. Earl Lazerson for his invaluable aid in proofreading.

W. J. L.

Ann Arbor, Michigan
November, 1955

CONTENTS

CHAPTER 1

INTRODUCTION

1-1 What is number theory? In number theory we are concerned with properties of certain of the integers

$$\ldots, -3, -2, -1, 0, 1, 2, 3, \ldots,$$

or sometimes with those properties of the real or complex numbers which depend rather directly on the integers. As in most branches of abstract thought, it is easier to characterize the theory of numbers extensively, by giving a large number of examples of problems which are usually considered parts of number theory, than to define it intensively, by saying that exactly those problems having certain characteristics will be included in the subject. Before considering such a list of types of problems, however, it might be worth while to make an exclusion.

In the opinion of the author, the theory of numbers does not include the axiomatic construction or characterization either of systems of numbers (integers, rational numbers, real numbers, or complex numbers) or of the fundamental operations and relations in these sets. Toward the end of this chapter, a few properties of the integers are mentioned which the student may not have considered explicitly before; aside from these, no properties will be assumed beyond what any high-school pupil knows. It is, of course, quite possible that the student will not have read a logical treatment of elementary arithmetic; if he wishes to do so, he might examine E. Landau's elegant *Foundations of Analysis* (New York: Chelsea Publishing Company, 1951), but he should not expect to find a treatment of this kind here. The contents of such a book are, in a sense, assumed to be known to the reader, but as far as understanding number theory is concerned, this assumption is of little consequence.

The problems treated in number theory can be divided into groups according to a more or less rough classification. First, there are multiplicative problems, concerned with divisibility properties of the integers. It will be proved later that any positive integer n greater than 1 can be represented uniquely, except for the order of the factors,

1

as a product of primes, a *prime* being any integer greater than 1 having no exact divisors except itself and 1. This might almost be termed the Fundamental Theorem of number theory, so manifold and varied are its applications. From the decomposition of n into primes, it is easy to determine the number of divisors of n. This number is called $\tau(n)$ by some writers and $d(n)$ by others; we shall use the former designation. The behavior of $\tau(n)$ is very erratic; the first few values are as follows:

n	$\tau(n)$		n	$\tau(n)$
1	1		13	2
2	2		14	4
3	2		15	4
4	3		16	5
5	2		17	2
6	4		18	6
7	2		19	2
8	4		20	6
9	3		21	4
10	4		22	4
11	2		23	2
12	6		24	8

If $n = 2^m$, the divisors of n are $1, 2, 2^2, \ldots, 2^m$, so that $\tau(2^m) = m + 1$. On the other hand, if n is a prime, then $\tau(n) = 2$. Since, as we shall see, there are infinitely many primes, it appears that the τ-function has arbitrarily large values, and yet has the value 2 for infinitely many n. Many questions might occur to anyone who thinks about the subject for a few moments and studies the above table. For example,

(a) Is it true that $\tau(n)$ is odd if and only if n is a square?

(b) Is it always true that if m and n have no common factor, then $\tau(m)\tau(n) = \tau(mn)$?

(c) Do the arguments of the form 2^m give the relatively largest values of the τ-function? That is, is the inequality

$$\tau(n) \leq \frac{\log n}{\log 2} + 1$$

correct for all n? If not, is there any better upper bound than the trivial one, $\tau(n) \leq n$?

(d) How large is $\tau(n)$ on the average? That is, what can be said about the quantity

$$\frac{1}{N} \sum_{n=1}^{N} \tau(n),$$

as N increases indefinitely?

(e) For large N, approximately how many solutions $n \leq N$ are there of the equation $\tau(n) = 2$? In other words, about how many primes are there among the integers $1, 2, \ldots, N$?

Of the above questions, which are fairly typical problems in multiplicative number theory, the first is very easy to answer in the affirmative. The next three are more difficult; they will be considered in Chapter 6. The last is very difficult indeed. It was conjectured by C. F. Gauss and A. Legendre, two of the greatest of number theorists, that the number $\pi(N)$ of primes not exceeding N is approximately $N/\log N$, in the sense that the relative error

$$\frac{|\pi(N) - (N/\log N)|}{N/\log N} = \left| \frac{\pi(N)}{N/\log N} - 1 \right|$$

is very small when N is sufficiently large. Many years later (1852–54), P. L. Chebyshev showed that if this relative error has any limiting value, it must be zero, but it was not until 1896 that J. Hadamard and C. de la Vallée Poussin finally proved what is now called the Prime Number Theorem, that

$$\lim_{N \to \infty} \frac{\pi(N)}{N/\log N} = 1.$$

In another direction, we have the problems of additive number theory: questions concerning the representability, and the number of representations, of a positive integer as a *sum* of integers of a specified kind. For instance, upon examination it appears that some integers, like $5 = 1^2 + 2^2$ and $13 = 2^2 + 3^2$, are representable as a sum of two squares, while others, like 3 or 12, are not. Which integers are so representable, and how many such representations are there?

A third category might include what are known as Diophantine equations, named after the Greek mathematician Diophantos, who first studied them. These are equations in one or more variables whose solutions are required to be integers, or at any rate rational numbers. For example, it is a familiar fact that $3^2 + 4^2 = 5^2$, which gives us a solution of the Diophantine equation $x^2 + y^2 = z^2$. Giving a particular solution is hardly of interest; what is desired is an explicit formula for *all* solutions. A very famous Diophantine equation is that known as Fermat's equation: $x^n + y^n = z^n$. P. Fermat asserted that this equation has no solution (in nonzero integers, of course) if $n \geq 3$; the assertion has never been proved or disproved for general n. There is at present practically no general theory of Diophantine equations, although there are many special methods, most of which were devised for the solution of particular equations.

Finally, there are problems in Diophantine approximations. For example, given a real number ξ and a positive integer N, find that rational number p/q for which $q \leq N$ and $|\xi - (p/q)|$ is minimal. The proofs that e and π are transcendental also fall in this category. This branch of number theory probably borrows the most from, and contributes the most to, other branches of mathematics.

The theorems of number theory can also be subdivided along entirely different lines—for example, according to the methods used in their proofs. Thus the dichotomies of elementary and nonelementary, analytic and synthetic. A proof is said to be elementary (although not necessarily simple!) if it makes no use of the theory of functions of a complex variable, and synthetic if it does not involve the usual concepts of analysis—limits, continuity, etc. Sometimes, but not always, the nature of the theorem shows that the proof will be in one or another of these categories. For example, the above-mentioned theorem about $\pi(x)$ is clearly a theorem of analytic number theory, but it was not until 1948 that an elementary proof was found. On the other hand, the following theorem, first proved by D. Hilbert, involves in its statement none of the concepts of analysis, yet the only proofs known prior to 1942 were analytic: Given any positive integer k, there is another integer s, depending only on k, such that every positive integer is representable as the sum of at most s kth powers, i.e., such that the equation

$$n = x_1^k + \cdots + x_s^k$$

is solvable in non-negative integers $x_1, \ldots, x_s$ for every n.

It may seem strange at first that the theory of functions of a complex variable is useful in treating arithmetic problems, since there is, *prima facie*, nothing common to the two disciplines. Even after we understand how function theory can be used, we must still reconcile ourselves to the rather disquieting thought that it apparently *must* be used in some problems—that there is, at present, simply no other way to deal with them. What is perhaps not a familiar fact to the general reader is that function theory is only one of many branches of mathematics which are at best only slightly related to number theory, but which enter in an essential way into number-theoretic considerations. This is true, for example, of abstract algebra, probability, Euclidean and projective geometry, topology, the theory of Fourier series, differential equations, and elliptic and other automorphic functions. In particular, it would appear that the rather common subsumption of number theory under algebra involves a certain distortion of the facts.

1-2 Proofs. It is a well-known phenomenon in mathematics that an excessively simple theorem frequently is difficult to prove (although the proof, in retrospect, may be short and elegant) just because of its simplicity. This is probably due in part to the lack of any hint in the statement of the theorem as to the machinery to be used in proving it, and in part to the lack of available machinery. Since many theorems of elementary number theory are of this kind, and since there is considerable diversity in the types of arguments used in their proofs, it might not be amiss to discuss the subject briefly.

First a psychological remark. If we are presented with a rather large number of theorems bearing on the same subject but proved by quite diverse means, the natural tendency is to regard the techniques used in the various proofs as special tricks, each applicable only to the theorem with which it is associated. A technique ceases to be a trick and becomes a method only when it has been encountered enough times to seem natural; correspondingly, a subject may be regarded as a "bag of tricks" if the relative number of techniques to results is too high. Unfortunately, elementary number theory has sometimes been regarded as such a subject. On working longer in the field, however, we find that many of the tricks become methods, and that there is more uniformity than is at first apparent. By making a conscious effort to abstract and retain the germs of the proofs that follow, the reader will begin to see patterns emerging sooner than he otherwise might.

Consider, for example, the assertion that $\tau(n)$ is even unless n is a square, i.e., the square of another integer. A proof of this is as follows: If d is a divisor of n, then so is the integer n/d. If n is not a square, then $d \neq n/d$, since otherwise $n = d^2$. Hence, if n is not a square, its divisors can be paired off into couples d, n/d, so that each divisor of n occurs just once as an element of some one of these couples. The number of divisors is therefore twice the number of couples and, being twice an integer, is even.

The principle here is that when we want to count the integers having a certain property (here "count" may also be replaced by "add"), it may be helpful first to group them in judicious fashion. There are several problems in the present book whose solutions depend on this idea.

In addition to the special methods appropriate to number theory, we shall have many occasions to use two quite general types of proof with which the student may not have had much experience: proofs by contradiction and proofs by induction.

An assertion P is said to have been proved by contradiction if it has been shown that, by assuming P false, we can deduce an assertion Q which is known to be incorrect or which contradicts the assumption that P is false. As an example, consider the theorem (known as early as the time of Euclid) that there are infinitely many prime numbers. To prove this by contradiction, we assume the opposite, namely, that there are only finitely many primes. Let these be $p_1, p_2, \ldots, p_k$; let N be the integer $p_1 p_2 \ldots p_k + 1$; and let Q be the assertion that N is divisible by some prime different from any of $p_1, p_2, \ldots, p_k$. Now N is divisible by some prime p (if N is itself prime, then $p = N$), and N is not divisible by any of the $p_1, p_2, \ldots, p_k$, since each of these primes leaves a remainder of 1 when divided into N. Hence Q is true. Since Q is not compatible with the falsity of the theorem, the theorem is true.

As for proofs by induction, let $P(n)$ be a statement involving an integral variable n; we wish to prove that $P(n)$ is valid for every integer n not less than a particular one, say n_0. The induction principle says that if $P(n_0)$ is valid, and if, for every $n \geq n_0$, one can deduce $P(n+1)$ by assuming that $P(n)$ is valid, then $P(n)$ is valid for every integer $n \geq n_0$. The statement $P(n_0)$ must, of course, be proved independently; usually, though not always, by direct verification. The difficulty, if any, normally lies in showing that $P(n)$ implies $P(n+1)$.

As an example, let us undertake to prove that the formula

$$1 + 2 + \cdots + n = \frac{n(n+1)}{2} \tag{1}$$

is correct, whatever the positive integer n may be. Here $n_0 = 1$. There are three steps:

(a) Show that the formula is correct when $n = 1$. This is trivial here.

(b) Show that if n is an integer for which (1) holds, the same is true of $n+1$. But if (1) holds, then adding $n+1$ to both sides gives

$$1 + 2 + \cdots + n + (n+1) = \frac{n(n+1)}{2} + n + 1 = \frac{(n+1)(n+2)}{2},$$

and this is simply (1) with n replaced by $n+1$, so that $P(n)$ implies $P(n+1)$.

(c) Use the principle of induction to deduce that (1) holds for every positive integer n.

As a second example, consider the Fibonacci sequence

$$1, 2, 3, 5, 8, 13, 21, \ldots,$$

in which every element after the second is the sum of the two numbers immediately preceding it. If we denote by u_n the nth element of this sequence, then the sequence is recursively defined by the conditions

$$u_1 = 1,$$

$$u_2 = 2,$$

$$u_n = u_{n-1} + u_{n-2}, \qquad n \geq 3. \tag{2}$$

We may verify, for as large n as we like, that

$$u_n < (\tfrac{7}{4})^n.$$

To prove this for all positive integral n, we take for $P(n)$ the following statement: the inequalities

$$u_n < (\tfrac{7}{4})^n \qquad \text{and} \qquad u_{n+1} < (\tfrac{7}{4})^{n+1} \tag{3}$$

hold. This is clearly equivalent to the earlier statement. We repeat the three steps.

(a) For $n = 1$, $P(n)$ reduces to the assertion that $1 < \tfrac{7}{4}$ and $2 < (\tfrac{7}{4})^2$.

(b) The induction hypothesis now is that $u_n < (\tfrac{7}{4})^n$ and $u_{n+1} < (\tfrac{7}{4})^{n+1}$, where n is a positive integer. Since $n + 2 \geq 3$, we have that

$$u_{n+2} = u_{n+1} + u_n,$$

by (2). Hence

$$u_{n+2} < (\tfrac{7}{4})^{n+1} + (\tfrac{7}{4})^n = (\tfrac{7}{4})^n(1 + \tfrac{7}{4}) < (\tfrac{7}{4})^n \cdot (\tfrac{7}{4})^2 = (\tfrac{7}{4})^{n+2},$$

and this inequality, together with the induction hypothesis, shows that

$$u_{n+1} < (\tfrac{7}{4})^{n+1} \qquad \text{and} \qquad u_{n+2} < (\tfrac{7}{4})^{n+2},$$

so that $P(n)$ implies $P(n + 1)$.

(c) By the induction principle, it follows that the inequality (3) holds for all positive integral n.

To avoid the artificial procedure of the last proof, it is frequently convenient to use the following formulation of the principle of induction, which can be shown to be equivalent to the first: if $P(n_0)$ is valid, and if for every $n \geq n_0$ the propositions $P(n_0)$, $P(n_0 + 1)$, ..., $P(n)$ together imply $P(n + 1)$, then $P(n)$ is valid for every $n \geq n_0$.

Using this formulation, we could have taken $P(n)$ to be the assertion $u_n < (\tfrac{7}{4})^n$ in the second example.

Besides the principle of induction, we shall have occasion to use three other properties of integers which the reader may not have encountered explicitly before:

(a) Every nonempty set of positive integers (or of non-negative integers) has a smallest element.

(b) If a and b are positive integers, there exists a positive integer n such that $na > b$.

(c) Let n be a positive integer. If a set of $n + 1$ elements is subdivided into n or fewer subsets, in such a way that each element belongs to precisely one subset, then some subset contains more than one element.

These assertions, which are consequences of the underlying axioms, will be assumed without proof.

1. Show that $\tau(n)$ is odd if n is a square.

2. Prove that

(a) $\displaystyle\sum_{m=1}^{n} m = \frac{n(n+1)}{2}$,   (b) $\displaystyle\sum_{m=1}^{n} m^2 = \frac{n(n+1)(2n+1)}{6}$,

(c) $\displaystyle\sum_{m=1}^{n} m^3 = \frac{n^2(n+1)^2}{4}$,

first by induction on n, and second by writing

$$\sum_{m=1}^{n+1} m^k = \sum_{m=1}^{n+1} ((m-1)+1)^k$$

and applying the binomial theorem to the summands on the right. Since the terms $\sum_{1}^{n} m^k$ drop out, this method can be used with $k = 2$ to prove (a), then with $k = 3$ to prove (b), etc.

3. Prove by induction that no two consecutive elements of the Fibonacci sequence $u_1, u_2, \ldots$ have a common divisor greater than 1.

4. Carry out the second proof of the inequality

$$u_n < (\tfrac{7}{4})^n$$

as indicated in the text.

5. Prove by induction that every integer greater than 1 can be represented as a product of primes.

6. Anticipating Theorem 1–1, suppose that every integer can be written in the form $6k + r$, where k is an integer and r is one of the numbers 0, 1, 2, 3, 4, 5.

(a) Show that if $p = 6k + r$ is a prime different from 2 and 3, then $r = 1$ or 5.

(b) Show that the product of numbers of the form $6k + 1$ is of the same form.

(c) Show that there exists a prime of the form $6k - 1 = 6(k - 1) + 5$.

(d) Show that there are infinitely many primes of the form $6k - 1$.

1–3 Radix representation. Although we have assumed a knowledge of the structure of the system of integers, we have said nothing about the method by which we will assign names to the integers. There are, of course, various ways of doing this, of which the Roman and decimal systems are probably the best known. While the decimal system has obvious advantages over Roman numerals, and the advantage of familiarity over any other method, it is not always the best system for theoretical purposes. A rather more general scheme is sometimes convenient, and it is the object of the following two theorems to show that this kind of representation is possible, i.e., that each integer can be given a unique name. **Here, and until Chapter 6, lower-case Latin letters will denote integers.**

THEOREM 1–1. *If a is positive and b is arbitrary, there is exactly one pair of integers q, r such that the conditions*

$$b = qa + r, \qquad 0 \leq r < a, \qquad (4)$$

hold.

Proof: First, we show that (4) has at least one solution.

Consider the set D of integers of the form $b - ua$, where u runs over all integers, positive and nonpositive. For the particular choice

$$u = \begin{cases} -1, & \text{if } b \geq 0, \\ b, & \text{if } b < 0, \end{cases}$$

the number $b - ua$ is non-negative, so that D contains non-negative elements. The subset consisting of the non-negative elements of D has a smallest element. Take r to be this number, and q the value of u which corresponds to it. Then

$$r = b - qa \geq 0, \qquad r - a = b - (q+1)a < 0,$$

so that (4) is satisfied.

To show the uniqueness, assume that also

$$b = q'a + r', \qquad 0 \leq r' < a.$$

Then if $q' < q$,

$$b - q'a = r' \geq b - (q-1)a = r + a \geq a,$$

while if $q' > q$,

$$b - q'a = r' \leq b - (q+1)a = r - a < 0.$$

Hence $q' = q, r' = r.$

THEOREM 1–2. *Let g be greater than 1. Then each a greater than 0 can be represented uniquely in the form*

$$a = c_0 + c_1 g + \cdots + c_n g^n,$$

where c_n is positive and $0 \leq c_m < g$ for $0 \leq m \leq n$.

Proof: We prove the representability by induction on a. For $a = 1$ we have $n = 0$, $c_0 = 1$.

Take a greater than 1 and assume that the theorem is true for 1, 2, $\ldots$, $a - 1$. Since g is larger than 1, the numbers g^0, g^1, g^2, $\ldots$ form an increasing sequence, and any positive integer lies between some pair of successive powers of g. More precisely, there is a unique $n \geq 0$ such that $g^n \leq a < g^{n+1}$. By Theorem 1–1,

$$a = c_n g^n + r, \qquad 0 \leq r < g^n.$$

Here $c_n > 0$, since $c_n g^n = a - r > g^n - g^n = 0$; moreover, $c_n < g$ because $c_n g^n \leq a < g^{n+1}$. If $r = 0$,

$$a = 0 + 0 \cdot g + \cdots + 0 \cdot g^{n-1} + c_n g^n;$$

while if r is positive, the induction hypothesis shows that r has a representation of the form

$$r = b_0 + b_1 g + \cdots + b_t g^t,$$

where b_t is positive and $0 \leq b_m < g$ for $0 \leq m \leq t$. Moreover, $t < n$. Thus

$$a = b_0 + b_1 g + \cdots + b_t g^t + 0 \cdot g^{t+1} + \cdots + 0 \cdot g^{n-1} + c_n g^n.$$

Now use the induction principle.

To prove uniqueness, assume that

$$a = c_0 + c_1 g + \cdots + c_n g^n = d_0 + d_1 g + \cdots + d_r g^r,$$

with $n \geq 0$, $c_n > 0$, and $0 \leq c_m < g$ for $0 \leq m \leq n$, and also $r \geq 0$, $d_r > 0$, and $0 \leq d_m < g$ for $0 \leq m \leq r$. Then, by subtraction, we have

$$0 = e_0 + e_1 g + \cdots + e_s g^s,$$

where $e_m = c_m - d_m$ and where s is the largest value of m for which $c_m \neq d_m$, so that $e_s \neq 0$. If $s = 0$, we have the contradiction $e_0 = e_s = 0$. If $s > 0$ we have, since

$$|e_m| = |c_m - d_m| \leq g - 1$$

and

$$e_s g^s = -(e_0 + \cdots + e_{s-1} g^{s-1}),$$

$$g^s \leq |e_s g^s| = |e_0 + \cdots + e_{s-1} g^{s-1}| \leq |e_0| + \cdots + |e_{s-1}| g^{s-1}$$
$$\leq (g - 1)(1 + g + \cdots + g^{s-1}) = g^s - 1,$$

which is also a contradiction. We conclude that $n = r$ and $c_m = d_m$ for $0 \le m \le n$, and the representation is unique.

By means of Theorem 1–2 we can construct a system of names or symbols for the positive integers in the following way. We choose arbitrary symbols to stand for the *digits* (i.e., the non-negative integers less than g) and replace the number

$$c_0 + c_1 g + \cdots + c_n g^n$$

by the simpler symbol $c_n c_{n-1} \cdots c_1 c_0$. For example, choosing g to be ten, and giving the smaller integers their customary symbols, we have the ordinary decimal system, in which, for example, 2743 is an abbreviation for the value of the polynomial $2x^3 + 7x^2 + 4x + 3$ when x is ten. But there is no reason why we must use ten as the *base*, or *radix*; if we used seven instead, we would write the integer whose decimal representation is 2743 as 10666, since

$$2743 = 6 + 6 \cdot 7 + 6 \cdot 7^2 + 0 \cdot 7^3 + 1 \cdot 7^4.$$

To indicate the base that is being used, we might write a subscript (in the decimal system), so that

$$(2743)_{10} = (10666)_7.$$

Of course, if the radix is larger than $(10)_{10}$, it will be necessary to invent symbols to replace $(10)_{10}, (11)_{10}, \ldots, g - 1$. For example, taking $g = (12)_{10}$ and putting $(10)_{10} = \alpha$, $(11)_{10} = \beta$, we have

$$(14)_{12} + (7)_{12} = (1\beta)_{12}$$

and

$$(31)_{12} \cdot (\alpha)_{12} = (37)_{10} \cdot (10)_{10} = (370)_{10} = (26\alpha)_{12}.$$

PROBLEMS

1. (a) Show that any integral weight less than 2^{n+1} can be weighed using only the standard weights $1, 2, 2^2, \ldots, 2^n$, by putting the unknown weight on one pan of the balance and a suitable combination of standard weights on the other pan.

(b) Prove that no other set of $n + 1$ weights will do this. [*Hint:* Name the weights so that $w_0 \le w_1 \le \cdots \le w_n$. Let k be the smallest index such that $w_k \ne 2^k$ and obtain a contradiction, using the fact that the number of nonempty subsets of a set of $n + 1$ elements is $2^{n+1} - 1$.]

2. Construct the addition and multiplication tables for the duodecimal digits, i.e., the digits in base twelve. Using these tables, evaluate

$$(21\alpha 9)_{12} \cdot (\beta 370)_{12}.$$

3. Let u_1, u_2, ... be the Fibonacci sequence defined in the preceding section.

(a) Prove by induction (or otherwise) that for $n > 0$,

$$u_{n-1} + u_{n-3} + u_{n-5} + \cdots < u_n,$$

the sum on the left continuing so long as the subscripts are positive.

(b) Show that every positive integer can be represented in a unique way in the form $u_{n_1} + u_{n_2} + \cdots + u_{n_k}$, where $k \geq 1$ and $n_{j-1} \geq n_j + 2$ for $j = 2, 3, \ldots, k$.

CHAPTER 2

THE EUCLIDEAN ALGORITHM AND
ITS CONSEQUENCES

2–1 Divisibility. Let a be different from zero, and let b be arbitrary. Then, if there is a c such that $b = ac$, we say that a *divides* b, and write $a|b$ (negation: $a \nmid b$). As usual, the letters involved represent integers.

The following statements are immediate consequences of this definition:

(a) For every $a \neq 0$, $a|0$ and $a|a$. For every b, $\pm 1|b$.

(b) If $a|b$ and $b|c$, then $a|c$.

(c) If $a|b$ and $a|c$, then $a|(bx + cy)$ for each x, y. (If $a|b$ and $a|c$, than a is said to be a *common divisor* of b and c.)

2–2 The Euclidean algorithm and greatest common divisor

THEOREM 2–1. *Given any two integers a, b not both zero, there is a unique integer d such that*

(a) $d > 0$;

(b) $d|a$ *and* $d|b$;

(c) *if* $d_1|a$ *and* $d_1|b$, *then* $d_1|d$.

Since $x|y$ implies that $|x| \leq |y|$, we call the d of Theorem 2–1 the *greatest common divisor* (abbreviated GCD) of a and b, and write $d = (a, b)$.

Proof: First let a and b be positive, and assume that $a \geq b$. Then, by Theorem 1–1, there are unique integers q_1, r_1 such that

$$a = bq_1 + r_1, \qquad 0 \leq r_1 < a.$$

Repeated application of this theorem shows the existence of unique pairs q_2, r_2; q_3, r_3; $\ldots$, such that

$$b = r_1 q_2 + r_2, \qquad 0 \leq r_2 < r_1,$$

$$r_1 = r_2 q_3 + r_3, \qquad 0 \leq r_3 < r_2,$$

$$\vdots \qquad\qquad\qquad \vdots$$

and this may be continued until we reach a remainder, say r_{k+1}, which is zero; the existence of such a k is assured because $r_1, r_2, \ldots$ is a decreasing sequence of non-negative integers. Thus the process terminates:

$$r_{k-3} = r_{k-2}q_{k-1} + r_{k-1}, \qquad 0 \leq r_{k-1} < r_{k-2},$$
$$r_{k-2} = r_{k-1}q_k + r_k, \qquad 0 \leq r_k < r_{k-1},$$
$$r_{k-1} = r_k q_{k+1}.$$

From the last equation we see that $r_k | r_{k-1}$; from the preceding equation, using statement (b) of Section 2–1, we see that $r_k | r_{k-2}$, etc. Finally, from the second and first equations, respectively, we have that $r_k | b$ and $r_k | a$. Thus r_k is a common divisor of a and b. Now let d_1 be any common divisor of a and b. From the first equation, $d_1 | r_1$; from the second, $d_1 | r_2$; etc.; from the equation before the last, $d_1 | r_k$. Thus we can take the d of the theorem to be r_k.

If $a < b$, interchange the names of a and b. If either a or b is negative, find the d corresponding to $|a|$, $|b|$. If a is zero, $(a, b) = |b|$.

If both d_1 and d_2 have the properties of the theorem, then d_1, being a common divisor of a and b, divides d_2. Similarly, $d_2 | d_1$. This clearly implies that $d_1 = d_2$, and the GCD is unique.

The chain of operations indicated by the above equations is known as the Euclidean algorithm; as will be seen, it is the cornerstone of multiplicative number theory. (In general, an algorithm is a systematic procedure which is applied repeatedly, each step depending on the results of the earlier steps. Other examples are the long division algorithm and the square root algorithm.) The Euclidean algorithm is actually quite practicable in numerical cases; for example, if we wish to find the GCD of 4147 and 10672, we have

$$10672 = 4147 \cdot 2 + 2378,$$
$$4147 = 2378 \cdot 1 + 1769,$$
$$2378 = 1769 \cdot 1 + 609,$$
$$1769 = 609 \cdot 2 + 551,$$
$$609 = 551 \cdot 1 + 58,$$
$$551 = 58 \cdot 9 + 29,$$
$$58 = 29 \cdot 2.$$

Hence $(4147, 10672) = 29$.

It is frequently important to know whether two integers a and b have a common factor larger than 1. If they have not, so that $(a, b) = 1$, we say that they are *relatively prime*, or *prime to each other*.

The following properties of the GCD are easily derived either from the definition or from the Euclidean algorithm.

(a) The GCD of more than two numbers, defined as that positive common divisor which is divisible by every common divisor, exists and can be found in the following way. Let there be n numbers $a_1, a_2, \ldots, a_n$, and define

$$D_1 = (a_1, a_2), \qquad D_2 = (D_1, a_3), \qquad \ldots, \qquad D_{n-1} = (D_{n-2}, a_n).$$

Then $(a_1, a_2, \ldots, a_n) = D_{n-1}$.

(b) $(ma, mb) = m(a, b)$, if $m \neq 0$.

(c) If $m|a$ and $m|b$, then $(a/m, b/m) = (a, b)/m$.

(d) If $(a, b) = d$, there exist integers x, y such that $ax + by = d$. (An important consequence of this is that if a and b are relatively prime, there exist x, y such that $ax + by = 1$. Conversely, if there is such a representation of 1, then clearly $(a, b) = 1$.)

(e) If a given integer is relatively prime to each of several others, it is relatively prime to their product. For if $(a, b) = 1$ and $(a, c) = 1$, there are x, y, t, and u such that $ax + by = 1$ and $at + cu = 1$, whence $ax + by(at + cu) = a(x + byt) + bc(yu) = 1$, and therefore $(a, bc) = 1$.

The Euclidean algorithm can be used to find the x and y of property (d). Thus, using the numerical example above, we have

$$
\begin{aligned}
29 &= 551 - 58 \cdot 9 & (58 &= 609 - 551 \cdot 1) \\
&= 551 - 9(609 - 551 \cdot 1) \\
&= 10 \cdot 551 - 9 \cdot 609 & (551 &= 1769 - 2 \cdot 609) \\
&= 10(1769 - 2 \cdot 609) - 9 \cdot 609 \\
&= 10 \cdot 1769 - 29 \cdot 609 & (609 &= 2378 - 1 \cdot 1769) \\
&= 10 \cdot 1769 - 29(2378 - 1 \cdot 1769) \\
&= 39 \cdot 1769 - 29 \cdot 2378 & (1769 &= 4147 - 2378) \\
&= 39(4147 - 2378) - 29 \cdot 2378 \\
&= 39 \cdot 4147 - 68 \cdot 2378 & (2378 &= 10672 - 2 \cdot 4147) \\
&= 175 \cdot 4147 - 68 \cdot 10672,
\end{aligned}
$$

so that $x = 175$, $y = -68$.

PROBLEMS

1. Evaluate (4655, 12075), and express the result as a linear combination of 4655 and 12075, that is, in the form $4655x + 12075y$.

2. Show that if $(a, b) = 1$, then $(a - b, a + b) = 1$ or 2.

3. Show that if $ax + by = m$, then $(a, b)|m$.

4. Show that no cancellation is possible in the fraction

$$\frac{a_1 + a_2}{b_1 + b_2}$$

if $a_1 b_2 - a_2 b_1 = \pm 1$.

5. Show that if $b|a$ and $c|a$, and $(b, c) = 1$, then $bc|a$.

6. Show that if $(b, c) = 1$, then $(a, bc) = (a, b)(a, c)$. [*Hint:* Prove that each member of the last equation divides the other. Use property (d) above, and the preceding problem.]

*7. Show that if $a + b \neq 0$, $(a, b) = 1$, and p is an odd prime, then

$$\left(a + b, \frac{a^p + b^p}{a + b}\right) = 1 \text{ or } p.$$

[*Hint:* If this GCD is d, then $a + b = kd$ and $(a^p + b^p)/(a + b) = ld$. Replace b and $a + b$ in the second equation by their values from the first, apply the binomial theorem, and show that $d|p$.]

*8. In the notation introduced in the proof of Theorem 2–1, show that each nonzero remainder r_m ($m \geq 2$) is less than $r_{m-2}/2$. (Consider separately the cases in which r_{m-1} is less than, equal to, and greater than $r_{m-2}/2$.) Deduce that the number of divisions in the Euclidean algorithm is less than

$$\frac{2 \log b}{\log 2} = 2.88 \ldots \log b,$$

where b is the larger of the two numbers whose GCD is being found. (Here and elsewhere, "log" means the natural logarithm.)

2–3 The Unique Factorization Theorem

THEOREM 2–2.   *Every integer $a > 1$ can be represented as a product of one or more primes.*

Proof: The theorem is true for $a = 2$. Assume it true for 2, 3, 4, $\ldots$, $a - 1$. If a is prime, we are through. Otherwise a has a divisor different from 1 and a, and we have $a = bc$, with $1 < b < a$, $1 < c < a$.

---

*Here and in all problems throughout the book, an asterisk is used to indicate a particularly difficult problem.

The induction hypothesis then implies that

$$b = \prod_{i=1}^{s} p_i', \qquad c = \prod_{i=1}^{t} p_i''$$

with p_i', p_i'' primes, and hence $a = p_1' p_2' \ldots p_s' p_1'' \ldots p_t''$.

Any positive integer which is not prime and which is different from unity is said to be *composite*. Hereafter p will be used to denote a prime number, unless otherwise specified.

THEOREM 2-3. *If $a|bc$ and $(a, b) = 1$, then $a|c$.*

Proof: If $(a, b) = 1$, there are integers x and y such that $ax + by = 1$, or $acx + bcy = c$. But a divides both ac and bc, and therefore divides c.

THEOREM 2-4. *If*

$$p \left| \prod_{m=1}^{n} p_m, \right.$$

then for at least one m, $p = p_m$.

Proof: Suppose that $p|p_1 p_2 \ldots p_n$ but that p is different from any of the $p_1, p_2, \ldots, p_{n-1}$. Then p is relatively prime to each of the $p_1, \ldots, p_{n-1}$, and so is relatively prime to their product. By Theorem 2-3, $p|p_n$, whence $p = p_n$.

THEOREM 2-5 (*Unique Factorization Theorem*). *The representation of $a > 1$ as a product of primes is unique up to the order of the factors.*

Proof: We must show exactly the following. From

$$a = \prod_{m=1}^{n_1} p_m = \prod_{m=1}^{n_2} p_m', \qquad (p_1 \leq p_2 \leq \cdots \leq p_{n_1}; \, p_1' \leq p_2' \leq \cdots \leq p_{n_2}'),$$

it follows that $n_1 = n_2$ and $p_m = p_m'$ for $1 \leq m \leq n_1$.
For $a = 2$ the assertion is true, since $n_1 = n_2 = 1$, $p_1 = p_1' = 2$.
Take $a > 2$ and assume the assertion correct for $2, 3, \ldots, a - 1$.
  (a) If a is prime, $n_1 = n_2 = 1$, $p_1 = p_1' = a$.
  (b) Otherwise $n_1 > 1$, $n_2 > 1$. From

$$p_1' \left| \prod_{m=1}^{n_1} p_m, \qquad p_1 \right| \prod_{m=1}^{n_2} p_m'$$

it follows by Theorem 2–4 that for at least one r and at least one s,

$$p_1' = p_r, \qquad p_1 = p_s'.$$

Since

$$p_1 \leq p_r = p_1' \leq p_s' = p_1,$$

we have $p_1 = p_1'$. Moreover, since $1 < p_1 < a$ and $p_1 | a$, we have

$$1 < \frac{a}{p_1} = \prod_{m=2}^{n_1} p_m = \prod_{m=2}^{n_2} p_m' < a,$$

and hence by the induction hypothesis,

$$n_1 - 1 = n_2 - 1 \qquad \text{and} \qquad p_m = p_m' \qquad \text{for } 2 \leq m \leq n_1.$$

Theorem 2–5, which appears natural enough when one is accustomed to working only with the ordinary integers, assumes greater significance when we encounter more general types of "integers" for which it is not true.

PROBLEMS

1. Show that if the reduced fraction a/b is a root of the equation

$$c_0 x^n + c_1 x^{n-1} + \cdots + c_n = 0,$$

where x is a real variable and $c_0, c_1, \ldots, c_n$ are integers with $c_0 \neq 0$, then $a | c_n$ and $b | c_0$. In particular, show that if k is an integer then $\sqrt[n]{k}$ is rational if and only if it is an integer.

2. The Unique Factorization Theorem shows that each integer $a > 1$ can be written uniquely as a product of powers of distinct primes. If the primes that do not divide a are included in this product with exponents 0, we can write

$$a = \prod_{i=1}^{\infty} p_i^{\alpha_i},$$

where p_i is the ith prime, $\alpha_i \geq 0$ for each i and $\alpha_i = 0$ for sufficiently large i, and the α_i's are uniquely determined by a. Show that if also

$$b = \prod_{i=1}^{\infty} p_i^{\beta_i},$$

then

$$(a, b) = \prod_{i=1}^{\infty} p_i^{\min(\alpha_i, \beta_i)},$$

where $\min(\alpha, \beta)$ is the smaller of α and β. Use this to give a different solution of Problem 6, Section 2–2.

3. Show that the Diophantine equation
$$x^2 - y^2 = N$$
is solvable in non-negative integers x and y if and only if N is odd or divisible by 4. Show further that the solution is unique if and only if N or $N/4$, respectively, is unity or a prime. [*Hint:* Factor the left side.]

4. Show that the following identity is formally correct:
$$\sum_{k=0}^{\infty} \frac{1}{2^{2k}} \cdot \sum_{k=0}^{\infty} \frac{1}{3^{2k}} \cdot \sum_{k=0}^{\infty} \frac{1}{5^{2k}} \cdots = \sum_{n=1}^{\infty} \frac{1}{n^2}.$$

The denominators occurring on the left are the even powers of the primes.

2–4 The linear Diophantine equation.   For simplicity, we consider only the equation in two variables

$$ax + by = c. \tag{1}$$

It is easy to devise a scheme for finding an infinite number of solutions of this equation in case any exist; it can best be explained by means of a numerical example, say $5x + 22y = 18$. Since x is to be an integer, $\frac{1}{5}(18 - 22y)$ must also be integral. Writing

$$x = \frac{18 - 22y}{5} = 3 - 4y + \frac{3 - 2y}{5},$$

we see that $\frac{1}{5}(3 - 2y)$ must also be an integer, say z. This gives

$$z = \frac{3 - 2y}{5}, \qquad 2y + 5z = 3.$$

We now repeat the argument, solving as before for the unknown which has the smaller coefficient:

$$y = \frac{3 - 5z}{2} = 1 - 2z + \frac{1 - z}{2},$$

$$\frac{1 - z}{2} = t, \qquad z = 1 - 2t.$$

Clearly, z will be an integer for any integral t, and we have

$$y = \frac{3 - 5(1 - 2t)}{2} = -1 + 5t,$$

$$x = \frac{18 - 22(-1 + 5t)}{5} = 8 - 22t.$$

Moreover, it is easily seen that any solution x, y of the original equation must be of this form, so that we have a *general solution* of the equation.

The same idea could be applied in the general case, but it is somewhat simpler to adopt a different approach. First of all, it should be noticed that (1) has no solution unless $d|c$, where $d = (a, b)$, and that if this requirement is satisfied, we can divide through in (1) by d to get a new equation

$$a'x + b'y = c', \qquad (2)$$

where now $(a', b') = 1$. We now use property (d) of Section 2–2 to assert the existence of numbers x_0', y_0' such that

$$a'x_0' + b'y_0' = 1,$$

so that $c'x_0'$, $c'y_0'$ is a solution of (2). Put $c'x_0' = x_0$, $c'y_0' = y_0$. If t is any integer, we have

$$a'(x_0 + b't) + b'(y_0 - a't) = a'x_0 + b'y_0 = c',$$

so that $x_0 + b't$, $y_0 - a't$ is a solution of (2) for each t. Finally, if x_1, y_1 is *any* solution of (2), we have

$$a'x_0 + b'y_0 = c', \qquad a'x_1 + b'y_1 = c',$$

and, by subtraction,

$$a'(x_0 - x_1) + b'(y_0 - y_1) = 0.$$

Thus $a'|(y_0 - y_1)$, $y_0 - y_1 = a't_1$, and $b'|(x_0 - x_1)$, $x_0 - x_1 = b't_2$. This gives $x_1 = x_0 - b't_2$, $y_1 = y_0 - a't_1$, and, requiring that these numbers satisfy (2), we have $t_2 = -t_1$. Hence every solution of (2) is of the form $x_0 + b't$, $y_0 - a't$, and every such pair constitutes a solution. Since every solution of (1) is a solution of (2) and conversely, we have the following theorem.

THEOREM 2–6.   *A necessary and sufficient condition that the equation*

$$ax + by = c$$

have a solution x, y in integers is that $d|c$, where $d = (a, b)$. If there is one solution, there are infinitely many; they are exactly the numbers of the form

$$x = x_0 + \frac{b}{d}t, \qquad y = y_0 - \frac{a}{d}t,$$

where t is an arbitrary integer.

There are various ways of getting a particular solution. Sometimes one can be found by inspection; if not, the method explained at the beginning of the section may be used or, what is almost the same thing, the Euclidean algorithm may be applied to find a solution of the equation which results from dividing the original equation through by (a, b). The latter process of successively eliminating the remainders in the Euclidean algorithm can be systematized, but this we shall not do at present. (See Section 9–2.)

PROBLEMS

1. Find a general solution of the linear Diophantine equation

$$2072x + 1813y = 2849.$$

2. Find all solutions of $19x + 20y = 1909$ with $x > 0$, $y > 0$.

3. Let m and n be positive integers, with $m \leq n$, and let $x_0, x_1, \ldots, x_k$ be all the distinct numbers among the two sequences

$$\frac{0}{m}, \frac{1}{m}, \ldots, \frac{m}{m} \quad \text{and} \quad \frac{0}{n}, \frac{1}{n}, \ldots, \frac{n}{n},$$

arranged so that $x_0 < x_1 < \cdots < x_k$. Describe k as a function of m and n. What is the shortest distance between successive x's?

*4. Let a and b be positive relatively prime integers. Then for certain non-negative integers n (which we shall refer to briefly as the *representable* integers), the equation $ax + by = n$ has a solution with $x \geq 0$, $y \geq 0$, while for other n it may not have. For example, if $n = 0, 3, 5$, or 6, or if $n \geq 8$, then $3x + 5y = n$ has such a solution. Show that this example is typical, in the following sense:

(a) There is always a number $N(a, b)$ such that for all $n \geq N(a, b)$, n is representable. (It may be helpful to combine the theory of the present section with the elementary analytic geometry of the line $ax + by = c$, interpreting x and y in the latter case as real variables. Note that so far it is only the existence of $N(a, b)$ that is in question, and not its size.)

(b) The minimal value of $N(a, b)$ is always $(a - 1)(b - 1)$.

(c) Exactly half the integers up to $(a - 1)(b - 1)$ are representable.

2–5 The least common multiple

THEOREM 2–7.   *The number* $\langle a, b \rangle = \dfrac{|ab|}{(a, b)}$ *has the following properties:*   (a) $\langle a, b \rangle \geq 0$;   (b) $a|\langle a, b \rangle$, $b|\langle a, b \rangle$;   (c) *If* $a|m$ *and* $b|m$, *then* $\langle a, b \rangle|m$.

Proof: (a) Obvious.

(b) Since $(a, b)|b$, we can write

$$\langle a, b \rangle = |a| \cdot \frac{|b|}{(a, b)} \, ,$$

and hence $a|\langle a, b \rangle$. Similarly,

$$\langle a, b \rangle = |b| \cdot \frac{|a|}{(a, b)} \, ,$$

and so $b|\langle a, b \rangle$.

(c) Let

$$m = ra = sb,$$

and put $d = (a, b)$, $a = a_1 d$, $b = b_1 d$. Then

$$m = r a_1 d = s b_1 d;$$

thus $a_1 | s b_1$, and since $(a_1, b_1) = 1$, it must be that $a_1 | s$. Thus $s = a_1 t$, and

$$m = t a_1 b_1 d = t \frac{ab}{d} \cdot$$

Because of the properties listed in Theorem 2–7, the number $\langle a, b \rangle$ is called the *least common multiple* (LCM) of a and b. The definition is easily extended to the case of more than two numbers, just as for the GCD. It is useful to remember that

$$ab = \pm (a, b)\langle a, b \rangle.$$

PROBLEMS

1. In the notation of Problem 2, Section 2–3, show that

$$\langle a, b \rangle = \prod_{i=1}^{\infty} p_i^{\max(\alpha_i, \beta_i)},$$

where max (α, β) is the larger of α and β.

*2. Show that

$$\min (\alpha, \max (\beta, \gamma)) = \max (\min (\alpha, \beta), \min (\alpha, \gamma)).$$

(By symmetry, one may suppose $\beta \geq \gamma$.) Deduce that

$$(a, \langle b, c \rangle) = \langle (a, b), (a, c) \rangle.$$

CHAPTER 3

CONGRUENCES

3-1 Introduction. The problem of solving the Diophantine equation $ax + by = c$ is just that of finding an x such that ax and c leave the same remainder when divided by b, since then $b|(c - ax)$ and we can take $y = (c - ax)/b$. As we shall see, there are many other instances also in which a comparison must be made of the remainders after dividing each of two numbers a and b by a third, say m. Of course, if the remainders are the same, then $m|(a - b)$, and conversely, and this might seem to be an adequate notation. But, as Gauss noticed, the following, for most purposes, is more suggestive: if $m|(a - b)$, then we write $a \equiv b \pmod{m}$, and say that a is *congruent to b modulo m*.

The use of the symbol "$\equiv$" is suggested by the similarity of the relation we are discussing to ordinary equality. Each of these two relations is an example of an *equivalence relation;* i.e., a relation $\mathcal{R}$ between elements of a set, such that if a and b are arbitrary elements, either a stands in the relation $\mathcal{R}$ to b (more briefly, $a \mathcal{R} b$) or not, and having the following properties:

(a) $a \mathcal{R} a$.

(b) If $a \mathcal{R} b$, then $b \mathcal{R} a$.

(c) If $a \mathcal{R} b$ and $b \mathcal{R} c$, then $a \mathcal{R} c$.

These are called the *reflexive, symmetric* and *transitive* properties, respectively. That ordinary equality between numbers is an equivalence relation is obvious (or it may be taken as an axiom): either $a = b$ or $a \neq b$; $a = a$; if $a = b$, then $b = a$; if $a = b$ and $b = c$, then $a = c$.

THEOREM 3-1. *Congruence modulo a fixed number m is an equivalence relation.*

Proof: (a) $m|(a - a)$, so that $a \equiv a \pmod{m}$.

(b) If $m|(a - b)$, then $m|(b - a)$; if $a \equiv b \pmod{m}$, then $b \equiv a \pmod{m}$.

(c) If $m|(a - b)$ and $m|(b - c)$, then $a - b = km$, $b - c = lm$, say, so that $a - c = (k + l)m$; if $a \equiv b \pmod{m}$ and $b \equiv c \pmod{m}$, then $a \equiv c \pmod{m}$.

24

Since we shall have occasion later to use several other equivalence relations, we pause to show a simple but important property enjoyed by all such relations. If $\mathcal{R}$ is an equivalence relation with respect to a set S, then corresponding to each element a of S there is a subset S_a of S which consists of exactly those elements of S which are equivalent to a, so that b is in S_a if and only if $a \mathcal{R} b$. Now if $a \mathcal{R} b$, then the sets S_a and S_b are identical: if c is in S_b, then $c \mathcal{R} b$, and since $a \mathcal{R} b$, also $c \mathcal{R} a$, so that c is in S_a. If, on the other hand, a is not equivalent to b, then S_a and S_b are disjoint; that is, they have no element in common. For if c is in S_a and in S_b, then $c \mathcal{R} a$ and $c \mathcal{R} b$, which entails $a \mathcal{R} b$. These disjoint sets, which jointly exhaust S, are called *equivalence classes;* an element of an equivalence class is sometimes called a *representative* of the class, and a *complete system of representatives* is any subset of S which contains exactly one element from each equivalence class.

Section 3–3 provides examples of all these notions, with somewhat different terminology.

PROBLEM

Decide whether each of the following is an equivalence relation. If it is, describe the equivalence classes.

(a) Congruence of triangles.

(b) Similarity of triangles.

(c) The relations "$\neq$", "$>$", and "$\geq$", relating real numbers.

(d) Parallelism of lines.

(e) Having the same mother.

(f) Having a parent in common.

3–2 Elementary properties of congruences. One reason for the superiority of the congruence notation is that congruences can be combined in much the same way as can equations.

THEOREM 3–2. *If $a \equiv b \,(\mathrm{mod}\, m)$ and $c \equiv d \,(\mathrm{mod}\, m)$, then $a + c \equiv b + d \,(\mathrm{mod}\, m)$, $ac \equiv bd \,(\mathrm{mod}\, m)$, and $ka \equiv kb \,(\mathrm{mod}\, m)$ for every integer k.*

Proof: These statements follow immediately from the definition. For if

$$m \,|\, (a - b) \qquad \text{and} \qquad m \,|\, (c - d),$$

then $m \,|\, (a - b + c - d) \qquad \text{and} \qquad m \,|\, ((a + c) - (b + d)).$

If $m|(a - b)$, then $m|k(a - b)$. Finally, if $m|(a - b)$ and $m|(c - d)$, then $m|(a - b)(c - d)$. But

$$(a - b)(c - d) = ac - bd + b(d - c) + d(b - a),$$

so that also $m|(ac - bd)$.

THEOREM 3–3. *If $f(x)$ is a polynomial with integral coefficients, and $a \equiv b \pmod{m}$, then $f(a) \equiv f(b) \pmod{m}$.*

Proof: Let $\qquad f(x) = c_0 + c_1 x + \cdots + c_n x^n.$

If $a \equiv b \pmod{m}$, then for every non-negative integer j,

$$a^j \equiv b^j \pmod{m},$$

and $\qquad\qquad c_j a^j \equiv c_j b^j \pmod{m},$

by Theorem 3–2. Adding these last congruences for $j = 0, 1, \ldots, n$, we have the theorem.

The situation is a little more complicated when we consider dividing both sides of a congruence by an integer. We cannot deduce from $ka \equiv kb \pmod{m}$ that $a \equiv b \pmod{m}$, for it may be that part of the divisibility of $ka - kb = k(a - b)$ by m is accounted for by the presence of the factor k. What is clearly necessary is that the part of m which does not divide k should divide $a - b$.

THEOREM 3–4. *If $ka \equiv kb \pmod{m}$ and $(k, m) = d$, then*

$$a \equiv b \left(\mod \frac{m}{d} \right).$$

Proof: Theorem 2–3.

PROBLEMS

1. Let $\qquad f(x) = a_0 x^n + a_1 x^{n-1} + \cdots + a_n,$

where $a_0, \ldots, a_n$ are integers. Show that if d consecutive values of f (i.e., values for consecutive integers) are all divisible by the integer d, then $d|f(x)$ for all integral x. Show by an example that this sometimes happens with $d > 1$ even when $(a_0, \ldots, a_n) = 1$.

2. In Theorem 3–3 take $a = 10$, $b = 1$, $m = 9$ to deduce the rule that an integer is divisible by 9 if and only if this is true of the sum of its digits. What is the corresponding rule for divisibility by 11? Use the fact that $7 \cdot 11 \cdot 13 = 1001$ to obtain a test for divisibility by any of the integers 7, 11, or 13.

3–3 Residue classes and Euler's φ-function. When dealing with congruences modulo a fixed integer m, the set of all integers breaks down into m classes, such that any two elements of the same class are congruent and two elements from two different classes are incongruent. For many purposes it is completely immaterial which element of one of these *residue classes* is used; for example, Theorem 3–3 shows this to be the case when one considers the values modulo m of a polynomial with integral coefficients. In these cases it suffices to consider an arbitrary set of representatives of the various residue classes; that is, a set consisting of one element of each residue class. Such a set a_1, a_2, . . . , a_m, called a *complete residue system modulo m*, is characterized by the following properties.

(a) If $i \neq j$, then $a_i \not\equiv a_j \pmod{m}$.

(b) If a is any integer, there is an index i with $1 \leq i \leq m$ for which $a \equiv a_i \pmod{m}$.

Examples of complete residue systems $\pmod{m}$ are the set of integers $0, 1, 2, \ldots, m - 1$, and the set $1, 2, \ldots, m$. The elements of a complete residue system need not be consecutive integers, however; for $m = 5$ we could take $1, 22, 13, -6, 2500$ as such a set.

THEOREM 3–5. *If $a_1, a_2, \ldots, a_m$ is a complete residue system $\pmod{m}$ and $(k, m) = 1$, then also $ka_1, ka_2, \ldots, ka_m$ is a complete residue system $\pmod{m}$.*

Proof: We show directly that properties (a) and (b) above hold for this new set.

(a) If $ka_i \equiv ka_j \pmod{m}$, then by Theorem 3–4, $a_i \equiv a_j \pmod{m}$, whence $i = j$.

(b) Theorem 2–6 shows that if $(k, m) = 1$, the congruence $kx \equiv a \pmod{m}$ has a solution for any fixed a. Let a solution be x_0. Since $a_1, \ldots, a_m$ is a complete residue system, there is an index i such that $x_0 \equiv a_i \pmod{m}$. Hence $kx_0 \equiv ka_i \equiv a \pmod{m}$.

The reason that we use the adjective "complete" when speaking of a residue system is that there is another kind which is frequently useful, called a *reduced residue system*. This is a set of integers $a_1, \ldots, a_h$, incongruent $\pmod{m}$, such that if a is any integer prime to m, there is an index i, $1 \leq i \leq h$, for which $a \equiv a_i \pmod{m}$. In other words, a reduced residue system is a set of representatives, one from each of the residue classes containing integers prime to m. (Clearly,

$(a, m) = (b, m)$ if $a \equiv b \pmod{m}$, since then $m | (a - b)$, so that $(a, m) | (a - b)$, and hence $(a, m) | b$; this implies that $(a, m) | (b, m)$, and also, by symmetry, that $(b, m) | (a, m)$.) The number h is the number of positive integers not exceeding m and prime to m. This function of m is customarily designated by $\varphi(m)$, and is called *Euler's φ-function* or the *totient* of m.

THEOREM 3–6. *If $a_1, \ldots, a_{\varphi(m)}$ is a reduced residue system* $(\bmod\ m)$ *and* $(k, m) = 1$, *then also* $ka_1, \ldots, ka_{\varphi(m)}$ *is a reduced residue system* $(\bmod\ m)$.

The proof is exactly parallel to that of Theorem 3–5.

Euler's φ-function has many interesting properties and, as we shall see, it occurs repeatedly in number-theoretic investigations.

THEOREM 3–7. *If* $(m, n) = 1$, *then* $\varphi(mn) = \varphi(m)\varphi(n)$.

(A function with this property is called a *multiplicative* function. For another example, see Problem 6, Section 2–2.)

Proof: Take integers m, n with $(m, n) = 1$, and consider the numbers of the form $mx + ny$. If we can so restrict the values which x and y assume that these numbers form a reduced residue system $(\bmod\ mn)$, there must be $\varphi(mn)$ of them. But also their number is then the product of the number of values which x assumes and the number of values which y assumes. Clearly, in order for $mx + ny$ to be prime to m, it is necessary that $(m, y) = 1$, and likewise we must have $(n, x) = 1$. Conversely, if these last two conditions are satisfied, then $(mx + ny, mn) = 1$. Hence let x range over a reduced residue system $(\bmod\ n)$, say $x_1, \ldots, x_{\varphi(n)}$, and let y run over a reduced residue system $(\bmod\ m)$, say $y_1, \ldots, y_{\varphi(m)}$. If for some indices i, j, k, l we have

$$mx_i + ny_j \equiv mx_k + ny_l \pmod{mn},$$

then

$$m(x_i - x_k) + n(y_j - y_l) \equiv 0 \pmod{mn}.$$

Since divisibility by mn implies divisibility by m, we have

$$m(x_i - x_k) + n(y_j - y_l) \equiv 0 \pmod{m},$$

$$n(y_j - y_l) \equiv 0 \pmod{m},$$

$$y_j \equiv y_l \pmod{m},$$

whence $j = l$. Similarly, $i = k$. Thus the numbers $mx + ny$ so formed are incongruent (mod mn). Now let a be any integer prime to mn; in particular, $(a, m) = 1$ and $(a, n) = 1$. Then Theorem 2–6 shows that there are integers X, Y (not necessarily in the chosen reduced residue systems) such that $mX + nY = a$, whence also $mX + nY \equiv a$ (mod mn). But there is an x_i such that $X \equiv x_i$ (mod n), and there is a y_j such that $Y \equiv y_j$ (mod m). This means that there are integers k, l such that $X = x_i + kn$, $Y = y_j + lm$. Hence

$$mX + nY = m(x_i + kn) + n(y_j + lm) \equiv mx_i + ny_j = a \ (\text{mod } mn).$$

Hence as x and y run over fixed reduced residue systems (mod n) and (mod m) respectively, $mx + ny$ runs over a reduced residue system (mod mn), and the proof is complete.

THEOREM 3–8. $$\varphi(m) = m \prod_{p \mid m} \left(1 - \frac{1}{p}\right),$$

where the notation indicates a product over all the distinct primes which divide m.

Proof: By Theorem 3–7, if $m = \prod_{i=1}^{r} p_i^{\alpha_i}$,
then

$$\varphi(m) = \prod_{i=1}^{r} \varphi(p_i^{\alpha_i}).$$

But we can easily evaluate $\varphi(p^\alpha)$ directly; all the positive integers not exceeding p^α are prime to p^α except the multiples of p, and there are just $p^{\alpha-1}$ of these. Hence

$$\varphi(p_i^{\alpha_i}) = p_i^{\alpha_i} - p_i^{\alpha_i - 1} = p_i^{\alpha_i} \left(1 - \frac{1}{p_i}\right),$$

and so

$$\varphi(m) = \prod_{i=1}^{r} p_i^{\alpha_i} \left(1 - \frac{1}{p_i}\right) = \prod_{i=1}^{r} p_i^{\alpha_i} \cdot \prod_{i=1}^{r} \left(1 - \frac{1}{p_i}\right)$$

$$= m \prod_{p \mid m} \left(1 - \frac{1}{p}\right).$$

For example, the integers 1, 5, 7, 11 are all those which do not exceed 12 and are prime to 12, and

$$\varphi(12) = 12(1 - \tfrac{1}{2})(1 - \tfrac{1}{3}) = 4.$$

THEOREM 3–9. $\sum_{d|n} \varphi(d) = n$.

Proof: Let $d_1, \ldots, d_k$ be the positive divisors of n. We separate the integers between 1 and n inclusive into classes $C(d_1), \ldots, C(d_k)$, putting an integer into the class $C(d_i)$ if its GCD with n is d_i. The number of elements in $C(d_i)$ is then

$$\sum_{\substack{a \leq n \\ (a,n)=d_i}} 1,$$

and since every integer up to n is in exactly one of the classes,

$$\sum_{d_i|n} \sum_{\substack{a \leq n \\ (a,n)=d_i}} 1 = n.$$

The number of integers a such that $a \leq n$ and $(a, n) = d_i$ is exactly equal to the number of integers b such that $b \leq n/d_i$ and $(b, n/d_i) = 1$; in fact, multiplying the b's by d_i, we get the a's. But from the definition of the Euler function, the number of b's is clearly $\varphi(n/d_i)$. Thus

$$\sum_{d_i|n} \varphi\left(\frac{n}{d_i}\right) = n,$$

which is equivalent to the theorem, since, as d_i runs over the divisors of n, n/d_i also runs over these divisors, but in reverse order.

To illustrate the theorem and its proof, take $n = 12$. Then

$$\varphi(1) + \varphi(2) + \varphi(3) + \varphi(4) + \varphi(6) + \varphi(12)$$
$$= 1 + 1 + 2 + 2 + 2 + 4 = 12,$$

$C(1) = \{1, 5, 7, 11\}, \qquad C(2) = \{2, 10\}, \qquad C(3) = \{3, 9\},$

$C(4) = \{4, 8\}, \qquad C(6) = \{6\}, \qquad C(12) = \{12\}.$

PROBLEMS

*1. Prove that if $(a, b) = d$, then

$$\varphi(ab) = \frac{d\varphi(a)\varphi(b)}{\varphi(d)}.$$

2. Show that if $n > 1$, then the sum of the positive integers less than n and prime to it is

$$\frac{n\varphi(n)}{2}.$$

[*Hint:* If m satisfies the conditions, so does $n - m$.]

3. Show that if $d|n$, then $\varphi(d)|\varphi(n)$.

4. Let n be positive. Show that any solution of the equation

$$\varphi(x) = 4n + 2$$

is of one of the forms p^α or $2p^\alpha$, where p is a prime of the form $4s - 1$. [*Hint:* Use the factorization of $\varphi(x)$ as given in Theorem 3–8.]

*5. Let $f(x)$ be a polynomial with integral coefficients, and let $\psi(n)$ denote the number of values

$$f(0), f(1), \ldots, f(n - 1)$$

which are prime to n.

(a) Show that ψ is multiplicative:

$$\psi(mn) = \psi(m) \cdot \psi(n) \qquad \text{if } (m, n) = 1.$$

(b) Show that

$$\psi(p^\alpha) = p^{\alpha-1}(p - b_p),$$

where b_p is the number of integers $f(0), f(1), \ldots, f(p - 1)$ which are divisible by the prime p.

6. How many fractions r/s are there satisfying the conditions

$$(r, s) = 1, \qquad 0 \leq r < s \leq n?$$

3–4 Linear congruences. Because of the analogy between congruences and equations, it is natural to ask about the solution of congruences involving one or more (integral) unknowns. In the case of an algebraic congruence $f(x) \equiv 0 \pmod{m}$, where $f(x)$ is a polynomial in x with integral coefficients, we see by Theorem 3–3 that if $x = a$ is a solution, so is every element of the residue class containing a. For this reason it is customary, for such congruences, to list only the solutions between 0 and $m - 1$, inclusive, with the understanding that any x congruent to one of those listed is also a solution. Similarly, when mention is made of the number of roots of a certain congruence, it is actually the number of residue classes that is meant.

The simplest case to treat is the linear congruence in one unknown; that is, the congruence

$$ax \equiv b \pmod{m}.$$

As we have already noticed, this is equivalent to the linear Diophantine equation

$$ax - my = b,$$

and by Theorem 2–6 this equation is solvable if and only if $(a, m)|b$.

If it is solvable, and if x_0, y_0 is a solution, then a general solution is

$$x \equiv x_0 \left(\operatorname{mod} \frac{m}{d} \right), \qquad y \equiv y_0 \left(\operatorname{mod} \frac{a}{d} \right),$$

where $d = (a, m)$. Among the numbers x satisfying the first of these congruences, the numbers

$$x_0, \; x_0 + \frac{m}{d}, \; x_0 + \frac{2m}{d}, \; \cdots, \; x_0 + \frac{(d-1)m}{d}$$

are incongruent $(\operatorname{mod} m)$, while every other such x is congruent $(\operatorname{mod} m)$ to one of these. Hence we have the following theorem:

THEOREM 3–10. *A necessary and sufficient condition that the congruence*

$$ax \equiv b \; (\operatorname{mod} m)$$

be solvable is that $(a, m) | b$. *If this is the case, there are exactly* (a, m) *solutions* $(\operatorname{mod} m)$.

While Theorem 3–10 gives assurance of the existence of a solution under appropriate circumstances and predicts the number of such solutions, it says nothing about finding them. For this purpose the simplest procedure, if no solution can be found by inspection, is to convert the congruence to an equation and solve by the method given at the beginning of Section 2–4.

Consider, for example, the congruence

$$34x \equiv 60 \; (\operatorname{mod} 98).$$

Since $(34, 98) = 2$ and $2 | 60$, there are just two solutions, to be found from

$$17x \equiv 30 \; (\operatorname{mod} 49).$$

This is equivalent to $17x - 49y = 30$, and we get

$$x = \frac{49y + 30}{17} = 3y + 2 - \frac{2y + 4}{17}, \qquad t = \frac{2y + 4}{17},$$

$$y = \frac{17t - 4}{2} = 8t - 2 + \frac{t}{2}, \qquad z = \frac{t}{2},$$

$$t = 2z.$$

Take $z = 0$; then $t = 0$, $y = -2$, $x = -4$. Hence

$$x \equiv -4 \ (\mathrm{mod}\ 49),$$

and the two solutions of the original congruence are

$$x \equiv -4,\ 45 \ (\mathrm{mod}\ 98).$$

The solution of a linear congruence in more than one unknown can be effected by the successive solution of a (usually large) number of congruences in a single unknown. Consider the congruence

$$a_1 x_1 + a_2 x_2 + \cdots + a_n x_n \equiv c \ (\mathrm{mod}\ m).$$

The obviously necessary condition for solvability, that $(a_1, \ldots, a_n, m)$ should divide c, is also sufficient, just as in the former case. For, assuming it satisfied, we can divide through by $(a_1, \ldots, a_n, m)$ to get

$$a_1' x_1 + \cdots + a_n' x_n \equiv c' \ (\mathrm{mod}\ m'),$$

where now $(a_1', \ldots, a_n', m') = 1$. If $(a_1', \ldots, a_{n-1}', m') = d'$, we must have

$$a_n' x_n \equiv c' \ (\mathrm{mod}\ d');$$

since $(a_n', d') = 1$, this has just one solution (mod d'). Thus there are m'/d' numbers x_n with $0 \le x_n < m'$ satisfying this congruence. Substituting these into the preceding congruence, we get m'/d' congruences in $n - 1$ unknowns, and the process can be repeated.

As an example, consider the congruence

$$2x + 7y \equiv 5 \ (\mathrm{mod}\ 12).$$

Here $(2, 7, 12) = 1$. Since $(2, 12) = 2$, we must have

$$7y \equiv 5 \ (\mathrm{mod}\ 2),$$

which clearly gives $y \equiv 1 \ (\mathrm{mod}\ 2)$, or $y \equiv 1, 3, 5, 7, 9, 11 \ (\mathrm{mod}\ 12)$. These give

$$2x \equiv 10, 8, 6, 4, 2, 0 \ (\mathrm{mod}\ 12)$$

respectively, or

$$x \equiv 5, 4, 3, 2, 1, 0 \ (\mathrm{mod}\ 6).$$

Thus the solutions (mod 12) are

$$x, y = 5, 1;\ \ 11, 1;\ \ 4, 3;\ \ 10, 3;\ \ 3, 5;\ \ 9, 5;$$
$$2, 7;\ \ 8, 7;\ \ 1, 9;\ \ 7, 9;\ \ 0, 11;\ \ 6, 11.$$

The general situation is given in the following theorem, which is easily proved by induction on the number of unknowns.

THEOREM 3-11. *The congruence*

$$a_1 x_1 + \cdots + a_n x_n \equiv c \pmod{m}$$

has just dm^{n-1} *or no solutions* $\pmod{m}$ *according as* $d|c$ *or* $d \nmid c$, *where* $d = (a_1, \ldots, a_n, m)$.

Turning now to the simultaneous solution of a system of linear congruences, we consider the system

$$\alpha_1 x \equiv \beta_1 \pmod{m_1}, \qquad \ldots, \qquad \alpha_n x \equiv \beta_n \pmod{m_n},$$
$$\alpha_i \text{ and } \beta_i \text{ integers.}$$

Clearly, no x satisfies all these congruences unless each is solvable separately. Assuming that this is so, we can restrict our attention to systems of the form

$$x \equiv c_1 \pmod{m_1}, \qquad \ldots, \qquad x \equiv c_n \pmod{m_n}.$$

It is clear that this system will have no solution unless every pair has. From the first of the congruences

$$x \equiv c_i \pmod{m_i}, \qquad x \equiv c_j \pmod{m_j},$$

we get $x = c_i + m_i y$; substituting in the second yields

$$m_i y \equiv c_j - c_i \pmod{m_j},$$

and consequently it must be true that

$$(m_i, m_j) | (c_i - c_j).$$

If this is the case, then y is unique $(\bmod\ m_j/(m_i, m_j))$, and x is unique $(\bmod\ m_i m_j/(m_i, m_j))$, that is, modulo the LCM of m_i and m_j. We have thus proved part of the following theorem.

THEOREM 3-12. *A necessary and sufficient condition that the system of congruences* $x \equiv c_i \pmod{m_i}$ $(i = 1, 2, \ldots, n)$ *be solvable is that for every pair of indices i, j between 1 and n inclusive,*

$$(m_i, m_j) | (c_i - c_j).$$

The solution, if it exists, is unique modulo the LCM of $m_1, \ldots, m_n$.

Proof: To prove the sufficiency we must show the following. If every pair from among the n congruences is solvable, and if any two of them are solved to give a single new congruence, then the $n - 1$ congruences consisting of this new one and the remaining $n - 2$ original congruences also have the property that every pair from among them is solvable. That is, assume that for all i and j with $1 \leq i \leq n$, $1 \leq j \leq n$, it is true that $(m_i, m_j)|(c_i - c_j)$, and let the solution of

$$x \equiv c_1 \ (\text{mod } m_1), \qquad x \equiv c_2 \ (\text{mod } m_2)$$

be

$$x \equiv f \ (\text{mod } \langle m_1, m_2 \rangle).$$

Then we must show that for $3 \leq i \leq n$,

$$(m_i, \langle m_1, m_2 \rangle)|(c_i - f).$$

This can easily be seen by considering the exponent α of any prime p which occurs in the prime-power factorization of $(m_i, \langle m_1, m_2 \rangle)$. Let the exponent of p in the factorization of m_j be β_j, for $j = 1, 2, \ldots, i$. Then p occurs in $\langle m_1, m_2 \rangle$ with exponent max (β_1, β_2), so that

$$\alpha = \min \ (\beta_i, \max \ (\beta_1, \beta_2)) = \max \ (\min \ (\beta_1, \beta_i), \min \ (\beta_2, \beta_i)).$$

But our assumption is that

$$p^{\min(\beta_1, \beta_i)}|(c_1 - c_i) \qquad \text{and} \qquad p^{\min(\beta_2, \beta_i)}|(c_2 - c_i),$$

and since $p^{\beta_1}|(c_1 - f)$ and $p^{\beta_2}|(c_2 - f)$ we see, by writing

$$c_1 - c_i = (c_1 - f) + (f - c_i),$$

$$c_2 - c_i = (c_2 - f) + (f - c_i),$$

that

$$p^{\min(\beta_1, \beta_i)}|(c_i - f) \qquad \text{and} \qquad p^{\min(\beta_2, \beta_i)}|(c_i - f),$$

so that also $p^{\alpha}|(c_i - f)$. Since p^{α} was an arbitrary prime-power factor of $(m_i, \langle m_1, m_2 \rangle)$, it follows that

$$(m_i, \langle m_1, m_2 \rangle)|(c_i - f),$$

and the sufficiency of the condition is proved.

Finally, solving the first two congruences simultaneously, we get a solution which is unique (mod $\langle m_1, m_2 \rangle$); solving this with the third, we get a solution unique (mod $\langle m_3, \langle m_1, m_2 \rangle \rangle$), that is, unique (mod $\langle m_1, m_2, m_3 \rangle$), etc.

As a consequence of Theorem 3–12, we have the following important result.

THEOREM 3–13 (*Chinese Remainder Theorem*). *Every system of linear congruences in which the moduli are relatively prime in pairs is solvable, the solution being unique modulo the product of the moduli.*

PROBLEMS

1. Solve the congruence $6x + 15y \equiv 9 \pmod{18}$.
2. Solve simultaneously:

$$x \equiv 1 \pmod 2,$$
$$x \equiv 1 \pmod 3,$$
$$x \equiv 3 \pmod 4,$$
$$x \equiv 4 \pmod 5.$$

3. Suppose that the system of congruences

$$x \equiv a_i \pmod{m_i}, \qquad i = 1, 2, \ldots, n,$$

is to be solved, where $(m_i, m_j) = 1$ for all i, j with $i \neq j$. Put

$$M = m_1 \ldots m_n,$$

and for $i = 1, \ldots, n$, let $y = b_i$ be a solution of the congruence

$$\frac{M}{m_i} y \equiv 1 \pmod{m_i}.$$

Then show that the solution x of the original system is given by

$$x \equiv \sum_{i=1}^{n} a_i b_i \frac{M}{m_i} \pmod{M}.$$

4. Show that given a, b, and n, with $(a, b) = 1$, there is an x such that

$$(ax + b, n) = 1.$$

[*Hint:* If $p|a$ and $p|n$, then $p \nmid (ax + b)$ for any x. If $p|n$ and $p \nmid a$, there is a solution of

$$ax + b \equiv 1 \pmod p.$$

Use the Chinese Remainder Theorem.]

3–5 Congruences of higher degree. We consider now the congruence

$$f(x) = a_0 x^n + a_1 x^{n-1} + \cdots + a_n \equiv 0 \pmod m,$$

where the a_i are not all congruent to zero $\pmod m$. If $m = \prod_{i=1}^{r} p_i^{\alpha_i}$, then clearly the given congruence is equivalent to the system of congruences

$$f(x) \equiv 0 \pmod{p_1^{\alpha_1}}, \qquad \ldots, \qquad f(x) \equiv 0 \pmod{p_r^{\alpha_r}}.$$

If for each i with $1 \leq i \leq r$, c_i is a root of $f(x) \equiv 0 \pmod{p_i^{\alpha_i}}$, then by the Chinese Remainder Theorem there is a solution x_0 of the system

$$x \equiv c_1 \pmod{p_1^{\alpha_1}}, \quad \ldots, \quad x \equiv c_r \pmod{p_r^{\alpha_r}},$$

and this x_0, which is unique modulo m, is a solution of the original congruence. Consequently, the number of solutions of the original congruence is the product of the numbers of roots of the congruences modulo the prime-power divisors of m. Hence we can restrict our attention to the case where the modulus is a power of a prime.

The reduction can easily be carried a step further, so that we have only to consider the higher degree congruence with prime modulus, together with a number of linear congruences with prime moduli. The idea is that the solutions of

$$f(x) \equiv 0 \pmod{p^\alpha} \tag{1}$$

are to be found among those of

$$f(x) \equiv 0 \pmod{p^\beta} \tag{2}$$

with $\beta < \alpha$. Suppose that for some $\beta < \alpha$ a solution of (2) is known, say a. (There may be others, of course.) Then every number $a + tp^\beta$ is a solution of (2); it is desired to determine t so that $a + tp^\beta$ is also a solution of

$$f(x) \equiv 0 \pmod{p^{\beta+1}}. \tag{3}$$

By Taylor's theorem,

$$f(a + tp^\beta) = f(a) + tp^\beta f'(a) + \frac{(tp^\beta)^2 f''(a)}{2!} + \cdots + \frac{(tp^\beta)^n f^{(n)}(a)}{n!}.$$

A term $c_j x^j$ in $f(x)$ leads to the term

$$\frac{j(j-1)\ldots(j-k+1)}{k!} c_j x^{j-k} = \binom{j}{k} c_j x^{j-k}$$

in $f^{(k)}(x)/k!$, so that the numbers $f^{(k)}(a)/k!$ are integers. Hence

$$f(a + tp^\beta) \equiv f(a) + tp^\beta f'(a) \pmod{p^{\beta+1}},$$

and (3) becomes

$$f(a) + tp^\beta f'(a) \equiv 0 \pmod{p^{\beta+1}}.$$

Now $p^\beta | f(a)$, so that this reduces to the linear congruence

$$f'(a) \cdot t \equiv -\frac{f(a)}{p^\beta} \pmod{p},$$

of which the number of solutions is

$$\begin{cases} 0, & \text{if } p|f'(a) \text{ but } p \nmid \dfrac{f(a)}{p^\beta}, \\[2ex] p, & \text{if } p|f'(a) \text{ and } p \mid \dfrac{f(a)}{p^\beta}, \\[2ex] 1, & \text{if } p \nmid f'(a). \end{cases}$$

The general procedure should now be clear, if all solutions of (2) with $\beta = 1$ are known. Choose one of them, say a_1. Corresponding to it there are 0, 1, or p solutions of (2) with $\beta = 2$, to be found by solving a linear congruence. If there are no solutions, start over with a different a_1. If there are solutions, choose one and find the corresponding solutions of (2) with $\beta = 3$. If all possibilities are explored in this way, all solutions of (1) can eventually be found.

Consider, for example, the congruence

$$f(x) = x^3 - 4x^2 + 5x - 6 \equiv 0 \pmod{27}.$$

We first search for roots of

$$x^3 - 4x^2 + 5x - 6 \equiv x^3 + 2x^2 + 2x \equiv 0 \pmod{3}.$$

Trying successively 0, 1, 2, we find the only solution of this congruence to be $x \equiv 0 \pmod{3}$. Putting $x = 0 + 3t$, we now wish to find t's for which

$$f(0 + 3t) \equiv 0 \pmod{9}.$$

As above, this reduces to

$$3f'(0)t \equiv -f(0) \pmod{9},$$

or

$$15t \equiv 6 \pmod{9},$$

or

$$5t \equiv 2 \pmod{3},$$

so that $t \equiv 1 \pmod{3}$. Putting $t = 1 + 3t_1$, we get $x = 3 + 9t_1$, and we ask that

$$f(3 + 9t_1) \equiv 0 \pmod{27}.$$

This gives
$$f(3) + 9t_1 f'(3) \equiv 0 \pmod{27},$$
or
$$9 \cdot 8 \cdot t_1 \equiv 0 \pmod{27},$$
$$t_1 \equiv 0 \pmod{3}.$$

Thus $t_1 = 3t_2$ and $x = 3 + 27t_2$, so that the only solution of the original congruence is
$$x \equiv 3 \pmod{27}.$$

If at any stage in the above argument there had been more than one possibility, each of them would have had to be followed through to obtain corresponding solutions.

PROBLEMS

1. Find all solutions of the congruence
$$x^3 - 3x^2 + 27 \equiv 0 \pmod{1125}.$$
[*Answer:* $x \equiv 51, 426, 801 \pmod{1125}$.]

2. If $f(x)$ is a nonconstant polynomial with integral coefficients, show that it assumes composite values for arbitrarily large x. [*Hint:* Apply Taylor's theorem to $f(m + k \cdot f(m))$.]

3. Suppose that the congruence $f(x) \equiv 0 \pmod{p}$ has as roots the s numbers $x_1, \ldots, x_s$, which are distinct $\pmod{p}$. Show that if $p \nmid f'(x_k)$ for $k = 1, \ldots, s$, then the congruence $f(x) \equiv 0 \pmod{p^\alpha}$ also has exactly s roots, for every $\alpha \geq 1$.

3–6 Congruences with prime moduli. If $f(x)$ and $f_1(x)$ are two polynomials whose corresponding (integral) coefficients are congruent modulo m, then we say that $f(x)$ and $f_1(x)$ are *identically congruent modulo m*, and write
$$f(x) \equiv f_1(x) \pmod{m}. \tag{4}$$

When there is no reference made to the numerical values of x in such a relation, it will always mean identical congruence. It should be noted that (4) is not equivalent to the assertion
$$f(x) \equiv f_1(x) \pmod{m} \qquad \text{for all } x,$$
since, for example, $x^3 \equiv x \pmod{3}$ for all x, but x^3 and x are not identically congruent modulo 3.

If $g(x)$ is also a polynomial with integral coefficients, and $g(x)$ has leading coefficient 1, then $f(x)$ can be divided by $g(x)$ in the usual fashion to obtain a quotient $q_1(x)$ and a remainder $r_1(x)$. Both q_1 and r_1 are polynomials with integral coefficients, and the degree of r_1 is less than that of g. If now

$$q_1(x) \equiv q(x) \pmod{m} \quad \text{and} \quad r_1(x) \equiv r(x) \pmod{m},$$

then

$$f(x) \equiv g(x)q(x) + r(x) \pmod{m}. \tag{5}$$

Such *division modulo m* is not always possible if the leading coefficient of $g(x)$ is not 1, since fractional coefficients may then be encountered. In the case of a prime modulus, however, it is possible to find an integer c such that $cg(x)$ has leading coefficient congruent to 1, and so to carry out the division.

If in (5),

$$r(x) \equiv 0 \pmod{m},$$

then $g(x)$ is said to *divide $f(x)$ modulo m*, or to *be a factor of $f(x)$ modulo m*, and we write

$$g(x)|f(x) \pmod{m}.$$

If $f(x)$ has no nonconstant factor (mod m) of lower degree than itself, it is said to be *irreducible* (mod m). If $(x - a)|f(x) \pmod{m}$, then a is said to be a *zero* of $f(x)$ (mod m), or a *root* of the congruence $f(x) \equiv 0 \pmod{m}$.

In the case of prime modulus, the Euclidean algorithm can easily be generalized, so that we can find the GCD (mod p) of any two polynomials. For example, if

$$f(x) = x^3 + 2x^2 - x + 1, \quad g(x) = x^2 - x + 1,$$

then

$$f(x) \equiv x \cdot g(x) + (x + 1) \pmod{3},$$

$$g(x) \equiv (x + 1)(x + 1) \pmod{3},$$

and so the GCD (mod 3) of $f(x)$ and $g(x)$ is the last nonvanishing remainder, namely $x + 1$. But

$$f(x) \equiv (x + 3)g(x) + (x - 2) \pmod{5},$$

$$g(x) \equiv (x + 1)(x - 2) + 3 \pmod{5},$$

$$x - 2 \equiv 3(2x + 1) \pmod{5},$$

so that $f(x)$ and $g(x)$ are relatively prime (i.e., have no common nonconstant divisor) modulo 5.

If the leading coefficient of $g(x)$ is not 1, it may be made so by multiplication by a suitable constant, and then one can find $(f(x), cg(x))$.

It is now possible to prove theorems analogous to Theorems 2–1 through 2–5, and so to show that every polynomial is congruent to a product of polynomials which are irreducible (mod p), and that this representation is unique except for the order of factors and the presence of a set of constant factors whose product is 1 (mod p). Notice that this result is not valid when the modulus is composite, for example,

$$(x - 1)x \equiv (x - 3)(x + 2) \pmod 6,$$

and each of the linear polynomials is of course irreducible.

Another assertion which holds only for prime modulus is that if

$$f(x)g(x) \equiv 0 \pmod p,$$

then either

$$f(x) \equiv 0 \qquad \text{or} \qquad g(x) \equiv 0 \pmod p.$$

For otherwise we may suppose, with no loss in generality, that the leading coefficients of $f(x)$ and $g(x)$ are 1. But then the leading coefficient of $f(x) \cdot g(x)$ is also 1, and therefore not 0.

THEOREM 3–14 (*Factor theorem*). *If a is a root of the congruence*

$$f(x) \equiv 0 \pmod m,$$

then

$$(x - a)|f(x) \pmod m,$$

and conversely.

Proof: Take $g(x) = x - a$ in equation (5). Then $r(x) = r$ is constant, and

$$f(x) \equiv (x - a)q(x) + r \pmod m.$$

Putting $x = a$, we see that $r \equiv 0 \pmod m$ if $f(a) \equiv 0 \pmod m$. Conversely, if

$$f(x) \equiv (x - a)g(x) \pmod m,$$

then

$$f(a) \equiv 0 \pmod m.$$

THEOREM 3–15 (*Lagrange's theorem*). *The congruence*

$$f(x) \equiv 0 \pmod{p}$$

in which

$$f(x) = a_0 x^n + \cdots + a_n, \qquad a_0 \not\equiv 0 \pmod{p},$$

has at most n roots.

Proof: For $n = 1$ this follows from Theorem 3–10. Assume that every congruence of degree $n - 1$ has at most $n - 1$ solutions, and that a is a root of the original congruence. Then

$$f(x) \equiv (x - a)q(x) \pmod{p},$$

where $q(x)$ is not identically zero $\pmod{p}$ and is of degree $n - 1$. It therefore has at most $n - 1$ zeros, say $c_1, \ldots, c_r$, where $r \leq n - 1$. Then if c is any number such that $f(c) \equiv 0 \pmod{p}$, then

$$(c - a)q(c) \equiv 0 \pmod{p},$$

so that either

$$c \equiv a \pmod{p}$$

or

$$q(c) \equiv 0 \pmod{p}, \qquad \text{that is, } c = c_i \text{ for some } i, 1 \leq i \leq r.$$

In other words, the original congruence has at most $r + 1 \leq n$ roots. The theorem now follows by the induction principle.

Again, this theorem is not valid for composite modulus.

PROBLEM

Let $f(x)$ be a polynomial of degree n, with integral coefficients. Show that if $n + 1$ consecutive values of $f(x)$ are divisible by a fixed prime p, then $p | f(x)$ for every integral x. Cf. Problem 1, Section 3–2.

3–7 The theorems of Fermat, Euler, and Wilson

THEOREM 3–16 (*Fermat's theorem*). *If $p \nmid a$, then*

$$a^{p-1} \equiv 1 \pmod{p}.$$

Since $\varphi(p) = p - 1$, this is a special case of

THEOREM 3–17 (*Euler's theorem*). *If $(a, m) = 1$, then*

$$a^{\varphi(m)} \equiv 1 \pmod{m}.$$

Proof: Let $c_1, \ldots, c_{\varphi(m)}$ be a reduced residue system (mod m), and let a be prime to m. Then $ac_1, \ldots, ac_{\varphi(m)}$ is also a reduced residue system (mod m), and

$$\prod_{i=1}^{\varphi(m)} ac_i = a^{\varphi(m)} \prod_{i=1}^{\varphi(m)} c_i \equiv \prod_{i=1}^{\varphi(m)} c_i \ (\text{mod } m).$$

Since $(m, \Pi c_i) = 1$, this implies that

$$a^{\varphi(m)} \equiv 1 \ (\text{mod } m).$$

We see from Euler's theorem that if we take the least positive remainders (mod m) of the sequence of powers $a, a^2, a^3, \ldots$ of a number a which is prime to m, we will have a periodic sequence, of period less than or equal to $\varphi(m)$. The period of this sequence—that is, the least positive exponent t such that $a^t \equiv 1$ (mod m)—is called the *order of a* (mod m), or the *exponent to which a belongs modulo m*, and we write $\text{ord}_m a = t$.

THEOREM 3–18. *If $a^u \equiv 1 (\text{mod } m)$, then $\text{ord}_m a | u$.*

Proof: Put $\text{ord}_m a = t$, and let $u = qt + r$, $0 \le r < t$. Then

$$a^u = a^{qt+r} = (a^t)^q \cdot a^r \equiv a^r \equiv 1 \ (\text{mod } m),$$

and if r were different from zero, there would be a contradiction with the definition of t.

THEOREM 3–19. *For every a prime to m, $\text{ord}_m a | \varphi(m)$.*

Proof: Follows immediately from Theorems 3–16 and 3–18.

As we shall see in the next chapter, the numbers a of order $\varphi(m)$ are of great importance.

The direct converse of Fermat's theorem does not hold; that is, it is not true that if for some a, $a^{m-1} \equiv 1$ (mod m), then m is prime. For example, the powers of 3, reduced modulo 91, are 3, 9, 27, 81, 61, 1, so that $\text{ord}_{91} 3 = 6$. Since $6|90$, $3^{90} \equiv 1$ (mod 91). But 91 is not prime. The clue to the proper converse lies in the observation that $\varphi(m) \le m - 1$ always, and $\varphi(m) = m - 1$ if and only if m is prime, so that m will certainly be prime if there is an a such that $\text{ord}_m a = m-1$.

THEOREM 3–20. *If there is an a for which $a^{m-1} \equiv 1$ (mod m), while none of the congruences $a^{(m-1)/p} \equiv 1$ (mod m) holds, where p runs over the prime divisors of $m - 1$, then m is prime.*

Proof: By the first hypothesis and Theorem 3–18, the exponent t to which a belongs (mod m) divides $m - 1$. On the other hand, since every proper divisor of $m - 1$ is a divisor of at least one of the numbers $(m - 1)/p$, the second hypothesis and Theorem 3–18 imply that t is not a proper divisor of $m - 1$. Consequently $t = m - 1$. By Theorem 3–19, $m - 1 | \varphi(m)$, and so $m - 1 = \varphi(m)$ and m is prime.

In a way, Theorem 3–20 is simply a restatement of the fact that $\varphi(m) = m - 1$ if and only if m is prime. But in distinction to this statement, it can actually be used to investigate the primality of large numbers.

Fermat's theorem exhibits congruences which have the maximum number of roots allowable by Lagrange's theorem. The following theorem gives another important example of such a situation.

THEOREM 3–21. *If p is prime and d divides $p - 1$, then there are exactly d roots of the congruence*

$$x^d \equiv 1 \pmod{p}.$$

Proof: Since $d | p - 1$,

$$x^{p-1} - 1 \equiv (x^d - 1)q(x) \pmod{p},$$

where $q(x)$ is a polynomial of degree $p - 1 - d$ in x. By Lagrange's theorem, the congruence

$$q(x) \equiv 0 \pmod{p}$$

has at most $p - 1 - d$ solutions. Since $x^{p-1} \equiv 1 \pmod{p}$ has exactly $p - 1$ solutions, $x^d \equiv 1 \pmod{p}$ must have at least $p - 1 - (p - 1 - d) = d$ solutions. Since it can have no more than this, it must have exactly d solutions.

As another consequence of Fermat's theorem, we have

THEOREM 3–22 (*Wilson's theorem*). *If p is prime, then*

$$(p - 1)! \equiv -1 \pmod{p}.$$

Proof: Fermat's theorem and Theorem 3–14 show that

$$x^{p-1} - 1 \equiv (x - 1)(x - 2) \cdots (x - p + 1) \pmod{p}$$

identically, so that the constant terms must be congruent:

$$-1 \equiv (-1)^{p-1}(p - 1)! \pmod{p}.$$

If p is odd, this gives the theorem. If $p = 2$, then we have

$$-1 \equiv 1 \equiv 1! \ (\text{mod } 2).$$

The converse of Wilson's theorem does hold.

THEOREM 3–23. *If $m > 1$ and $(m - 1)! \equiv -1 \ (\text{mod } m)$, then m is prime.*

Proof: If m is composite, it has a proper divisor $d > 1$. But then

$$(m - 1)! \equiv 0 \not\equiv -1 \ (\text{mod } d),$$

and *a fortiori*,

$$(m - 1)! \not\equiv -1 \ (\text{mod } m).$$

There is another way of obtaining Wilson's theorem which also throws some light on a subject to be considered in much more detail in Chapter 5. Let a be any integer not divisible by the odd prime p, and let b be one of the numbers $1, \ldots, p - 1$. Then we know that there is a unique solution $(\text{mod } p)$ of the congruence $bx \equiv a \ (\text{mod } p)$. Let b', called the *associate* of b, be that positive solution which is less than p. We must distinguish two cases, according as some b is associated with itself or not. If $b = b'$, then $b^2 \equiv a \ (\text{mod } p)$, so that the congruence $x^2 \equiv a \ (\text{mod } p)$ has a solution; in this case a is said to be a *quadratic residue* of p. If the congruence $x^2 \equiv a \ (\text{mod } p)$ has no solution, a is called a *quadratic nonresidue* of p. (Similar definitions hold for nth power residues and nonresidues.)

If a is a quadratic residue of p, and if $b_1{}^2 \equiv a \ (\text{mod } p)$, then clearly also $(p - b_1)^2 \equiv a \ (\text{mod } p)$; by Lagrange's theorem there are no other solutions. Thus in this case the numbers $1, \ldots, p - 1$ can be grouped into $(p - 3)/2$ pairs of associates, the product of each pair being congruent to $a \ (\text{mod } p)$, together with the two numbers b_1 and $p - b_1$. Thus

$$(p - 1)! = \prod_{b=1}^{p-1} b \equiv a^{(p-3)/2} \cdot b_1(p - b_1) \equiv -a^{(p-1)/2} \ (\text{mod } p). \quad (6)$$

On the other hand, if a is a quadratic nonresidue of p, the numbers $1, 2, \ldots, p - 1$ can be grouped into $(p - 1)/2$ pairs of associates, and

$$(p - 1)! = \prod_{b=1}^{p-1} b \equiv a^{(p-1)/2} \ (\text{mod } p). \quad (7)$$

In order to give a uniform statement of (6) and (7), we define the *Legendre symbol* (a/p) (also frequently written $\left(\dfrac{a}{p}\right)$ or $(a|p)$) to

CHAPTER 4

PRIMITIVE ROOTS AND INDICES

4-1 Integers belonging to a given exponent (mod p)

THEOREM 4-1. *If* $\operatorname{ord}_m a = t$, *then* $\operatorname{ord}_m a^n = t/(n, t)$.

Proof: Let $(n, t) = d$. Then, since $a^t \equiv 1 \pmod{m}$, we have

$$(a^t)^{n/d} = (a^n)^{t/d} \equiv 1 \pmod{m},$$

so that if $\operatorname{ord}_m a^n = t'$, then

$$t' \left| \frac{t}{d} \right. \tag{1}$$

But from the congruence

$$(a^n)^{t'} \equiv 1 \pmod{m},$$

we have that $t \mid nt'$, or

$$\frac{t}{d} \left| \frac{n}{d} t' \right.$$

Since

$$\left(\frac{t}{d}, \frac{n}{d} \right) = 1,$$

this gives

$$\frac{t}{d} \left| t' \right. \tag{2}$$

Combining (1) and (2), we have

$$t' = \frac{t}{d} \cdot$$

THEOREM 4-2. *If any integer belongs to* t (mod p), *then exactly* $\varphi(t)$ *incongruent numbers belong to* t (mod p).

Proof: Assume that $\operatorname{ord}_p a = t$. Then by Theorem 3–19, $t \mid (p - 1)$, so that by Theorem 3–21 there are exactly t roots of the congruence

48

$x^t \equiv 1 \pmod{p}$. But all the numbers $a, a^2, \ldots, a^t$ are roots of this congruence and they are incongruent $\pmod{p}$, so that they are the only roots. By Theorem 4–1, the powers of a which belong to $t \pmod{p}$ are the numbers a^n with $(n, t) = 1$, $1 \leq n \leq t$, and there are just $\varphi(t)$ of these numbers.

THEOREM 4–3. *If $t \mid (p - 1)$, there are $\varphi(t)$ incongruent numbers $\pmod{p}$ which belong to $t \pmod{p}$.*

Proof: Let d run over the divisors of $p - 1$, and for each such d let $\psi(d)$ be the number of integers among $1, 2, \ldots, p - 1$ of order $d \pmod{p}$. By Theorem 3–19 and Fermat's theorem, each of the integers $1, 2, \ldots, p - 1$ belongs to exactly one of the d. Hence

$$\sum_{d \mid p-1} \psi(d) = p - 1.$$

But also

$$\sum_{d \mid p-1} \varphi(d) = p - 1,$$

by Theorem 3–9, so that

$$\sum_{d \mid p-1} \psi(d) = \sum_{d \mid p-1} \varphi(d).$$

By Theorem 4–2, the value of $\psi(d)$ is either zero or $\varphi(d)$ for each d, and we deduce from the last equation that $\psi(d) = \varphi(d)$ for each d dividing $p - 1$.

If $\operatorname{ord}_m a = \varphi(m)$, then a is said to be a *primitive root* of m. The importance of this notion lies in the fact that if g is such a primitive root, then its powers

$$g, g^2, \ldots, g^{\varphi(m)}$$

are distinct $\pmod{m}$ and are all relatively prime to m; they therefore constitute a reduced residue system modulo m. Thus we have a convenient way of representing all the elements of a reduced residue system, some of the implications of which are to be found later in this chapter and in the problems.

It follows immediately from Theorem 4–1 that the other primitive roots of m are those powers g^k for which $(k, \varphi(m)) = 1$. Either from this remark or from Theorem 4–3 we have

THEOREM 4–4. *There are exactly $\varphi(\varphi(p))$ primitive roots of a prime p.*

1. Show that if $\operatorname{ord}_p a = t$, $\operatorname{ord}_p b = u$, and $(t, u) = 1$, then $\operatorname{ord}_p (ab) = tu$.

2. Show that if $p \equiv 1 \pmod 4$ and g is a primitive root of p, then so is $-g$. Show by a numerical example that this need not be the case if $p \equiv 3 \pmod 4$.

3. Show that if p is of the form $2^m + 1$ and $(a/p) = -1$, then a is a primitive root of p.

4. Show that if p is an odd prime and $\operatorname{ord}_p a = t > 1$, then

$$\sum_{k=1}^{t-1} a^k \equiv -1 \pmod p.$$

4–2 Primitive roots of composite moduli. Theorem 4–4 immediately brings the following questions to mind: Do all numbers have primitive roots? If not, which do, and how many are there? The first question is easily answered in the negative, since 8 has none: $\varphi(8) = 4$, but

$$\operatorname{ord}_8 1 = 1, \qquad \operatorname{ord}_8 3 = 2, \qquad \operatorname{ord}_8 5 = 2, \qquad \operatorname{ord}_8 7 = 2.$$

On the other hand, since 5 is a primitive root of 6, there are composite numbers having primitive roots. The answer to the second question is, therefore, not just the set of primes, as one might think.

After the primes themselves, the simplest moduli to treat are the prime powers. We need a preliminary result.

THEOREM 4–5.   (a) *If p is prime, then*

$$a \equiv b \pmod{p^n} \qquad implies \qquad a^{p^s} \equiv b^{p^s} \pmod{p^{n+s}} \qquad (3)$$

for every pair of positive integers n, s.

(b) *If p is an odd prime and $p \nmid b$, then*

$$a^{p^s} \equiv b^{p^s} \pmod{p^{n+s}} \qquad implies \qquad a \equiv b \pmod{p^n} \qquad (4)$$

for every pair of positive integers n, s.

Proof: (a) We use induction on s. Assume that $a \equiv b \pmod{p^n}$. Then

$$a = hp^n + b,$$

and

$$a^p = (hp^n)^p + \binom{p}{1}(hp^n)^{p-1}b + \cdots + \binom{p}{p-1} hp^n b^{p-1} + b^p.$$

Now p occurs in the numerator of the binomial coefficient

$$\binom{p}{k} = \frac{p!}{k!(p-k)!},$$

but it is not present in the denominator if $0 < k < p$; hence $p \,\Big|\, \binom{p}{k}$ for $0 < k < p$, and for such k,

$$p^{n+1} \,\Big|\, \binom{p}{k} p^{n(p-k)}.$$

But also $p^{n+1} | p^{np}$, so that $a^p \equiv b^p \pmod{p^{n+1}}$. Hence (3) is correct for $s = 1$ and every n.

Suppose that (3) is valid for $s = 1, 2, \ldots, s'$, for every n, and suppose that $a \equiv b \pmod{p^n}$. (This congruence is now to be regarded as the premise of (3) with $s = s' + 1$.) Then the induction hypothesis with $s = 1$ gives

$$a^p \equiv b^p \pmod{p^{n+1}}. \tag{5}$$

Using (5) as the premise of (3) with $s = s'$ gives

$$(a^p)^{p^{s'}} \equiv (b^p)^{p^{s'}} \pmod{p^{n+1+s'}},$$

or

$$a^{p^{s'+1}} \equiv b^{p^{s'+1}} \pmod{p^{n+(s'+1)}},$$

which is the conclusion of (3) with $s = s' + 1$. Hence (3) holds for every pair of positive integers n, s.

(b) We first prove (4) for $s = 1$, by induction on n; we suppose throughout that $p \neq 2$ and $p \nmid b$. Thus we wish to show that

$$a^p \equiv b^p \pmod{p^{n+1}} \qquad \text{implies} \qquad a \equiv b \pmod{p^n}.$$

If $a^p \equiv b^p \pmod{p^2}$, then also $a^p \equiv b^p \pmod{p}$, and, by Fermat's theorem, $a \equiv b \pmod{p}$. Now assume that

$$a^p \equiv b^p \pmod{p^{n'}} \qquad \text{implies} \qquad a \equiv b \pmod{p^{n'-1}}$$

and that

$$a^p \equiv b^p \pmod{p^{n'+1}}.$$

Then

$$a^p \equiv b^p \pmod{p^{n'}},$$

so that

$$a \equiv b \pmod{p^{n'-1}}.$$

But if $a = up^{n'-1} + b$, then

$$a^p \equiv b^p + up^{n'}b^{p-1} \pmod{p^{n'+1}}$$

if $p > 2$, and so $p|u$, whence $a = u_1 p^{n'} + b$ and

$$a \equiv b \pmod{p^{n'}},$$

and the implication follows by induction on n.

To complete the proof of (4), we use induction on s. Assume that

$$a^{p^{s'-1}} \equiv b^{p^{s'-1}} \pmod{p^{n+s'-1}} \qquad \text{implies} \qquad a \equiv b \pmod{p^n}$$

for every n, and assume that

$$a^{p^{s'}} \equiv b^{p^{s'}} \pmod{p^{n+s'}}.$$

Then

$$(a^p)^{p^{s'-1}} \equiv (b^p)^{p^{s'-1}} \pmod{p^{n+s'}},$$

whence

$$a^p \equiv b^p \pmod{p^{n+1}},$$

so that, by what we have just proved,

$$a \equiv b \pmod{p^n}.$$

The result follows by induction on s.

Let p be a prime. Then if $p^n|a$ and $p^{n+1} \nmid a$, we will write for brevity $p^n \| a$.

THEOREM 4–6. *If p is an odd prime, $\mathrm{ord}_p\, a = t$, and $p^z \| (a^t - 1)$, then*

$$\mathrm{ord}_{p^n}\, a = t \cdot p^{\max(0, n-z)}.$$

Proof: Assume the hypotheses of the theorem are satisfied. If $n \leq z$, then $p^n|(a^t - 1)$. This is not true for any exponent $t' < t$, since if $p^n|(a^{t'} - 1)$, then $p|(a^{t'} - 1)$, so that $t|t'$. Hence in this case $\mathrm{ord}_{p^n}\, a = t$, which proves the theorem for $n \leq z$.

If $n > z$, we get from Theorem 4–5 and the last hypothesis of the present theorem that

$$a^{tp^{n-z}} \equiv 1 \pmod{p^n}.$$

We must show that $a^d \not\equiv 1 \pmod{p^n}$ if d is a proper divisor of tp^{n-z}. Let $d = t_1 p^r$, where $r \leq n - z$ and $t_1|t$, and assume that $a^{t_1 p^r} \equiv 1 \pmod{p^n}$.

By Theorem 4–5 again, $a^{t_1} \equiv 1 \pmod{p^{n-r}}$,

whence $$a^{t_1} \equiv 1 \ (\mathrm{mod}\ p),$$

so that $t|t_1$, and $t = t_1$. Since $p^z \| (a^t - 1)$ and $p^{n-r}|(a^t - 1)$, we have $n - r \leq z$, whence $n - z = r$.

We can use Theorem 4-6 to construct primitive roots of p^n, where p is an odd prime; that is, numbers which belong to $p^{n-1}(p - 1)$ modulo p^n. Let g be a primitive root of p. Then if $p^2 \nmid (g^{p-1} - 1)$, Theorem 4-6 shows that

$$\mathrm{ord}_{p^n}\, g = (p - 1)p^{n-1},$$

and g is also a primitive root of p^n for all positive n. If $p^2|(g^{p-1} - 1)$, then $g + p$ is also a primitive root of p, and

$$(g + p)^{p-1} - 1 \equiv g^{p-1} + (p - 1)g^{p-2}p - 1$$
$$\equiv (p - 1)pg^{p-2} \not\equiv 0 \ (\mathrm{mod}\ p^2),$$

so that by Theorem 4-6,

$$\mathrm{ord}_{p^n}\, (g + p) = (p - 1)p^{n-1},$$

and $g + p$ is a primitive root of p^n for all positive n. We have thus proved

THEOREM 4-7.   *Any power of an odd prime has a primitive root.*

Turning now to other composite numbers, it is convenient to define a function $\lambda(m)$, called the *universal exponent* of m:

$$\lambda(1) = 1,$$

$$\lambda(2^\alpha) = \begin{cases} \varphi(2^\alpha) = 2^{\alpha-1} \text{ if } \alpha = 1, 2, \\ \frac{1}{2}\varphi(2^\alpha) = 2^{\alpha-2} \text{ if } \alpha > 2, \end{cases}$$

$$\lambda(p^\alpha) = \varphi(p^\alpha), \qquad p \text{ an odd prime},$$

$$\lambda(2^\alpha \cdot p_1{}^{\alpha_1} \ldots p_r{}^{\alpha_r}) = \langle \lambda(2^\alpha), \lambda(p_1{}^{\alpha_1}), \ldots, \lambda(p_r{}^{\alpha_r}) \rangle,$$

$$p_1, \ldots, p_r \text{ distinct odd primes}.$$

Euler's theorem can now be strengthened somewhat.

THEOREM 4-8.   *If $(a, m) = 1$, then*

$$a^{\lambda(m)} \equiv 1 \ (\mathrm{mod}\ m).$$

Proof: (a) If $m = 2^\alpha$ with $\alpha \leq 2$, this is Euler's theorem.
(b) If $m = 2^\alpha$ with $\alpha > 2$, a must be odd, so that $a^2 \equiv 1 \ (\mathrm{mod}\ 2^3)$.

By Theorem 4–5, $(a^2)^{2^{\alpha-3}} = a^{2^{\alpha-2}} \equiv 1 \pmod{2^\alpha}$.

(c) If $m = p^\alpha$, where p is odd, we have Euler's theorem again.

(d) Finally, suppose that $m = 2^\alpha \cdot p_1{}^{\alpha_1} \cdots p_r{}^{\alpha_r}$. By (a), (b) and (c), each of the congruences

$$a^{\lambda(2^\alpha)} \equiv 1 \pmod{2^\alpha},$$

$$a^{\lambda(p_i{}^{\alpha_i})} \equiv 1 \pmod{p_i{}^{\alpha_i}} \qquad i = 1, 2, \ldots, r,$$

holds. Since all the exponents of a divide $\lambda(m)$, it follows that

$$a^{\lambda(m)} \equiv 1 \pmod{2^\alpha},$$

$$a^{\lambda(m)} \equiv 1 \pmod{p_i{}^{\alpha_i}}, \qquad i = 1, 2, \ldots, r,$$

and hence

$$a^{\lambda(m)} \equiv 1 \pmod{m}.$$

As a complement to Theorem 4–8, we have

THEOREM 4–9. $\lambda(m)$ *is the smallest positive value of x such that $a^x \equiv 1 \pmod{m}$ for every a prime to m. That is, there is always an integer which belongs to $\lambda(m) \pmod{m}$.*

Proof: (a) If $m = 1$: $\lambda(1) = 1$ and $\mathrm{ord}_1 1 = 1$.

(b) If $m = 2$: $\lambda(2) = 1$ and $\mathrm{ord}_2 1 = 1$.

(c) If $m = 4$: $\lambda(4) = 2$ and $\mathrm{ord}_4 3 = 2$.

(d) If $m = 2^\alpha$, $\alpha > 2$: $\lambda(2^\alpha) = 2^{\alpha-2}$ and $\mathrm{ord}_{2^\alpha} 5 = 2^{\alpha-2}$. For if $\mathrm{ord}\, 5 = d$, then $d \mid 2^{\alpha-2}$, so that $d = 2^\beta$, where $\beta \leq \alpha - 2$. But it is easily proved by induction on α that for $\alpha \geq 3$,

$$5^{2^{\alpha-3}} = 1 + 2^{\alpha-1} h_\alpha,$$

where h_α is an odd number. Hence $5^{2^{\alpha-3}} \not\equiv 1 \pmod{2^\alpha}$ and $\beta = \alpha - 2$.

(e) If $m = p^\alpha$ with p odd: $\lambda(p^\alpha) = \varphi(p^\alpha)$, and by Theorem 4–7, p^α has a primitive root.

(f) If m is arbitrary: Let $m = p_1{}^{\alpha_1} \cdots p_r{}^{\alpha_r}$, with $2 \leq p_1 < \cdots < p_r$. By the first five steps of the proof, there are numbers $a_1, \ldots, a_r$ such that $\mathrm{ord}_{p_i{}^{\alpha_i}} a_i = \lambda(p_i{}^{\alpha_i})$ for $i = 1, \ldots, r$. By the Chinese Remainder Theorem, there is a single integer a such that $a \equiv a_i \pmod{p_i{}^{\alpha_i}}$, for $i = 1, \ldots, r$, and the order of $a \pmod{p_i{}^{\alpha_i}}$ is the same as that of a_i, for each i. Hence if $a^x \equiv 1 \pmod{m}$, then $\lambda(p_i{}^{\alpha_i}) \mid x$ for each i, and so $\lambda(m)$, since it is the LCM of the numbers $\lambda(p_i{}^{\alpha_i})$, also divides x. By Theorem 4–8, $\mathrm{ord}_m a = \lambda(m)$.

An integer whose order (mod m) is $\lambda(m)$ is called a *primitive λ-root* of m. Theorem 4–9 says in effect that every modulus has a primitive λ-root.

As a combination of Theorems 4–2 and 4–9, we have

THEOREM 4–10. *There are $\varphi(\lambda(m))$ primitive λ-roots of m congruent to powers of any given primitive λ-root.*

Notice that in general it is not the case that all the primitive λ-roots are congruent to powers of a single one. For example, if $m = 2^4$, $g = 5$, then the only other primitive λ-root congruent to a power of 5 is 13, while 3 and 11 are also primitive λ-roots.

Moreover, we can now deduce

THEOREM 4–11. *The numbers having primitive roots are*

$$1, 2, 4, p^\alpha, 2p^\alpha,$$

where p is any odd prime.

Proof: We already know that 1, 2, 4, and p^α have primitive roots. Since

$$\lambda(2p^\alpha) = \langle \lambda(2), \lambda(p^\alpha) \rangle = \lambda(p^\alpha) = \varphi(p^\alpha) = \varphi(2p^\alpha),$$

every number $2p^\alpha$ has primitive roots. On the other hand, if $m = 2^\alpha \cdot p_1^{\alpha_1} \cdots p_r^{\alpha_r}$ with $\alpha > 2$, p_i odd, $r \geq 1$, then

$$\lambda(m) \leq \tfrac{1}{2}\varphi(2^\alpha)\varphi(p_1^{\alpha_1}) \cdots \varphi(p_r^{\alpha_r}) \leq \tfrac{1}{2}\varphi(m),$$

and if

$$m = p_1^{\alpha_1} \cdots p_r^{\alpha_r} \qquad \text{with } r > 1$$

or if

$$m = 4p_1^{\alpha_1} \cdots p_r^{\alpha_r} \qquad \text{with } r \geq 1,$$

then each of the numbers $\lambda(4)$, $\lambda(p_i^{\alpha_i})$ is even, so that again

$$\lambda(m) \leq \tfrac{1}{2}\varphi(m).$$

This completes the proof.

The problem of efficiently finding a primitive root of a given large modulus q is not simple. It is, of course, a finite problem, and for specific modulus can be solved by successively testing the elements of a reduced residue system. A slightly more rapid method is indicated in Problem 4 at the end of the next section, but it also is laborious for large q, particularly if $\varphi(q)$ has many distinct prime divisors.

1. Show that if q has primitive roots, there are $\varphi(\varphi(q))$ of them, and their product is congruent to 1 $(\mathrm{mod}\ q)$ if $q > 6$. [*Hint:* Represent all the primitive roots in terms of a single one.]

2. Find all the primitive roots of 25.

*3. It is an unproved conjecture that no two consecutive integers, except 8 and 9, are perfect powers. Show that at any rate the only pair x, y satisfying the conditions

$$3^x - 2^y = 1, \qquad x > 1, y > 1$$

is 2, 3. [*Hint:* Use Theorem 4–6 and Problem 3, Section 3–7 to show that $3^{x-1}|y$.]

4. Show that if g is a primitive root of p^2, then the roots of the congruence

$$x^{p-1} \equiv 1\ (\mathrm{mod}\ p^2)$$

are g^{np}, $n = 1, 2, \ldots, p - 1$; that is, that these numbers are distinct roots, and there are no others. [*Hint:* Show that the congruence has only $p - 1$ roots. Cf. Problem 9, Section 3–7.]

4–3 Indices. Let q be a number having primitive roots and let g be one of them. Then the numbers $g, g^2, \cdots, g^{\varphi(q)}$ are distinct $(\mathrm{mod}\ q)$, and they are all prime to q; therefore they constitute a reduced residue system $(\mathrm{mod}\ q)$. The relation between a number a and the exponent of a power of g which is congruent to a $(\mathrm{mod}\ q)$ is very similar to the relation between an ordinary positive real number x and its logarithm. This exponent is called an *index* of a to the base g, and written "$\mathrm{ind}_g a$." That is, $\mathrm{ind}_g a$ will stand for any number t such that $g^t \equiv a$ $(\mathrm{mod}\ q)$; it is defined only if $(a, q) = 1$, and is unique modulo $\varphi(q)$. The following facts are immediate consequences of the definition.

THEOREM 4–12. *If g is a primitive root of q and $a \equiv b$ $(\mathrm{mod}\ q)$, then*

$$\mathrm{ind}_g a \equiv \mathrm{ind}_g b\ (\mathrm{mod}\ \varphi(q)),$$

$$\mathrm{ind}_g (ab) \equiv \mathrm{ind}_g a + \mathrm{ind}_g b\ (\mathrm{mod}\ \varphi(q)),$$

and

$$\mathrm{ind}_g a^n \equiv n\ \mathrm{ind}_g a\ (\mathrm{mod}\ \varphi(q)).$$

The procedure for finding the indices of the elements of a reduced residue system is quite simple if a primitive root is known. If g is a primitive root of q, construct a table of two rows and $\varphi(q)$ columns, of which the second row consists of the integers 1, 2, ..., $\varphi(q)$ in order. In the first row enter g in the first column. Multiply this by g and reduce modulo q for the element in the second column, multiply this result by g and reduce modulo q for the element in the third column, etc. (When the table is complete, the last element in the first row should be 1.) Then the index of any element of the first row appears directly below that element.

If, for example, $q = 17$ and $g = 3$, we have the table

a:	3	9	10	13	5	15	11	16	14	8	7	4	12	2	6	1
ind a:	1	2	3	4	5	6	7	8	9	10	11	12	13	14	15	16

while if $q = 18$ and $g = 5$, we have

a:	5	7	17	13	11	1
ind a:	1	2	3	4	5	6

By Theorem 4–1, if $\text{ord}_m g = \varphi(m)$, then

$$\text{ord}_m g^n = \frac{\varphi(m)}{(n, \varphi(m))},$$

so that a is a primitive root of m if and only if $(\text{ind } a, \varphi(m)) = 1$. Thus in the above table we see that the primitive roots of 18 are 5 and 11, since the only numbers less than $\varphi(18) = 6$ and prime to it are 1 and 5.

Indices are quite useful in solving binomial congruences. For example, the congruence

$$10x \equiv 8 \pmod{18}$$

implies

$$5x \equiv 4 \pmod 9,$$

which implies

$$\text{ind } 5 + \text{ind } x \equiv \text{ind } 4 \pmod 6,$$

$$\text{ind } x \equiv \text{ind } 4 - \text{ind } 5 \pmod 6.$$

Since 2 is a primitive root of 9, we construct the table as before:

n:	2	4	8	7	5	1
ind n:	1	2	3	4	5	6

Thus $\text{ind } x \equiv 2 - 5 \equiv 3 \pmod{6}$,

whence $x \equiv 8 \pmod 9$,

so that $x \equiv 8 \text{ or } 17 \pmod{18}$.

The investigation of the congruence $x^n \equiv c \pmod m$, where $(m, c) = 1$, can be reduced to the study of the solutions of

$$x^n \equiv c \pmod p$$

by previously explained methods. But the latter is entirely equivalent to

$$n \cdot \text{ind } x \equiv \text{ind } c \pmod{p - 1},$$

which has solutions if and only if $(n, p - 1) | \text{ind } c$; if this condition is satisfied there are $d = (n, p - 1)$ roots. This criterion has the disadvantage that it requires knowledge of the value of $\text{ind } c$; the following is more useful.

THEOREM 4–13. *Let $(c, q) = 1$, q being any number which has primitive roots. Then a necessary and sufficient condition that the congruence*

$$x^n \equiv c \pmod q \tag{6}$$

be solvable is that

$$c^{\varphi(q)/d} \equiv 1 \pmod q,$$

where $d = \big(n, \varphi(q)\big)$.

Proof: By an argument similar to that just given for prime modulus, a necessary and sufficient condition for the solvability of (6) is that $\text{ind } c \equiv 0 \pmod d$. This is equivalent to

$$\frac{\varphi(q)}{d} \text{ ind } c \equiv 0 \pmod{\varphi(q)},$$

or, what is the same thing,

$$c^{\varphi(q)/d} \equiv 1 \pmod q.$$

If $x^n \equiv c \pmod m$ is solvable, and $(m, c) = 1$, c is said to be an *nth power residue* of m, otherwise a *nonresidue*.

THEOREM 4–14. *The number of incongruent nth power residues of q is $\varphi(q)/d$, and these residues are the roots of the congruence*

$$x^{\varphi(q)/d} \equiv 1 \pmod q.$$

Proof: The second statement is a paraphrase of Theorem 4–13. Since q has a primitive root g, the roots of the congruence $x^{\varphi(q)/d} \equiv 1 \pmod{q}$ are the numbers g^t for which

$$g^{t\varphi(q)/d} \equiv 1 \pmod{q},$$

and this requires that $d|t$. But the number of multiples t of d with $1 \leq t \leq \varphi(q)$ is exactly $\varphi(q)/d$. (Note that this is a generalization of Theorem 3–24.)

PROBLEMS

1. Show that if g and h are primitive roots of p, then

$$\text{ind}_h\, a \equiv \text{ind}_g\, a \cdot \text{ind}_h\, g \pmod{p-1}.$$

2. Given that 2 is a primitive root of 29, construct a table of indices, and use it to solve the following congruences:

(a) $17x \equiv 10 \pmod{29}$ (b) $17x^2 \equiv 10 \pmod{29}$.

3. Develop a method for solving the congruence

$$Ax^2 + Bx + C \equiv 0 \pmod{p}$$

by use of indices, when p is an odd prime which does not divide A. (First show that the given congruence can be replaced by one in which the coefficient of x^2 is 1; then, after suitable modifications, complete the square.) Apply your method to

(a) $17x^2 - 3x + 10 \equiv 0 \pmod{29}$ (b) $17x^2 - 4x + 10 \equiv 0 \pmod{29}$.

4. Let q be a number having primitive roots. Show that h is a primitive root of q if and only if h is an rth power nonresidue of q for every prime r dividing $\varphi(q)$. [*Hint:* Write $h = g^k$, where g is a primitive root of q, and show that each of the allegedly equivalent statements is equivalent to the equation $(k, \varphi(q)) = 1$.] By eliminating all the appropriate powers of the elements of a reduced residue system, find all the primitive roots of (a) 13, (b) 29. (Cf. Problem 3, Section 4–1.)

*5. Show that for $x > 1$ the quantity

$$f(x) = \frac{y^q - 1}{y - 1} = y^{q-1} + \cdots + y + 1$$

$$= (y-1)^{q-1} + \binom{q}{1}(y-1)^{q-2} + \cdots + \binom{q}{2}(y-1) + q,$$

where q is prime, $q^n > 2$, and $y = x^{q^{n-1}}$, has the following properties:
  (a) For $x \equiv 1 \pmod{q}$, $q\|f(x)$, and for $x \not\equiv 1 \pmod{q}$, $q{\nmid}f(x)$.
  (b) $f(x) > q$.

(c) $(f(x), x) - 1$.

(d) If $p \neq q$ and $p|f(x)$, then $p \equiv 1 \pmod{q^n}$.

Deduce that there exists a prime $p \equiv 1 \pmod{q^n}$, and then by taking $x = p_1 \ldots p_r$, where each $p_i \equiv 1 \pmod{q^n}$, that there are infinitely many primes $p \equiv 1 \pmod{q^n}$. (Cf. Problem 4, Section 3–7, where $q = 2$.)

4–4 An application to Fermat's conjecture. A simple way of attempting to show that the equation

$$x^n + y^n = z^n \tag{7}$$

has no nonzero solutions for $n \geq 3$ is to show that the infinitely many congruences

$$x^n + y^n \equiv z^n \pmod{p}, \qquad p = 2, 3, 5, \ldots$$

impose absurd conditions on the variables. For example, in the case $n = 3$ the congruence

$$x^3 + y^3 \equiv z^3 \pmod{7}$$

implies that $7|xyz$. For if $7 \nmid u$, then $u^6 \equiv 1 \pmod{7}$, so that $u^3 \equiv \pm 1 \pmod{7}$, and for no choice of signs is $\pm 1 \pm 1 \equiv \pm 1 \pmod{7}$. If we could find infinitely many primes p such that

$$x^3 + y^3 \equiv z^3 \pmod{p}$$

implies $p|xyz$, then clearly equation (7) could have no nonzero solution for $n = 3$. We shall show that this cannot be done, either for $n = 3$ or for larger n. The proof depends on the following combinatorial lemma.

THEOREM 4–15. *If the numbers $1, 2, \ldots, N$ are distributed into m disjoint classes, and if* $N > m!e$, then at least one class contains the difference of two of its elements.*

Proof: Suppose that the numbers $1, 2, \ldots, N$ have been put into m disjoint classes so that no class contains the difference of any two of its elements. Let a class having the largest number of elements be called Z_1; then if Z_1 is composed of $x_1, \ldots, x_{n_1}$, we have $N \leq n_1 m$. If the names are so chosen that $x_1 < x_2 < \cdots < x_{n_1}$, the $n_1 - 1$ differences

$$x_2 - x_1, x_3 - x_1, \ldots, x_{n_1} - x_1 \tag{8}$$

are also integers between 1 and N, inclusive, and by assumption they

---

*Here, contrary to our convention, the number $e = 2.718\ldots$ is not an integer.

lie in the remaining $m - 1$ classes. Let Z_2 be a class in which the largest number of differences (8) lie. If Z_2 contains the n_2 differences

$$x_\alpha - x_1, x_\beta - x_1, \ldots, \tag{9}$$

then clearly $n_1 - 1 \leq n_2 (m - 1)$. Now the $n_2 - 1$ differences

$$x_\beta - x_\alpha, x_\gamma - x_\alpha, \ldots \tag{10}$$

do not lie in either Z_1 or Z_2, so they must be distributed among the remaining $m - 2$ classes. If n_3 is the largest number of differences (10) in any single class, then $n_2 - 1 \leq n_3 (m - 2)$. Continuing in this way, we have

$$n_\mu - 1 \leq n_{\mu+1} (m - \mu), \tag{11}$$

for $\mu = 1, 2, \ldots, m_1$, where m_1 is such that $n_{m_1} = 1$. From (11), we have

$$\frac{n_\mu}{(m - \mu)!} \leq \frac{1}{(m - \mu)!} + \frac{n_{\mu+1}}{(m - \mu - 1)!}, \qquad \mu = 1, 2, \ldots, n_1$$

and adding all these inequalities gives

$$\frac{n_1}{(m - 1)!} \leq \frac{1}{(m - 1)!} + \frac{1}{(m - 2)!} + \cdots + \frac{1}{(m - m_1)!} < e.$$

Hence

$$N \leq n_1 m < m! e,$$

and the proof is complete.

THEOREM 4–16. *There are only finitely many primes p for which every solution of the congruence*

$$x^n + y^n \equiv z^n \pmod{p} \tag{12}$$

is such that $p | xyz$. More precisely, if $p > n! e + 1$, then (12) has solutions such that $p \nmid xyz$.

Proof: First suppose that $n | (p - 1)$, so that $p - 1 = nr$ for suitable r. Let g be a primitive root of p, and let s_m be the smallest positive residue (mod p) of g^m. Then the numbers $s_1, \ldots, s_{p-1}$ are the integers $1, 2, \ldots, p - 1$ in some order. We now classify the numbers s_m according to the residue classes of their subscripts (mod n), so that for each t with $0 \leq t \leq n - 1$, the numbers

$$s_t, s_{t+n}, \ldots, s_{t+(r-1)n}$$

form a single class, there being n classes altogether. By Theorem

4–15, if $p - 1 > n!e$, then some class contains three elements, say s_{t+jn}, s_{t+kn}, s_{t+ln}, such that

$$s_{t+jn} - s_{t+kn} = s_{t+ln}.$$

But then

$$g^{t+jn} \equiv g^{t+kn} + g^{t+ln} \pmod{p},$$

whence

$$g^{jn} \equiv g^{kn} + g^{ln} \pmod{p},$$

and the numbers $x = g^k$, $y = g^l$, $z = g^j$ give the desired solution of (12).

If $n \nmid (p - 1)$, let $d = (n, p - 1)$. Then by what we have just proved, the congruence

$$x^d + y^d \equiv z^d \pmod{p}$$

is solvable with $p \nmid xyz$ if $p - 1 > d!e$. But by Theorem 4–13, any dth power is an nth power residue of p, since

$$\left(u^d\right)^{(p-1)/(p-1,n)} = \left(u^d\right)^{(p-1)/d} = u^{p-1} \equiv 1 \pmod{p}.$$

Thus there exist an x_1, y_1, and z_1 such that

$$x_1{}^n \equiv x^d, \qquad y_1{}^n \equiv y^d, \qquad z_1{}^n \equiv z^d \qquad \pmod{p},$$

and hence

$$x_1{}^n + y_1{}^n \equiv z_1{}^n \pmod{p}.$$

PROBLEM

Show that if $x^3 + y^3 \equiv z^3 \pmod{9}$, then $3 \mid xyz$. Use the result of this section, together with the method of Section 3–5, to show that this is an atypical phenomenon: that for fixed n, the congruence

$$x^n + y^n \equiv z^n \pmod{p^\alpha}$$

has a solution such that $p \nmid xyz$, if p is sufficiently large and $\alpha \geq 1$.

REFERENCES

Section 1–4

The main theorem was first proved by L. E. Dickson, *Journal für die Reine und Angewandte Mathematik* (Berlin) **135**, 134–141, 181–188 (1909). The proof given here is due to I. Schur, *Jahresbericht der Deutschen Matematikervereinigung* (Leipzig) **25**, 114–117 (1917). Dickson's proof is more difficult, but shows that equation (12) is solvable if $p > cn^4$ for suitable c.

CHAPTER 5

QUADRATIC RESIDUES

5–1 Introduction. The subject of nth power residues is a large and difficult one. It happens, however, that for the case $n - 2$ many elegant and important results can be obtained by elementary considerations, and it is to these that we now turn our attention. A fundamental tool in the investigation of quadratic residues is Euler's criterion, Theorem 3–24, which was generalized somewhat in Theorem 4–13, namely that a necessary and sufficient condition for a number a prime to q to be a quadratic residue of q is that $a^{\varphi(q)/2} \equiv 1 \pmod{q}$. (Here q is a number greater than 2 having primitive roots.) The two problems with which we shall deal are, first, to extend this criterion to general composite moduli (in so doing we shall find that it suffices to restrict our attention to odd prime moduli); and, second, to find an efficient method for determining all the primes of which a given integer a is a quadratic residue.

The prime 2 plays a rather special role in the theory of quadratic residues, not so much because of an intrinsic difference between it and the odd primes (which does exist, as we saw in the discussion of primitive roots of composite moduli) as because the congruences are quadratic; in a similar fashion, 3 must be treated separately when considering cubic congruences. On account of this, we shall use the symbol p to represent an *odd* prime throughout this chapter.

5–2 Composite moduli

THEOREM 5–1. *A number a prime to m is a quadratic residue of m if and only if it is a quadratic residue of all odd prime divisors of m and is congruent to 1 (mod 4) if $m \equiv 4 \pmod 8$, and congruent to 1 (mod 8) if $8|m$.*

Proof: Let

$$m = 2^\alpha \prod_{i=1}^{r} p_i^{\alpha_i}.$$

Then the congruence

$$x^2 \equiv a \pmod{m}$$

is equivalent to the system of congruences

$$x^2 \equiv a \pmod{2^\alpha}$$

$$x^2 \equiv a \pmod{p_i^{\alpha_i}}, \qquad i = 1, \ldots, r,$$

so that a is a quadratic residue of m if and only if it is a quadratic residue of every prime-power divisor of m.

(a) If a is a quadratic residue of p, it is a quadratic residue of p^α, and conversely. (The converse is trivial.) For if a is a residue of p, it follows from Euler's criterion that

$$a^{(p-1)/2} \equiv 1 \pmod{p}.$$

By Theorem 4–5,

$$a^{p^{\alpha-1}(p-1)/2} \equiv 1 \pmod{p^\alpha},$$

and so a is a quadratic residue of p^α by Euler's criterion again. If a is a quadratic residue of p and $x_1^2 \equiv a \pmod{p}$, then also $(-x_1)^2 \equiv a \pmod{p}$, and these are the only solutions, by Lagrange's theorem. Using the method of Section 3–5, it is easily seen that if $p \nmid f'(x_1)$ for each root x_1 of $f(x) \equiv 0 \pmod{p}$, then the congruence $f(x) \equiv 0 \pmod{p^\alpha}$ has exactly as many roots as the congruence with prime modulus. In this case $f(x) = x^2 - a$, $f'(x) = 2x$, and since p is odd and $p \nmid x_1$, it follows that $x^2 \equiv a \pmod{p^\alpha}$ has exactly two solutions $\pmod{p^\alpha}$ if a is a quadratic residue of p.

(b) For modulus 2^α the situation is more complicated: If a is odd,

   (i) $x^2 \equiv a \pmod{2}$ is always uniquely solvable,
   (ii) $x^2 \equiv a \pmod{4}$ is solvable if and only if $a \equiv 1 \pmod{4}$; it then has two roots,
   (iii) $x^2 \equiv a \pmod{2^\alpha}$, for $\alpha \geq 3$, is solvable if and only if $a \equiv 1 \pmod{8}$; it then has four roots.

The first statement is obvious, and the second follows immediately upon noting that any odd square is congruent to 1 (mod 4). (The two solutions are, of course, ± 1.) For the case $\alpha \geq 3$, recall that it was shown in the proof of Theorem 4–9 that 5 is a primitive λ-root of 2^α, and so the numbers $5, 5^2, 5^3, \ldots, 5^{2^{\alpha-2}}$ are distinct (mod 2^α). Since by the binomial theorem $(1 + 2^2)^n \equiv 1 + 4n \pmod{8}$, $5^n \equiv 1 \pmod{8}$ if and only if n is even. Thus $5^2, 5^4, 5^6, \ldots, 5^{2^{\alpha-2}}$ are $2^{\alpha-3}$ numbers which are distinct (mod 2^α) and which are all congruent to 1 (mod 8). But since there are exactly $2^{\alpha-3}$ numbers in a complete residue system (mod 2^{α}) which are congruent to 1 (mod 8), it follows that every number congruent to 1 (mod 8) is congruent to 5^{2n} for some n, so that every $a \equiv 1$ (mod 8) is a quadratic residue of 2^{α}. (If $5^{2n} \equiv a$ (mod 2^{α}), the congruence $x^2 \equiv a$ (mod 2^{α}) has the solution $x = 5^n$.) On the other hand, every odd square is of the form $8k + 1$, so that $x^2 \equiv a$ (mod 2^{α}) is certainly not solvable unless $a \equiv 1$ (mod 8).

Assume that $b^2 \equiv a$ (mod 2^{α}), and let x be any other solution of this congruence, so that also $x^2 \equiv a$ (mod 2^{α}). Then $x^2 - b^2 = (x - b)(x + b) \equiv 0$ (mod 2^{α}). Both x and b are odd, so $x - b$ and $x + b$ are even; since $(x - b, x + b) = 2(x, b)$, one of them has 2 as a simple factor. Since

$$\frac{x - b}{2} \cdot \frac{x + b}{2} \equiv 0 \pmod{2^{\alpha-2}},$$

one factor must be divisible by $2^{\alpha-2}$, that is,

$$\frac{x \pm b}{2} \equiv 0 \pmod{2^{\alpha-2}}.$$

Hence $x \equiv \pm b$ (mod $2^{\alpha-1}$), and x must be congruent to one of $\pm b$, $\pm b + 2^{\alpha-1}$ (mod 2^{α}). It is immediately verified that each of these four numbers is a solution.

Combining the results of (a) and (b) gives Theorem 5-1. By the Chinese Remainder Theorem, the number of roots of $x^2 \equiv a$ (mod m) is the product of the numbers of roots of the congruences with prime-power moduli. As shown above, if a is a quadratic residue of m, this number is 2 for each odd prime-power factor, and 1, 2 or 4 according as α is 0 or 1, 2, or more than two, where $2^{\alpha} \| m$. Hence we have

THEOREM 5-2. *If $(a, m) = 1$ and the congruence $x^2 \equiv a$ (mod m) is solvable, it has exactly $2^{\sigma+\tau}$ solutions, where σ is the number of distinct odd prime divisors of m and τ is 0, 1, or 2 according as $4 \nmid m$, $2^2 \| m$, or $8 | m$.*

PROBLEMS

1. Decide whether 5 is a quadratic residue of 44.

2. Show that the product of the quadratic residues of a prime p is congruent to 1 or -1 (mod p) according as $p \equiv -1$ or 1 (mod 4). [*Hint:* Write the residues of p in terms of a primitive root.]

*3. Prove the following generalization of Wilson's theorem: The product of the positive integers less than m and prime to m is congruent to $-1 \pmod{m}$ if $m = 4$, p^α, or $2p^\alpha$, and to 1 otherwise. [*Hint:* Proceed as in the second proof of Wilson's theorem, associating a and a' if $aa' \equiv 1 \pmod{m}$. Use Theorem 5-2 to count the elements associated with themselves.]

5-3 Quadratic residues of primes, and the Legendre symbol.

As was seen in Section 5-2, the quadratic residues of powers of 2 can be given explicitly, and the quadratic residues of powers of an odd prime are identical with those of the prime itself. Consequently, there remains only the investigation of quadratic residues of odd primes. Hereafter we shall make use of the simplifying notation of the Legendre symbol (a/p), introduced at the end of Chapter 3. It will be recalled that for $(a, p) = 1$, we put

$$(a/p) = \begin{cases} 1, & \text{if } a \text{ is a quadratic residue of } p, \\ -1, & \text{if } a \text{ is a quadratic nonresidue of } p. \end{cases}$$

For completeness we put $(a/p) = 0$ if $p|a$, so that (a/p) is now defined for every odd prime p.

THEOREM 5-3. *The Legendre symbol (a/p) has the following properties:*

(a) $(ab/p) = (a/p)(b/p)$. *Thus the product of two residues or two nonresidues is a residue; the product of a residue and a nonresidue is a nonresidue.*

(b) *If $a \equiv b \pmod{p}$, then $(a/p) = (b/p)$.*

(c) $(a^2/p) = 1$ *if $p \nmid a$.*

(d) $(-1/p) = (-1)^{(p-1)/2}$.

Proof: The first two parts are obvious if $p|ab$, so suppose that $p \nmid ab$. In the proof of Theorem 3-24 it was shown that $(a/p) \equiv a^{(p-1)/2} \pmod{p}$. Hence

$$(ab/p) \equiv (ab)^{(p-1)/2} = a^{(p-1)/2}b^{(p-1)/2} \equiv (a/p)(b/p) \pmod{p},$$

and since (a/p) assumes only the values ± 1, it follows that $(ab/p) = (a/p)(b/p)$. Property (d) also follows immediately from this congruence. Properties (b) and (c) are obvious.

It follows from Theorem 5-3 that in investigating the Legendre symbol (a/p), there will be no loss in generality in assuming that a is a positive prime. For example, Theorem 5–3 shows that

$$(-48/31) = (-1/31)(48/31) = (-1/31)(3/31)(16/31)$$
$$= (-1/31)(3/31)$$
$$= (30/31)(3/31) = (2/31)(3/31)(5/31)(3/31)$$
$$= (2/31)(5/31),$$

so that $(-48/31)$ can be evaluated either from

$$(-48/31) = (-1)^{\frac{1}{2}(31-1)}(3/31) = -(3/31)$$

or from

$$(-48/31) = (2/31)(5/31).$$

In general, (a/p) can be written as the product of Legendre symbols, in which the first entries are the distinct prime divisors of a which divide a to an odd power.

Although it will be used only in the case where a is prime, the following theorem is valid for all a's for which $p \nmid a$.

THEOREM 5–4 (*Gauss's lemma*). *If μ is the number of elements of the set a, $2a$, $\ldots$, $\frac{1}{2}(p-1)a$ whose numerically least residues (mod p) are negative, then*

$$(a/p) = (-1)^{\mu}.$$

Example: If $a = 3$, $p = 31$, the numerically least residues (mod 31) of $3 \cdot 1, 3 \cdot 2, \ldots, 3 \cdot 15$ are $3, 6, 9, 12, 15, -13, -10, -7, -4, -1, 2, 5, 8, 11, 14$; thus $\mu = 5$, $(3/31) = -1$, and from the above numerical example, $(-48/31) = 1$.

Proof: Replace the numbers of the set a, $2a$, $\ldots$, $\frac{1}{2}(p-1)a$ by their numerically smallest residues (mod p); denote the positive ones by $r_1, r_2, \ldots$ and the negative ones by $-r_1', -r_2', \ldots$. Clearly no two r_i's are equal, and no two r_i''s are equal. If $m_1a \equiv r_i$ and $m_2a \equiv -r_j'$ (mod p), then $r_i = r_j'$ would imply $a(m_1 + m_2) \equiv 0$ (mod p), which implies $m_1 + m_2 \equiv 0$ (mod p), and this is impossible because the m's are strictly between 0 and $p/2$. Hence the $(p-1)/2$ numbers r_i, r_i' are distinct integers between 1 and $(p-1)/2$ inclusive, and are therefore exactly the numbers $1, 2, \ldots, (p-1)/2$ in some order. Hence,

$$a \cdot 2a \cdots \frac{p-1}{2}a \equiv (-1)^{\mu}\frac{p-1}{2}! \text{ (mod } p\text{)},$$

$$a^{(p-1)/2} \equiv (-1)^{\mu} \text{ (mod } p\text{)}.$$

Since also $a^{(p-1)/2} \equiv (a/p) \pmod{p}$, it follows that

$$(a/p) \equiv (-1)^{\mu} \pmod{p},$$

and finally,

$$(a/p) = (-1)^{\mu}.$$

In distinction to Euler's criterion, Gauss's lemma can be used to characterize the primes of which a given integer a is a quadratic residue. For example, if $a = 2$, then μ is the number of numbers $2m$, with $1 \leq m \leq (p-1)/2$, which are greater than $p/2$; this is clearly true if and only if $m > p/4$. Thus if we write $[x]$ to stand for the largest integer not exceeding x, it follows that

$$\mu = \frac{p-1}{2} - \left[\frac{p}{4}\right].$$

If now

$p = 8k+1$,   then $\mu = 4k - [2k + \frac{1}{4}] = 4k - 2k \equiv 0 \pmod 2$,

$p = 8k+3$,   then $\mu = 4k + 1 - [2k + \frac{3}{4}] = 4k + 1 - 2k \equiv 1 \pmod 2$,

$p = 8k+5$,   then $\mu = 4k + 2 - [2k + 1 + \frac{1}{4}] = 2k + 1 \equiv 1 \pmod 2$,

$p = 8k+7$,   then $\mu = 4k + 3 - [2k + 1 + \frac{3}{4}] = 2k + 2 \equiv 0 \pmod 2$,

and we deduce that 2 is a quadratic residue of primes of the form $8k \pm 1$ and a nonresidue of primes $8k \pm 3$. Since it happens that the quantity $(p^2 - 1)/8$ satisfies exactly the same congruences as μ above, this result can be stated in the following form.

THEOREM 5–5. $(2/p) = (-1)^{(p^2-1)/8}$.

As an application of Theorem 5–5, we have

THEOREM 5–6.   (a) *2 is a primitive root of the prime $p = 4q + 1$ if q is an odd prime.*

   (b)   *2 is a primitive root of $p = 2q + 1$ if q is a prime of the form $4k + 1$.*

   (c)   *-2 is a primitive root of $p = 2q + 1$ if q is a prime of the form $4k - 1$.*

Proof: (a) If $\mathrm{ord}_p 2 = t$, then $t | p - 1$, which is equivalent to saying that $t | 4q$. Aside from 4, every proper divisor of $4q$ is also a divisor of $2q$, and if $2^4 \equiv 1 \pmod p$, then p is 5 and q is not prime. Hence it suffices to show that $2^{2q} \not\equiv 1 \pmod p$. But

$$2^{2q} = 2^{(p-1)/2} \equiv (2/p) \equiv (-1)^{(p^2-1)/8} \equiv (-1)^{2q^2+q} \equiv -1 \pmod p.$$

Parts (b) and (c) can be proved in a similar fashion. Part (a) shows that 2 is a primitive root of 13, 29, 53, . . . ; part (b) shows that 2 is a primitive root of 11, 59, 83, . . . , and part (c) that −2 is a primitive root of 7, 23, 47, ... It is an unproved conjecture that 2 is a primitive root of infinitely many primes, which would follow from Theorem 5–5 if it could be shown that there are infinitely many primes p of the kinds described in (a) and (b).

Referring to (a), this requires a proof that the function $4x + 1$ assumes prime values for infinitely many prime arguments. Unfortunately, there is no nonconstant rational function known to have this property. If one could prove that the function $x + 2$ has it, one would have proved a conjecture which is one of the outstanding problems in additive number theory: that there are infinitely many "twin primes," such as 17 and 19, or 101 and 103.

PROBLEMS

1. Apply Gauss's lemma to determine the primes of which −2 is a quadratic residue, and show that your result is consistent with Theorem 5–3, parts (a) and (d), and Theorem 5–5.

2. Complete the proof of Theorem 5–6.

*3. Show that 7 is a primitive root of any prime of the form $2^{4n} + 1$ with $n > 0$. [*Hint:* Show first that it suffices to prove that $(7/p) = -1$, and then show that any *prime* of the specified form is congruent to 3 or 5 (mod 7). Note that $2^4 \equiv 2$ (mod 7).]

4. Show that the numbers $6k - 1$ and $6k + 1$ are twin primes if and only if the equation $k = 6xy \pm x \pm y$ has no solution in positive integers x and y for any of the four choices of sign. [Note that if $6k + 1 = mn$, then $m \equiv n \equiv \pm 1$ (mod 6).] Show that this characterizes all the twin primes except 3 and 5.

5–4 The law of quadratic reciprocity. Gauss's lemma can be used to establish a deep property of the Legendre symbol which is an essential tool both in determining the quadratic character of a prime q (mod p) and in finding the primes p of which q is a quadratic residue.

THEOREM 5–7 (*Quadratic reciprocity law*). *If p and q are distinct odd primes, then*

$$(p/q)(q/p) = (-1)^{\frac{1}{2}(p-1)\cdot\frac{1}{2}(q-1)}.$$

In other words, $(p/q) = (q/p)$ unless both p and q are of the form $4k - 1$, in which case $(p/q) = -(q/p)$.

Proof: By Gauss's lemma, the numbers μ and ν in the equations

$$(q/p) = (-1)^\mu, \qquad (p/q) = (-1)^\nu$$

are the numbers of the multiples

$$q, 2q, \ldots, \frac{p-1}{2} q,$$

and

$$p, 2p, \ldots, \frac{q-1}{2} p$$

whose absolutely smallest residues (mod p) and (mod q) respectively are negative, and we need only show that

$$\mu + \nu \equiv \frac{p-1}{2} \cdot \frac{q-1}{2} \pmod 2.$$

If y is chosen so that

$$-\frac{p}{2} < qx - py < \frac{p}{2},$$

then clearly $qx - py$ is the numerically smallest residue of qx (mod p). From this inequality we get

$$\frac{qx}{p} - \frac{1}{2} < y < \frac{qx}{p} + \frac{1}{2}.$$

Thus y is unique and non-negative; if $y = 0$ then $qx - py = qx > 0$, and there is no contribution to μ in this case. Moreover, we see that for $x \le (p-1)/2$,

$$\frac{qx}{p} - \frac{1}{2} < \frac{q-1}{2},$$

so that also $y \le (q-1)/2$. The number μ denotes therefore the number of combinations of x and y from the sequences

$(p) \qquad\qquad\qquad 1, 2, \ldots, \dfrac{p-1}{2}$

and

$(q) \qquad\qquad\qquad 1, 2, \ldots, \dfrac{q-1}{2},$

respectively, for which

$$0 > qx - py > -\frac{p}{2}.$$

Similarly, ν is the number of pairs x and y from the sequences (p) and (q) respectively, for which

$$0 > py - qx > -\frac{q}{2}.$$

For any other pair x and y from (p) and (q) respectively, either

$$py - qx > \frac{p}{2}$$

or

$$py - qx < -\frac{q}{2};$$

let there be λ of the former and ρ of the latter. Then clearly

$$\frac{p-1}{2} \cdot \frac{q-1}{2} = \mu + \nu + \lambda + \rho.$$

Finally, as x and y run through (p) and (q) respectively, the numbers

$$x' = \frac{p+1}{2} - x \quad \text{and} \quad y' = \frac{q+1}{2} - y$$

run through the same sequences, but in the opposite order. And if $py - qx > p/2$, then

$$py' - qx' = p\left(\frac{q+1}{2} - y\right) - q\left(\frac{p+1}{2} - x\right)$$

$$= \frac{p-q}{2} - (py - qx) < \frac{p-q}{2} - \frac{p}{2} = -\frac{q}{2}.$$

Hence $\lambda = \rho$, and

$$\frac{p-1}{2} \cdot \frac{q-1}{2} = \mu + \nu + 2\lambda \equiv \mu + \nu \pmod{2}.$$

By combining the law of quadratic reciprocity with the properties of the Legendre symbol mentioned in Theorem 5–3, it is easy to evaluate (q/p) if p and q do not lie beyond the extent of the available tables of factorizations of integers. For example, 2819 and 4177 are both primes and $4177 \equiv 1 \pmod 4$, so that

$$\begin{aligned}
(2819/4177) &= (4177/2819) = (1358/2819) = (2 \cdot 7 \cdot 97/2819) \\
&= (2/2819)(7/2819)(97/2819) \\
&= -1 \cdot - (2819/7)(2819/97) = (5/7)(6/97) \\
&= (7/5)(2/97)(97/3) \\
&= (2/5)(1/3) = -1,
\end{aligned}$$

and so 2819 is not a quadratic residue of 4177.

Moreover, the quadratic reciprocity law can be used to determine the primes p of which a given prime q is a quadratic residue. This result, which is contained in the next theorem, has sometimes been taken as the quadratic reciprocity law, rather than Theorem 5–7.

THEOREM 5–8. *Every $p \neq q$ can be uniquely represented in the form $4qk \pm a$, where $0 < a < 4q$ and $a \equiv 1 \pmod 4$. For a fixed odd prime q, the solutions of the equation $(q/p) = 1$ are exactly the primes $p \neq q$ such that the corresponding a is a quadratic residue of q; that is, $(q/p) = (a/q)$. The numbers a such that*

$$0 < a < 4q, \qquad a \equiv 1 \pmod 4 \qquad and \qquad (a/q) = 1, \qquad (1)$$

are given by the least positive residues $\pmod{4q}$ of the numbers $1^2, 3^2, 5^2, \ldots, (q-2)^2$.

Proof: Clearly every odd number can be written in the form $4qk' + a'$, where $1 \le a' < 4q$, and a' is odd. If $a' \equiv 1 \pmod 4$, take $a = a'$ and $k = k'$, while if $a' \equiv -1 \pmod 4$, take $a = 4q - a'$ and $k = k' + 1$. Thus every odd number, and therefore every p, has a representation either as $4qk + a$ (if $a' \equiv 1 \pmod 4$) or as $4qk - a$ (if $a' \equiv -1 \pmod 4$). This proves the first sentence.

If $p \equiv a \pmod{4q}$, then $p \equiv 1 \pmod 4$, so that

$$(q/p) = (p/q) = (a/q).$$

If, on the other hand, $p \equiv -a \pmod{4q}$, then $p \equiv -1 \pmod 4$, and

$$\begin{aligned}
(q/p) &= (-1)^{\frac{1}{2}(p-1)(q-1)}(p/q) = (-1)^{\frac{1}{2}(p-1)(q-1)}(-a/q) \\
&= (-1)^{\frac{1}{2}(p-1) \cdot \frac{1}{2}(q-1)}(-1)^{\frac{1}{2}(q-1)}(a/q) \\
&= (-1)^{\frac{1}{2}(p+1) \cdot \frac{1}{2}(q-1)}(a/q) = (a/q).
\end{aligned}$$

Thus always $(q/p) = (a/q)$, which proves the second sentence.

Finally, if $(a/q) = 1$, there is an x such that

$$x^2 \equiv a \pmod{q} \qquad \text{and} \qquad 1 \leq x \leq q - 1,$$

whence also

$$(q - x)^2 \equiv a \pmod{q} \qquad \text{and} \qquad 1 \leq q - x \leq q - 1.$$

Since either x or $q - x$ is odd—say x'—we have

$$x'^2 \equiv a \pmod{q}, \qquad 1 \leq x' \leq q - 2, \qquad x' \equiv 1 \pmod{2}.$$

But then

$$x'^2 \equiv 1 \equiv a \pmod{4},$$

so that

$$x'^2 \equiv a \pmod{4q},$$

and the proof is complete.

To illustrate, take $q = 3$. Then the only integer satisfying the conditions (1) is 1, so that 3 is a quadratic residue of primes $12k \pm 1$. Every other odd number is of one of the forms $12k \pm 3$ or $12k \pm 5$, and no prime except 3 occurs in the progressions $12k \pm 3$. Hence $(3/p)$ is completely determined by the equations

$$(3/p) = \begin{cases} 1, & \text{if } p \equiv \pm 1 \pmod{12}, \\ -1, & \text{if } p \equiv \pm 5 \pmod{12}. \end{cases}$$

Similarly, taking $q = 17$ we consider the squares

$$1^2, 3^2, 5^2, 7^2, 9^2, 11^2, 13^2, 15^2,$$

which reduce (mod 68) to

$$1, 9, 25, 49, 13, 53, 33, 21.$$

We have. that 17 is a quadratic residue of primes of the forms

$$68k \pm 1, 9, 13, 21, 25, 33, 49, \text{ and } 53,$$

and a nonresidue of primes of the forms

$$68k \pm 5, 29, 37, 41, 45, 57, 61, \text{ and } 65;$$

17 itself is the only prime of the forms $68k \pm 17$.

In general, out of the $2q$ progressions $4qk \pm a$, $q - 1$ contain only primes of which q is a residue, $q - 1$ contain only primes of which q is a nonresidue, and two (either $4qk \pm q$ or $4qk \pm 3q$, according as $q \equiv 1$ or 3 (mod 4)) contain no primes besides q itself.

Determining the primes of which a composite number is a quadratic residue is somewhat more complicated. To illustrate, consider the problem of finding the primes p for which $(10/p) = 1$. This requires that either $(2/p) = (5/p) = 1$ or $(2/p) = (5/p) = -1$, so that either

$$p \equiv \pm 1 \pmod 8 \quad \text{and} \quad p \equiv \pm 1 \pmod{10}$$

or

$$p \equiv \pm 3 \pmod 8 \quad \text{and} \quad p \equiv \pm 3 \pmod{10},$$

all combinations of signs being allowed. Thus we have the following pairs of congruences, each pair to be solved simultaneously:

$p \equiv 1 \pmod 8$	$p \equiv -1 \pmod 8$	$p \equiv 1 \pmod 8$
$p \equiv 1 \pmod{10}$	$p \equiv -1 \pmod{10}$	$p \equiv -1 \pmod{10}$
$p \equiv -1 \pmod 8$	$p \equiv 3 \pmod 8$	$p \equiv -3 \pmod 8$
$p \equiv 1 \pmod{10}$	$p \equiv 3 \pmod{10}$	$p \equiv -3 \pmod{10}$
$p \equiv 3 \pmod 8$	$p \equiv -3 \pmod 8$	
$p \equiv -3 \pmod{10}$	$p \equiv 3 \pmod{10}.$	

Solving (by the method of Problem 3, Section 3–4, for example), we obtain

$$p \equiv 1, -1, 9, 31, 3, -3, 27, 13 \pmod{40};$$

that is, 10 is a quadratic residue of the primes $40k \pm 1, 3, 9, 13$, and a nonresidue of the others.

PROBLEMS

1. Evaluate the Legendre symbols $(503/773)$ and $(501/773)$.

2. Characterize the primes of which 5 is a quadratic residue; those of which 6 is a quadratic residue.

3. Show that if $p = 4m + 1$ and $d|m$, then $(d/p) = 1$. [*Hint:* Let q be a prime divisor of m, and consider separately the cases $q = 2$ and $q > 2$.]

4. Deduce from the representation $N = 6119 = 82^2 - 5 \cdot 11^2$ that if $p|N$, then $(5/p) = 1$. Use this to find the factorization of N. (It suffices to consider $p < 80$.) Use similar ideas to factor $43993 = 211^2 - 2^4 \cdot 33$.

5. Prove that 4751 is prime.

5–5 An application. It is clear that if a given integer a is congruent to 1 $\pmod p$ for every prime p, then $a = 1$, since $p|(a - 1)$ implies $p \leq |a| + 1$ unless $a - 1 = 0$. Here we have an instance of the following principle: if an assertion involving a congruence holds for every prime modulus p, then the statement with the congruence replaced by the corresponding equation may be implied. With this in mind, it is natural to ask whether it is true that if, for fixed integers a and n, a is an nth power modulo p for every p, then a must be an nth power. (Saying that a is an nth power (mod p) means, of course, that a is congruent to the nth power of some integer; in other words, that a is an nth power residue of p.) Unfortunately, this is not quite the case: if the congruence $x^n \equiv a \pmod{p}$ is solvable for every p, then $a = b^n$ for some b if $8 \nmid n$, but if $8 \mid n$, either $a = b^n$ or $a = 2^{n/2} b^n$. Powers of 2 higher than the second cause difficulty here, just as they did in the study of primitive roots. (Cf. Problem 1 at the end of this section.)

At the present time, the theorem just stated cannot be proved in a simple way. Even in the special case $n = 2$ which we now treat, it is necessary to use a rather deep result about the existence of primes in certain arithmetic progressions.

THEOREM 5–9. *A fixed integer is a quadratic residue of every prime if and only if it is a square.*

Proof: If $a = b^2$, the congruence $x^2 \equiv a \pmod{p}$ has the solution $x \equiv b \pmod{p}$ for every p.

Suppose, on the other hand, that a is not a square. Then it can be written as $\pm m^2 p_1 p_2 \cdots p_r$, where $r \geq 1$ and $p_i \neq p_j$ if $i \neq j$. Suppose first that a is positive; then we wish to show the existence of a prime p such that

$$(a/p) = (m^2 p_1 \cdots p_r/p) = (p_1/p) \cdots (p_r/p) = -1.$$

We attempt to find a p such that $(p_i/p) = 1$ if $1 \leq i < r$, while $(p_r/p) = -1$. Here, of course, one of the primes $p_1, \ldots, p_r$ may be 2. But since 2 is a quadratic residue of primes $8k \pm 1$, and a nonresidue of primes $8k \pm 5$, the following statement is true for *every* prime q:

If $p \equiv 1 \pmod{4q}$, then $(q/p) = 1$. On the other hand, for each q there is a u such that $q \nmid u$, $u \equiv 1 \pmod{4}$, and if $p \equiv u \pmod{4q}$, then $(q/p) = -1$.

The first part is obvious. When $q = 2$, u may be taken to be 5 in the second part, while if $q > 2$, u may be taken as any of the N numbers remaining out of the q integers between 1 and $4q$ which are congruent to 1 (mod 4), after the removal of (a) the least positive residues (mod $4q$) of the $(q-1)/2$ squares $1^2, 3^2, \ldots, (q-2)^2$, and (b) that one of q, $3q$ which is congruent to 1 (mod 4). Since

$$N = q - \frac{q-1}{2} - 1 = \frac{q-1}{2} \geq 1,$$

such an integer u exists.

Now consider the system of congruences

$$x \equiv 1 \ (\text{mod } 4p_1)$$
$$\vdots$$
$$x \equiv 1 \ (\text{mod } 4p_{r-1})$$
$$x \equiv u \ (\text{mod } 4p_r),$$

when $r > 1$, or the single congruence

$$x \equiv u \ (\text{mod } 4p_1)$$

when $r = 1$, where u is the number characterized above, with $q = p_r$ or p_1. For $r > 1$, the necessary and sufficient condition that the system be solvable is, by Theorem 3–12, that for all i and j,

$$(4p_i, 4p_j) | (c_i - c_j),$$

where $c_i = 1$ if $i < r$ and $c_i = u$ if $i = r$. Since $c_i \equiv 1$ (mod 4) for every i, this requirement is clearly satisfied, so that the system can be replaced by a single congruence

$$x \equiv u' \ (\text{mod } 4p_1 \cdots p_r),$$

where $(u', 4p_1 \cdots p_r) = 1$. If now $p = 4p_1 \cdots p_r k + u'$ or $p = 4p_1 k + u$, in the cases $r > 1$ and $r = 1$, respectively, then

$$(a/p) = 1 \cdots 1 \cdot (-1) = -1,$$

and it is seen that in the case $a > 0$, the theorem is a consequence of the famous

DIRICHLET'S THEOREM. *If s and t are relatively prime, there are infinitely many primes of the form $sk + t$.*

Proofs of special cases of Dirichlet's theorem have been indicated in Problem 4 of Section 3–7 and Problem 5 of Section 4–3. The general theorem is proved in Volume II of this work.

If $a = -m^2$, then $(a/p) = -1$ if $p \nmid a$ and $p \equiv -1$ (mod 4). If $p_1, \ldots, p_k$ is any set of primes of the form $4k - 1$, then the number

$4p_1 \ldots p_k - 1$ has a prime divisor of this same form distinct from $p_1, \ldots, p_k$, so there are infinitely many primes of this form, and in particular there is one which does not divide a. For this p, $(a/p) = -1$.

If $a = -m^2 p_1 \cdots p_r$, where $r \geq 1$ and $p_i \neq p_j$ if $i \neq j$, then we must find a p such that $p \nmid a$ and

$$(-1/p)(p_1/p) \ldots (p_r/p) = -1.$$

But if p is a prime for which $(-a/p) = -1$, as determined above, then $p \equiv 1 \pmod 4$, so that $(a/p) = (-a/p)$. The proof is complete.

PROBLEMS

1. Show that the congruence

$$x^{2^\alpha} \equiv 2^{2^{\alpha-1}} \pmod p$$

has a solution for every prime p, if $\alpha \geq 3$. [*Hint:* Consider the factorization

$$x^{2^\alpha} - 2^{2^{\alpha-1}}$$

$$= (x^2-2)(x^2+2)((x-1)^2+1)((x+1)^2+1)(x^{2^3}+2^{2^2}) \cdots (x^{2^{\alpha-1}}+2^{2^{\alpha-2}}),$$

and show that every p divides one of the first three factors for suitable x.]

2. Show that if the congruence $x^n \equiv a \pmod m$ is solvable for every m, then a is an nth power. [*Hint:* Consider the moduli $p^{\alpha+1}$, where $p^\alpha \| a$ and α is positive.]

5–6 The Jacobi symbol. As was pointed out at the end of the proof of the law of quadratic reciprocity, it is necessary to have available rather extensive factorization tables if one is to evaluate Legendre symbols with large entries. Partly to obviate such a list, and partly for theoretical purposes, it has been found convenient to extend the definition of the Legendre symbol (a/p) so as to give meaning to (a/b) when b is not a prime. This is done in the following way: put $(a/1) = 1$, and if b is greater than 1 and odd, put

$$(a/b) = (a/p_1)(a/p_2) \cdots (a/p_r), \tag{2}$$

where $p_1 p_2 \cdots p_r$ is the prime factorization of b, and the symbols on the right in (2) are Legendre symbols. Then the symbol on the left in (2) is called a *Jacobi symbol;* like the Legendre symbol, it is undefined for even second entry. As we shall see, many more of its properties are similar to those of the Legendre symbol, but there is one crucial point at which the analogy breaks down: it may happen that $(a/b) = 1$ even when a is not a quadratic residue of b. For it is clearly necessary that each of the Legendre symbols (a/p_i) have the value 1 in order for a to be a residue of b, while $(a/b) = 1$ if an even number of the factors in (2) are -1 while the remainder are $+1$. On the other hand, a is certainly not a quadratic residue of b if $(a/b) = -1$.

The following theorem lists properties of the Jacobi symbol which were proved for the Legendre symbol in Theorems 5–3, 5–5, and 5–7, together with one (the second) which is peculiar to the extended function.

THEOREM 5–10.   *The Jacobi symbol has these properties:*
   (a) $(a_1a_2/b) = (a_1/b)(a_2/b)$.
   (b) $(a/b_1b_2) = (a/b_1)(a/b_2)$.
   (c) *If* $a_1 \equiv a_2 \pmod{b}$, *then* $(a_1/b) = (a_2/b)$.
   (d) $(-1/b) = (-1)^{(b-1)/2}$.
   (e) $(2/b) = (-1)^{\frac{1}{8}(b^2-1)}$.
   (f) *If* $(a, b) = 1$, *then* $(a/b)(b/a) = (-1)^{\frac{1}{2}(a-1)\cdot\frac{1}{2}(b-1)}$.
Here the second entry in each symbol is a positive odd number.

Proof: (a) Put $b = p_1 \cdots p_r$. Then

$$(a_1a_2/b) = (a_1a_2/p_1) \cdots (a_1a_2/p_r),$$

and since these are Legendre symbols,

$$(a_1a_2/b) = (a_1/p_1) \cdots (a_1/p_r)(a_2/p_1) \cdots (a_2/p_r) = (a_1/b)(a_2/b).$$

   (b) Put $b_1 = p_1 \cdots p_r$ and $b_2 = p_1' \cdots p_s'$. Then

$$
\begin{aligned}
(a/b_1b_2) &= (a/p_1 \cdots p_r p_1' \cdots p_s') \\
&= ((a/p_1) \cdots (a/p_r))((a/p_1') \cdots (a/p_s')) \\
&= (a/b_1)(a/b_2).
\end{aligned}
$$

   (c) If $a_1 \equiv a_2 \pmod{b}$ and $b = p_1 \cdots p_r$, then $a_1 \equiv a_2 \pmod{p_i}$ for $i = 1, \ldots, r$. Hence $(a_1/p_i) = (a_2/p_i)$, and

$$(a_1/b) = (a_1/p_1) \cdots (a_1/p_r) = (a_2/p_1) \cdots (a_2/p_r) = (a_2/b).$$

   (d) Put $b = p_1 \cdots p_r$. Then

$$(-1/b) = \prod_{i=1}^{r} (-1/p_i) = \prod_{i=1}^{r} (-1)^{(p_i-1)/2}$$

or $$(-1/b) = (-1)^{\sum_{i=1}^{r} \frac{p_i-1}{2}} \tag{3}$$

But if m and n are odd, then

$$(m - 1)(n - 1) \equiv 0 \ (\text{mod } 4),$$

$$mn - 1 \equiv m + n - 2 \ (\text{mod } 4),$$

$$\frac{mn - 1}{2} \equiv \frac{m - 1}{2} + \frac{n - 1}{2} \ (\text{mod } 2).$$

Repeated application of this fact shows that

$$\sum_{i=1}^{r} \frac{p_i - 1}{2} \equiv \frac{p_1 \cdots p_r - 1}{2} \ (\text{mod } 2),$$

so that $(-1/b) = (-1)^{(b-1)/2}$, by (3).

(e) The proof of this is the same as that just given, except that, using the fact that $m^2 \equiv 1 \ (\text{mod } 8)$ if m is odd, we deduce from the congruence

$$(m^2 - 1)(n^2 - 1) \equiv 0 \ (\text{mod } 64)$$

that

$$\frac{m^2 - 1}{8} + \frac{n^2 - 1}{8} \equiv \frac{(mn)^2 - 1}{8} \ (\text{mod } 2).$$

(f) Put $a = p_1 \cdots p_r$, $b = p_1' \cdots p_s'$. Then

$$(a/b)(b/a) = \prod_{i=1}^{s} (a/p_i') \prod_{j=1}^{r} (b/p_j)$$

$$= \prod_{j=1}^{r} \prod_{i=1}^{s} (p_j/p_i') \cdot \prod_{j=1}^{r} \prod_{i=1}^{s} (p_i'/p_j)$$

$$= \prod_{j=1}^{r} \prod_{i=1}^{s} (p_j/p_i')(p_i'/p_j)$$

$$= (-1)^{\sum_{j=1}^{r} \sum_{i=1}^{s} \frac{p_j-1}{2} \cdot \frac{p_i'-1}{2}} = (-1)^{\sum_{j=1}^{r} \frac{p_j-1}{2} \cdot \sum_{i=1}^{s} \frac{p_i'-1}{2}}$$

$$= (-1)^{\frac{1}{2}(a-1) \cdot \frac{1}{2}(b-1)}.$$

Because the laws of operation and combination are the same for the two types, Jacobi symbols can be used (and according to the same rules) in evaluating Legendre symbols, even though they do not give complete information about the quadratic character of a modulo b; all that is required is that one begin with a Legendre symbol. This means that the first entry in each symbol does not have to be factored, except that powers of 2 must be removed. Thus, using the numerical example considered earlier, we have

$$(2819/4177) = (4177/2819) = (1358/2819) = (2/2819)(679/2819)$$
$$= -(679/2819) = (2819/679) = (103/679)$$
$$= -(679/103) = -(61/103) = -(103/61)$$
$$= -(42/61) = -(2/61)(21/61) = (61/21)$$
$$= (19/21) = (21/19) = (2/19) = -1,$$

and we can again conclude that 2819 is a nonresidue of 4177.

PROBLEM

Evaluate (751/919), both with and without the use of Jacobi symbols. The entries are primes.

REFERENCES

Section 5–5

The general theorem stated in the first paragraph is due to E. Trost, *Nieuw Archief voor Wiskunde* (Amsterdam) **18**, 58–61 (1934). It has been generalized by H. Flanders, *Annals of Mathematics* **57**, 392–400 (1953).

CHAPTER 6

NUMBER-THEORETIC FUNCTIONS AND THE DISTRIBUTION OF PRIMES

6–1 Introduction. A number-theoretic function is any function which is defined for positive integral argument or arguments. Euler's φ-function is such, as are $n!$, n^2, e^n, etc. The functions which are interesting from the point of view of number theory are, of course, those like φ whose value depends in some way on the arithmetic nature of the argument, and not simply on its size. Two of the most interesting of such functions are $\tau(n)$, the number of positive divisors of n, and $\sigma(n)$, the sum of these divisors. These functions have been treated extensively in the literature, partly because of their simplicity and partly because they occur in a natural way in the investigation of many other problems. For this reason we shall pause briefly to demonstrate some of their fundamental properties. Recall that, as noted in Chapter 3, a number-theoretic function which is not identically zero is said to be multiplicative if $f(mn) = f(m)f(n)$ whenever $(m, n) = 1$.

THEOREM 6–1. *The functions σ and τ are multiplicative.*

Proof: Assume that $(m, n) = 1$. Then by the Unique Factorization Theorem, every divisor of mn can be represented uniquely as the product of a divisor of m and a divisor of n, and conversely, every such product is a divisor of mn. Clearly this implies that τ is multiplicative, and that

$$\sum_{d|m} d \cdot \sum_{d'|n} d' = \sum_{d''|mn} d'',$$

so that also $\sigma(m)\sigma(n) = \sigma(mn)$.

If f is any multiplicative function and the prime-power factorization of n is

$$n = \prod_{i=1}^{r} p_i^{\alpha_i},$$

then clearly

$$f(n) = f\left(\prod_{i=1}^{r} p_i^{\alpha_i}\right) = \prod_{i=1}^{r} f(p_i^{\alpha_i}),$$

and so the function is completely determined when its value is known for every prime-power argument. In the cases at hand, we have

$$\tau(p^\alpha) = \alpha + 1$$

and $$\sigma(p^\alpha) = 1 + p + \cdots + p^\alpha = \frac{p^{\alpha+1} - 1}{p - 1}.$$

Thus we have proved

THEOREM 6-2. *If $n = p_1{}^{\alpha_1} \ldots p_r{}^{\alpha_r}$, then*

$$\tau(n) = \prod_{i=1}^{r} (\alpha_i + 1) \quad and \quad \sigma(n) = \prod_{i=1}^{r} \frac{p_i{}^{\alpha_i+1} - 1}{p_i - 1}.$$

There is another way of proving the multiplicativity of σ and τ which uses a basic property of all multiplicative functions:

THEOREM 6-3. *If f is multiplicative and F is the function defined by the equation*

$$F(n) = \sum_{d \mid n} f(d),$$

then F is also multiplicative.

Remark: The multiplicativity of σ and τ follows immediately from the relations

$$\sigma(n) = \sum_{d \mid n} d, \qquad \tau(n) = \sum_{d \mid n} 1,$$

since the functions f_1 and f_2 defined by the equations

$$f_1(n) = n \quad and \quad f_2(n) = 1 \quad for all n$$

are obviously multiplicative.

Proof: Let $(m, n) = 1$. Then every divisor d of mn can be written uniquely as the product of a divisor d_1 of m and a divisor d_2 of n, and $(d_1, d_2) = 1$. Hence

$$F(mn) = \sum_{d \mid mn} f(d) = \sum_{\substack{d_1 \mid m \\ d_2 \mid n}} f(d_1 d_2) = \sum_{\substack{d_1 \mid m \\ d_2 \mid n}} f(d_1) f(d_2)$$

$$= \sum_{d_1 \mid m} f(d_1) \cdot \sum_{d_2 \mid n} f(d_2) = F(m) F(n).$$

We shall see in the next section that the converse of Theorem 6–3 also holds.

A problem that was of great interest to the Greeks was that of determining all the *perfect numbers*, that is, numbers such as 6 which are equal to the sum of their proper divisors. In our notation this amounts to asking for all solutions of the equation

$$\sigma(n) = 2n.$$

It was known as early as Euclid's time that every number of the form

$$n = 2^{p-1}(2^p - 1),$$

in which both p and $2^p - 1$ are primes, is perfect. This is easy to verify:

$$\sigma(n) = \frac{2^p - 1}{2 - 1} \cdot \frac{(2^p - 1)^2 - 1}{(2^p - 1) - 1} = (2^p - 1) \cdot 2^p = 2n.$$

It happens that a partial converse also holds: every *even* perfect number is of the Euclid type. To see this we put $n = 2^{k-1} \cdot n'$, where $k \geq 2$. Then

$$\sigma(n) = \sigma(2^{k-1})\sigma(n') = (2^k - 1)\sigma(n'),$$

so that if n is perfect, it must be that

$$(2^k - 1)\sigma(n') = 2n = 2^k n'.$$

This implies that $(2^k - 1)|n'$, so we put $n' = (2^k - 1)n''$ and obtain

$$\sigma(n') = 2^k n''.$$

Since n' and n'' are divisors of n' whose sum is

$$n'' + (2^k - 1)n'' = 2^k n'' = \sigma(n'),$$

it must be that they are the only divisors of n', so that n' must be prime, and so $n'' = 1$, $n' = 2^k - 1$. Thus $n = 2^{k-1}(2^k - 1)$, where $2^k - 1$ is prime; this can happen only if k itself is prime.

There are two problems connected with perfect numbers which have not yet been solved. One is whether there are any odd perfect numbers; various necessary conditions are known for an odd number to be perfect, which show that any such number must be extremely large, but no conclusive results have been obtained. The other question is about the primes p for which $2^p - 1$ is prime. These *Mersenne primes* $2^p - 1$ are completely known for $p < 2300$ (the corresponding p's are 2, 3, 5, 7, 13, 17, 19, 31, 61, 89, 107, 127, 521, 607, 1279, 2203, 2281), but it is not known whether there are infinitely many such primes.

Aside from φ, σ, and τ, the function with which we shall be most concerned in this chapter is $\pi(x)$, already defined in Chapter 1 as the number of primes not exceeding x. (We now drop the restriction that all variables are integer-valued.) It was shown there that $\pi(x)$ increases indefinitely with x, that is, that *there are infinitely many primes*. We now give another proof, which depends on the Unique Factorization Theorem.

Assume that there are only k primes, say $p_1, \ldots, p_k$. By the Unique Factorization Theorem, every integer larger than 1 can be written uniquely as the product of a square-free number (that is, an integer which is the product of distinct primes) and a square. But with only k primes at our disposal, there are only

$$\binom{k}{1} + \cdots + \binom{k}{k-1} + 1 = 2^k - 1 < 2^k$$

square-free numbers, and there are not more than $\sqrt{n}$ perfect squares less than or equal to n. This means that there are fewer than $2^k \cdot \sqrt{n}$ positive integers not exceeding n, which is obviously false if $n \geq 2^k \sqrt{n}$, that is if $\sqrt{n} \geq 2^k$. Actually, this argument proves a little more, namely that

$$2^{\pi(n)} > \sqrt{n}, \quad \text{or} \quad \pi(n) > \frac{\log n}{2 \log 2}.$$

For later use in this chapter we now prove a general combinatorial theorem of very wide applicability. (The product representation for the φ-function, for example, is a special case.) The result is sometimes called the principle of cross-classification.

THEOREM 6–4. *Let S be a set of N distinct elements, and let $S_1, \ldots, S_r$ be arbitrary subsets of S containing $N_1, \ldots, N_r$ elements, respectively. For $1 \leq i < j < \cdots < l \leq r$, let $S_{ij\ldots l}$ be the intersection of $S_i, S_j, \ldots, S_l$, that is, the set of all elements of S common to $S_i, S_j, \ldots, S_l$; and let $N_{ij\ldots l}$ be the number of elements of $S_{ij\ldots l}$. Then the number of elements of S not in any of $S_1, \ldots, S_r$ is*

$$K = N - \sum_{1 \leq i \leq r} N_i + \sum_{1 \leq i < j \leq r} N_{ij} - \sum_{1 \leq i < j < k \leq r} N_{ijk} + \cdots$$
$$+ (-1)^r N_{12\cdots r}.$$

Remark: To obtain the product formula for the φ-function, take S to be the set of integers $1, \ldots, n$, and for $1 \leq k \leq r$, take S_k to be the set of elements of S which are divisible by p_k, where $n = p_1^{\alpha_1} \cdots p_r^{\alpha_r}$.

If $d|n$, the number of integers $s \le n$ such that $d|s$ is n/d; hence

$$\varphi(n) = n - \sum_{1 \le i \le r} \frac{n}{p_i} + \sum_{1 \le i < j \le r} \frac{n}{p_i p_j} - \cdots = n \prod_{p|n} \left(1 - \frac{1}{p}\right).$$

Proof: Let a certain element s of S belong to exactly m of the sets $S_1, \ldots, S_r$. If $m = 0$, s is counted just once, in N itself. If $0 < m \le r$, then s is counted once, or $\binom{m}{0}$ times, in N, $\binom{m}{1}$ times in the terms N_i, $\binom{m}{2}$ times in the terms N_{ij}, etc. Hence the total contribution to K arising from the element s is

$$\binom{m}{0} - \binom{m}{1} + \binom{m}{2} - \cdots + (-1)^m \binom{m}{m} = (1 - 1)^m = 0.$$

PROBLEMS

1. Find an expression for $\sigma_k(n)$, the sum of the kth powers of the divisors of n.

2. Prove that

$$\sum_{d|n} \tau^3(d) = \left(\sum_{d|n} \tau(d)\right)^2.$$

[*Hint:* Both sides are multiplicative functions, so it suffices to consider the case $n = p^\alpha$. Cf. Problem 2, Section 1-2.]

3. Show that, if $\sigma(n)$ is odd, then n is a square or the double of a square.

4. Show that the number of representations of an integer n as a sum of one or more consecutive positive integers is $\tau(n_1)$, where n_1 is the largest odd divisor of n. [*Hint:* If

$$n = (r+1) + (r+2) + \cdots + (r+s) = \sum_{k=1}^{r+s} k - \sum_{k=1}^{r} k = \tfrac{1}{2}s(s + 2r + 1),$$

then either s or $s + 2r + 1$ divides n_1.]

*5. Show that the number of ordered pairs of integers whose LCM is n is $\tau(n^2)$.

6. Show that

$$\sum_{n=1}^{\infty} \frac{\tau(n)}{n^s} = \left(\sum_{n=1}^{\infty} \frac{1}{n^s}\right)^2$$

if $s > 1$. [The series involved converge absolutely, and therefore can be rearranged in any order.]

*7. (a) Show that the sum of the odd divisors of n is

$$-\sum_{d|n} (-1)^{n/d} d.$$

[*Hint:* Let d_1 be an odd divisor of n, and find the total contribution to this sum from all divisors of n of the form $2^k d_1$.]

(b) Show that if n is even, then

$$\sum_{d|n} (-1)^{n/d} d = 2\sigma(n/2) - \sigma(n).$$

*8. Show that, if $d|n$ and $(n, r) = 1$, then the number of solutions (mod n)
of

$$x \equiv r \pmod{d}, \qquad (x, n) = 1$$

is

$$\frac{\varphi(n)}{\varphi(d)} = \frac{n}{d} \prod_{\substack{p|n \\ p\nmid d}} \left(1 - \frac{1}{p}\right).$$

[*Hint:* Take S of Theorem 6–4 to be the n/d numbers

$$x = r + td, \qquad 1 \leq t \leq n/d.$$

If $p|d$, then $p\nmid x$. Let the subsets consist of those elements of S divisible by the various primes which divide n but not d.]

6–2 The Möbius function. As we saw in Theorem 6–3, if f is any multiplicative function and F is its sum function, so that

$$F(n) = \sum_{d|n} f(d),$$

then F is also multiplicative. We now ask whether the converse is true—whether the multiplicativity of F implies the multiplicativity of f. To this end we attempt to express $f(n)$ as a sum, over the divisors of n, of terms involving $F(d)$. Assuming that F is multiplicative, it is enough to consider $F(p^n)$, and if the converse in question is valid we can also restrict attention to $f(p^n)$. Since

$$f(p^n) = F(p^n) - F(p^{n-1}),$$

we can write

$$f(p^n) = \sum_{\alpha=0}^{n} \mu(p^{n-\alpha}) F(p^\alpha) = \sum_{d|p^n} \mu\left(\frac{p^n}{d}\right) F(d),$$

if we define the function μ in the following way:

$$\mu(1) = 1,$$
$$\mu(p) = -1,$$
$$\mu(p^n) = 0 \qquad \text{for } n > 1.$$

If we now require in addition that μ be multiplicative, then $\mu(n)$ is defined for all positive integral n, and it is easily seen that

$$\mu(n) = \begin{cases} 1 & \text{if } n = 1, \\ 0 & \text{if } n \text{ is divisible by a square larger than 1,} \\ (-1)^\nu & \text{if } n = p_1 \cdots p_\nu, \text{ where the } p_i \text{ are distinct primes.} \end{cases}$$

This function μ is commonly called the *Möbius function*; it plays an important role in the theory of numbers. On the basis of the heuristic argument above, it is reasonable to conjecture that, for any n,

$$f(n) = \sum_{d|n} \mu\left(\frac{n}{d}\right) F(d),$$

and that from this formula one might be able to deduce the multiplicativity of f from that of F. We now substantiate these conjectures.

THEOREM 6-5. $\displaystyle\sum_{d|n} \mu(d) = \begin{cases} 1 & \text{if } n = 1, \\ 0 & \text{if } n > 1. \end{cases}$

Proof: By Theorem 6-3, the function

$$M(n) = \sum_{d|n} \mu(d)$$

is multiplicative, and since

$$M(p^\alpha) = \begin{cases} 1, & \text{if } \alpha = 0, \\ 1 - 1 + 0 + \cdots + 0, & \text{if } \alpha \geq 1, \end{cases}$$

we see that $M(n) = 0$ if n is divisible by any prime, that is, if $n > 1$.

THEOREM 6-6 (*Möbius inversion formula*). *If f is any number-theoretic function (not necessarily multiplicative) and*

$$F(n) = \sum_{d|n} f(d),$$

then

$$f(n) = \sum_{d|n} F(d)\mu\left(\frac{n}{d}\right) = \sum_{d|n} F\left(\frac{n}{d}\right)\mu(d).$$

Proof: We have

$$\sum_{d|n} \mu(d) F\left(\frac{n}{d}\right) = \sum_{d_1 d_2 = n} \mu(d_1) F(d_2) = \sum_{d_1 d_2 = n} \mu(d_1) \sum_{d|d_2} f(d)$$
$$= \sum_{d_1 d|n} \mu(d_1) f(d) = \sum_{d|n} f(d) \sum_{d_1 | \frac{n}{d}} \mu(d_1),$$

and, by Theorem 6-5, the coefficient of $f(d)$ is zero unless $n/d = 1$ (that is, unless $d = n$), when it is 1, so that this last sum is equal to $f(n)$.

As an example of Theorem 6-6, we have

THEOREM 6-7. $\displaystyle\varphi(n) = n \sum_{d|n} \frac{\mu(d)}{d}.$

This follows immediately from Theorem 3–9:

$$\sum_{d|n} \varphi(d) = n.$$

It can also be obtained directly from the product representation of $\varphi(n)$:

$$\varphi(n) = n \prod_{p|n} \left(1 - \frac{1}{p}\right) = n \left(1 + \sum_{\substack{p_1 \cdots p_\nu | n \\ p_1 < \cdots < p_\nu}} \frac{(-1)^\nu}{p_1 \cdots p_\nu}\right) = n \sum_{d|n} \frac{\mu(d)}{d} .$$

THEOREM 6–8. *If*

$$F(n) = \sum_{d|n} f(d)$$

and F is multiplicative, so is f.

Proof: If $(m, n) = 1$, then

$$f(mn) = \sum_{\substack{d_1|m \\ d_2|n}} F(d_1 d_2)\mu\left(\frac{mn}{d_1 d_2}\right)$$

$$= \sum_{\substack{d_1|m \\ d_2|n}} F(d_1)F(d_2)\mu\left(\frac{m}{d_1}\right)\mu\left(\frac{n}{d_2}\right)$$

$$= \sum_{d_1|m} F(d_1)\mu\left(\frac{m}{d_1}\right) \sum_{d_2|n} F(d_2)\mu\left(\frac{n}{d_2}\right) = f(m)f(n).$$

PROBLEMS

*1. Show that

$$\frac{1}{\varphi(n)} = \frac{1}{n} \sum_{d|n} \frac{\mu^2(d)}{\varphi(d)} .$$

2. Show that

$$\sum_{d^2|n} \mu(d) = |\mu(n)|.$$

3. Let f be any number-theoretic function of two variables. Show that if F is defined by the equation

$$F(m, n) = \sum_{\substack{d_1|m \\ d_2|n}} f(d_1, d_2),$$

then

$$f(m, n) = \sum_{\substack{d_1|m \\ d_2|n}} \mu(d_1)\mu(d_2)F\left(\frac{m}{d_1}, \frac{n}{d_2}\right).$$

*4. Let $J_k(n)$ be the number of ordered sets of k equal or distinct positive integers, none of which exceeds n and whose GCD is prime to n. Show, in the order indicated, that

(a) $\sum\limits_{d|n} J_k(d) = n^k$,

(b) J_k is multiplicative,

(c) $J_k(n) = n^k \prod\limits_{p|n} \left(1 - \dfrac{1}{p^k}\right).$

5. Let
$$\Lambda(n) = \begin{cases} \log p & \text{if } n \text{ is a power of any prime } p, \\ 0 & \text{otherwise.} \end{cases}$$

Show that
$$\log n = \sum_{d|n} \Lambda(d),$$

and deduce that
$$\sum_{d|n} \mu(d) \log d = -\Lambda(n).$$

6. If ϑ is any multiplicative function, then the function ϑ' defined by the equation
$$\sum_{d|n} \vartheta(d)\vartheta'\left(\frac{n}{d}\right) = \begin{cases} 1 & \text{if } n = 1, \\ 0 & \text{if } n > 1, \end{cases}$$

is also multiplicative. In this notation, find μ' and τ'.

7. If ϑ and ϑ' have the relation specified in Problem 6 and if
$$F(n) = \sum_{d|n} f(d)\vartheta\left(\frac{n}{d}\right),$$

then
$$f(n) = \sum_{d|n} F(d)\vartheta'\left(\frac{n}{d}\right).$$

8. Show that if f is multiplicative, then
$$\sum_{d|n} \mu(d)f(d) = \prod_{p|n} (1 - f(p)).$$

[*Hint:* Show that the function on the left is multiplicative.]

6–3 The function $[x]$. Another function which is of importance in number theory is the function $[x]$, introduced in the last chapter to represent the largest integer not exceeding x. In other words, for each real x, $[x]$ is the unique integer such that
$$x - 1 < [x] \leq x < [x] + 1.$$

For later purposes we list some of the properties of $[x]$.

(a) $x = [x] + \vartheta$, where $0 \le \vartheta < 1$. ϑ is called the *fractional part* of x.

(b) $[x + n] = [x] + n$, if n is an integer.

(c) $[x] + [-x] = \begin{cases} 0 & \text{if } x \text{ is an integer,} \\ -1 & \text{otherwise.} \end{cases}$

(d) $[x_1] + [x_2] \le [x_1 + x_2]$.

(e) $[x/n] = [[x]/n]$ if n is a positive integer.

(f) $0 \le [x] - 2[x/2] \le 1$. (Equivalently, $[x] - 2[x/2]$ assumes only the values 0 and 1.)

(g) The number of integers m for which $x_1 < m \le x_2$ is $[x_2] - [x_1]$.

(h) The number of multiples of m which do not exceed x is $[x/m]$.

(i) The least non-negative residue of a, modulo m, is the number a' defined by the equation

$$a = m \left[\frac{a}{m} \right] + a'.$$

These properties may easily be proved using the definition of $[x]$ and the first property above.

Another quantity closely related to $[x]$ is the nearest integer to x, which is $[x + \frac{1}{2}]$. Sometimes the quantity $-[-x]$ is also useful; it is the smallest integer not less than x.

In order to simplify the notation, summation signs will sometimes be used with the real variable x as upper limit. In these cases, it is understood that the summation variable takes values up to $[x]$; in other words,

$$\sum_{k=a}^{x} f(k) = \sum_{k=a}^{[x]} f(k).$$

The following relation between the greatest integer function and the factorial function will be of importance later.

THEOREM 6–9. *If n is a positive integer, the exponent of the highest power of a prime p which divides $n!$ is*

$$\left[\frac{n}{p} \right] + \left[\frac{n}{p^2} \right] + \left[\frac{n}{p^3} \right] + \cdots.$$

That is, if we set
$$\sum_{k=1}^{\infty} \left[\frac{n}{p^k}\right] = E(p, n),$$

then
$$p^{E(p,n)} \big\| n!.$$

Remark: The sum has, of course, only finitely many nonzero terms.

Proof: The multiples of p from among the numbers $1, 2, \ldots, n$ are counted once each in $[n/p]$, those which are also multiples of p^2 are counted again in $[n/p^2]$, etc. Thus if $p^r \| m$, the total contribution to the sum

$$\left[\frac{n}{p}\right] + \left[\frac{n}{p^2}\right] + \cdots$$

from the number m is exactly r, as it should be.

PROBLEMS

1. Carry out the proofs of the properties of $[x]$ listed in the text.

2. Prove that $[2x] + [2y] \geq [x] + [y] + [x + y]$, where x and y are arbitrary real numbers. [*Hint:* Consider separately the cases that neither, one, or both of $x - [x]$, $y - [y]$ are greater than $\frac{1}{2}$.]

3. Let $f(x, n)$ be the number of integers less than or equal to x and prime to n. Show that

(a) $\sum_{d|n} f\left(\frac{x}{d}, \frac{n}{d}\right) = [x]$. [Parallel the proof of Theorem 3–9.]

(b) $f(x, n) = \sum_{d|n} \mu(d) \cdot \left[\frac{x}{d}\right]$.

4. Let x be a number between 0 and 1. Let a_1 be the smallest positive integer such that

$$x_1 = x - \frac{1}{a_1} \geq 0,$$

let a_2 be the smallest positive integer such that

$$x_2 = x_1 - \frac{1}{a_2} \geq 0,$$

etc. Show that this leads to a finite expansion

$$x = \frac{1}{a_1} + \frac{1}{a_2} + \cdots + \frac{1}{a_n}$$

(that is, that $x_{n+1} = 0$ for some n) if and only if x is rational.

5. (a) Show that

$$\frac{(ab)!}{a!\,(b!)^a}$$

is an integer if a and b are positive integers. [*Hint:* Use induction on a.]

(b) Show that

$$\frac{(2a)!\,(2b)!}{a!\,b!\,(a+b)!}$$

is an integer. [*Hint:* Use Problem 2.]

6–4 The symbols "O", "o", and "$\sim$". If we construct tables of values of the common number-theoretic functions, we are immediately struck by how erratically they behave. Thus $\tau(n)$ can be arbitrarily large, since for example $\tau(2^m) = m + 1$, and yet $\tau(n) = 2$ whenever n is prime. Neither φ nor σ varies quite so wildly, in the sense that each of them definitely grows with n, but they are still far from monotonic. It is one of the objects of this chapter to see what can be said about the size of these and other functional values simply in terms of the size of their arguments.

A very convenient notation has been introduced by Landau for use in this connection. Let $g(x)$ be defined and positive for all positive x. Then if $f(x)$ is any function defined on some unbounded set S of positive numbers (which in all applications here will be either the set of positive integers or the set of positive real numbers), and if there is a number M such that

$$\frac{|f(x)|}{g(x)} < M$$

for all sufficiently large $x \in S$, then we write $f(x) = O(g(x))$. (The symbol $\in$ means "is an element of.") If

$$\lim_{\substack{x \to \infty \\ x \in S}} \frac{f(x)}{g(x)} = 0,$$

we write $f(x) = o\big(g(x)\big)$, and if

$$\lim_{\substack{x \to \infty \\ x \in S}} \frac{f(x)}{g(x)} = 1,$$

we write $f(x) \sim g(x)$, and say that $f(x)$ is asymptotically equal to $g(x)$. For example,

$$\sin x = O(x),$$
$$\sin x = o(x),$$
$$\sin x = O(1),$$
$$\varphi(n) = O(n),$$
$$\sqrt{x} = o(x),$$

$x^k = o(e^x)$ for every constant k,

$\log^k x = o(x^\alpha)$ for every pair of constants $\alpha > 0$ and k,

$$[x] \sim x.$$

Here each of the second and third equations gives more information than the one preceding it; the first says that $\sin x$ does not grow any faster than x itself, the second that it does not grow as fast, and the third that $\sin x$ remains bounded as x increases. In the fourth equation, $O(n)$ could not be replaced by $o(n)$, since $\varphi(p) = p - 1 \sim p$.

The purpose of introducing these symbols is that, by their use, a complicated expression can be replaced by its principal or largest term, plus a remainder or error term whose possible size is indicated. Retaining an estimate for the error term is necessary because if several such expressions are combined, one has eventually to show that the sum of the error terms is still of smaller order of magnitude than the principal term. This in turn makes it necessary to combine terms involving "O" and "o". The following abbreviated rules apply:

(a) $O(O(g(x))) = O(g(x))$,

(b) $O(o(g(x))) = o(O(g(x))) = o(o(g(x))) = o(g(x))$,

(c) $O(g(x)) \pm O(g(x)) = O(g(x)) \pm o(g(x)) = O(g(x))$,

(d) $o(g(x)) \pm o(g(x)) = o(g(x))$,

(e) $\{O(g(x))\}^2 = O(g^2(x))$,

(f) $O(g(x)) \cdot o(g(x)) = \{o(g(x))\}^2 = o(g^2(x))$.

The meaning of the first statement, for example, is that if $f(x) = O(g(x))$ and $h(x) = O(f(x))$, then $h(x) = O(g(x))$; this follows from the fact that, if $0 < f(x) < M_1 g(x)$ and $|h(x)| < M_2 f(x)$, then $|h(x)| < M_1 M_2 g(x)$. The other assertions are equally straightforward; they need not be remembered explicitly, but are listed here to help orient the student, who should analyze all of them. Notice that suitable combinations of these rules give more general ones; for example, rules (a) and (c) show that

$$O(f(x)) \pm O(g(x)) = O(\max (f(x), g(x))).$$

A useful fact to remember is that the implication

$$f(x) = O(g(x)) \qquad \text{implies} \qquad h(f(x)) = O(h(g(x)))$$

does not hold in general; a sufficient condition is that $h(kx) = O(h(x))$ for every positive constant k, if $h(x), f(x) \to \infty$ as $x \to \infty$. Thus if $f(x)$ is larger than some positive constant for every $x > 0$, then $f(x) = O(g(x))$ implies that $\log f(x) = O (\log g(x))$, but it does not imply that

$$e^{f(x)} = O(e^{g(x)}),$$

since, for example, $\log x = O (\log \sqrt{x})$ but $x \neq O(\sqrt{x})$.

The situation is quite different for the "o" symbol. If $f(x) = o(g(x))$, then

$$e^{f(x)} = o(e^{g(x)})$$

if $f(x)$ increases indefinitely with x, but the relation $\log f(x) = o(\log g(x))$ may be false, e.g., if $f(x) = \sqrt{x}$, $g(x) = x$.

Another important point arises when we want to add together a set of error terms, the number $a(x)$ of such terms being an increasing function of x. It is not true without restriction that

$$\sum_{k=1}^{a(x)} O(g_k(x)) = O \left(\sum_{k=1}^{a(x)} g_k(x) \right),$$

since, for example,

$$x = O(x), \qquad 2x = O(x), \qquad \dots,$$

but

$$\sum_{k=1}^{[x]} kx \neq O \left(\sum_{k=1}^{[x]} x \right).$$

What is needed here, of course, is that the constants implied in the symbols $O(g_k(x))$ all be bounded above by some number independent of k. The corresponding principle for the "o" symbol is this: if $f_k(x) = o(g_k(x))$, then we can write $f_k(x) = \epsilon_k(x)g_k(x)$, where $\epsilon_k(x) \to 0$ as $x \to \infty$, for fixed k, and if $\max (|\epsilon_1(x)|, \dots, |\epsilon_{a(x)}(x)|) \to 0$ as $x \to \infty$, then

$$\sum_{k=1}^{a(x)} f_k(x) = o \left(\sum_{k=1}^{a(x)} g_k(x) \right).$$

Turning now to the relation $f(x) \sim g(x)$, notice first that it is equivalent to the equation $f(x) = g(x) + o(g(x))$. Hence if $g(x) \to \infty$ as x increases indefinitely, the difference $f(x) - g(x)$ need not remain bounded; all that is asserted is that it is of smaller order of magnitude than $g(x)$ itself.

To give more precise information about $f(x)$, we must consider not $f(x)$ but $f(x) - g(x)$. As an example of this, consider the following theorem, which is not strictly a number-theoretic result, but which will be useful in what follows:

THEOREM 6–10. *There is a constant $\gamma = 0.57721 \ldots$ (called Euler's constant) such that*

$$\sum_{k=1}^{n} \frac{1}{k} = \log n + \gamma + O\left(\frac{1}{n}\right). \tag{1}$$

Remark: The relation

$$\sum_{k=1}^{n} \frac{1}{k} - \log n \sim \gamma, \quad \text{or} \quad \lim_{n \to \infty} \left(\sum_{k=1}^{n} \frac{1}{k} - \log n \right) = \gamma, \tag{2}$$

is weaker than (1), since it says nothing about the error except that it approaches zero. Notice that (2) is not equivalent to

$$\sum_{k=1}^{n} \frac{1}{k} \sim \log n + \gamma \tag{3}$$

(that is, terms may not be "transposed" in an asymptotic relation), for (3) has no more content than the simpler relation

$$\sum_{k=1}^{n} \frac{1}{k} \sim \log n.$$

Proof: Put

$$\alpha_k = \log k - \log (k - 1) - \frac{1}{k}, \quad k = 2, 3, \ldots,$$

and put

$$\gamma_n = \sum_{k=1}^{n} \frac{1}{k} - \log n, \quad n = 1, 2, \ldots,$$

so that

$$1 - \gamma_n = \sum_{k=2}^{n} \alpha_k, \quad n = 2, 3, \ldots.$$

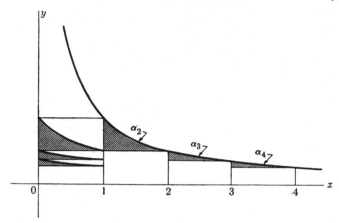

FIGURE 6-1

Geometrically, the number α_k represents the difference between the area of the region between the x-axis and the curve $y = 1/x$ in the interval $k - 1 \leq x \leq k$, and the area of the rectangle inscribed in this region; it is therefore positive. The regions having areas α_2, α_3, and α_4 are shaded in Fig. 6–1. If the regions having areas $\alpha_2, \ldots, \alpha_n$ are translated parallel to the x-axis into the interval $0 \leq x \leq 1$, it becomes obvious that $0 < 1 - \gamma_n < 1$ and that $1 - \gamma_{n+1} > 1 - \gamma_n$, for $n = 1, 2, \ldots$. Since every bounded increasing sequence is convergent, we have that $\lim_{n \to \infty} (1 - \gamma_n)$ exists; we call the limit $1 - \gamma$. Referring again to the square $0 \leq x \leq 1$, $0 \leq y \leq 1$, we see that the region whose area is

$$\gamma_n - \gamma = (1 - \gamma) - (1 - \gamma_n) = \sum_{k=n+1}^{\infty} \alpha_k$$

is contained in the rectangle $0 \leq x \leq 1$, $0 \leq y \leq 1/n$, of area $1/n$, so that

$$\gamma - \gamma_n = O\left(\frac{1}{n}\right).$$

The proof is complete.

If $f(x) \sim g(x)$ and $g(x) \to \infty$ as $x \to \infty$, then $\log f(x) \sim \log g(x)$. The relation $e^{f(x)} \sim e^{g(x)}$ is usually false, however; it is true only when $f(x) - g(x) = o(1)$. Finally, under the above suppositions, together with that of the continuity of f and g, one may deduce that

$$\int_a^x f(t)dt \sim \int_a^x g(t)dt$$

for sufficiently large fixed a, by applying L'Hôpital's rule, but the corresponding relation $f'(x) \sim g'(x)$ is not always valid.

PROBLEMS

1. Carry out the proofs of all the unproved statements in this section.
2. Show that

$$\sum_{\substack{i,j=1 \\ i \neq j}}^{\infty} \left[\frac{x}{p_i p_j} \right] = x \sum_{\substack{p_i p_j \leq x \\ i \neq j}} \frac{1}{p_i p_j} + O(x),$$

where p_i is the ith prime.

3. Show that if $f(x)$ tends to zero monotonically as x increases without limit, and is continuous for $x > 0$, and if the series

$$\sum_{k=1}^{\infty} f(k)$$

diverges, then

$$\sum_{k=1}^{n} f(k) \sim \int_1^n f(x)dx.$$

What can be said if the infinite series converges?

4. It is known that for every n, the nth prime p_n is greater than $n \log n$. Use this to show that if B_n is defined by the equation

$$\sum_{i=1}^{n} \frac{1}{p_i} - \log \log n = B_n, \qquad n = 3, 4, \ldots,$$

then $B_3, B_4, \ldots$ is a decreasing sequence.

6–5 The sieve of Eratosthenes. We now turn to the study of $\pi(x)$, and shall obtain many of the classical elementary results concerning the distribution of primes. None of these estimates is the best of its kind that is known, but to obtain more accurate results would require either too long a discussion to be worth while or the use of tools not available here, as, for example, the theory of functions of a complex variable. For many purposes our results are quite as useful as the better estimates.

One method of estimating $\pi(x)$ is based upon the observation that if n is less than or equal to x and is not divisible by any prime less than or equal to $\sqrt{x}$, then it is prime. Thus if we eliminate from the integers between 1 and x first all multiples of 2, then all multiples of 3, then all multiples of 5, etc., until all multiples of all primes less than or equal to $\sqrt{x}$ have been eliminated, then the numbers remaining are prime. This method of eliminating the composite numbers is known as the *sieve of Eratosthenes*; it has been adapted by Viggo Brun and others into a powerful method of estimating the number of integers in a certain interval having specified divisibility properties with respect to a certain set of primes.

We can modify the process just described by striking out the multiples of the first r primes $p_1, \ldots, p_r$, retaining r as an independent variable until the best choice for it can be clearly seen. If p_r is not the largest prime less than or equal to $\sqrt{x}$, but is some smaller prime, then of course it is no longer the case that all the integers remaining are primes, but certainly none of the primes except $p_1, \ldots, p_r$ have been removed. Thus if $A(x, r)$ is the number of integers remaining after all multiples of $p_1, \ldots, p_r$ (including $p_1, \ldots, p_r$ themselves, of course) have been removed from the integers less than or equal to x, then

$$\pi(x) \leq r + A(x, r).$$

In order to estimate $A(x, r)$ we use Theorem 6–4. If we take the $N = [x]$ objects to be the positive integers $\leq x$ and take S_k, for $1 \leq k \leq r$, to be the set of elements of S divisible by p_k, then

$$N_1 = \left[\frac{x}{2}\right], \qquad N_2 = \left[\frac{x}{3}\right], \qquad \cdots, \qquad N_{12} = \left[\frac{x}{2 \cdot 3}\right], \qquad \cdots,$$

and so

$$A(x, r) = [x] - \sum_{i=1}^{r}\left[\frac{x}{p_i}\right] + \sum_{1 \leq i < j \leq r}\left[\frac{x}{p_i p_j}\right] - \sum_{1 \leq i < j < k \leq r}\left[\frac{x}{p_i p_j p_k}\right]$$
$$+ \cdots + (-1)^r\left[\frac{x}{p_1 p_2 \cdots p_r}\right].$$

The difference between this expression and

$$x - \sum_{1 \leq i \leq r}\frac{x}{p_i} + \sum_{1 \leq i < j \leq r}\frac{x}{p_i \cdot p_j} - \cdots + (-1)^r\frac{x}{p_1 p_2 \cdots p_r}$$

does not exceed

$$1 + \binom{r}{1} + \binom{r}{2} + \cdots + \binom{r}{r} = 2^r,$$

and consequently

$$\pi(x) \leq r + x \prod_{i=1}^{r} \left(1 - \frac{1}{p_i}\right) + 2^r.$$

We need an estimate for the product occurring here.

THEOREM 6–11. *If* $x \geq 2$, *then*

$$\prod_{p \leq x} \left(1 - \frac{1}{p}\right) < \frac{1}{\log x}.$$

Proof: We have

$$\prod_{p \leq x} \frac{1}{1 - 1/p} = \prod_{p \leq x} \left(1 + \frac{1}{p} + \frac{1}{p^2} + \cdots\right),$$

and, by the Unique Factorization Theorem, this is the sum of the reciprocals of all integers having only the primes not exceeding x as prime divisors. In particular, all integers less than or equal to x are of this form, and so

$$\prod_{p \leq x} \frac{1}{1 - 1/p} > \sum_{k=1}^{x} \frac{1}{k} > \int_{1}^{[x]+1} \frac{du}{u} > \log x,$$

and the theorem follows.

We can now prove

THEOREM 6–12.

$$\pi(x) = O\left(\frac{x}{\log \log x}\right).$$

Proof: As above,

$$\pi(x) \leq r + 2^r + x \cdot \prod_{i=1}^{r} \left(1 - \frac{1}{p_i}\right) \leq 2^{r+1} + x \prod_{i=1}^{r} \left(1 - \frac{1}{p_i}\right),$$

and by Theorem 6–11,

$$\pi(x) \leq 2^{r+1} + \frac{x}{\log p_r}.$$

But $p_r \geq r$, and so

$$\pi(x) < \frac{x}{\log r} + 2^{r+1}.$$

Taking $r = [\log x]$, this becomes

$$\pi(x) < \frac{x}{\log \log x} + 2 \cdot 2^{\log x} = \frac{x}{\log \log x} + O(x^{\log 2}).$$

The last term is $O(x^{1-\epsilon})$ for some $\epsilon > 0$, and this is $o\left(\dfrac{x}{\log \log x}\right)$.
Hence

$$\pi(x) = O\left(\frac{x}{\log \log x}\right) + o\left(\frac{x}{\log \log x}\right) = O\left(\frac{x}{\log \log x}\right).$$

*PROBLEM

Show by a sieve argument that the number of square-free integers not exceeding x is less than

$$x \prod \left(1 - \frac{1}{p^2}\right) + o(x).$$

6–6 Sums involving primes. Theorems 6–11 and 6–12 bear a rather peculiar relation to each other: Theorem 6–11 was used in the proof of Theorem 6–12, yet the import of Theorem 6–11 is that the primes are not too infrequent, while that of Theorem 6–12 is that they are not too frequent. For if, for example, the primes were so scarce that $p_n > cn^2$ for some positive number c, and for all n, the product

$$\prod_{p \leq x} \left(1 - \frac{1}{p}\right)$$

would be bounded away from zero as $x \to \infty$, which it is not. It follows from Theorem 6–12, however, that there is no constant c' such that $p_n < c'n$ for all n. The following theorem is, in its implications, analogous to Theorem 6–11.

THEOREM 6–13. *The series* $\sum\limits_{p} \dfrac{1}{p}$ *diverges.*

Proof: By Theorem 6–11,

$$\log \prod_{p \leq x} \left(1 - \frac{1}{p}\right) = \sum_{p \leq x} \log \left(1 - \frac{1}{p}\right) < -\log \log x.$$

But since the curve $y = \log(1 + x)$ lies entirely above the curve $y = 2x$ in the interval $-\frac{1}{2} \leq x < 0$ (see Fig. 6–2), and since $p \geq 2$, we have

$$-\frac{2}{p} < \log\left(1 - \frac{1}{p}\right)$$

for all primes p, and so

$$\sum_{p \leq x} \frac{2}{p} > \log\log x.$$

In order to get more precise information about the behavior of the sum

$$\sum_{p \leq x} \frac{1}{p},$$

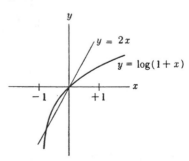

FIGURE 6–2

we proceed in a rather roundabout way, making use of the connection established in Theorem 6–9 between $n!$ and the primes not exceeding n.

THEOREM 6–14. $$\sum_{p \leq x} \frac{\log p}{p} \sim \log x.$$

Proof: By Theorem 6–9,

$$n! = \prod_{p \leq n} p^{[n/p] + [n/p^2] + \cdots},$$

and so

$$\log n! = \sum_{p \leq n} \left[\frac{n}{p}\right] \log p + \sum_{p \leq n} \left(\left[\frac{n}{p^2}\right] + \left[\frac{n}{p^3}\right] + \cdots\right) \log p.$$

Now

$$\sum_{p \leq n} \left[\frac{n}{p}\right] \log p \leq \sum_{p \leq n} \frac{n}{p} \log p,$$

and

$$\sum_{p \leq n} \left[\frac{n}{p}\right] \log p \geq \sum_{p \leq n} \left(\frac{n}{p} - 1\right) \log p = \sum_{p \leq n} \frac{n}{p} \log p - \sum_{p \leq n} \log p$$

$$\geq \sum_{p \leq n} \frac{n}{p} \log p - \log n \sum_{p \leq n} 1.$$

Moreover,

$$0 \leq \sum_{p \leq n} \left(\left[\frac{n}{p^2}\right] + \left[\frac{n}{p^3}\right] + \cdots\right) \log p \leq \sum_{p \leq n} \left(\frac{n}{p^2} + \frac{n}{p^3} + \cdots\right) \log p.$$

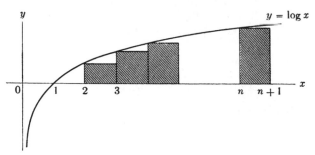

FIGURE 6–3

Thus

$$\log n! = \sum_{p \le n} \frac{n}{p} \log p + O(\pi(n) \log n)$$

$$+ O\left(n \sum_{p \le n} \left(\frac{1}{p^2} + \frac{1}{p^3} + \cdots\right) \log p\right)$$

$$= n \sum_{p \le n} \frac{\log p}{p} + O(\pi(n) \log n) + O\left(n \sum_p \frac{\log p}{p(p-1)}\right),$$

and by Theorem 6–12, and the fact that the series

$$\sum_k \frac{\log k}{k(k-1)}$$

converges, this gives

$$\log n! = n \sum_{p \le n} \frac{\log p}{p} + O\left(\frac{n \log n}{\log \log n}\right) + O(n)$$

$$= n \sum_{p \le n} \frac{\log p}{p} + O\left(\frac{n \log n}{\log \log n}\right).$$

On the other hand, by comparing the area under the curve $y = \log x$ with the total area of the inscribed rectangles (see Fig. 6–3), we see easily that

$$\log n! = \sum_{m=1}^{n} \log m = \int_1^n \log t \, dt + O(\log n) = n \log n + O(n).$$

Combining the two estimates obtained for $\log n!$, we have

$$n \sum_{p \le n} \frac{\log p}{p} + O\left(\frac{n \log n}{\log \log n}\right) = n \log n + O(n),$$

so that

$$\sum_{p \le n} \frac{\log p}{p} = \log n + O\left(\frac{\log n}{\log \log n}\right).$$

This proves the theorem when x is an integer n. But if $n \le x < n+1$, then

$$\sum_{p \le x} \frac{\log p}{p} = \sum_{p \le n} \frac{\log p}{p} = \log n + O\left(\frac{\log n}{\log \log n}\right)$$

$$= \log x + O\left(\frac{\log x}{\log \log x}\right). \tag{4}$$

THEOREM 6–15. *Suppose that $\lambda_1, \lambda_2, \ldots$ is a nondecreasing sequence with limit infinity, that $c_1, c_2, \ldots$ is an arbitrary sequence of real or complex numbers, and that $f(x)$ has a continuous derivative for $x \ge \lambda_1$. Put*

$$C(x) = \sum_{\lambda_n \le x} c_n,$$

where the summation is over all n for which $\lambda_n \le x$. Then for $x \ge \lambda_1$,

$$\sum_{\lambda_n \le x} c_n f(\lambda_n) = C(x)f(x) - \int_{\lambda_1}^{x} C(t)f'(t)\,dt.$$

Proof: We have

$$\sum_{\lambda_n \le x} c_n f(\lambda_n) = C(\lambda_1)f(\lambda_1) + (C(\lambda_2) - C(\lambda_1))f(\lambda_2) + \cdots$$
$$+ (C(\lambda_\nu) - C(\lambda_{\nu-1}))f(\lambda_\nu),$$

where λ_ν is the greatest λ_n which does not exceed x. ·Regrouping the terms, we have

$$\sum_{\lambda_n \le x} c_n f(\lambda_n) = C(\lambda_1)(f(\lambda_1) - f(\lambda_2)) + \cdots$$
$$+ C(\lambda_{\nu-1})(f(\lambda_{\nu-1}) - f(\lambda_\nu))$$
$$+ C(\lambda_\nu)(f(\lambda_\nu) - f(x)) + C(\lambda_\nu)f(x)$$
$$= -\int_{\lambda_1}^{x} C(t)f'(t)\,dt + C(x)f(x),$$

since $C(t)$ is a step function, constant over each of the intervals $(\lambda_{i-1}, \lambda_i)$ and over the interval (λ_ν, x).

THEOREM 6–16. $\displaystyle \sum_{p \le x} \frac{1}{p} \sim \log \log x.$

Proof: Take $\lambda_n = p_n$, $\quad c_n = \log p_n / p_n$, $\quad f(t) = 1/\log t$ in Theorem 6–15. By (4),

$$\sum_{p \le x} \frac{1}{p} = \frac{1}{2} + \sum_{2 < p \le x} \left(\frac{\log p}{p} \cdot \frac{1}{\log p} \right)$$

$$= \frac{1}{\log x} \sum_{2 < p \le x} \frac{\log p}{p} - \int_3^x \left(\sum_{p \le t} \frac{\log p}{p} \right) \frac{-dt}{t \log^2 t} + \frac{1}{2}$$

$$= \frac{1}{\log x} \left\{ \log x + O\left(\frac{\log x}{\log \log x} \right) \right\}$$

$$+ \int_3^x \left\{ \log t + O\left(\frac{\log t}{\log \log t} \right) \right\} \frac{dt}{t \log^2 t} + \frac{1}{2}$$

$$= O(1) + \int_3^x \frac{dt}{t \log t} + \int_3^x O\left(\frac{1}{t \log t \log \log t} \right) dt$$

$$= \log \log x + O(1) + \int_3^x O\left(\frac{1}{t \log t \log \log t} \right) dt.$$

Now for some constant M,

$$\left| \int_3^x O\left(\frac{1}{t \log t \log \log t} \right) dt \right| < M \int_3^x \frac{dt}{t \log t \log \log t}$$

$$= O(\log \log \log x),$$

so that

$$\sum_{p \le x} \frac{1}{p} = \log \log x + O(\log \log \log x),$$

which proves the theorem.

Use Theorems 6–15 and 6–16 to show that

$$\int_2^x \frac{\pi(t)}{t^2} dt = \sum_{p \le x} \frac{1}{p} + o(1) \sim \log \log x,$$

and deduce that for no positive constant δ is there a $T = T(\delta)$ such that for all $t > T$,

$$\pi(t) > (1 + \delta) \frac{t}{\log t},$$

and that for no $\delta > 0$ is there a $T = T(\delta)$ such that for all $t > T$,

$$\pi(t) < (1 - \delta) \frac{t}{\log t}.$$

This implies that, if

$$\lim_{x \to \infty} \frac{\pi(x)}{x/\log x}$$

exists, it must be 1.

6–7 The order of $\pi(x)$. We now show that the actual order of $\pi(x)$ is $x/\log x$.

THEOREM 6–17. *There are positive finite constants c_1 and c_2 such that for $x \geq 2$,*

$$c_1 \frac{x}{\log x} < \pi(x) < c_2 \frac{x}{\log x}.$$

Proof: Take $n \geq 2$. Corresponding to each $p \leq 2n$ there is a unique integer r_p such that $p^{r_p} \leq 2n < p^{r_p+1}$. We first prove that

$$\prod_{n < p \leq 2n} p \;\Big|\; \frac{(2n)!}{n!n!} \quad \text{and} \quad \frac{(2n)!}{n!n!} \;\Big|\; \prod_{p \leq 2n} p^{r_p}. \tag{5}$$

The first part is obvious, since any prime between n and $2n$ occurs as a factor of $(2n)!$ but does not occur in the denominator $(n!)^2$. For the second part, we have that the highest power of p which divides the numerator $(2n)!$, by Theorem 6–9, has exponent

$$\sum_{m=1}^{r_p} \left[\frac{2n}{p^m}\right],$$

while the highest power of p which divides the denominator has exponent

$$2 \sum_{m=1}^{r_p} \left[\frac{n}{p^m}\right],$$

so that the highest power of p dividing $\binom{2n}{n}$ has exponent

$$\sum_{m=1}^{r_p} \left\{\left[\frac{2n}{p^m}\right] - 2\left[\frac{n}{p^m}\right]\right\} \leq \sum_{m=1}^{r_p} 1 = r_p.$$

Here we have used property (f) of $[x]$, from the list in Section 6–3. From (5) we get

$$n^{\pi(2n)-\pi(n)} \leq \prod_{n < p \leq 2n} p \leq \binom{2n}{n} \leq \prod_{p \leq 2n} p^{r_p} \leq (2n)^{\pi(2n)},$$

whence $(\pi(2n) - \pi(n)) \log n \leq \log \binom{2n}{n} \leq \pi(2n) \log 2n.$

Clearly $\binom{2n}{n} \leq 2^{2n}$, and also

$$\binom{2n}{n} = \frac{(n + 1) \cdots (2n)}{1 \cdots n} = \prod_{a=1}^{n} \frac{n + a}{a} \geq \prod_{a=1}^{n} 2 = 2^n;$$

thus $(\pi(2n) - \pi(n)) \log n \leq 2n \log 2,$

or* $\pi(2n) - \pi(n) \leq c_3 \dfrac{n}{\log n},$ (6)

and $\pi(2n) \log 2n \geq n \log 2,$

or $\pi(2n) > c_4 \dfrac{n}{\log n} \cdot$ (7)

If $x \geq 4$, we get from (7) that

$$\pi(x) \geq \pi\left(2\left[\frac{x}{2}\right]\right) > c_4 \frac{[x/2]}{\log [x/2]} > c_5 \frac{x}{\log x},$$

and since $\pi(x) \geq 1$ for $2 \leq x < 4$, $\pi(x) > c_1 (x/\log x)$ for $x \geq 2$.

If $y \geq 4$, we get from (6) that

$$\pi(y) - \pi\left(\frac{y}{2}\right) = \pi(y) - \pi\left(\left[\frac{y}{2}\right]\right) \leq 1 + \pi\left(2\left[\frac{y}{2}\right]\right) - \pi\left(\left[\frac{y}{2}\right]\right)$$

$$< 1 + c_3 \frac{[y/2]}{\log [y/2]} < c_6 \frac{y}{\log y},$$

and so for $y \geq 2$, $\pi(y) - \pi\left(\dfrac{y}{2}\right) < c_7 \dfrac{y}{\log y} \cdot$

Using the trivial bound $\pi(y/2) \leq y/2$, we get

$$\pi(y) \log y - \pi\left(\frac{y}{2}\right) \log \frac{y}{2} = \left\{\pi(y) - \pi\left(\frac{y}{2}\right)\right\} \log y + \pi\left(\frac{y}{2}\right) \log 2$$

$$< \log y \cdot c_7 \frac{y}{\log y} + \frac{y}{2} < c_8 y.$$

---

*Here $c_3, c_4, \cdots$ will denote certain positive constants, whose exact values will be of no concern.

If we put $y = x/2^m$ with $2^m \leq x/2$ and $m \geq 0$, this becomes

$$\pi\left(\frac{x}{2^m}\right) \log \frac{x}{2^m} - \pi\left(\frac{x}{2^{m+1}}\right) \log \frac{x}{2^{m+1}} < c_8 \frac{x}{2^m},$$

and summing over all such m's, we have

$$\pi(x) \log x - \pi\left(\frac{x}{2^{\mu+1}}\right) \log \frac{x}{2^{\mu+1}} < c_2 x,$$

where $2^\mu \leq x/2 < 2^{\mu+1}$. But $x/2^{\mu+1} < 2$, so that $\pi(x/2^{\mu+1}) = 0$, and we have

$$\pi(x) \log x < c_2 x,$$

which completes the proof.

THEOREM 6–18. *There are positive constants c_9, c_{10} such that for $r > 1$,*

$$c_9 r \log r < p_r < c_{10} r \log r.$$

Proof: Taking x to be p_r in Theorem 6–17, we get

$$c_1 \frac{p_r}{\log p_r} < r < c_2 \frac{p_r}{\log p_r}.$$

The right-hand inequality gives immediately

$$p_r > c_9 r \log p_r > c_9 r \log r.$$

Using the other inequality and the fact that $\log u = o(\sqrt{u})$, we have that for $r > c_{11}$,

$$\frac{\log p_r}{\sqrt{p_r}} < c_1 < \frac{r \log p_r}{p_r},$$

$$p_r < r^2,$$

$$\log p_r < 2 \log r,$$

and so for $r > c_{11}$,

$$p_r < \frac{1}{c_1} r \cdot 2 \log r,$$

whence finally $p_r < c_{10} r \log r$ for all $r > 1$.

We can use Theorem 6–17 to improve Theorems 6–14 and 6–16. Examining the proof of Theorem 6–14, we see that the error term can now be reduced to $O(n)$, since by Theorem 6–17, $\pi(n) \log n = O(n)$. Thus we have

THEOREM 6–19. $\displaystyle \sum_{p \leq x} \frac{\log p}{p} = \log x + O(1)$.

Following the argument used for Theorem 6–14, we now have

$$\sum_{p \leq x} \frac{1}{p} = \frac{1}{\log x} \left(\log x + O(1) \right) + \int_2^x \log t \cdot \frac{dt}{t \log^2 t}$$

$$+ \int_2^x \left(\sum_{p \leq t} \frac{\log p}{p} - \log t \right) \frac{dt}{t \log^2 t}$$

$$= 1 + O\left(\frac{1}{\log x} \right) + \log \log x - \log \log 2$$

$$+ \int_2^\infty \left(\sum_{p \leq t} \frac{\log p}{p} - \log t \right) \frac{dt}{t \log^2 t} - \int_x^\infty \frac{O(1) dt}{t \log^2 t}.$$

Here the first integral is convergent, and the second is clearly $O(1/\log x)$. This proves

THEOREM 6–20. *There is a constant C such that*

$$\sum_{p \leq x} \frac{1}{p} = \log \log x + C + O\left(\frac{1}{\log x} \right).$$

PROBLEM

Apply Theorem 6–20 to show that for some constant B,

$$\sum_{p \leq x} \log \left(1 - \frac{1}{p} \right) = -\log \log x - B + O\left(\frac{1}{\log x} \right).$$

Deduce that

$$\prod_{p \leq x} \left(1 - \frac{1}{p} \right) = \frac{e^{-B}}{\log x} + O\left(\frac{1}{\log^2 x} \right).$$

By Theorem 6–11, $B \geq 0$; although we do not prove it, B is Euler's constant. [Use the law of the mean to show that if $f(x) \to 0$ as $x \to \infty$, then $e^{f(x)} = 1 + O(f(x))$.]

6–8 Bertrand's conjecture. In 1845 J. Bertrand showed empirically that there is a prime between n and $2n$ for all n greater than 1 and less than six million, and predicted that this is true for all positive integral n. Chebyshev proved this in 1850, and indeed that for every

$\epsilon > \frac{1}{5}$ there is a ξ such that for every $x > \xi$ there is a prime between x and $(1 + \epsilon)x$. Since that time, analytic methods have been used to show that this last theorem is true for every $\epsilon > 0$. We shall content ourselves here with a proof of Bertrand's original conjecture. The proof given is due to P. Erdös.

It is worth noting that Theorem 6–19 implies a weak form of the theorem.

THEOREM 6–21. *There exists a positive constant c_{12} such that there is a prime between n and $c_{12}n$ for all n.*

Proof: By Theorem 6–19, there is a constant A such that

$$\log n - A < \sum_{p \leq n} \frac{\log p}{p} < \log n + A$$

for all n. Suppose that there is no prime between n and ne^{2A}. Then

$$\sum_{p \leq n} \frac{\log p}{p} = \sum_{p \leq ne^{2A}} \frac{\log p}{p},$$

and so by Theorem 6–19 again,

$$\sum_{p \leq n} \frac{\log p}{p} > \log(ne^{2A}) - A = \log n + A.$$

With this contradiction, the theorem is proved with $c_{12} = e^{A}$.

For the proof of the more exact theorem we need two lemmas.

THEOREM 6–22. $\prod_{p \leq n} p < 4^{n}.$

Proof: We use induction on n. The theorem is obvious if $n = 1$ or 2. Suppose it is true for $1, 2, \ldots, n - 1$, where $n \geq 3$. Then we can restrict attention to odd n, since otherwise

$$\prod_{p \leq n} p = \prod_{p \leq n-1} p < 4^{n-1} < 4^{n},$$

so we can put $n = 2m + 1$. From its definition, the binomial coefficient

$$\binom{2m + 1}{m} = \frac{(2m + 1)!}{m!(m + 1)!}$$

is divisible by every prime p with $m + 2 \leq p \leq 2m + 1$. Hence

$$\prod_{p \leq 2m+1} p \leq \binom{2m + 1}{m} \cdot \prod_{p \leq m+1} p < \binom{2m + 1}{m} 4^{m+1}.$$

But the numbers

$$\binom{2m+1}{m} \quad \text{and} \quad \binom{2m+1}{m+1}$$

are equal, and both occur in the expansion of $(1+1)^{2m+1}$, so that

$$\binom{2m+1}{m} \leq \frac{1}{2} \cdot 2^{2m+1} = 4^m,$$

and so

$$\prod_{p \leq 2m+1} p < 4^m \cdot 4^{m+1} = 4^{2m+1}.$$

The theorem follows by induction on n.

THEOREM 6–23. *If $n \geq 3$ and $\frac{2}{3}n < p \leq n$, then $p \nmid \binom{2n}{n}$.*

Proof: The restrictions on n and p are such that

(a) p is greater than 2,

(b) p and $2p$ are the only multiples of p which are less than or equal to $2n$, since $3p$ is greater than $2n$,

(c) p itself is the only multiple of p which is less than or equal to n.

From (a) and (b), $p^2 \| (2n)!$, and from (c), $p^2 \| (n!)^2$, so that $p \nmid (2n)!/(n!)^2$.

THEOREM 6–24. *For any positive integer n, there is a prime p such that $n < p \leq 2n$.*

Proof: This is true for $n = 1, 2, 3$. Assume the theorem false for a certain integer $n \geq 4$. Then by Theorem 6–23, every prime which divides $\binom{2n}{n}$ must be less than or equal to $2n/3$. Let p be such a prime, and suppose that $p^\alpha \| \binom{2n}{n}$. Then by the proof of Theorem 6–17, since

$$\binom{2n}{n} \Bigg| \prod_{p \leq 2n} p^{r_p}, \qquad (p^{r_p} \leq 2n < p^{r_p+1})$$

it follows that $p^\alpha \leq 2n$. Thus if $\alpha \geq 2$, then $p \leq \sqrt{2n}$, and so there are at most $[\sqrt{2n}]$ primes appearing in the prime-power factorization of $\binom{2n}{n}$ with exponent larger than 1. Hence

$$\binom{2n}{n} \leq (2n)^{[\sqrt{2n}]} \cdot \prod_{p \leq 2n/3} p.$$

But $\binom{2n}{n}$ is the largest of the $2n + 1$ terms in the expansion of $(1 + 1)^{2n}$, so that

$$4^n < (2n + 1) \binom{2n}{n},$$

and so

$$\frac{4^n}{2n + 1} < (2n)^{\sqrt{2n}} \cdot \prod_{p \leq 2n/3} p.$$

By Theorem 6–22, this implies that

$$\frac{4^n}{2n + 1} < (2n)^{\sqrt{2n}} \cdot 4^{2n/3},$$

and since $2n + 1 < 4n^2$, this gives

$$4^n < (2n)^{\sqrt{2n}+2} \cdot 4^{2n/3}, \qquad \text{or} \qquad 4^{n/3} < (2n)^{\sqrt{2n}+2}.$$

Taking logarithms, we have

$$\frac{n \log 4}{3} < (\sqrt{2n} + 2) \log 2n.$$

This inequality is false for $n > 512$, so the theorem is true for $n > 512$. But in the sequence of primes

$$2, 3, 5, 7, 13, 23, 43, 83, 163, 317, 557,$$

each number is smaller than twice the one preceding it, and the theorem is also true for all $n \leq 512$. It is therefore true for all n.

PROBLEMS

1. It follows from the Problem of Section 6–6 that in Theorem 6–17, $c_1 < 1 < c_2$. If estimates had been made of c_1 and c_2 in the proof of Theorem 6–17 (which would be simple to do), we would know, as a consequence, two particular constants c_1 and c_2 for which the inequality of Theorem 6–13 holds. Suppose that this is the case, and that $c_2/c_1 = \beta > 1$. Show that if $\epsilon > 0$, then

$$\frac{\pi((1 + \epsilon)x) - \pi(x)}{x/\log x} > c_1(1 + \epsilon - \beta) + O\left(\frac{1}{\log x}\right).$$

Deduce that if $\epsilon > \beta - 1$, the number of primes between x and $(1 + \epsilon)x$ tends to infinity with x.

2. Show that there is a constant $A > 0$ such that

$$\sum_{x < p \le x^2} \frac{1}{p} > A$$

for all sufficiently large x. Deduce that for each $\epsilon > 0$, there are infinitely many pairs p_n and p_{n+1} of consecutive primes such that

$$p_{n+1} < (1 + \epsilon)p_n.$$

6–9 The order of magnitude of φ, σ, and τ. The quantity $\pi(x)$ is reasonably well-behaved, and so one can make fairly precise statements about its size as a function of x. This is not true of the other functions we have considered, which vary much too wildly to permit asymptotic approximations. There are, however, various weaker statements which can be made about their size which still yield considerable information.

Consider, for example, the quantity $\tau(n)$. A moment's thought shows that the number of divisors of n is much smaller than n itself, for large n; it is to be expected that $\tau(n) = o(n)$. And while $\tau(n) = 2$ infinitely many times, it is also possible to make $\tau(n)$ arbitrarily large for suitable n. Thus if the points $(n, \tau(n))$ are plotted in a coordinate system, as in Fig. 6–4, there is a unique "lowest" polygonal path extending upward and to the right from $(1, 1)$ which is concave downward and is such that every point $(n, \tau(n))$ lies on or below it. Suppose that this path is described by the equation $y = T(x)$. While we shall not obtain an asymptotic estimate for $T(x)$, the following theorem shows that it increases more rapidly than any power of $\log x$, and less rapidly than any positive power of x.

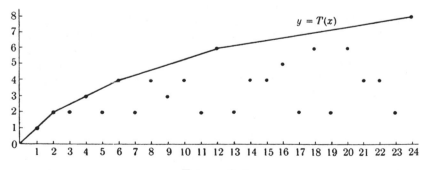

FIGURE 6–4

THEOREM 6–25.   (a) *The relation* $\tau(n) = O(\log^h n)$ *is false for every constant h.*

(b) *The relation* $\tau(n) = O(n^\delta)$ *is true for every fixed* $\delta > 0$.

Proof: (a) Let n be any of the numbers $(2 \cdot 3 \cdots p_r)^m$, $m = 1, 2, \ldots$; here r is arbitrary but fixed. Then

$$\tau(n) = \prod_1^r (m + 1) = (m + 1)^r > m^r.$$

But $m = \dfrac{\log n}{\log (2 \cdot 3 \cdots p_r)}$, so that

$$\tau(n) > \frac{\log^r n}{(\log (2 \cdot 3 \cdots p_r))^r} > c_{13} \log^r n,$$

where $c_{13} > 0$ is a constant depending only on r, and not on n.

(b) Let $$f(n) = \frac{\tau(n)}{n^\delta} \, ;$$

then f is multiplicative. But

$$f(p^m) = \frac{m + 1}{p^{m\delta}} \le \frac{2m}{p^{m\delta}} = \frac{2}{\log p} \, \frac{\log p^m}{p^{m\delta}} \le \frac{2}{\log 2} \cdot \frac{\log p^m}{p^{m\delta}} ,$$

so that $f(p^m) \to 0$ as $p^m \to \infty$, i.e., as either p or m, or both, increases. This clearly implies that $f(n) \to 0$ as $n \to \infty$, which is the assertion.

Alternative proof of (b): Let δ be positive, and let

$$n = \prod_1^r p_i^{\alpha_i}.$$

Then $\dfrac{\tau(n)}{n^\delta} = \dfrac{\alpha_1 + 1}{p_1^{\alpha_1 \delta}} \cdots \dfrac{\alpha_r + 1}{p_r^{\alpha_r \delta}} \le \prod_{p_i | n} \max_{x \ge 0} \left(\dfrac{x + 1}{p_i^{\delta x}} \right).$

For fixed δ, the quantity

$$\max_{x \ge 0} \frac{x + 1}{p^{\delta x}}$$

is equal to 1 for sufficiently large p, and is never smaller than 1. Hence

$$\frac{\tau(n)}{n^\delta} \le \prod_p \max_{x \ge 0} \left(\frac{x + 1}{p^{\delta x}} \right) = c_\delta,$$

and c_δ is a finite constant independent of n. Hence $\tau(n) = O(n^\delta)$.

By actually evaluating the constant c_5, one obtains inequalities such as

$$\tau(n) \leq \sqrt{3n}, \qquad \tau(n) < 4\sqrt[3]{n}, \qquad \ldots,$$

which of course are very poor estimates for large n, but which are sometimes more useful than the statement of Theorem 6–25, where nothing is said about the behavior of $\tau(n)$ for all n, but only for large n.

As regards the φ-function, we have the trivial upper bound $\varphi(n) \leq n - 1$ for $n > 1$, equality being attained whenever n is prime. The corresponding lower bound is less obvious.

THEOREM 6–26. *There is a positive constant* c_{14} *such that for all* $n > 3$,

$$\varphi(n) > \frac{c_{14}n}{\log \log n}.$$

Proof: We have

$$\frac{\varphi(n)}{n} = \prod_{p|n} \left(1 - \frac{1}{p}\right),$$

so that

$$\log \frac{\varphi(n)}{n} = \sum_{p|n} \log \left(1 - \frac{1}{p}\right) = -\sum_{p|n} \frac{1}{p} + \sum_{p|n} \left\{\log \left(1 - \frac{1}{p}\right) + \frac{1}{p}\right\}$$

$$> -\sum_{p|n} \frac{1}{p} - c_{15},$$

since

$$\sum_{p} \left\{\log \left(1 - \frac{1}{p}\right) + \frac{1}{p}\right\} > \sum_{p} \left(\frac{1}{p} - \frac{1}{p-1}\right)$$

$$> -\sum_{n=2}^{\infty} \left(\frac{1}{n-1} - \frac{1}{n}\right) = -1.$$

Now let $p_1, \ldots, p_{r-\rho}$ be the primes less than $\log n$ which divide n, so that

$$\sum_{p|n} \frac{1}{p} = \sum_{k=1}^{r-\rho} \frac{1}{p_k} + \sum_{k=r-\rho+1}^{r} \frac{1}{p_k} = S_1 + S_2.$$

Then

$$\log^\rho n \leq p^\rho_{r-\rho+1} \leq \prod_{k=r-\rho+1}^{r} p_k \leq n,$$

so that
$$\rho \le \frac{\log n}{\log \log n},$$

and
$$S_2 \le \frac{1}{\log n} \cdot \frac{\log n}{\log \log n} < c_{16}.$$

By Theorem 6–20,
$$S_1 < \log \log p_{r-\rho} + c_{17} < \log \log \log n + c_{17}.$$

Combining these results, we get
$$\log \frac{\varphi(n)}{n} > -\log \log \log n - c_{18},$$

and so
$$\frac{\varphi(n)}{n} > \frac{c_{14}}{\log \log n}.$$

We can use Theorem 6–26 to obtain a corresponding upper bound for $\sigma(n)$, with the help of the following theorem.

THEOREM 6–27. *There is a positive constant c_{19} such that*

$$c_{19} < \frac{\sigma(n)\varphi(n)}{n^2} < 1.$$

Proof: If $n = \prod p^\alpha$, then

$$\sigma(n)\varphi(n) = \prod_{p|n} \left(\frac{p^{\alpha+1} - 1}{p - 1} \right) n \prod_{p|n} \left(1 - \frac{1}{p} \right)$$

$$= n \prod_{p|n} \frac{1 - p^{-(\alpha+1)}}{1 - 1/p} \cdot n \prod_{p|n} \left(1 - \frac{1}{p} \right)$$

$$= n^2 \prod_{p|n} (1 - p^{-(\alpha+1)}).$$

Here the coefficient of n^2 is clearly less than 1 and greater than or equal to

$$\prod_{p|n} \left(1 - \frac{1}{p^2} \right) > \prod_{k=2}^{n} \left(1 - \frac{1}{k^2} \right).$$

Now

$$\log \left[\prod_{k=2}^{n} \left(1 - \frac{1}{k^2} \right) \right] = \sum_{k=2}^{n} \log \left(1 - \frac{1}{k^2} \right),$$

and for $x > 0$, $\log (1 - x) > \dfrac{-x}{1 - x}$,

so that $-\displaystyle\sum_{k=2}^{n} \frac{1}{k^2 - 1} < \sum_{k=2}^{n} \log\left(1 - \frac{1}{k^2}\right)$.

Since the first sum in this inequality tends to a limit as $n \to \infty$, it follows that the above coefficient of n^2 is bounded away from zero, and the theorem follows.

THEOREM 6–28. $\sigma(n) = O(n \log \log n)$.

Proof: By Theorem 6–26,

$$\frac{\varphi(n)}{n} > \frac{c_{14}}{\log \log n},$$

and by Theorem 6–27,

$$\frac{\sigma(n)}{n} < \frac{n}{\varphi(n)} < \frac{\log \log n}{c_{14}},$$

$$\frac{\sigma(n)}{n} = O(\log \log n),$$

$$\sigma(n) = O(n \log \log n).$$

PROBLEM

Show that there is an infinite sequence of positive integers $n_1, n_2, \ldots$ such that

$$\varphi(n_k) < \frac{cn_k}{\log \log n_k}, \qquad k = 1, 2, \ldots,$$

for some constant c.

6–10 Average order of magnitude. Another way of describing the behavior of a number-theoretic function is in terms of its average order, that is, in terms of the quantity

$$\frac{1}{n} \sum_{m=1}^{n} f(m).$$

Summing the values of a function has the effect of smoothing out its irregularities, so that it is frequently possible to make quite precise statements about the size of the sum.

THEOREM 6–29. *If*

$$F(n) = \sum_{d|n} f(d),$$

then

$$\sum_{m=1}^{n} F(m) = \sum_{m=1}^{n} \left[\frac{n}{m}\right] f(m).$$

Proof:

$$\sum_{m=1}^{n} F(m) = \sum_{m=1}^{n} \sum_{d|m} f(d) = \sum_{d=1}^{n} \sum_{k=1}^{[n/d]} f(d) = \sum_{d=1}^{n} \left[\frac{n}{d}\right] f(d).$$

THEOREM 6–30. *If*

$$F(n) = \sum_{d|n} f(d),$$

then

$$\sum_{m=1}^{n} F(m) = \sum_{m=1}^{t} \left[\frac{n}{m}\right] f(m) + \sum_{m=1}^{n/t} G\left(\frac{n}{m}\right) - \left[\frac{n}{t}\right] G(t),$$

where t is any positive integer not exceeding n, and

$$G(\xi) = G([\xi]) = \sum_{m=1}^{\xi} f(m).$$

Proof: By the definition of G, $f(m) = G(m) - G(m-1)$, and so by partial summation (cf. the proof of Theorem 6–15),

$$\sum_{m=1}^{n} F(m) = \sum_{m=1}^{t} \left[\frac{n}{m}\right] f(m) + \sum_{m=t+1}^{n} \left[\frac{n}{m}\right] f(m)$$

$$= \sum_{m=1}^{t} \left[\frac{n}{m}\right] f(m) + \sum_{m=t+1}^{n} \left[\frac{n}{m}\right] (G(m) - G(m-1))$$

$$= \sum_{m=1}^{t} \left[\frac{n}{m}\right] f(m) + \sum_{m=t}^{n} \left\{ G(m) \cdot \left(\left[\frac{n}{m}\right] - \left[\frac{n}{m+1}\right] \right) \right\}$$

$$- \left[\frac{n}{t}\right] G(t).$$

As was noted earlier, $[n/m] - [n/(m+1)]$ is the number of integers u such that

$$\frac{n}{m+1} < u \leq \frac{n}{m}.$$

For each such u, $n/u - 1 < m \leq n/u$, so that $m = [n/u]$. Hence

$$G(m) \left(\left[\frac{n}{m} \right] - \left[\frac{n}{m+1} \right] \right) = \sum_{n/(m+1) < u \leq n/m} G\left(\frac{n}{u} \right),$$

$$\sum_{m=t}^{n} G(m) \left(\left[\frac{n}{m} \right] - \left[\frac{n}{m+1} \right] \right) = \sum_{m=t}^{n} \sum_{n/(m+1) < u \leq n/m} G\left(\frac{n}{u} \right)$$

$$= \sum_{u=1}^{n/t} G\left(\frac{n}{u} \right),$$

and the proof is complete.

THEOREM 6–31. $\displaystyle\sum_{m=1}^{n} \tau(m) = n \log n + (2\gamma - 1)n + O(n^{\frac{1}{2}})$,

where γ is Euler's constant.

Proof: Take $F = \tau$, $f = 1$, $t = [\sqrt{n}]$ in Theorem 6–30; then $G(\xi) = [\xi]$ and

$$\sum_{m=1}^{n} \tau(m) = \sum_{m=1}^{[\sqrt{n}]} \left[\frac{n}{m} \right] + \sum_{m=1}^{n/[\sqrt{n}]} \left[\frac{n}{m} \right] - \left[\frac{n}{[\sqrt{n}]} \right] [\sqrt{n}]$$

$$= 2 \sum_{m=1}^{\sqrt{n}} \left[\frac{n}{m} \right] - n + O(\sqrt{n})$$

$$= 2 \sum_{m=1}^{\sqrt{n}} \frac{n}{m} + O(\sqrt{n}) - n + O(\sqrt{n})$$

$$= 2n \sum_{m=1}^{\sqrt{n}} \frac{1}{m} - n + O(\sqrt{n})$$

$$= 2n \left(\log \sqrt{n} + \gamma + O(1/\sqrt{n}) \right) - n + O(\sqrt{n})$$

$$= n \log n + n(2\gamma - 1) + O(\sqrt{n}).$$

The term $O(\sqrt{n})$ in Theorem 6–31 is not the best possible estimate of the error. The problem of increasing the accuracy of the estimate, usually called Dirichlet's divisor problem, has received a large amount of study. It is known that $O(n^{\frac{1}{2}})$ can be replaced by $O(n^{\frac{1}{3}})$, but not by $O(n^{\frac{1}{4}})$. The exact exponent, if such exists, is still unknown.

For the purpose of illustrating the methods available for estimating averages, we give a second proof of Theorem 6–31. By Theorem 6–29,

$$\sum_{m=1}^{n} \tau(m) = \sum_{m=1}^{n} \left[\frac{n}{m} \right].$$

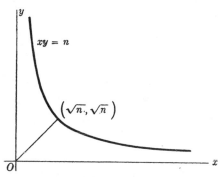

FIGURE 6–5

Geometrically, this last sum represents the number of *lattice points* (x, y) (that is, points such that x and y are integers) with positive coordinates, on or below the hyperbola $xy = n$, since for fixed x the number of integers y such that $1 \leq y \leq n/x$ is exactly $[n/x]$.

By symmetry, the number of lattice points (x, y) with $0 < xy \leq n$, $y > x$, is equal to the number with $0 < xy \leq n$, $y < x$ (see Fig. 6–5). Hence the number of points (x, y) with $0 < xy \leq n$ is twice the number of those with $y > x$, plus the number with $y = x$:

$$\sum_{m=1}^{n} \tau(m) = 2 \sum_{x=1}^{\sqrt{n}} \left(\left[\frac{n}{x} \right] - x \right) + [\sqrt{n}]$$

$$= 2n \sum_{1}^{n} \frac{1}{x} + O(\sqrt{n}) - 2 \frac{[\sqrt{n}]([\sqrt{n}] + 1)}{2} + O(\sqrt{n})$$

$$= 2n \left(\log \sqrt{n} + \gamma + O(1/\sqrt{n}) \right) - n + O(\sqrt{n})$$

$$= n \log n + (2\gamma - 1)n + O(\sqrt{n}).$$

To get an asymptotic estimate for the sum of the first n values of the φ-function, we need a preliminary result concerning the famous *Riemann ζ-function*, which is defined for $s > 1$ by the equation

$$\zeta(s) = \sum_{n=1}^{\infty} \frac{1}{n^s}.$$

For $s \leq 1$ the function is, for the time being, undefined, since the series fails to converge for such s. It is a well-known result, which we shall use without proof (see Problem 6 below), that

$$\zeta(2) = \sum_{n=1}^{\infty} \frac{1}{n^2} = \frac{\pi^2}{6}.$$

THEOREM 6-32. *For s > 1,*

$$\frac{1}{\zeta(s)} = \sum_{n=1}^{\infty} \frac{\mu(n)}{n^s}.$$

Proof: The series of the theorem and that for $\zeta(s)$ converge absolutely for $s > 1$, so that they may be multiplied together by adding all possible products of a term from one series and a term from the other, and the resulting terms may be arranged in any convenient order. Hence

$$\sum_{m=1}^{\infty} \frac{1}{m^s} \sum_{n=1}^{\infty} \frac{\mu(n)}{n^s} = \sum_{m,n=1}^{\infty} \frac{\mu(n)}{(mn)^s} = \sum_{t=1}^{\infty} \frac{1}{t^s} \sum_{d|t} \mu(d) = 1.$$

THEOREM 6-33. $\displaystyle\sum_{m=1}^{n} \varphi(m) = \frac{3n^2}{\pi^2} + O(n \log n).$

Proof: Since

$$\varphi(m) = m \sum_{d|m} \frac{\mu(d)}{d},$$

we have

$$\sum_{m=1}^{n} \varphi(m) = \sum_{m=1}^{n} m \sum_{d|m} \frac{\mu(d)}{d} = \sum_{dd' \le n} d'\mu(d) = \sum_{d=1}^{n} \mu(d) \sum_{d'=1}^{n/d} d'$$

$$= \sum_{d=1}^{n} \mu(d) \frac{[n/d]^2 + [n/d]}{2} = \frac{1}{2} \sum_{d=1}^{n} \mu(d) \left[\frac{n}{d}\right]^2$$

$$+ O\left(\sum_{d=1}^{n} \left[\frac{n}{d}\right]\right)$$

$$= \frac{1}{2} \sum_{d=1}^{n} \mu(d) \frac{n^2}{d^2} + O\left(\sum_{d=1}^{n} \frac{n}{d}\right) + O(n \log n)$$

$$= \frac{n^2}{2} \left(\sum_{d=1}^{\infty} \frac{\mu(d)}{d^2} - \sum_{d=n+1}^{\infty} \frac{\mu(d)}{d^2}\right) + O(n \log n)$$

$$= \frac{n^2}{2} \frac{1}{\zeta(2)} + O\left(n^2 \sum_{n+1}^{\infty} \frac{1}{d^2}\right) + O(n \log n)$$

$$= \frac{3n^2}{\pi^2} + O(n) + O(n \log n)$$

$$= \frac{3n^2}{\pi^2} + O(n \log n).$$

THEOREM 6–34. $\displaystyle\sum_{m=1}^{n} \sigma(m) = \frac{\pi^2 n^2}{12} + O(n \log n).$

Proof:

$$\sum_{m=1}^{n} \sigma(m) = \sum_{m=1}^{n} \sum_{d|m} d = \sum_{m=1}^{n} \sum_{d=1}^{n/m} d = \frac{1}{2} \sum_{m=1}^{n} \left(\left[\frac{n}{m}\right]^2 + \left[\frac{n}{m}\right] \right)$$

$$= \frac{1}{2} n^2 \sum_{1}^{n} \frac{1}{m^2} + O(n \log n)$$

$$= \frac{n^2}{2} \left(\zeta(2) - O(1/n) \right) + O(n \log n)$$

$$= \frac{n^2 \zeta(2)}{2} + O(n \log n).$$

PROBLEMS

1. Show that $\displaystyle\sum_{n \leq x} \frac{\tau(n)}{n} = \frac{\log^2 x}{2} + 2\gamma \log x + O(1).$ [*Hint:* Use Theorems 6–15 and 6–31.]

2. Let $\delta(n)$ be the largest odd divisor of n. Show that

$$\sum_{n \leq x} \delta(n) = \frac{x^2}{3} + O(x) \qquad \text{and} \qquad \sum_{n \leq x} \frac{\delta(n)}{n} = \frac{2x}{3} + O(1).$$

[*Hint:* Classify the numbers less than or equal to x according to the exponents of the powers of 2 dividing them, and show that

$$\sum_{n \leq x} \delta(n) = \sum_{n=0}^{(x-1)/2} (2n+1) + \sum_{n=0}^{(x-2)/4} \frac{4n+2}{2} + \sum_{n=0}^{(x-4)/8} \frac{8n+4}{4} + \cdots .]$$

*3. Show that $\displaystyle\sum_{n \leq x} \frac{\varphi(n)}{n} \sim \frac{6x}{\pi^2}.$

Deduce that the numbers $\varphi(n)/n$ are not uniformly distributed in the interval $(0, 1)$. [A sequence $\{a_n\}$ of numbers in $(0, 1)$ is said to be uniformly distributed if, for every α and β with $0 \leq \alpha < \beta < 1$,

$$\lim_{N \to \infty} \frac{1}{N} \sum_{\substack{n \leq N \\ \alpha \leq a_n < \beta}} 1 = \beta - \alpha.]$$

4. Prove that $\displaystyle\sum_{d=1}^{n} \varphi(d) \left[\frac{n}{d}\right] = \frac{n(n+1)}{2}.$

(Use Theorem 6–29.)

*5. Using the result of Problem 1, Section 6–2, show that

$$\sum_{n \leq x} \frac{1}{\varphi(n)} \sim \log x \sum' \frac{1}{n\varphi(n)},$$

where the accent designates summation over the square-free integers.

*6. Prove that $\zeta(2) = \pi^2/6$ by evaluating the double integral

$$I = \int_0^1 \int_0^1 \frac{dx \, dy}{1 - xy}$$

in two ways. [*Hint:* Obtain $I = \zeta(2)$ from the expansion

$$\frac{1}{1 - xy} = 1 + xy + x^2y^2 + x^3y^3 + \cdots,$$

which is valid for $|xy| < 1$. Then evaluate the integral directly by rotating the coordinate system about the origin through 45° to obtain

$$I = 4 \int_0^{1/\sqrt{2}} du \int_0^u \frac{dv}{2 - u^2 + v^2} + 4 \int_{1/\sqrt{2}}^{\sqrt{2}} du \int_0^{\sqrt{2}-u} \frac{dv}{2 - u^2 + v^2},$$

integrating with respect to v, and making the substitution $u = \sqrt{2} \cos \vartheta$.]

6–11 An application. As was pointed out earlier, it is not known whether 2 is a primitive root of infinitely many primes, nor has the same question been settled for any other fixed integer. It is therefore natural to ask, what can be said about the size of the smallest primitive root g_p, as a function of p? Unfortunately, the little that is known (such as the theorem that g_p is less than $\sqrt{p} \log^{17} p$ for large p) cannot be developed here; we content ourselves with an estimate for the smallest quadratic nonresidue n_p of p. Since g_p is certainly a nonresidue of p, any upper bound for g_p would imply the same bound for n_p, but not conversely.

THEOREM 6–35. *There is a quadratic nonresidue of p between 1 and $\sqrt{p}$, for all sufficiently large primes p.*

Proof: Corresponding to each pair of integers x, y with $(x, y) = 1$, $0 < x < \sqrt{p}$, $0 < y < \sqrt{p}$, there corresponds an integer z, unique modulo p, such that

$$x \equiv yz \pmod{p}. \tag{8}$$

Different pairs x, y yield different z's. For if

$$x_1 \equiv y_1 z \pmod{p} \qquad \text{and} \qquad x_2 \equiv y_2 z \pmod{p},$$

then

$$x_1 y_2 \equiv x_2 y_1 \pmod{p},$$

whence $x_1 y_2 = x_2 y_1$. But this, together with the hypothesis that $(x_1, y_1) = (x_2, y_2) = 1$, clearly implies that $x_1 = x_2$ and $y_1 = y_2$.

Now there are $2\varphi(m)$ ordered pairs x, y of relatively prime positive integers whose largest element is m, if $1 < m < \sqrt{p}$, and one pair both of whose elements are 1, so that the total number of pairs is

$$1 + 2 \sum_{m=2}^{\sqrt{p}} \varphi(m) = 2 \sum_{m=1}^{\sqrt{p}} \varphi(m) - 1.$$

If this number is larger than $p/2$, then there are more than $(p-1)/2$ different residue classes z, and since there are only $(p-1)/2$ quadratic residues of p, at least one z must be a nonresidue. But then it follows from (8) that one of x and y (each of which is smaller than $\sqrt{p}$) is also a quadratic nonresidue of p. Thus the proof will be complete when it is shown that, for all large p,

$$2 \sum_{m=1}^{\sqrt{p}} \varphi(m) - 1 > \frac{p}{2}. \tag{9}$$

Using the estimate of Theorem 6–33, we have that

$$2 \sum_{m=1}^{\sqrt{p}} \varphi(m) - 1 = \frac{6}{\pi^2} [\sqrt{p}]^2 + O(\sqrt{p} \log p)$$

$$= \frac{6}{\pi^2} p + O(\sqrt{p} \log p) \sim \frac{6p}{\pi^2}.$$

Since $6/\pi^2 > \frac{1}{2}$, it is clear that (9) holds for sufficiently large p, where the lower bound of validity depends upon the implied constant in the term $O(\sqrt{p} \log p)$.

By refining the argument slightly, it can be shown that the error term in Theorem 6–33 is numerically less than $1 \cdot n \log n$. Using this, the phrase "sufficiently large p" of Theorem 6–35 can be replaced by "$p > 10^4$", and finally, by reference to tables of the smallest primitive roots of primes less than 10^4, it can be shown that $n_p < \sqrt{p}$ for every $p \neq 2, 3, 7, 23$.

REFERENCES

Section 6-1

J. J. Sylvester (*Mathematical Papers*, vol. 4, New York: Cambridge University Press, 1912, pp. 588, 625–629) proved that an odd perfect number must have at least five distinct prime factors, and conjectured that it must have at least six. His conjecture was verified by U. Kühnel, *Mathematische Zeitschrift* (Berlin) **52**, 202–211 (1949). It follows that if there is an odd perfect number, it cannot be smaller than $2.2 \cdot 10^{12}$.

For a complete account of the Mersenne numbers, see R. C. Archibald, *Scripta Mathematica* **3**, 112–119 (1935), and H. S. Uhler, *Scripta Mathematica* **18**, 122–131, (1952).

Section 6-8

The proof of Bertrand's conjecture given here is a modification of that given by P. Erdös, *Acta Universitatis Szegediensis* (Szeged, Hungary) **5**, 194–198 (1932). The proof of Theorem 6–22 is simpler than that originally given; it was found independently by Erdös and L. Kalmár in 1939, but was not published.

Section 6-11

The inequality $g_p < \sqrt{p} \log^{17} p$ is due to Erdös, *Bulletin of the American Mathematical Society* **51**, 131–132 (1945). The inequality

$$\left| \sum_{m=1}^{n} \varphi(m) - \frac{3n^2}{\pi^2} \right| < n \log n$$

is due to R. Tambs-Lyche, *Kongelige Norske Videnskabers Selskabs Forhandlinger* (Trondhjem, Norway) **9**, 58–61 (1936). For further results on this error term, see Erdös and H. N. Shapiro, *Canadian Journal of Mathematics* **3**, 375–385 (1951), and A. Z. Valfish, *Akademiya Nauk Gruzinskoi SSR Trudy Tbilisskogo Matematicheskogo Instituta imeni A. N. Razmadze* **19**, 1–31 (1953). [American Mathematical Society Translations, Series 2, **4**, 1–30.]

CHAPTER 7

SUMS OF SQUARES

7–1 An approximation theorem. In this chapter we consider the
following questions: Given k, what integers can be represented as a
sum of k squares? If an integer is so representable, how many representations are there? Both problems will be completely solved for
$k = 2$, a partial answer to the first will be given for $k = 3$, and it will
be shown that every integer is a sum of four squares (and hence of
k squares, if $k \geq 4$).

We shall need the following approximation theorem.

THEOREM 7–1. *If ξ is a real number and t is a positive integer, there
are integers x and y such that*

$$\left| \xi - \frac{x}{y} \right| \leq \frac{1}{y(t+1)}, \qquad 1 \leq y \leq t.$$

Proof: The $t + 1$ numbers

$$0 \cdot \xi - [0 \cdot \xi], \qquad 1 \cdot \xi - [1 \cdot \xi], \qquad \ldots, \qquad t\xi - [t\xi]$$

all lie in the interval $0 \leq u < 1$. Call them, in increasing order of
magnitude, $\alpha_0, \alpha_1, \ldots, \alpha_t$. Mark the numbers $\alpha_0, \ldots, \alpha_t$ on a circle
of unit circumference, that is, a unit interval on which 0 and 1 are
identified. Then the $t + 1$ differences

$$\alpha_1 - \alpha_0, \qquad \alpha_2 - \alpha_1, \qquad \ldots, \qquad \alpha_t - \alpha_{t-1}, \qquad \alpha_0 - \alpha_t + 1$$

are the lengths of the arcs of the circle between successive α's, and so
they are non-negative and

$$(\alpha_1 - 0) + (\alpha_2 - \alpha_1) + \cdots + (1 - \alpha_t) = 1.$$

It follows that at least one of these $t + 1$ differences does not exceed
$(t + 1)^{-1}$. But each difference is of the form

$$g_1 \xi - g_2 \xi - N,$$

where N is an integer, and we can take $y = |g_1 - g_2|$, $x = \pm N$.

125

7-2 Sums of two squares. A representation of the positive integer n as a sum of two squares, say $n = x^2 + y^2$, will be termed *proper* or *improper* according as $(x, y) = 1$ or $(x, y) > 1$. Throughout this section, "representable" will mean "representable as a sum of two squares," with an analogous meaning for "properly representable."

THEOREM 7-2. *If $p \equiv 3 \pmod 4$ and $p|n$, then n has no proper representation.*

Proof: If $p|n$, $n = x^2 + y^2$, and $(x, y) = 1$, then $p{\nmid}x$ and $p{\nmid}y$. Hence there is an integer u such that $y \equiv ux \pmod p$, and

$$x^2 + y^2 \equiv x^2 + u^2x^2 \equiv x^2(1 + u^2) \equiv 0 \pmod p,$$

so that

$$u^2 \equiv -1 \pmod p.$$

It follows that -1 is a quadratic residue of p, and so either $p = 2$ or $p \equiv 1 \pmod 4$, by Theorem 5-3.

THEOREM 7-3. *An integer $n = \prod p_i{}^{\alpha_i}$ is representable if and only if α_i is even for every i for which $p_i \equiv 3 \pmod 4$.*

Proof: Suppose first that $p^{2k+1}||n$, where $p \equiv 3 \pmod 4$, and suppose that $n = x^2 + y^2$, where $(x, y) = d$ and $p^j||d$. Then $x = dx_1$ and $y = dy_1$, where $(x_1, y_1) = 1$. But if $x_1{}^2 + y_1{}^2 = N$, then $p^{2k-2j+1}|N$, and $2k - 2j + 1 > 0$; this contradicts Theorem 7-2.

It remains only to show that if $n = n_1 n_2{}^2$, where n_1 is square-free and without divisors congruent to 3 (mod 4), then n is representable. It suffices to prove n_1 representable. Since

$$(x_1{}^2 + y_1{}^2)(x_2{}^2 + y_2{}^2) = (x_1 x_2 + y_1 y_2)^2 + (x_1 y_2 - x_2 y_1)^2, \quad (1)$$

the product of representable numbers is representable; this, together with the fact that $1 = 1^2 + 0^2$ is representable, shows that we need only consider the various prime factors of n_1. Since $2 = 1^2 + 1^2$ is representable, it suffices to show that if $p \equiv 1 \pmod 4$, then p is representable. For later purposes, however, we prove a more precise result.

THEOREM 7-4. *If $n > 1$ and $u^2 \equiv -1 \pmod n$, there are unique integers x and y such that*

$$n = x^2 + y^2, \qquad x > 0, \qquad y > 0, \qquad (x, y) = 1,$$
$$y \equiv ux \pmod n.$$

Remark: In case n is a prime $p \equiv 1 \pmod 4$, the congruence $u^2 \equiv -1 \pmod p$ is solvable, and so p is representable.

Proof: The idea of the proof is to replace the equation $x^2 + y^2 = n$ by the equivalent conditions

$$x^2 + y^2 \equiv 0 \pmod n, \qquad 0 < x^2 + y^2 < 2n.$$

To satisfy these conditions, we require that x be one of the numbers $1, 2, \ldots, [\sqrt{n}]$, and then seek a y such that $y \equiv ux \pmod n$ (so that $x^2 + y^2 \equiv 0 \pmod n$) and $1 \leq y < \sqrt{n}$. But if $y \equiv ux \pmod n$, then $y = ux + an$, and so we want a linear combination of u and n to be small.

We apply Theorem 7–1, with

$$\xi = -\frac{u}{n}, \qquad t = [\sqrt{n}],$$

and see that there are integers a and x_1 such that $1 \leq x_1 \leq [\sqrt{n}]$ and

$$\left| -\frac{u}{n} - \frac{a}{x_1} \right| \leq \frac{1}{x_1(1 + [\sqrt{n}])} < \frac{1}{x_1\sqrt{n}},$$

so that

$$|ux_1 + na| < \sqrt{n}.$$

Put $y_1 = ux_1 + na$. If $y_1 > 0$, put $y' = y_1$, $x' = x_1$. If $y_1 < 0$, then $-y_1 \equiv -ux_1 \pmod n$, and since $u^2 \equiv -1 \pmod n$ we have

$$u^2 y_1 \equiv -ux_1 \pmod n,$$

$$u(-y_1) \equiv x_1 \pmod n,$$

and we take $x' = -y_1$, $y' = x_1$. In either case,

$$y' \equiv ux' \pmod n, \qquad x'^2 + y'^2 = n, \qquad x' > 0, \qquad y' > 0.$$

From the relation

$$n = x_1{}^2 + y_1{}^2 = x_1{}^2 + u^2 x_1{}^2 + 2ux_1 na + n^2 a^2$$
$$= x_1{}^2(1 + u^2) + ux_1 an + na(ux_1 + an),$$

we obtain

$$1 = \left(\frac{1 + u^2}{n} x_1 + ua \right) x_1 + ay_1,$$

so that $(x_1, y_1) = 1$, whence $(x', y') = 1$.

Finally, to prove the uniqueness, suppose that besides x', y', there is a pair x'', y'' satisfying all conditions of the theorem. Then by equation (1),

$$n^2 = (x'^2 + y'^2)(x''^2 + y''^2) = (x'x'' + y'y'')^2 + (x'y'' - x''y')^2.$$

But

$$x'x'' + y'y'' \equiv x'x'' + u^2x'x'' = x'x''(1 + u^2) \equiv 0 \pmod{n},$$

and since $x'x'' + y'y'' > 0$, it follows that $x'x'' + y'y'' = n$, $x'y'' - x''y' = 0$. Hence $x'' = kx'$, $y'' = ky'$, and it is clear that $k = 1$.

THEOREM 7–5. *The number $P_2(n)$ of proper representations of n is four times the number of solutions of the congruence $u^2 \equiv -1 \pmod{n}$. Hence (by Theorems 5–1 and 5–2),*

$$P_2(n) = \begin{cases} 0 & \text{if } 4|n \text{ or if some } p \equiv 3 \pmod 4 \text{ divides } n; \\ 2^{s+2} & \text{if } 4{\nmid}n, \text{ no } p \equiv 3 \pmod 4 \text{ divides } n, \text{ and } s \text{ is the} \\ & \text{number of distinct odd prime divisors of } n. \end{cases}$$

Proof: The theorem is trivial if $n = 1$. If $n > 1$, then $xy \neq 0$, and the number of representations is four times the number of positive representations. To each u such that $u^2 \equiv -1 \pmod{n}$, there corresponds exactly one proper representation with $x > 0$, $y > 0$, $y \equiv ux \pmod{n}$. Conversely, if $x^2 + y^2 = n$ and $(x, y) = 1$, then $(x, n) = 1$, so that the congruence $y \equiv ux \pmod{n}$ has a unique solution $\pmod{n}$, and

$$x^2 + y^2 = x^2(1 + u^2) \equiv 0 \pmod{n},$$

which implies that $u^2 \equiv -1 \pmod{n}$.

COROLLARY. *A prime $p \equiv 1 \pmod 4$ can be represented uniquely (up to order and sign) as a sum of two squares.*

This follows immediately from the theorem, for in this case $P_2(p)$ is 8, so that p has essentially only one proper representation. It clearly has no improper representation.

PROBLEM

Show that if n is a positive odd number of which -2 is a quadratic residue, then there are integers x and y such that $2x^2 + y^2 = n$ and $(x, y) = 1$.

7-3 The Gaussian integers. In order to obtain an expression for the *total* number of representations of an integer as a sum of two squares, we turn our attention momentarily to the arithmetic of the so-called Gaussian integers: the complex numbers $a + bi$, where a and b are ordinary (or rational) integers. In this section, Greek letters will be used exclusively to designate Gaussian integers, and the set of all such integers will be denoted by $R[i]$.

It is clear that if α and β are in $R[i]$, then so also are $\alpha \pm \beta$ and $\alpha\beta$. If $\alpha = a + bi$, then $(a + bi)(a - bi) = a^2 + b^2$ is called the *norm* of α, and designated by $\mathbf{N}\alpha$. It is easily verified that $\mathbf{N}\alpha\mathbf{N}\beta = \mathbf{N}\alpha\beta$.

An integer α whose reciprocal is also an integer is called a *unit*; since

$$\frac{1}{\alpha} = \frac{1}{a + bi} = \frac{a - bi}{\mathbf{N}\alpha},$$

α is a unit if and only if

$$(a^2 + b^2)|a \qquad \text{and} \qquad (a^2 + b^2)|b.$$

If $a \neq 0$ and $b \neq 0$, then $a^2 + b^2 > \max(|a|, |b|)$; hence either a or b must be zero. If $a = 0$, then $b^2|b$, whence $b = \pm 1$. If $b = 0$, then $a = \pm 1$. Thus the units are ± 1, $\pm i$. The numbers $\pm \alpha$ and $\pm \alpha i$ are called the *associates* of α. An integer is a unit if and only if its norm is 1.

We say that α *divides* β, and write $\alpha|\beta$, if there is an integer γ such that $\beta = \alpha\gamma$. If $\alpha|\beta$, then $\mathbf{N}\alpha|\mathbf{N}\beta$. A unit divides any integer; if an integer has no divisors other than its associates and the units, it is said to be *prime*. Thus $1 + i$ is prime, since the equation

$$1 + i = (a + bi)(c + di)$$

implies

$$\mathbf{N}(1 + i) = 2 = \mathbf{N}(a + bi)\mathbf{N}(c + di),$$

which shows that either $\mathbf{N}(a + bi)$ or $\mathbf{N}(c + di)$ is 1, so that either $a + bi$ or $c + di$ is a unit. More generally, this argument shows that if $\mathbf{N}\alpha$ is a rational prime, then α is a prime of $R[i]$. Thus, corresponding to the representation $p = x^2 + y^2$ of a rational prime $p \equiv 1 \pmod 4$, we have the decomposition

$$p = (x + iy)(x - iy)$$

into primes of $R[i]$. In this case the factors are not associated: x and

y are relatively prime and numerically larger than 1, and are therefore distinct, so that the supposition that a relation of the form

$$i^n(x + iy) = x - iy$$

holds, yields

$$i^n = 1 \quad \text{and} \quad i^{n+1} = -i,$$

which is impossible.

The primes $p \equiv 3 \pmod 4$ do not split further in $R[i]$; that is, they are also prime in the larger set. For if

$$p = (a + bi)(c + di),$$

then

$$p^2 = (a^2 + b^2)(c^2 + d^2).$$

But the only factorizations of p^2 are $p \cdot p$ and $1 \cdot p^2$, and it is impossible that $a^2 + b^2 = c^2 + d^2 = p$, by Theorem 7–3; hence one of the numbers $a + bi$, $c + di$ is a unit.

If α is not prime, it can be represented as a product of primes. For then $\alpha = \beta\gamma$, where $\mathbf{N}\beta > 1$ and $\mathbf{N}\gamma > 1$, and consequently $\mathbf{N}\beta < \mathbf{N}\alpha$, $\mathbf{N}\gamma < \mathbf{N}\alpha$. If β and γ are primes, we are through; if one is not, it can be factored, with the factors having still smaller norms. The process cannot continue indefinitely, since the norms are strictly decreasing positive rational integers, and so we come eventually to a prime factorization.

To show the uniqueness of this factorization, we use the following analog of Theorem 1–1.

THEOREM 7–6. *If α and β are integers of $R[i]$, and $\beta \neq 0$, then there are integers ρ, $\varkappa$ such that*

$$\alpha = \beta\varkappa + \rho, \qquad \mathbf{N}\rho < \mathbf{N}\beta.$$

Proof: Since $\beta \neq 0$, we can write

$$\frac{\alpha}{\beta} = \frac{a + bi}{c + di} = \frac{(a + bi)(c - di)}{c^2 + d^2} = A + Bi,$$

where A and B are rational numbers. Let x be the nearest integer to A, and y the nearest integer to B, so that

$$|A - x| \leq \tfrac{1}{2},$$

$$|B - y| \leq \tfrac{1}{2}.$$

Then

$$\left| \frac{\alpha}{\beta} - (x + iy) \right| = |(A - x) + (B - y)i|$$
$$= ((A - x)^2 + (B - y)^2)^{\frac{1}{2}} \leq (\tfrac{1}{4} + \tfrac{1}{4})^{\frac{1}{2}} < 1.$$

Hence if we put

$$x + iy = \varkappa, \qquad \alpha - \beta(x + iy) = \rho,$$

then

$$\mathbf{N}\rho = \mathbf{N}(\alpha - \varkappa\beta) = \mathbf{N}\beta \cdot \mathbf{N} \left(\frac{\alpha}{\beta} - \varkappa \right) < \mathbf{N}\beta,$$

and $\varkappa, \rho \in R[i]$. The proof is complete.

Starting from Theorem 7–6, the development in Chapter 2 leading to the Unique Factorization Theorem for rational integers can now be paralleled, and we obtain a Unique Factorization Theorem for $R[i]$.

THEOREM 7–7. *Every integer of $R[i]$ can be represented as a product of primes. This representation is unique, aside from the order of the factors and the presence of a set of units whose product is 1.*

It can be shown that the primes of $R[i]$ are exactly the ones we have already found, i.e., the associates of the following numbers:

(a) $1 + i$,
(b) $a \pm bi$, where $a^2 + b^2 = p \equiv 1 \pmod 4$,
(c) q, where $q \equiv 3 \pmod 4$.

PROBLEMS

1. Use the ideas of this section to give a new proof of the Corollary to Theorem 7–5.

2. Find the GCD (in $R[i]$) of $21 + 49i$ and $78 + 8i$.

3. Show that if $\alpha = a + bi$ is prime in $R[i]$, then either $(a, b) = 1$ or $ab = 0$. Use this to deduce that the only primes in $R[i]$ are those listed above. [*Hint:* If $(a, b) = 1$, show that $\mathbf{N}\alpha = 2^t p_1 \ldots p_r$, where $t = 0$ or 1 and $p_i \equiv 1 \pmod 4$ for $i = 1, 2, \ldots, r$. Note also that $\mathbf{N}\alpha = \mathbf{N}\bar{\alpha}$.]

7–4 The total number of representations. Suppose that n has the factorization

$$n = 2^u \cdot \prod_{p_j \equiv 1 (\mathrm{mod}\ 4)} p_j^{t_j} \cdot \prod_{q_j \equiv 3 (\mathrm{mod}\ 4)} q_j^{s_j}.$$

Put

$$\prod_{p_j \equiv 1 (\mathrm{mod}\ 4)} p_j^{t_j} = n', \qquad \prod_{q_j \equiv 3 (\mathrm{mod}\ 4)} q_j^{s_j} = m.$$

THEOREM 7-8. *If $n \geq 1$, then the number $r_2(n)$ of representations of n as a sum of two squares is zero if m is not a square, and is $4\tau(n')$ if m is a square.*

Proof: The case in which m is not a square is covered by Theorem 7-3. If m is a square, each s_j is even, and we can put $s_j = 2r_j$. In this case we shall prove the theorem by establishing, by means of the identity $x^2 + y^2 = (x + iy)(x - iy)$, a one-to-one correspondence between the various representations of n on the one hand, and the factorizations of n as a product of two conjugate Gaussian integers, on the other. We must count these factorizations. Since $1 + i = i(1 - i)$, and $2 = i(1 - i)^2$, we can write the prime decomposition of n in $R[i]$ in the form

$$n = i^u (1 - i)^{2u} \prod ((a + bi)(a - bi))^t \prod q^{2r},$$

where the subscripts in the products have been omitted for clarity, and where

$$a > 0, \qquad b > 0, \qquad p = a^2 + b^2.$$

Then every divisor of n in $R[i]$ is of the form

$$x + iy = i^v (1 - i)^{u_1} \prod ((a + bi)^{t_1}(a - bi)^{t_2}) \prod q^{r_1},$$

where

$$0 \leq v \leq u, \qquad 0 \leq u_1 \leq 2u, \qquad 0 \leq t_1 \leq t, \qquad 0 \leq t_2 \leq t,$$
$$0 \leq r_1 \leq 2r.$$

Not every such divisor leads to a representation; we must also require that the complex conjugate,

$$x - iy = (-i)^v (1 + i)^{u_1} \prod ((a + bi)^{t_2}(a - bi)^{t_1}) \prod q^{r_1}$$
$$= i^{u_1 - v}(1 - i)^{u_1} \prod ((a + bi)^{t_2}(a - bi)^{t_1}) \prod q^{r_1},$$

be such that $(x + iy)(x - iy) = n$. It is clear that this is the case if and only if $u_1 = u$, $t_1 + t_2 = t$, $r_1 = r$. Since the powers of i are periodic, with period 4, we obtain all the distinct factorizations of n into conjugate factors by listing the numbers

$$i^v (1 - i)^u \prod ((a + bi)^{t_1}(a - bi)^{t - t_1}) \prod q^r,$$

where u, t, and r are fixed, v is one of the integers 0, 1, 2, 3, and t_1 is one of $0, 1, \ldots, t$. Their total number is $4 \prod (t + 1) = 4\tau(n')$.

PROBLEM

Show that $\sum\limits_{m=1}^{n} r_2(m) = \pi n + O(\sqrt{n})$.

[*Hint:* The sum on the left is the number of lattice points inside or on the circle $x^2 + y^2 = n$. Associate each such point with the unit square of which it is the lower left corner. The resulting region has a polygonal boundary, no point of which is at distance greater than $\sqrt{2}$ from the circle.]

7–5 Sums of three squares. The problem of the solvability of the equation

$$n = x^2 + y^2 + z^2 \tag{2}$$

is much more difficult than the corresponding question for the sum of either two or four squares. The result is this: (2) is solvable if and only if n is not of the form $4^t(8k + 7)$. We prove here only the trivial half of this theorem, that if n is of the specified form then (2) has no integral solutions.

Since a square can have only the values 0, 1, or 4 (mod 8), the sum of three squares is congruent to 0, 1, 2, 3, 4, 5, or 6 (mod 8), so that no $n \equiv 7 \pmod 8$ is so representable. If $4|n$ and (2) holds, then x, y, and z must all be even, so that $n/4$ must also be a sum of three squares. Therefore n cannot be a power of 4 times a nonrepresentable number.

It might be mentioned that one reason that problems concerning three squares are more difficult than those concerning either two or four is that there is no composition identity in this case analogous to (1) or to that given below for four squares. Indeed, the fact that 3 and 5 are sums of three squares, while 15 is not, shows that no such identity·is possible.

7–6 Sums of four squares

THEOREM 7–9. *Every positive integer can be represented as a sum of four squares.*

Since

$$(x_1{}^2 + x_2{}^2 + x_3{}^2 + x_4{}^2)(y_1{}^2 + y_2{}^2 + y_3{}^2 + y_4{}^2)$$
$$= (x_1y_1 + x_2y_2 + x_3y_3 + x_4y_4)^2 + (x_1y_2 - x_2y_1 + x_3y_4 - x_4y_3)^2$$
$$+ (x_1y_3 - x_3y_1 + x_4y_2 - x_2y_4)^2 + (x_1y_4 - x_4y_1 + x_2y_3 - x_3y_2)^2,$$

the product of representable numbers is representable. Since 1 is also representable, it suffices to prove that every prime p is representable. The proof, which uses the same idea as the proof of Theorem 7–4, depends on the following theorem.

THEOREM 7–10. *Let r, s, and m be positive integers with $r < s$, and let λ_σ ($\sigma = 1, \ldots, s$) be positive numbers (not necessarily integers) smaller than m, such that*

$$\lambda_1 \ldots \lambda_s > m^r.$$

Then the system of r linear congruences

$$\sum_{\sigma=1}^{s} a_{\rho\sigma} x_\sigma \equiv 0 \ (\mathrm{mod}\ m), \qquad \rho = 1, \ldots, r,$$

where the a's are integers, has a solution in integers $x_1, \ldots, x_s$, not all zero, such that $|x_\sigma| \le \lambda_\sigma$ for $\sigma = 1, \ldots, s$.

Proof: Put

$$y_\rho = \sum_{\sigma=1}^{s} a_{\rho\sigma} x_\sigma, \qquad \text{for } \rho = 1, \ldots, r.$$

For each σ, let x_σ range over the integers $0, 1, \ldots, [\lambda_\sigma]$; this gives $1 + [\lambda_\sigma]$ choices for x which are distinct (mod m), since $\lambda_\sigma < m$, and there are

$$\prod_{\sigma=1}^{s} (1 + [\lambda_\sigma])$$

different s-tuples $x_1, \ldots, x_s$. Corresponding to each s-tuple $x_1, \ldots, x_s$ there is an r-tuple $y_1, \ldots, y_r$, and so we have found

$$\prod_{\sigma=1}^{s} (1 + [\lambda_\sigma]) > \prod_{\sigma=1}^{s} \lambda_\sigma > m^r$$

r-tuples $y_1, \ldots, y_r$. But there are only m^r integral r-tuples which are distinct (mod m), so that there must be sets $y_1', \ldots, y_r'$ and $y_1'', \ldots, y_r''$ such that $y_\rho' \equiv y_\rho''$ (mod m) for $\rho = 1, \ldots, r$. If these r-tuples correspond to $x_1', \ldots, x_s'$ and $x_1'', \ldots, x_s''$ respectively, then

$$y_\rho' - y_\rho'' = \sum_{\sigma=1}^{s} a_{\rho\sigma}(x_\sigma' - x_\sigma'') \equiv 0 \ (\mathrm{mod}\ m), \qquad \rho = 1, \ldots, r,$$

and not all of $x_\sigma' - x_\sigma''$ are zero, while $|x_\sigma' - x_\sigma''| \le \lambda_\sigma$ for $\sigma = 1, \ldots, s$.

Proof of Theorem 7-9: If p is a prime, then the congruence

$$x^2 + y^2 + 1 \equiv 0 \pmod{p}$$

has a solution. For if x and y range independently over the numbers $0, 1, \ldots, (p-1)/2$ (this for odd p; the assertion is clearly correct for $p = 2$), then all the numbers x^2 are distinct $\pmod{p}$, and the same is true of the numbers $-(1 + y^2)$. For if $x_i^2 \equiv x_j^2 \pmod{p}$, then $p \mid (x_i - x_j)(x_i + x_j)$. But $0 < x_i + x_j < p$, unless $x_i = x_j = 0$, so $p \mid (x_i - x_j)$, $x_i \equiv x_j \pmod{p}$, and so $x_i = x_j$. But we have altogether

$$\frac{p+1}{2} + \frac{p+1}{2} = p + 1$$

numbers x^2 and $-1 - y^2$, so some x^2 is congruent to some $-1 - y^2$, modulo p, which is the assertion.

Suppose that $a^2 + b^2 + 1 \equiv 0 \pmod{p}$. By Theorem 7-10, the congruences

$$x \equiv az + bt \pmod{p},$$

$$y \equiv bz - at \pmod{p}$$

have a nontrivial solution x, y, z, t with

$$\max(|x|, |y|, |z|, |t|) \leq \sqrt{p} + \epsilon;$$

here $r = 2$, $s = 4$, $m = p$, and we have chosen $\lambda_\sigma = \sqrt{p} + \epsilon$, where $\epsilon > 0$ is so small that $\sqrt{p} + \epsilon < p$. Now x, y, z, and t are integers, while $\sqrt{p}$ is not; if ϵ is chosen so small that $\sqrt{p} + \epsilon < 1 + [\sqrt{p}]$, it follows that

$$\max(|x|, |y|, |z|, |t|) < \sqrt{p}.$$

We have

$$x^2 + y^2 \equiv (a^2 + b^2)(z^2 + t^2) \equiv -(z^2 + t^2) \pmod{p},$$

while

$$0 < x^2 + y^2 + z^2 + t^2 < p + p + p + p = 4p,$$

so that

$$x^2 + y^2 + z^2 + t^2 = Ap,$$

where $A = 1, 2$, or 3. If $A = 1$, we are finished. If $A = 2$, then x is congruent to y, z, or $t \pmod{2}$. If $x \equiv y \pmod{2}$, then $z \equiv t \pmod{2}$, and $\quad p = \left(\dfrac{x+y}{2}\right)^2 + \left(\dfrac{x-y}{2}\right)^2 + \left(\dfrac{z+t}{2}\right)^2 + \left(\dfrac{z-t}{2}\right)^2,$

where the quantities in parentheses are integers.

In the case $A = 3$, we note first that $p = 3$ has a representation: $3 = 1^2 + 1^2 + 1^2$, so that we need only consider $p \neq 3$. The square of an integer is congruent to 0 or 1 (mod 3), and the equation

$$x^2 + y^2 + z^2 + t^2 = 3p$$

implies that

$$x^2 + y^2 + z^2 + t^2 \equiv 0 \pmod{3},$$

while

$$x^2 + y^2 + z^2 + t^2 \not\equiv 0 \pmod{9}.$$

By the congruence, one of the quantities—say x—is divisible by 3, and either all the others are, or all are not, divisible by 3. Because of the incongruence, $3 \nmid yzt$, so that y, z, and t are all congruent to ± 1 (mod 3). Let z' be that one of $\pm z$ such that $z' \equiv y$ (mod 3), and let t' be that one of $\pm t$ such that $t' \equiv y$ (mod 3). Then

$$p = \left(\frac{y + z' + t'}{3}\right)^2 + \left(\frac{x + z' - t'}{3}\right)^2 + \left(\frac{x - y + t'}{3}\right)^2$$
$$+ \left(\frac{x + y - z'}{3}\right)^2,$$

where the quantities in parentheses are integers. The proof is complete.

REFERENCES

Section 7–5

A brief proof of the theorem that every number not of the form $4^t(8k + 7)$ can be represented as the sum of three squares is to be found in Landau, *Handbuch der Lehre von der Verteilung der Primzahlen*, Leipzig: Teubner Gesellschaft, 1909, vol. 1, pp. 550–555.

Section 7–6

Theorem 7–10, and its application to Theorem 7–9, are due to A. Brauer and T. L. Reynolds, *Canadian Journal of Mathematics* **3**, 367–373 (1951).

CHAPTER 8

PELL'S EQUATION AND SOME APPLICATIONS

8-1 Introduction. The Diophantine equation $x^2 - dy^2 = N$ (where N and d are integers), commonly known as Pell's equation, was actually never considered by Pell; it was because of a mistake on Euler's part that his name has been attached to it. The early Greek and Indian mathematicians had considered special cases, but Fermat was the first to deal systematically with it. He said that he had shown, in the special case where $N = 1$ and $d > 0$ is not a perfect square, that there are infinitely many integral solutions x, y; as usual, he did not give a proof. The first published proof was given by Lagrange, using the theory of continued fractions. Prior to this, Euler had shown that there are infinitely many solutions if there is one.

Before beginning a systematic investigation, it might be worth while to indicate some of the ways in which the equation arises and some of the reasons, therefore, for its importance. On the one hand, knowledge of the solutions of Pell's equation is essential in finding integral solutions of the general quadratic equation

$$ax^2 + bxy + cy^2 + dx + ey + f = 0,$$

in which a, b, $\ldots$, f are integers. For, writing the left side as a polynomial in x,

$$ax^2 + (by + d)x + cy^2 + ey + f = 0,$$

it is clear that, if the equation is solvable for a certain y, the discriminant

$$(by + d)^2 - 4a(cy^2 + ey + f),$$

or, what is the same thing,

$$(b^2 - 4ac)y^2 + (2bd - 4ae)y + d^2 - 4af,$$

must be a perfect square, say z^2. Putting

$$b^2 - 4ac = p, \qquad 2bd - 4ae = q, \qquad d^2 - 4af = r,$$

we have

$$py^2 + qy + r - z^2 = 0.$$

Again, the discriminant of this quadratic in y must be a perfect square, say

$$q^2 - 4p(r - z^2) = w^2.$$

Thus we are led to consider the Pell equation

$$w^2 - 4pz^2 = q^2 - 4pr;$$

knowing solutions of it, we can, at any rate, obtain rational solutions of the original equation.

As a second example, consider the *real quadratic field* $R(\sqrt{d})$, consisting of all the numbers of the form

$$a + b\sqrt{d}, \qquad d > 1, \ d \text{ square-free}$$

where a and b are rational numbers—positive, negative or zero. Each of these numbers with $b \neq 0$ is a zero of a unique quadratic polynomial with relatively prime integral coefficients, that of x^2 being positive. If the leading coefficient of the polynomial is 1, the corresponding number is said to be an *integer* of the field. (Notice the close analogy between the present discussion and that in the first portion of Section 7–3. There, of course, we were working with the integers of the nonreal quadratic field $R(\sqrt{-1})$.) Starting with this notion of integer, it is possible to construct an arithmetic very similar to that developed in Chapter 2 for the ordinary, or rational, integers. Denote by $R[\sqrt{d}]$ the set of all integers of $R(\sqrt{d})$. If α, β, and α/β are in $R[\sqrt{d}]$, then we say that β *divides* α, and write $\beta|\alpha$. If $\alpha|1$, then α is a *unit* of $R[\sqrt{d}]$. If every factorization $\alpha = \beta\gamma$ into the product of integers of $R[\sqrt{d}]$ is such that either β or γ is a unit, then α is *prime* in $R[\sqrt{d}]$. Finally, the *norm* $\mathbf{N}\alpha$ of $\alpha = a + b\sqrt{d}$ is the product of α and its algebraic conjugate $\bar{\alpha} = a - b\sqrt{d}$, namely $a^2 - db^2$. It is a rational integer if α is an integer of the field, and $\mathbf{N}\alpha \cdot \mathbf{N}\beta = \mathbf{N}(\alpha\beta)$ always.

Two complications now arise, however, which must be dealt with. The more serious, with which we shall not be concerned for the time being, is that the analog of the Unique Factorization Theorem does not hold for every d, and it is necessary to introduce a rather sophisticated mechanism to deal with this problem. The other complication is that, in distinction to the set of rational integers where there are only the two units ± 1, a real quadratic field has infinitely many, as will follow from the theorems of this chapter. For it is easily seen that α is a unit of $R(\sqrt{d})$ if and only if $\mathbf{N}\alpha = \pm 1$, that is, if and only if

$$a^2 - db^2 = \pm 1.$$

Since it can be shown that $a + b\sqrt{d}$ is a quadratic integer if and only if a and b are both rational integers (or in case $d \equiv 1 \pmod 4$, a and b may also be halves of odd integers), the infinitude of units follows from Lagrange's theorem concerning Pell's equation. Clearly, knowledge of the structure of this set of units will depend on a thorough analysis of that equation for $N = \pm 1$ and ± 4.

We shall give a third application of Pell's equation, this time to the minimum of an indefinite quadratic form, later in this chapter. There will be others in Volume II.

PROBLEMS

*1. Let d be greater than 1 and square-free, and let α be in $R[\sqrt{d}]$. Show that if $d \not\equiv 1 \pmod 4$, then

$$\alpha = a + b\sqrt{d}, \qquad a,\ b \text{ integers},$$

while if $d \equiv 1 \pmod 4$, then

$$\alpha = \frac{a + b\sqrt{d}}{2}, \qquad a,\ b \text{ integers such that } a \equiv b \pmod 2.$$

[*Hint:* First show that if $\bar{\alpha}$ is the conjugate of α in $R(\sqrt{d})$, then $\alpha + \bar{\alpha}$ and $\alpha\bar{\alpha}$, and therefore also $4\alpha\bar{\alpha} - (\alpha + \bar{\alpha})^2$, must be in $R[\sqrt{d}]$.]

2. Show that α is a unit of $R[\sqrt{d}]$ if and only if $\mathbf{N}\alpha = \pm 1$.

3. Find some solutions of the Diophantine equation

$$x^2 + 6xy - 4y^2 - 4x - 12y - 19 = 0.$$

8–2 The case $N = \pm 1$. For the present we shall concern ourselves with the equation

$$x^2 - dy^2 = 1. \tag{1}$$

The case in which d is a negative integer is easily dealt with: if $d = -1$, then the only solutions are $\pm 1, 0$ and $0, \pm 1$, while if $d < -1$, the only solutions are $\pm 1, 0$. So from now on we may restrict attention to equations of the form (1) with $d > 0$. If d is a square, then (1) can be written as

$$x^2 - (d'y)^2 = 1,$$

and since the only two squares which differ by 1 are 0 and 1, the only solutions in this case are $\pm 1, 0$. Suppose then that d is not a square.

THEOREM 8–1. *For any irrational number ξ, the inequality*

$$|x - \xi y| < \frac{1}{y} \qquad (2)$$

has infinitely many solutions.

Proof: According to Theorem 7–1, if ξ is irrational the inequalities

$$0 < |x - \xi y| < \frac{1}{t}, \qquad 1 \le y \le t, \qquad (3)$$

have a solution for each positive integer t. It is clear that each solution of (3) is also a solution of (2). Taking $t = 1$ in (3) gives a solution x_1, y_1 of (2). Then for suitable $t_1 > 1$,

$$|x_1 - \xi y_1| > \frac{1}{t_1},$$

and taking $t = t_1$ in (3) gives a solution x_2, y_2 of (2). Since

$$|x_2 - \xi y_2| < |x_1 - \xi y_1|,$$

the two solutions so far found are distinct. Now choose $t_2 > t_1$ so that

$$|x_2 - \xi y_2| > \frac{1}{t_2},$$

and for $t = t_2$ find x_3, y_3. Clearly this procedure can be continued indefinitely, yielding infinitely many solutions of (2).

THEOREM 8–2. *There are infinitely many solutions of the equation*

$$x^2 - dy^2 = k \qquad (4)$$

in positive integers x, y for some k with $|k| < 1 + 2\sqrt{d}$.

Proof: If x, y is a solution of (2), then

$$|x + y\sqrt{d}| = |x - y\sqrt{d} + 2y\sqrt{d}| < \frac{1}{y} + 2y\sqrt{d} \le (1 + 2\sqrt{d})y,$$

and so

$$|x^2 - dy^2| < \frac{1}{y}(1 + 2\sqrt{d})y = 1 + 2\sqrt{d}.$$

Since there are infinitely many distinct pairs x, y available, but only finitely many integers numerically smaller than $1 + 2\sqrt{d}$, infinitely many of the numbers $x^2 - dy^2$ must have a common value, which is the theorem.

THEOREM 8–3. *Equation* (1) *has at least one solution with $y \neq 0$.*

Proof: Separate the infinitely many solutions of (4) into k^2 classes, putting two solutions x_1, y_1 and x_2, y_2 in the same class if and only if $x_1 \equiv x_2 \pmod{k}$ and $y_1 \equiv y_2 \pmod{k}$. Then some class contains at least two different solutions, say x_1, y_1 and x_2, y_2, with $x_1 x_2 > 0$. Put

$$x = \frac{x_1 x_2 - d y_1 y_2}{k}, \qquad y = \frac{x_1 y_2 - x_2 y_1}{k};$$

we shall show that x and y are integers with $y \neq 0$ for which $x^2 - dy^2 = 1$. It follows immediately from the congruences

$$x_1 \equiv x_2 \pmod{k}, \qquad y_1 \equiv y_2 \pmod{k},$$

that

$$x_1 y_2 \equiv x_2 y_1 \pmod{k},$$

and so y is an integer. Also, from these congruences and from (4),

$$x_1 x_2 - d y_1 y_2 \equiv x_1{}^2 - d y_1{}^2 = k \equiv 0 \pmod{k},$$

and so x is an integer. Furthermore

$$x^2 - dy^2 = \frac{1}{k^2} \left((x_1 x_2 - d y_1 y_2)^2 - d(x_1 y_2 - x_2 y_1)^2 \right)$$

$$= \frac{1}{k^2} \left(x_1{}^2 x_2{}^2 - d x_1{}^2 y_2{}^2 + d^2 y_1{}^2 y_2{}^2 - d x_2{}^2 y_1{}^2 \right)$$

$$= \frac{1}{k^2} \left(x_1{}^2 - d y_1{}^2 \right) \left(x_2{}^2 - d y_2{}^2 \right) = 1.$$

Finally, if $y = 0$, then

$$x_1 y_2 = x_2 y_1,$$

so that for some a, $x_1 = a x_2$ and $y_1 = a y_2$. But since x_1, y_1 and x_2, y_2 are both solutions of (4), it must be that $a = 1$, contrary to the assumption that x_1, y_1 and x_2, y_2 are different solutions.

THEOREM 8–4. *If x_1, y_1 and x_2, y_2 are solutions of the Pell equation* (1), *then so also are the integers x, y defined by the equation*

$$(x_1 + y_1 \sqrt{d})(x_2 + y_2 \sqrt{d}) = x + y\sqrt{d}. \tag{5}$$

Proof: It follows from (5) that also

$$(x_1 - y_1 \sqrt{d})(x_2 - y_2 \sqrt{d}) = x - y\sqrt{d},$$

and multiplying corresponding sides of these equations gives

$$x^2 - dy^2 = (x_1{}^2 - dy_1{}^2)(x_2{}^2 - dy_2{}^2) = 1.$$

In particular, it follows from Theorems 8–3 and 8–4, taking $x_1 = x_2$, $y_1 = y_2$, that the numbers x, y defined by

$$(x_1 + y_1\sqrt{d})^n = x + y\sqrt{d}$$

form a solution for every positive value of n; that this is also true for negative values of n follows from the fact that

$$\frac{1}{x_1 + y_1\sqrt{d}} = x_1 - y_1\sqrt{d}.$$

We shall now show that a general solution can be obtained in this fashion. For brevity, we shall refer to $x + y\sqrt{d}$, as well as x, y, as a solution of equation (1). It will be called *positive* if $x > 0$ and $y > 0$. The positive solutions will be ordered by the size of x, or what is the same thing, by the size of $x + y\sqrt{d}$, since if $x_1 + y_1\sqrt{d}$ and $x_2 + y_2\sqrt{d}$ are positive solutions, and $x_1 > x_2$, then

$$x_1 + y_1\sqrt{d} > x_2 + y_2\sqrt{d},$$

and conversely.

THEOREM 8–5. *If x_1, y_1 is the minimal positive solution of equation* (1), *then a general solution is given by the equation*

$$x + y\sqrt{d} = \pm(x_1 + y_1\sqrt{d})^n, \tag{6}$$

where n can assume any integral value, positive, negative or zero.

Remark: Because of this theorem, the minimal positive solution of (1) is sometimes called the *fundamental solution.*

Proof: That (6) actually furnishes a solution for each $n > 0$, we have just seen. Since

$$(x + y\sqrt{d})^{-n} = (x - y\sqrt{d})^n,$$

(6) also gives a solution for each $n < 0$. Since the solutions with $y = 0$ correspond to $n = 0$, it suffices to show that (6) gives every solution with $y \neq 0$. Furthermore, if x_1, y_1, and n are positive and

$$x + y\sqrt{d} = (x_1 + y_1\sqrt{d})^n > 1,$$

then
$$-x + (-y)\sqrt{d} = -(x_1 + y_1\sqrt{d})^n < 1,$$
$$x + (-y)\sqrt{d} = (x_1 + y_1\sqrt{d})^{-n} < 1,$$
$$-x + y\sqrt{d} = -(x_1 + y_1\sqrt{d})^{-n} < 1,$$

so that it suffices to show that every solution of (1) with both x and y positive (so that $x + y\sqrt{d} > 1$) satisfies the equation
$$x + y\sqrt{d} = (x_1 + y_1\sqrt{d})^n, \qquad n > 0.$$

Put $x_1 + y_1\sqrt{d} = \alpha$; then if x, y is any positive solution of (1), $x + y\sqrt{d} \geq \alpha$, since α is minimal. Hence there is an $n > 0$ such that
$$\alpha^n \leq x + y\sqrt{d} < \alpha^{n+1}.$$
But then
$$1 \leq (x + y\sqrt{d})\alpha^{-n} = (x + y\sqrt{d})(x_1 - y_1\sqrt{d})^n < \alpha,$$

and this, by Theorem 8–4, contradicts the minimality of α unless
$$(x + y\sqrt{d})(x_1 - y_1\sqrt{d})^n = 1,$$
whence
$$x + y\sqrt{d} = (x_1 + y_1\sqrt{d})^n.$$

Turning now to the case $N = -1$, we find a somewhat similar situation, with the essential difference that the equation is not always solvable. This is the case, for example, when $d = 3$, for the expression $x^2 - 3y^2$ assumes only the values 0, 1, and 2 (mod 4). However, it is again true that all solutions can be expressed in terms of a single one, when such exists.

THEOREM 8–6. *Let d be a positive nonsquare integer. Then if the equation*
$$z^2 - dt^2 = -1 \tag{7}$$
is solvable, and if $z_1 + t_1\sqrt{d}$ is the minimal positive solution, a general solution is given by
$$z + t\sqrt{d} = \pm(z_1 + t_1\sqrt{d})^{2n+1}, \qquad n = 0, \pm 1, \ldots.$$

With the earlier notation,
$$\alpha = x_1 + y_1\sqrt{d} = (z_1 + t_1\sqrt{d})^2.$$

Proof: We prove the second assertion first. It is clear that

$(z_1 + t_1\sqrt{d})^2$ is a solution of (1), so that

$$1 < x_1 + y_1\sqrt{d} \le (z_1 + t_1\sqrt{d})^2. \tag{8}$$

This gives

$$-z_1 + t_1\sqrt{d} < -z_1x_1 + dy_1t_1 + (-z_1y_1 + t_1x_1)\sqrt{d} \le z_1 + t_1\sqrt{d},$$

where the number in the center of this inequality (which we will call $z + t\sqrt{d}$, for the moment) is again a solution of (7), so that in particular $t \ne 0$. But if a number lies between the minimal positive solution of (7) and its reciprocal, the same is true of the reciprocal of that number, so that either

$$1 < z + t\sqrt{d} \le z_1 + t_1\sqrt{d}$$

or

$$1 < -z + t\sqrt{d} < z_1 + t_1\sqrt{d}.$$

Using the minimality of $z_1 + t_1\sqrt{d}$, we conclude that

$$z + t\sqrt{d} = z_1 + t_1\sqrt{d}.$$

Now suppose that $z + t\sqrt{d}$ is any solution of (7), where we can again restrict attention to the case $z, t > 0$. Then as in the proof of Theorem 8–5, we can find an n such that

$$1 \le (z + t\sqrt{d})\alpha^{-n} < \alpha = (z_1 + t_1\sqrt{d})^2,$$

or, dividing through by $z_1 + t_1\sqrt{d}$,

$$-z_1 + t_1\sqrt{d} \le x' + y'\sqrt{d} < z_1 + t_1\sqrt{d},$$

where $x' + y'\sqrt{d}$ satisfies (1). This inequality implies that

$$\alpha^{-1} < x' + y'\sqrt{d} < \alpha,$$

so that $x' + y'\sqrt{d} = 1$, and

$$z + t\sqrt{d} = (z_1 + t_1\sqrt{d})\alpha^n = (z_1 + t_1\sqrt{d})^{2n+1}.$$

PROBLEMS

1. Find a general solution of the equation $x^2 - 2y^2 = 1$.
2. Describe all the integral solutions of the equation

$$x^2 + 6xy + 7y^2 + 8x + 24y + 15 = 0.$$

3. Show that

$$\lim_{n \to \infty} \inf \left(n(n\sqrt{2} - [n\sqrt{2}]) \right) = \frac{1}{2\sqrt{2}} \cdot$$

[The assertion means simply that if a_n stands for the quantity in parentheses, and if $\epsilon > 0$, then

$$a_n < \frac{1 + \epsilon}{2\sqrt{2}}$$

for infinitely many n, while

$$a_n > \frac{1 - \epsilon}{2\sqrt{2}}$$

for all sufficiently large n.]

4. Show that a necessary condition that the equation $x^2 - dy^2 = -1$ be solvable is that d have a proper representation as a sum of two squares.

8-3 The case $|N| > 1$. Because of its special interest in connection with the units of real quadratic fields, we consider separately the case $|N| = 4$.

THEOREM 8-7. *Let d be positive and not a square. If $r_1 + s_1\sqrt{d}$ is the minimal positive solution of the equation*

$$r^2 - ds^2 = 4, \tag{9}$$

then a general solution is given by

$$r + s\sqrt{d} = \pm 2 \left(\frac{r_1 + s_1\sqrt{d}}{2} \right)^n, \qquad n = 0, \pm 1, \ldots. \tag{10}$$

If the equation

$$r'^2 - ds'^2 = -4 \tag{11}$$

is solvable, and its minimal positive solution is $r_1' + s_1'\sqrt{d}$, then a general solution is given by

$$r' + s'\sqrt{d} = \pm 2 \left(\frac{r_1' + s_1'\sqrt{d}}{2} \right)^{2n+1}, \qquad n = 0, \pm 1, \ldots.$$

Proof: Clearly, the double of any solution of (1) is a solution of (9). While this remark shows that (9) is always solvable, not all the solutions can necessarily be found in this way, since, for example, $3^2 - 5 \cdot 1^2 = 4$, and 3 and 1 are odd.

If $r_2 + s_2\sqrt{d}$ and $r_3 + s_3\sqrt{d}$ are any solutions of (9), then

$$r + s\sqrt{d} = 2\frac{r_2 + s_2\sqrt{d}}{2} \cdot \frac{r_3 + s_3\sqrt{d}}{2}$$

is another integral solution. For, from (9), $r_i^2 \equiv ds_i^2 \pmod{2}$, so that $r_i \equiv ds_i \pmod{2}$. Hence

$$2r = r_2r_3 + ds_2s_3 \equiv d^2s_2s_3 + ds_2s_3 \equiv d(d+1)s_2s_3 \equiv 0 \pmod{2}$$

and

$$2s = r_2s_3 + r_3s_2 \equiv ds_2s_3 + ds_2s_3 \equiv 2ds_2s_3 \equiv 0 \pmod{2},$$

so that r and s are integers. Also,

$$(r + s\sqrt{d})(r - s\sqrt{d}) = r^2 - ds^2 = 4\frac{r_2^2 - ds_2^2}{4} \cdot \frac{r_3^2 - ds_3^2}{4} = 4.$$

It follows that the numbers $r + s\sqrt{d}$ defined in (10) are solutions of (9), for every n. The remainder of the proof for the case $N = 4$ is an easy modification of the proof of Theorem 8–5. The proof for $N = -4$ is a straightforward combination of the above considerations and the proof of Theorem 8–6.

For general N, the situation is rather complicated. The following theorem gives a partial result.

THEOREM 8–8. *If $d > 0$ is not a square, and if the Pell equation*

$$u^2 - dv^2 = N \tag{12}$$

has one solution, it has infinitely many. In particular, if x_1, y_1 is a solution of equation (1) and u_1, v_1 is a solution of equation (12), then the integers u, v determined by

$$u + v\sqrt{d} = (x_1 + y_1\sqrt{d})(u_1 + v_1\sqrt{d}), \tag{13}$$

form a solution of equation (12).

Proof: The second statement is proved in exactly the same way as was Theorem 8–4. The first statement follows immediately from the second, making use of Theorem 8–5.

Notice that it may not be possible to obtain *all* solutions of (12) from one solution and the set of all solutions of (1). For example, the equation $u^2 - 2v^2 = 49$ has the solutions 7 and $9 + 4\sqrt{2}$, and neither can be obtained from the other by multiplying by a solution of $x^2 - 2y^2 = 1$.

Theorem 8–8 can be used to obtain a finite criterion for the solvability of (12). If two solutions of (12) are related as in (13), we say that they belong to the same *class*. We now find bounds on the smallest element of each class, where the solutions are ordered by the size of u. (We can require that u be positive, since $u + v\sqrt{d}$ and $-u - v\sqrt{d}$ are in the same class.) The investigation is carried out only for $N > 0$; the case $N < 0$ is similar.

To do this we ask, given a solution $u_1 + v_1\sqrt{d}$ of (12), with $u_1 > 0$, when is it possible to find a smaller solution $u + v\sqrt{d}$, with $u > 0$, in the same class? That is, we want to find u and v such that

$$u + v\sqrt{d} = (x + y\sqrt{d})(u_1 + v_1\sqrt{d}), \qquad 0 < u < u_1,$$
$$x^2 - dy^2 = 1.$$

Let $\alpha = x_1 + y_1\sqrt{d}$ be the minimal positive solution of (1). If $v_1 > 0$, take $x + y\sqrt{d} = \alpha^{-1} = x_1 - y_1\sqrt{d}$; while if $v_1 < 0$, take $x + y\sqrt{d} = \alpha$; in either case, we get

$$u = u_1 x_1 - y_1|v_1|d = u_1\left(x_1 - y_1\sqrt{d}\,\frac{|v_1|\sqrt{d}}{u_1}\right)$$

$$= u_1\left\{x_1 - y_1\sqrt{d} + y_1\sqrt{d}\left(1 - \sqrt{1 - \frac{N}{u_1^2}}\right)\right\}.$$

Here $0 < N/u_1^2 < 1$. Since

$$0 < 1 - \sqrt{1 - t} = \frac{t}{1 + \sqrt{1 - t}} < \frac{t}{2 - t}$$

for $0 < t < 1$, we have

$$0 < u < u_1\left(\alpha^{-1} + \frac{y_1\sqrt{d}\,N}{2u_1^2 - N}\right).$$

A little manipulation shows that the coefficient of u_1 is smaller than 1, so that $u < u_1$, if

$$u_1 > \sqrt{\frac{\beta y_1\sqrt{d} + 1}{2}\,N}, \qquad \text{where } \beta = \frac{\alpha}{\alpha - 1}.$$

Since $y_1\sqrt{d} = \sqrt{x_1^2 - 1} < x_1$, we have proved

THEOREM 8–9. *If equation* (12) *is solvable, it has a solution with*

$$0 < u < \sqrt{\frac{\beta x_1 + 1}{2}\cdot N}, \tag{14}$$

where $\alpha = x_1 + y_1\sqrt{d}$ is the minimal positive solution of equation (1) *and $\beta = \alpha/(\alpha - 1)$. If there are two or more classes of solutions of equation* (12), *each contains an element for which equation* (14) *holds.*

This reduces the question of the solvability of (12) to a finite problem, once the minimal positive solution of (1) is known; it suffices to decide whether any of the numbers $(u^2 - N)/d$ is a square, for u in the interval (14). If there are two or more such values of u, it is a simple matter to decide whether the corresponding solutions are in the same class.

For example, when $d = 2$ we have the minimal positive solution $3^2 - 2 \cdot 2^2 = 1$, and it is easily seen that (14) holds if $0 < u < \frac{3}{2}\sqrt{N}$. Since also $N = u^2 - 2v^2 \leq u^2$, we need only examine the integers u between $\sqrt{N}$ and $\frac{3}{2}\sqrt{N}$, for each N.

PROBLEMS

1. Complete the proof of Theorem 8–7.

2. The statement obtained from Theorem 8–7 by replacing 2 and 4 by 7 and 49, respectively, is false, as is seen by considering the numerical example immediately following Theorem 8–8. Where would the analogous proof break down?

3. Show that if $N < 0$, Theorem 8–9 remains correct if the inequality (14) is replaced by

$$0 < u < \sqrt{\frac{\alpha x_1 |N|}{2(\alpha + 1)}}.$$

[*Hint:* Prove and use the fact that for $t > 0, \sqrt{1 + t} - 1 < t/2$.]

4. Describe all the units of $R(\sqrt{2})$; of $R(\sqrt{5})$. Cf. Problem 1, Section 8–1.

8–4 An application. We showed in Theorem 8–1 that if ξ is irrational, the inequality

$$\left| \xi - \frac{x}{y} \right| < \frac{1}{y^2}$$

has infinitely many solutions in integers x, y. It is the object of the present section to make a more detailed examination of the approximability of a quadratic irrationality (that is, an irrational root of a quadratic equation with integral coefficients) by rationals, making use of the preceding results concerning Pell's equation.

It is easy to see that when ξ is a quadratic irrationality, there is a constant $g_0 = g_0(\xi)$ such that the inequality

$$\left| \xi - \frac{x}{y} \right| < \frac{1}{gy^2}$$

does not hold for any x, y if $g > g_0$. For if ξ is defined by the equation

$$f(\xi) = a\xi^2 + b\xi + c = 0, \qquad a, b, c \text{ integers,}$$

and if $f(x)$ factors as

$$f(x) = a(x - \xi)(x - \xi'),$$

then

$$\left| \xi - \frac{x}{y} \right| = \frac{|ax^2 + bxy + cy^2|}{y^2 |a| \cdot |\xi' - x/y|} \geq \frac{1}{y^2 |a| \cdot |\xi' - x/y|},$$

since $ax^2 + bxy + cy^2$ is an integer different from zero. Since ξ is irrational, $\xi \neq \xi'$. Hence from the above inequality, either

$$\left| \xi - \frac{x}{y} \right| > \frac{|\xi - \xi'|}{2} \qquad \text{or} \qquad \left| \xi - \frac{x}{y} \right| \geq \frac{1}{3y^2 |a| \cdot |(\xi - \xi')/2|},$$

and we can take

$$g(\xi) = \min\left(\frac{2}{|\xi - \xi'|}, \quad \frac{2}{3|a| \cdot |(\xi - \xi')/2|} \right)$$

for any $\epsilon > 0$.

We are thus led to consider the quantity $M(\xi)$, which is the upper limit of those numbers λ for which the inequality

$$\left| \xi - \frac{x}{y} \right| < \frac{1}{\lambda y^2}$$

has infinitely many solutions. It was first treated by A. Markov, who made an extensive investigation of $M(\xi)$ in connection with the problem of determining an upper bound for the minimum value assumed by an *indefinite quadratic form*, i.e., an expression

$$Ax^2 + Bxy + Cy^2$$

in which $D = B^2 - 4AC > 0$, D is not a square, and x, y are integral variables. Markov made use of the theory of continued fractions, but we shall derive certain of his results using only the theorems just proved concerning Pell's equation.

In order to avoid interrupting the argument later, we first prove a lemma.

THEOREM 8–10. *Let*

$$f(x, y) = ax^2 + bxy + cy^2$$

have integral coefficients such that $d = b^2 - 4ac > 0$ *and* d *is not a square. Then if the equation*

$$f(x, y) = k \qquad (15)$$

has one solution in integers, it has infinitely many, and each of the two quantities

$$|2ax + by + y\sqrt{d}| \qquad and \qquad |2ax + by - y\sqrt{d}| \qquad (16)$$

is less than any prescribed positive number for infinitely many such solutions.

Proof: (a) In the case $k = a$, let X, Y be integers such that $X^2 - dY^2 = 1$, so that

$$4a^2X^2 - 4a^2dY^2 = 4a^2.$$

If we put

$$2aX = 2ax + by, \qquad 2aY = y,$$

in which case x and y, given by

$$x = X - bY, \qquad y = 2aY, \qquad (17)$$

are integers, then

$$(2ax + by)^2 - dy^2 = 4af(x, y) = 4a^2,$$

or

$$f(x, y) = a. \qquad (18)$$

Since by Theorem 8–5 there are infinitely many pairs X, Y, there are infinitely many solutions of (18). Since

$$\lim_{y \to \infty} \left\{ \left(\frac{2ax + by}{y} \right)^2 - d \right\} = \lim_{y \to \infty} \frac{4a^2}{y^2} = 0,$$

it is clear that one of the quantities in (16) is smaller than any prescribed $\epsilon > 0$ for sufficiently large y. Moreover, if the integers x, y determined by X, Y in (17) are such that one of the quantities in (16) is small, then the numbers x', y' determined by X, $-Y$ are such that the other quantity is small.

(b) In the case $k \neq a$, let x_1, y_1 be a solution of (15), and x_2, y_2 a solution of (18). Then the integers x, y defined by the equation

$$2ax + by + y\sqrt{d} = \frac{(2ax_1 + by_1 + y_1\sqrt{d})(2ax_2 + by_2 - y_2\sqrt{d})}{2a}$$

again satisfy (15). Since there are infinitely many solutions of (18), the same is true of (15). Furthermore, for fixed x_1, y_1, the first quantity in (16) will be small if x_2, y_2 ranges over those solutions of (18) for which $|2ax_2 + by_2 - y_2\sqrt{d}|$ is small, while the second will be small for the remaining solutions of (18).

THEOREM 8–11. *Let ξ be a real quadratic irrationality of discriminant d:*

$$a\xi^2 + b\xi + c = 0, \qquad d = b^2 - 4ac > 0, \qquad d \text{ not a square,}$$
$$(a, b, c) = 1, \qquad a > 0.$$

Then if k is the smallest positive integer for which the equation

$$|ax^2 + bxy + cy^2| = k$$

has an integral solution,

$$M(\xi) = \frac{\sqrt{d}}{k}.$$

Proof: (a) $M(\xi)$ must be less than or equal to $\sqrt{d}/k$. For assume on the contrary that

$$M(\xi) = \frac{\sqrt{d}}{(1 - \delta)k},$$

where $0 < \delta < 1$. Then the inequality

$$\left| \frac{\sqrt{d} - b}{2a} - \frac{x}{y} \right| < \frac{(1 - \delta)k}{\sqrt{d}y^2}$$

holds for an infinite sequence S of distinct fractions x/y. (The case that $\xi = (-b - \sqrt{d})/2a$ is treated similarly.) Then, multiplying through by $\sqrt{d}$, we have

$$\left| \left(\frac{b}{2a} + \frac{x}{y} \right)\sqrt{d} - \frac{d}{2a} \right| < \frac{(1 - \delta)k}{y^2},$$

or

$$\left| \sqrt{d} - \frac{dy}{2ax + by} \right| < \frac{2a(1 - \delta)k}{y|2ax + by|}.$$

Hence

$$\frac{2a(1-\delta)k}{y|2ax+by|}\left|\sqrt{d}+\frac{dy}{2ax+by}\right| > \left|d-\frac{d^2y^2}{(2ax+by)^2}\right|$$

$$=\frac{4ad}{(2ax+by)^2}|ax^2+bxy+cy^2| \geq \frac{4akd}{(2ax+by)^2},$$

and, therefore,

$$1-\delta.> \frac{2dy}{|2ax+by|\cdot|\sqrt{d}+dy/(2ax+by)|}$$

$$=\frac{2d}{|b+2ax/y|\cdot|\sqrt{d}+dy/(2ax+by)|}. \qquad (19)$$

But as x/y runs through the sequence S, y increases without limit and

$$b+2a\frac{x}{y}\to\sqrt{d},$$

$$\sqrt{d}+\frac{dy}{2ax+by}\to 2\sqrt{d},$$

so that

$$\lim_{\substack{y\to\infty\\x/y\in S}}\frac{2d}{|b+2ax/y|\cdot|\sqrt{d}+dy/(2ax+by)|}=1,$$

which contradicts (19).

(b) $M(\xi)$ must be greater than or equal to $\sqrt{d}/k$. For from the definition of k, and Theorem 8–10, we have that the equation

$$|(2ax+by)^2-dy^2|=4ak$$

has infinitely many integral solutions x, y. The left side factors into

$$|2ax+by-y\sqrt{d}|\cdot|2ax+by+y\sqrt{d}|=4ak. \qquad (20)$$

By Theorem 8–10, each of these factors is small for infinitely many pairs x, y. Henceforth we restrict x, y to the set T of solutions of (20) for which the first factor is smaller than the second. (The proof in the alternate case proceeds similarly.) Then

$$\frac{x}{y}\to\frac{-b+\sqrt{d}}{2a}$$

as $|y|$ tends to infinity. Furthermore,

$$\left| \frac{x}{y} - \frac{-b + \sqrt{d}}{2a} \right| = \frac{4ak}{4a^2|x/y + (b + \sqrt{d})/2a|y^2}$$

$$= \frac{k}{a|x/y + (b + \sqrt{d})/2a|y^2}$$

and

$$\frac{x}{y} + \frac{b + \sqrt{d}}{2a} \to \frac{\sqrt{d}}{a}$$

as $|y|$ tends to infinity. Hence, given $\epsilon > 0$, the inequality

$$\left| \frac{x}{y} - \frac{-b + \sqrt{d}}{2a} \right| < \frac{k(1 + \epsilon)}{y^2 \sqrt{d}}$$

holds for all $(x, y) \in T$ with $|y| > y_0(\epsilon)$. Hence $M(\xi) \geq \sqrt{d}/k$.

The proof is now complete, since if $M(\xi) \geq \sqrt{d}/k$ and also $M(\xi) \leq \sqrt{d}/k$, it must be true that $M(\xi) = \sqrt{d}/k$.

COROLLARY. *If ξ is defined as in Theorem 8–11, then*

$$\frac{\sqrt{d}}{a} \leq M(\xi) \leq \sqrt{d}.$$

For clearly $k \geq 1$, and $k \leq a$ since $a \cdot 1^2 + b \cdot 1 \cdot 0 + c \cdot 0^2 = a$.

PROBLEM

Generalize the result of Problem 3, Section 8–2, evaluating

$$\lim_{n \to \infty} \inf n(n\sqrt{m} - [n\sqrt{m}]),$$

where m is a positive square-free integer.

8–5 The minima of indefinite quadratic forms. So far we have used Theorem 8–11 to obtain information concerning the quantity $M(\xi)$; it can also be used, in conjunction with the following well-known theorem of A. Hurwitz, to obtain information about the numerically smallest value assumed by an indefinite quadratic form.

THEOREM 8–12 (*Hurwitz' theorem*). *If ξ is any irrational number, then there are infinitely many integral solutions x, y of the inequality*

$$\left| \xi - \frac{x}{y} \right| < \frac{1}{\sqrt{5}\, y^2}.$$

Consequently, $M(\xi) \geq \sqrt{5}$ for every irrational ξ.

We defer the proof for a moment. Assuming the theorem to be correct, a comparison of it and Theorem 8–11 yields the following result.

THEOREM 8–13. *If $f(x, y)$ is an indefinite binary quadratic form of nonsquare discriminant d, then*

$$0 < |f(x, y)| \leq \frac{\sqrt{d}}{\sqrt{5}}$$

for suitable integers x, y.

The coefficient $1/\sqrt{5}$ occurring here is best possible, in the sense that the theorem becomes false (for some quadratic forms) if $1/\sqrt{5}$ is replaced by a smaller constant. For the form $k(x^2 + xy - y^2)$ has discriminant $5k^2$, and it is clear that this form assumes no nonzero value numerically smaller than $|k(1^2 + 1 \cdot 0 - 0^2)| = |k|$.

8–6 Farey sequences, and a proof of Hurwitz' theorem. A very simple proof of Hurwitz' theorem can be deduced from the well-known properties of the so-called *Farey sequences* F_n, which are the sequences of rational numbers a/b with $0 < b \leq n$, $(a, b) = 1$, arranged in increasing order of magnitude. Thus for the first few values of n we have

$$F_1: \quad \ldots, \frac{-1}{1}, \frac{0}{1}, \frac{1}{1}, \frac{2}{1}, \ldots$$

$$F_2: \quad \ldots, \frac{-1}{2}, \frac{0}{1}, \frac{1}{2}, \frac{1}{1}, \frac{3}{2}, \ldots$$

$$F_3: \quad \ldots, \frac{-1}{3}, \frac{0}{1}, \frac{1}{3}, \frac{1}{2}, \frac{2}{3}, \frac{1}{1}, \frac{4}{3}, \ldots$$

$$F_4: \quad \ldots, \frac{-1}{4}, \frac{0}{1}, \frac{1}{4}, \frac{1}{3}, \frac{1}{2}, \frac{2}{3}, \frac{3}{4}, \frac{1}{1}, \frac{5}{4}, \ldots$$

$$F_5: \quad \ldots, \frac{-1}{5}, \frac{0}{1}, \frac{1}{5}, \frac{1}{4}, \frac{1}{3}, \frac{2}{5}, \frac{1}{2}, \frac{3}{5}, \frac{2}{3}, \frac{3}{4}, \frac{4}{5}, \frac{1}{1}, \frac{6}{5}, \ldots$$

$$\vdots$$

Clearly the number of elements of F_n which lie between 0 and 1 inclusive is $1 + \varphi(1) + \varphi(2) + \cdots + \varphi(n)$.

The rational numbers p/q and r/s are said to be *adjacent in F_n* if they are successive elements of F_n.

THEOREM 8–14.   (a) *If p/q and r/s are adjacent in F_n, then* $|ps - qr| = 1$.

(b) *If $|ps - qr| = 1$, then p/q and r/s are adjacent in F_n for*

$$\max (q, s) \leq n < q + s,$$

and they are separated by the single element $(p + r)/(q + s)$ in F_{q+s}.

Remark: This theorem on the one hand gives necessary and sufficient conditions that p/q and r/s be adjacent in F_n, and on the other hand gives the law of formation of the new elements that appear in going from F_n to F_{n+1}. The number $(p + r)/(q + s)$ is called the *mediant* of p/q and r/s.

Proof: Suppose that p/q and r/s are elements of F_n such that $qr - ps = 1$, so that $r/s > p/q$. As t varies continuously from zero to infinity, the number

$$f(t) = \frac{p + tr}{q + ts}$$

increases steadily from p/q to r/s, so that there is a one-to-one correspondence between the positive real numbers t and the points of the interval

$$\frac{p}{q} < x < \frac{r}{s}. \tag{21}$$

Moreover, it is clear that $f(t)$ is rational if and only if t is rational; since we are interested only in the rational numbers in the interval, we put $t = u/v$, where $(u, v) = 1$ and $u > 0$, $v > 0$. This gives

$$f\left(\frac{u}{v}\right) = \frac{vp + ur}{vq + us}.$$

Since

$$q(vp + ur) - p(vq + us) = u(qr - ps) = u,$$

$$s(vp + ur) - r(vq + us) = v(ps - qr) = -v,$$

we have $(vp + ur, vq + us) = 1$. Thus we have shown that as u and v run over all pairs of relatively prime positive integers, the reduced fraction $(vp + ur)/(vq + us)$ runs over all rational numbers between p/q and r/s.

Among these fractions, the one with $u = v = 1$ is clearly the unique one of smallest denominator; it is the mediant of p/q and r/s, and

$$|(p + r)q - (q + s)p| = 1, \qquad |r(q + s) - s(p + r)| = 1.$$

Since $q + s > \max(q, s)$, part (b) of the theorem follows. To prove (a), we proceed inductively. F_1 consists of the integers $\ldots, -1/1, 0/1, 1/1, \ldots$, and $|a \cdot 1 - (a + 1) \cdot 1| = 1$, so that (a) is true for $n = 1$. If it is true for $n = m$, it is also true for $n = m + 1$, since the only elements of F_{m+1} not in F_m are certain mediants of adjacent elements of F_m. The assertion follows by the induction principle.

Proof of Hurwitz' theorem: If a/b is a reduced fraction and c is a positive real number, designate by $I_c(a/b)$ the closed interval

$$\left[\frac{a}{b} - \frac{1}{cb^2}, \frac{a}{b} + \frac{1}{cb^2}\right].$$

Hurwitz' theorem says that if ξ is irrational, there are infinitely many fractions x/y such that $\xi \in I_{\sqrt{5}}(x/y)$.

For each n, ξ lies between some two adjacent elements of F_n, say

$$\frac{p}{q} < \xi < \frac{r}{s}.$$

We divide the interval $[p/q, r/s]$ into left and right halves:

$$J_L = \left[\frac{p}{q}, \frac{p + r}{q + s}\right], \qquad J_R = \left[\frac{p + r}{q + s}, \frac{r}{s}\right].$$

We now ask, how large may c be if it is required that the three intervals $I_c(p/q)$, $I_c((p + r)/(q + s))$, $I_c(r/s)$ together completely cover the interval J_L? If this is the case, and $\xi \in J_L$, then ξ must be an interior point of one of these intervals I_c, and we have a solution of the inequality $|\xi - x/y| < 1/cy^2$.

Clearly $I_c(p/q)$ and $I_c(r/s)$ overlap (or abut) if and only if

$$\frac{p}{q} + \frac{1}{cq^2} \geq \frac{r}{s} - \frac{1}{cs^2},$$

and this reduces, with the aid of the relation $rq - ps = 1$, to

$$c \leq qs\left(\frac{1}{q^2} + \frac{1}{s^2}\right) = \frac{s}{q} + \frac{q}{s};$$

or, putting $f(x) = x + 1/x$, to

$$c \leq f\left(\frac{s}{q}\right).$$

Similarly, $I_c(p/q)$ and $I_c((p + r)/(q + s))$ overlap if and only if

$$c \leq f\left(\frac{q + s}{q}\right) = f\left(1 + \frac{s}{q}\right),$$

and so J_L is certainly covered by the intervals I_c if

$$c \leq \max\left(f(s/q), f(1 + s/q)\right),$$

and *a fortiori* if

$$c \leq \min_{x>0}\left\{\max\left(f(x), f(1 + x)\right)\right\} = c_0.$$

But a glance at the curves $y = f(x)$ and $y = f(1 + x)$ shows that the curve $y = \max\left(f(x), f(1+x)\right)$ is concave upward for $x > 0$, and has its minimum c_0 at $x = x_0$, where $f(x_0) = f(1 + x_0)$. (See Fig. 8–1.) A simple calculation gives

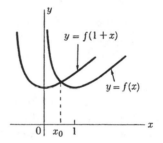

$$x_0 = \frac{\sqrt{5} - 1}{2}, \qquad c_0 = \sqrt{5}.$$

FIGURE 8–1

The proof can now be completed in either of two ways. The simpler is to note that ξ, being irrational, must lie, for infinitely many n, in the left half J_L of the interval between its surrounding Farey points; for if not, it would have to lie in all the intervals

$$\left[\frac{p}{q}, \frac{r}{s}\right], \qquad \left[\frac{p + r}{q + s}, \frac{r}{s}\right], \qquad \left[\frac{p + 2r}{q + 2s}, \frac{r}{s}\right], \qquad \ldots,$$

and the only point common to all these intervals is r/s itself. And whenever $\xi \in J_L$, the above argument shows that at least one of the numbers

$$\frac{p}{q}, \qquad \frac{p + r}{q + s}, \qquad \frac{r}{s}$$

affords a solution of Hurwitz' inequality. Finally, this gives infinitely many solutions, because ξ lies in infinitely many J_L's and infinitely many J_R's, so that only finitely many of these intervals have a common end point.

Alternatively, one can also examine the conditions under which the intervals $I_c(p/q)$, $I_c((p + r)/(q + s))$, and $I_c(r/s)$ completely cover the interval J_R; an argument similar to that given above shows that this is the case if

$$c \leq \min_{x > 0} \left(\max \left\{ f(x), f\left(\frac{x}{1 + x}\right) \right\} \right) = \sqrt{5}.$$

It is then not necessary to distinguish the cases $\xi \in J_L$, $\xi \in J_R$.

Again using the fact that ξ lies in infinitely many left half-intervals and infinitely many right half-intervals, we deduce the following stronger form of Hurwitz' theorem.

THEOREM 8–15. *If ξ is irrational there are infinitely many solutions of the inequality*

$$\left| \xi - \frac{x}{y} \right| < \frac{1}{\sqrt{5}\, y^2}. \tag{22}$$

If, for arbitrary n, ξ lies between the adjacent elements p/q and r/s of F_n, then at least one of the three numbers p/q, $(p + r)/(q + s)$, r/s is a solution of the inequality (22).

CHAPTER 9

RATIONAL APPROXIMATIONS TO REAL NUMBERS

9–1 Introduction. In the investigation of the solvability of the equation $n = x^2 + y^2$ in Chapter 7, it was convenient to use the fact that if x is real and t is a positive integer, there are integers p and q such that

$$|qx - p| \leq \frac{1}{t + 1}, \qquad 1 \leq q \leq t.$$

In connection with Pell's equation we used an easy consequence of this theorem, that if x is irrational, the inequality

$$|qx - p| < \frac{1}{q}$$

has infinitely many integral solutions q and p with $q > 0$. Finally, the investigation, in Chapter 8, of the numerically smallest nonzero value assumed by an indefinite binary quadratic form hinged on Hurwitz' theorem, which states that the inequality

$$|qx - p| < \frac{1}{\sqrt{5}q}$$

has infinitely many integral solutions q and p with $q > 0$, if x is irrational. These theorems, while quantitatively different, all tell something about how small the absolute value of the linear form $qx - p$ can be made if the integers q and p are not both zero. Several generalizations of this problem come to mind at once, involving either a larger number of variables, or more than one such form, or both. The investigation of the behavior of such sets of forms is a central problem in the theory of Diophantine approximations; while many results have been obtained, few of them have the quantitative precision of Hurwitz' theorem, which becomes false if $\sqrt{5}$ is replaced by any larger constant. (This statement has not yet been proved; it is a consequence of Theorem 9–9.) One reason for this is that it is only in the simple case of one linear form in two variables, that a simple algorithm can be constructed which yields all the pairs p, q for which $|qx - p|$ is "small," in a sense which will be made explicit below. Naturally, it is much easier to investigate the small values of a function if one knows what values to use for the arguments. One of the objects of this chapter is to develop this algorithm.

Rewriting Hurwitz' inequality in the form

$$\left| x - \frac{p}{q} \right| < \frac{1}{\sqrt{5}q^2} ,$$

we see that we are here concerned with a notion of "good" rational approximations to an irrational number which differs essentially from that generally understood in analysis. There, we say that p/q is a better approximation to x than is r/s if

$$\left| x - \frac{p}{q} \right| < \left| x - \frac{r}{s} \right| .$$

The question of finding this kind of good approximation is rather uninteresting arithmetically, although of course it may be necessary to use approximate values of irrational numbers in arithmetic investigations. What is involved in the theorems we are now discussing is a comparison of the exactness of the approximation with the size of the denominator of the fraction used, the comparison being effected by taking the product

$$q \left| x - \frac{p}{q} \right| .$$

At least if x is irrational, the first factor in this product gets large as the second approaches zero; for out of all the elements of an arbitrary Farey sequence F_N there is one which is nearest to x, so to find a nearer rational number it is necessary to consider fractions whose denominators exceed N. To require the above product to be small is therefore a much stronger condition than that imposed in analysis.

Instead of searching immediately for appropriate fractions corresponding to a given x, it is fruitful (and indeed necessary, to make precise the meaning of "appropriate") to make x, instead of p/q, the unknown quantity. That is, we fix a rational number p/q, and ask what numbers should be considered as having p/q as a good approximation. Put so crudely, the question is unanswerable; we must decide what other rational numbers are competing with p/q. It seems natural to consider just the elements of some Farey sequence F_N which contains p/q, and to say that p/q is a "good" approximation to x if, for some $N \geq q$, $|qx - p| \leq |sx - r|$ for all r/s in F_N. This is perhaps most easily thought of this way: we measure distance from p/q in F_N, not by the usual expression $|x - p/q|$, but by $q|x - p/q|$, so that there is an individual measuring rod (or, more briefly, a *metric*) associated with each element of F_N. It is clear that "distances" increase more rapidly (in comparison with ordinary length) when measured from an element of F_N with large denominator than from an element with small denominator. We now associate with p/q all those points x such that the "distance" $|qx - p|$ from p/q is less than or equal to the "distance" $|sx - r|$ from an arbitrary element r/s of F_N. Call this set of points $R_N(p, q)$; formally, $R_N(p, q)$ is the set of x such that

$$\min_{r/s \in F_N} (|sx - r|) = |qx - p|.$$

Clearly, p/q itself is in $R_N(p, q)$.

Each of these sets $R_N(p, q)$ consists of a single interval. To see this, we first prove that if p/q and r/s are adjacent in F_N, then no point x between them belongs to any $R_N(t, u)$, if t/u is neither p/q nor r/s. This is obvious if x is either p/q or r/s. Suppose that

$$\frac{t}{u} < \frac{p}{q} < x < \frac{r}{s} \, ;$$

the other possible order, in which $t/u > r/s$, is treated similarly. If $q \leq u$, then

$$0 < qx - p = q\left(x - \frac{p}{q}\right) < q\left(x - \frac{t}{u}\right) \leq u\left(x - \frac{t}{u}\right) = ux - t,$$

so if the assertion is false, it must be that $q > u$. But then if

$$qx - p \geq ux - t,$$

so that

$$x \geq \frac{p - t}{q - u} \, ,$$

we have

$$0 < r - sx \leq r - s\,\frac{p - t}{q - u} = \frac{(qr - sp) - (ur - st)}{q - u} \leq 0,$$

since $qr - sp = 1$ while $ur - st \geq 1$. This contradiction shows that

$R_N(p, q)$ does not extend past the two elements to which p/q is adjacent in F_N. But the condition

$$|qx - p| \leq |sx - r|$$

gives

$$qx - p \leq r - sx,$$

or

$$x \leq \frac{p + r}{q + s},$$

so that $R_N(p, q)$ consists of all x between the two points which are the mediants of p/q and its immediate neighbors in F_N.

In particular, it follows that the new points which appear in going from F_N to F_{N+1} always appear at end points of intervals R_N.

We now adopt the following convention: if for some N, the number x is a point of $R_N(p, q)$ then p/q will be called a *best approximation* to x. For this N, $|qx - p|$ is less than or equal to any expression $|sx - r|$ with $s \leq N$; a fortiori, $|x - p/q|$ is less than or equal to $|x - r/s|$ if $s \leq q$, so that if p/q is a best approximation to x in our present sense, it is also the rational number closest to x (in the ordinary sense) out of all those with denominators not exceeding q.

There is thus associated with a fixed x a unique sequence of best approximations; the sequence will be infinite unless x is rational, in which case x lies inside its own interval R_N, for N greater than or equal to the denominator of x. (For rational x, the sequence is not quite unique, since x is a common end point of two intervals R_N for some N.) If $N \geq q$, $R_{N+1}(p, q)$ is contained in $R_N(p, q)$, so that if p/q is a best approximation to x, then certainly x is in $R_q(p, q)$. If h is the largest non-negative integer for which $x \in R_{q+h}(p, q)$, then $x \in R_N(p, q)$ for $q \leq N \leq q + h$. Thus, if for fixed x we define a_N/b_N for $N = 1, 2, \ldots$ as that rational number such that $x \in R_N(a_N, b_N)$, then for suitable $N_0, N_1, \ldots$ we have $1 = N_0 < N_1 < N_2 < \cdots$ and

$$\frac{a_{N_0}}{b_{N_0}} = \frac{a_1}{b_1} = \frac{a_2}{b_2} = \cdots = \frac{a_{N_1-1}}{b_{N_1-1}} \neq \frac{a_{N_1}}{b_{N_1}},$$

$$\frac{a_{N_1}}{b_{N_1}} = \frac{a_{N_1+1}}{b_{N_1+1}} = \cdots = \frac{a_{N_2-1}}{b_{N_2-1}} \neq \frac{a_{N_2}}{b_{N_2}},$$

$$\frac{a_{N_2}}{b_{N_2}} = \frac{a_{N_2+1}}{b_{N_2+1}} = \cdots = \frac{a_{N_3-1}}{b_{N_3-1}} \neq \frac{a_{N_3}}{b_{N_3}},$$

$$\vdots$$

Since a_{Nk-1}/b_{Nk-1} and a_{Nk}/b_{Nk} are adjacent elements of F_{Nk}, we have

$$|b_{Nk}a_{Nk-1} - b_{Nk-1}a_{Nk}| = 1, \qquad k = 1, 2, \ldots. \qquad (1)$$

Now consider the following problem: *given a real number x, to find a systematic method for determining the sequence of best approximations to x.* We begin by reducing x by its greatest integer $[x] = \lambda_0$; the new number $x' = x - [x]$ is then in the interval $(0, 1)$. Put $P_0 = \lambda_0$, $Q_0 = 1$, so that P_0/Q_0 is or is not a_1/b_1 according as the fractional part x' of x is less than or greater than $\frac{1}{2}$. (In what follows, we shall assume that if x is rational, its denominator is sufficiently large that equality does not occur in statements such as the preceding one. This point will be considered in detail later.) If $P_0/Q_0 = a_1/b_1$, we put $P_k/Q_k = a_{Nk}/b_{Nk}$, while if $P_0/Q_0 \neq a_1/b_1$ we put $P_k/Q_k = a_{Nk-1}/b_{Nk-1}$, for $k = 1, 2, \ldots$. Thus the sequence $\{P_k/Q_k\}$ coincides with $\{a_{Nk}/b_{Nk}\}$, except that P_0/Q_0 may not be a best approximation. If we also put $P_{-1} = 1$, $Q_{-1} = 0$, then

$$Q_0 P_{-1} - Q_{-1} P_0 = 1. \qquad (2)$$

The numbers P_1/Q_1, P_2/Q_2, ... are now to be determined. It turns out that this can be done using an algorithm, closely related to the Euclidean algorithm, of considerable importance in many branches of mathematics. Unfortunately, the deduction of this algorithm is necessarily somewhat complicated, since one must obtain the sequences $\{P_k\}$ and $\{Q_k\}$ from three others yet to be defined: $\{\alpha_k\}$, $\{x_k\}$, and $\{\lambda_k\}$. The final result, however, is quite simple.

If $P_0/Q_0 = a_1/b_1$, the relation

$$|Q_k P_{k-1} - Q_{k-1} P_k| = 1 \qquad k = 0, 1, \ldots, \qquad (3)$$

holds, by (1) and (2). If $P_0/Q_0 \neq a_1/b_1$, then

$$\frac{P_1}{Q_1} - \frac{P_0}{Q_0} = 1, \qquad Q_0 = Q_1 = 1,$$

and

$$|Q_1 P_0 - Q_0 P_1| = 1, \qquad (4)$$

so that (3) again holds, by (1), (2), and (4). The relation (3) is therefore always valid.

The numbers P_k and Q_k are now defined recursively, as follows: $P_{-1} = 1$, $Q_{-1} = 0$, $P_0 = [x]$, $Q_0 = 1$, and, for $k \geq 1$, P_k and Q_k constitute that solution p, q of the inequality

$$|qx - p| < |Q_{k-1}x - P_{k-1}|$$

for which q is positive and minimal. If we put $\alpha_k = Q_k x - P_k$ for $k = 0, 1, \ldots$, we must find the minimal solution of the inequality

$$|qx - p| < |\alpha_{k-1}|.$$

Fortunately, we need not consider all pairs p, q, but only those for which

$$|P_{k-1}q - Q_{k-1}p| = 1,$$

on account of (3). Since we know that one solution of this equation is $q = Q_{k-2}$, $p = P_{k-2}$, it follows that every solution is of the form

$$q = \epsilon(Q_{k-2} + \lambda Q_{k-1}), \qquad p = \epsilon(P_{k-2} + \lambda P_{k-1}), \tag{5}$$

where $\epsilon = \pm 1$ and λ is an integer, so that

$$|qx - p| = |\lambda(Q_{k-1}x - P_{k-1}) + (Q_{k-2}x - P_{k-2})| = |\lambda\alpha_{k-1} + \alpha_{k-2}|.$$

Thus we can rephrase the definition of P_k and Q_k: if $k \geq 1$ and P_l and Q_l are known for $l < k$, then P_k, Q_k are the p and q of equations (5) if λ and ϵ are so determined that

$$|\lambda\alpha_{k-1} + \alpha_{k-2}| < |\alpha_{k-1}|; \quad \epsilon(\lambda Q_{k-1} + Q_{k-2}) \text{ is positive and minimal.}$$

Since $Q_{k-1} > 0$, these conditions are equivalent to the following:

$$\left|\lambda - \left(-\frac{\alpha_{k-2}}{\alpha_{k-1}}\right)\right| < 1; \quad \epsilon\left\{\lambda - \left(-\frac{Q_{k-2}}{Q_{k-1}}\right)\right\} \text{ positive and minimal.} \tag{6_k}$$

To see how to solve (6_k), let us consider the case $k = 1$. We have

$$\alpha_{-1} = 0 \cdot x - 1 = -1, \qquad \alpha_0 = 1 \cdot x - [x],$$

so that (6_k) becomes

$$\left|\lambda - \frac{1}{x - [x]}\right| < 1; \quad \epsilon\left(\lambda - \frac{0}{1}\right) \text{ positive and minimal.} \tag{6_1}$$

The number

$$x_1 = \frac{1}{x - [x]}$$

is larger than 1, so that the two integral solutions λ of the inequality of (6_1) are positive; the solution of (6_1) is clearly

$$\lambda = \lambda_1 = [x_1], \qquad \epsilon = +1.$$

This gives $P_1 = \lambda_1 P_0 + P_{-1}, \qquad Q_1 = \lambda_1 Q_0 + Q_{-1},$

$$Q_1 P_0 - Q_0 P_1 = (\lambda_1 Q_0 + Q_{-1}) P_0 - Q_0 (\lambda_1 P_0 + P_{-1})$$
$$= -(Q_0 P_{-1} - Q_{-1} P_0),$$

and since $Q_0 P_{-1} - Q_{-1} P_0 = 1 \cdot 1 - 0 \cdot \lambda_0 = 1,$

we have $Q_1 P_0 - Q_0 P_1 = -1.$

These calculations provide a basis for an inductive proof of the following theorem.

THEOREM 9 1. *Put $x_0 = x$, and define*

$$x_k = \frac{1}{x_{k-1} - [x_{k-1}]} \qquad for\ 1 \le k < n + 1,$$

where n is the smallest index for which $x_n - [x_n] = 0$, if there is such, and is infinity otherwise. Then for $1 \le k < n + 1$,

$$x_k = -\frac{\alpha_{k-2}}{\alpha_{k-1}}, \tag{7}$$

and the solution λ, ϵ of (6_k) is $\lambda = \lambda_k = [x_k]$, $\epsilon = +1$. Hence $\{P_k/Q_k\}$ is recursively defined by the equations

$$P_{-1} = 1, \quad P_0 = \lambda_0, \quad P_k = P_{k-1}\lambda_k + P_{k-2}$$
$$Q_{-1} = 0, \quad Q_0 = 1, \quad Q_k = Q_{k-1}\lambda_k + Q_{k-2} \qquad for\ 1 \le k < n + 1, \tag{8}$$

and $Q_k P_{k-1} - Q_{k-1} P_k = (-1)^k \qquad for\ 0 \le k < n + 1.$ (9)

Proof: As we have just seen, all the assertions of the theorem are true when $k = 1$, and (9) holds for $k = 0$. Suppose the assertions true for some $k < n + 1$, and for all indices smaller than this k. We wish to determine P_{k+1} and Q_{k+1} by solving (6_{k+1}).

If n is finite and $k + 1 = n + 1$, then $x_k - [x_k] = x_k - \lambda_k = 0$. Reversing the argument that led to (6_k), this means that $Q_k x - P_k = 0$, or that $x = P_n/Q_n$, and the sequence $\{P_k/Q_k\}$ terminates with P_n/Q_n. Thus the entire sequence of best approximations has already been determined when $k = n$, so that we need only consider the case $k < n$.

If $k < n$, we must solve

$$\left| \lambda - \left(-\frac{\alpha_{k-1}}{\alpha_k} \right) \right| < 1; \quad \epsilon \left\{ \lambda - \left(-\frac{Q_{k-1}}{Q_k} \right) \right\} \text{ positive and minimal.} \tag{6_{k+1}}$$

From the induction hypothesis and the definition of x_{k+1}, we have

$$-\frac{\alpha_{k-1}}{\alpha_k} = -\frac{\alpha_{k-1}}{Q_k x - P_k} = -\frac{\alpha_{k-1}}{(Q_{k-1}\lambda_k + Q_{k-2})x - (P_{k-1}\lambda_k + P_{k-2})}$$

$$= -\frac{\alpha_{k-1}}{\alpha_{k-1}\lambda_k + \alpha_{k-2}} = \frac{1}{-\alpha_{k-2}/\alpha_{k-1} - \lambda_k} = \frac{1}{x_k - [x_k]} = x_{k+1} > 1.$$

Since $-Q_{k-1}/Q_k < 0$, the solution of (6_{k+1}) is clearly

$$\lambda = \lambda_{k+1} = [x_{k+1}], \qquad \epsilon = +1,$$

whence $Q_{k+1} = Q_k\lambda_{k+1} + Q_{k-1}, \qquad P_{k+1} = P_k\lambda_{k+1} + P_{k-1}.$
Moreover,

$$\begin{aligned} Q_{k+1}P_k - Q_k P_{k+1} &= (Q_k\lambda_{k+1} + Q_{k-1})P_k - Q_k(P_k\lambda_{k+1} + P_{k-1}) \\ &= -(Q_k P_{k-1} - Q_{k-1}P_k) = (-1)^{k+1}, \end{aligned}$$

by the induction hypothesis, and the proof is complete.

To see how Theorem 9–1 solves the problem of finding the best approximations to x, take $x = \sqrt{2}$. Then $\lambda_0 = [\sqrt{2}] = 1$ and

$$x_1 = \frac{1}{x - [x]} = \frac{1}{\sqrt{2} - 1} = \sqrt{2} + 1, \qquad \lambda_1 = [\sqrt{2} + 1] = 2,$$

$$x_2 = \frac{1}{x_1 - [x_1]} = \frac{1}{\sqrt{2} - 1} = \sqrt{2} + 1, \qquad \lambda_2 = [\sqrt{2} + 1] = 2,$$
$$\vdots$$

and in general, $x_k = \sqrt{2} + 1$ and $\lambda_k = 2$, for $k \geq 1$. Hence

$$P_{-1} = 1, \qquad P_0 = 1, \qquad P_k = 2P_{k-1} + P_{k-2},$$

so that $\{P_k\} = 1, 1, 3, 7, 17, \ldots$, and

$$Q_{-1} = 0, \qquad Q_0 = 1, \qquad Q_k = 2Q_{k-1} + Q_{k-2},$$

so that $\{Q_k\} = 0, 1, 2, 5, 12, \ldots.$
Thus the best approximations to $\sqrt{2}$ are

$$1, \frac{3}{2}, \frac{7}{5}, \frac{17}{12}, \cdots.$$

Of course, not every x will give a constant sequence of λ's, as happens with $\sqrt{2}$. In general, while arbitrarily many λ's can be determined, no explicit (i.e., nonrecursive) formula for the entire sequence can be given.

THEOREM 9–2. *If $\{x_k\}$, $\{\lambda_k\}$, $\{P_k\}$, and $\{Q_k\}$ are determined as in Theorem 9–1, then the following relations hold:*

$$x = \frac{P_{k-1}x_k + P_{k-2}}{Q_{k-1}x_k + Q_{k-2}}, \qquad 1 \le k < n+1, \qquad (10)$$

$$x = \lambda_0 + \cfrac{1}{\lambda_1 + \cfrac{1}{\lambda_2 + \cfrac{}{\ddots \; + \cfrac{1}{\lambda_{k-1} + \cfrac{1}{x_k}}}}}, \qquad 1 \le k < n+1, \quad (11)$$

$$\frac{P_k}{Q_k} = \lambda_0 + \cfrac{1}{\lambda_1 + \cfrac{1}{\lambda_2 + \cfrac{}{\ddots \; + \cfrac{1}{\lambda_{k-1} + \cfrac{1}{\lambda_k}}}}}, \qquad 0 \le k < n+1. \quad (12)$$

Proof: From (7) and the definition of the α's, we obtain

$$x_k = -\frac{Q_{k-2}x - P_{k-2}}{Q_{k-1}x - P_{k-1}},$$

which yields (10). The definition of $\{x_k\}$ and $\{\lambda_k\}$ gives

$$x \quad = \lambda_0 + \frac{1}{x_1},$$

$$x_1 \quad = \lambda_1 + \frac{1}{x_2}, \qquad (13)$$

$$\vdots$$

$$x_{k-1} = \lambda_{k-1} + \frac{1}{x_k},$$

and if we successively eliminate $x_1, x_2 \ldots, x_{k-1}$, equation (11) results. To obtain (12), consider the equations (13) with x_k as an independent variable which assumes values greater than 1, with fixed $\lambda_0, \lambda_1, \ldots, \lambda_{k-1}$. Then x is a function of x_k, given more briefly by (11). Since $P_{k-1}, P_{k-2}, Q_{k-1}$, and Q_{k-2} depend only on $\lambda_0, \ldots, \lambda_{k-1}$, (10) and (11) are different expressions for the same functional relation. If in (10) x_k is given the value λ_k, then $x = P_k/Q_k$, and substituting these values in (11) gives (12).

PROBLEMS

1. Carry through the procedure described in this section to find the first few best approximations to $\sqrt{3}$.

2. Find all the best approximations to $339/62$.

3. Show that if $x = \frac{1}{2}(1 + \sqrt{5})$, then each λ_i is 1.

9–2 The rational case. We now suppose that x is rational. If x is an interior point of $R_{Q_k}(P_k, Q_k)$, we have the strict inequality $|Q_k x - P_k| < |Q_{k-1} x - P_{k-1}|$, or $|\lambda_k - x_k| < 1$. It may happen, however, that for some r/s with $s > Q_{k-1}$, x is the common end point of the abutting intervals $R_s(P_{k-1}, Q_{k-1})$ and $R_s(r, s)$, while x is an interior point of $R_{s-1}(P_{k-1}, Q_{k-1})$. In this case, it is a matter of choice whether r/s is to be included among the best approximations to x; it has not been included up to now, since we have required the strict inequality $|\lambda_k - x_k| < 1$. Fortunately, this ambiguity occurs only once for each x, for we know from earlier calculations that x is the mediant of P_{k-1}/Q_{k-1} and r/s:

$$x = \frac{P_{k-1} + r}{Q_{k-1} + s}, \tag{14}$$

so that x is the first rational number to appear between P_{k-1}/Q_{k-1} and r/s in the sequences $F_s, F_{s+1}, \ldots$. Hence $k = n$, and if r/s is to be included among the best approximations, and if we put $r/s = P_n/Q_n$, then $P_{n+1}/Q_{n+1} = x$. Comparing equations (8) and (14), we see that $\lambda_{n+1} = 1$, and by (12),

$$x = \lambda_0 + \cfrac{1}{\lambda_1 + \cfrac{}{\ddots \cfrac{}{+ \cfrac{1}{\lambda_n + \cfrac{1}{1}}}}}.$$

If r/s is not included, then $x = P_n/Q_n$ and

$$x = \lambda_0 + \cfrac{1}{\lambda_1 + \cfrac{}{\ddots}}$$

$$+ \cfrac{1}{(\lambda_n + 1)}$$

To illustrate, take $x = \frac{2}{3}$. Then

$$x_1 = \frac{1}{\frac{2}{3} - [\frac{2}{3}]} = \frac{3}{2}, \qquad x_2 = \frac{1}{\frac{3}{2} - [\frac{3}{2}]} = 2,$$

$$\lambda_0 = [\tfrac{2}{3}] = 0, \qquad \lambda_1 = [x_1] = 1, \qquad \lambda_2 = [x_2] = 2,$$

and we have

$$\frac{2}{3} = 0 + \frac{1}{1 + \frac{1}{2}}, \qquad \frac{P_0}{Q_0} = \frac{0}{1}, \qquad \frac{P_1}{Q_1} = \frac{1}{1}, \qquad \frac{P_2}{Q_2} = \frac{2}{3}.$$

But we could also write

$$\frac{2}{3} = 0 + \cfrac{1}{1 + \cfrac{1}{1 + \cfrac{1}{1}}},$$

with $\qquad \dfrac{P_0}{Q_0} = \dfrac{0}{1}, \qquad \dfrac{P_1}{Q_1} = \dfrac{1}{1}, \qquad \dfrac{P_2}{Q_2} = \dfrac{1}{2}, \qquad \dfrac{P_3}{Q_3} = \dfrac{2}{3}.$

In this case, $\frac{2}{3}$ is the right end point of $R_2(1, 2)$ and the left end point of $R_2(1, 1)$; with the normal procedure, $\frac{1}{2}$ would not be included among the best approximations to $\frac{2}{3}$.

An expression

$$a_0 + \cfrac{1}{a_1 + \cfrac{1}{a_2 + \cfrac{}{\ddots}}} \tag{15}$$

$$+ \cfrac{1}{a_n}$$

is called a *finite regular continued fraction;* it is finite because there are only finitely many a's, and regular because the a's are integers, $a_1, \ldots, a_n$ are positive, and the numerators are all $+1$. We shall deal only with regular continued fractions in this book. For typographical simplicity, we write

$$a_0 + \frac{1}{a_1 +} \frac{1}{a_2 +} \cdots \frac{1}{a_n} \tag{16}$$

in place of (15). The numbers

$$\frac{p_0}{q_0} = a_0, \qquad \frac{p_1}{q_1} = a_0 + \frac{1}{a_1}, \qquad \frac{p_2}{q_2} = a_0 + \frac{1}{a_1 +} \frac{1}{a_2}, \qquad \cdots$$

are called the *convergents* of the continued fraction (16). If (16) has the value x, we can put

$$x = a_0 + \frac{1}{a_1 +} \cdots \frac{1}{a_{k-1} +} \frac{1}{x_k'},$$

where

$$x_k' = a_k + \frac{1}{a_{k+1} +} \cdots \frac{1}{a_n}, \qquad \text{or} \qquad x_k' = a_k + \frac{1}{x_{k+1}'},$$

for $k = 1, 2, \ldots, n$. Since $x_k' \geq 1$, we have

$$x_1' = \frac{1}{x - a_0} = \frac{1}{x - [x]}, \qquad a_1 = [x_1'],$$

$$x_2' = \frac{1}{x_1' - a_1} = \frac{1}{x_1' - [x_1']}, \qquad a_2 = [x_2'],$$

$$\vdots$$

Thus the sequence $\{x_k'\}$ is identical with the sequence $\{x_k\}$ defined in Theorem 9–1, and $\{a_k\}$ is therefore identical with $\{\lambda_k\}$. Hence we have the following theorem.

THEOREM 9–3. *The convergents (possibly excepting p_0/q_0) of any finite regular continued fraction are the best approximations to the value of the continued fraction. Every rational number can be expanded into a finite regular continued fraction, and this expansion is unique, except for the variation indicated by the identity*

$$a_0 + \frac{1}{a_1 +} \cdots \frac{1}{a_n} = a_0 + \frac{1}{a_1 +} \cdots \frac{1}{(a_n - 1) +} \frac{1}{1}, \qquad a_n > 1.$$

Moreover, the identities

(a) $p_0 = a_0$, $p_k = p_{k-1}a_k + p_{k-2}$,

(b) $q_0 = 1$, $q_k = q_{k-1}a_k + q_{k-2}$,

(c) $q_k p_{k-1} - q_{k-1}p_k = (-1)^k$,

(d) $x = \dfrac{p_{k-1}x_k + p_{k-2}}{q_{k-1}x_k + q_{k-2}}$

hold for $k = 1, 2, \ldots n$, *if we define* $p_{-1} = 1$, $q_{-1} = 0$, *and*

$$x = a_0 + \frac{1}{a_1 +} \cdots \frac{1}{a_n}, \qquad a_n \geq 1.$$

It might be worth mentioning that the continued fraction expansion of $x = p/q$ can be deduced immediately from the Euclidean algorithm as applied to a and b. For if

$$a = ba_0 + r_0,$$
$$b = r_0 a_1 + r_1,$$
$$r_0 = r_1 a_2 + r_2,$$
$$\vdots$$
$$r_{n-3} = r_{n-2}a_{n-1} + r_{n-1},$$
$$r_{n-2} = r_{n-1}a_n,$$

then

$$\frac{a}{b} = a_0 + \frac{r_0}{b} = a_0 + \frac{1}{b/r_0},$$

$$\frac{b}{r_0} = a_1 + \frac{r_1}{r_0} = a_1 + \frac{1}{r_0/r_1},$$

$$\frac{r_0}{r_1} = a_2 + \frac{r_2}{r_1} = a_2 + \frac{1}{r_1/r_2},$$

$$\vdots$$

$$\frac{r_{n-3}}{r_{n-2}} = a_{n-1} + \frac{r_{n-1}}{r_{n-2}} = a_{n-1} + \frac{1}{r_{n-2}/r_{n-1}},$$

$$\frac{r_{n-2}}{r_{n-1}} = a_n,$$

so that

$$\frac{a}{b} = a_0 + \frac{1}{a_1 +} \frac{1}{a_2 +} \cdots \frac{1}{a_n}.$$

1. Prove that the convergents p_k/q_k are in reduced form, i.e., that $(p_k, q_k) = 1$.

2. If $x = a/b$, where $(a, b) = 1$, and

$$x = a_0 + \frac{1}{a_1 +} \frac{1}{a_2 +} \cdots \frac{1}{a_n},$$

then $p_n = a$, $q_n = b$. Use this and an identity of Theorem 9–3 to find a solution of the linear Diophantine equation $ax + by = c$. In particular, find a general solution of $247x + 77y = 31$.

3. Show that for $k \leq n$,

$$\frac{q_k}{q_{k-1}} = a_k + \frac{1}{a_{k-1} +} \cdots \frac{1}{a_2 +} \frac{1}{a_1}.$$

9–3 The irrational case. Now consider the case that x is an irrational number ξ. The sequences $\{x_k\}$, $\{\lambda_k\}$, and $\{P_k/Q_k\}$ are now infinite, and we write

$$\xi = \lambda_0 + \frac{1}{\lambda_1 +} \frac{1}{\lambda_2 +} \cdots. \tag{17}$$

This equation must be understood as an abbreviation for the equation

$$\xi = \lim_{n \to \infty} \left(\lambda_0 + \frac{1}{\lambda_1 +} \cdots \frac{1}{\lambda_n} \right) = \lim_{n \to \infty} \frac{P_n}{Q_n} ; \tag{18}$$

the convergents P_n/Q_n play a role here analogous to that of the partial sums of an infinite series.

Conversely, if we start with an arbitrary infinite regular continued fraction

$$a_0 + \frac{1}{a_1 +} \frac{1}{a_2 +} \cdots, \tag{19}$$

we can show that the convergents

$$\frac{p_0}{q_0} = a_0, \qquad \frac{p_1}{q_1} = a_0 + \frac{1}{a_1} , \qquad \frac{p_2}{q_2} = a_0 + \frac{1}{a_1 +} \frac{1}{a_2} , \cdots$$

always converge to an irrational number ξ. For take $n > 2$ and put

$$\frac{p}{q} = \frac{p_n}{q_n} = a_0 + \frac{1}{a_1 +} \cdots \frac{1}{a_n}.$$

Then by Theorem 9–1, the numbers $a_0, a_1, \ldots, a_{n-2}$ are uniquely determined by p/q, and the convergents of p/q, which are also convergents of (19), satisfy the usual recursion relations:

$$p_{-1} = 1, \qquad p_0 = a_0, \qquad p_k = p_{k-1}a_k + p_{k-2},$$
$$q_{-1} = 0, \qquad q_0 = 1, \qquad q_k = q_{k-1}a_k + q_{k-2}, \tag{20}$$

for $k = 1, 2, \ldots, n - 2$. Moreover,

$$\begin{aligned}
q_k p_{k-1} - q_{k-1} p_k &= (q_{k-1}a_k + q_{k-2})p_{k-1} - q_{k-1}(p_{k-1}a_k + p_{k-2}) \\
&= -(q_{k-1}p_{k-2} - q_{k-2}p_{k-1}) = \cdots \\
&= (-1)^k (q_0 p_{-1} - q_{-1} p_0),
\end{aligned}$$

so that

$$q_k p_{k-1} - q_{k-1} p_k = (-1)^k \tag{21}$$

for $k = 0, 1, \ldots, n - 2$. Since n is arbitrary, the relations (20) and (21) hold for all $k \geq 1$. By (21),

$$\begin{aligned}
q_k p_{k-2} - q_{k-2} p_k &= (q_{k-1}a_k + q_{k-2})p_{k-2} - q_{k-2}(p_{k-1}a_k + p_{k-2}) \\
&= a_k(q_{k-1}p_{k-2} - q_{k-2}p_{k-1}) \\
&= (-1)^{k-1} a_k. \tag{22}
\end{aligned}$$

From (22), we see that

$$\frac{p_{2k-2}}{q_{2k-2}} < \frac{p_{2k}}{q_{2k}}, \qquad \frac{p_{2k-1}}{q_{2k-1}} > \frac{p_{2k+1}}{q_{2k+1}},$$

so that

$$\frac{p_0}{q_0} < \frac{p_2}{q_2} < \frac{p_4}{q_4} < \cdots, \qquad \frac{p_1}{q_1} > \frac{p_3}{q_3} > \frac{p_5}{q_5} > \cdots.$$

By (21),

$$\frac{p_{2k}}{q_{2k}} < \frac{p_{2k+1}}{q_{2k+1}},$$

so that

$$\frac{p_{2k}}{q_{2k}} < \frac{p_{2l+1}}{q_{2l+1}}$$

for every $l \geq k$. Hence

$$\frac{p_0}{q_0} < \frac{p_2}{q_2} < \frac{p_4}{q_4} < \cdots < \frac{p_5}{q_5} < \frac{p_3}{q_3} < \frac{p_1}{q_1},$$

so that the sequences $\{p_{2k}/q_{2k}\}$ and $\{p_{2k+1}/q_{2k+1}\}$, being monotonic and bounded, are convergent. But (21) can be rewritten in the form

$$\frac{p_{k-1}}{q_{k-1}} - \frac{p_k}{q_k} = \frac{(-1)^k}{q_{k-1}q_k},$$

and since $q_k \to \infty$ as $k \to \infty$ we see that

$$\lim_{k \to \infty} \left(\frac{p_{2k}}{q_{2k}} - \frac{p_{2k+1}}{q_{2k+1}} \right) = 0,$$

and consequently $\lim p_k/q_k$ exists. Call this limit ξ, and put

$$\xi = a_0 + \frac{1}{a_1 +} \cdots \frac{1}{a_{n-1} +} \frac{1}{\xi_n}.$$

It follows, just as in the rational case, that

$$\xi_1 = \frac{1}{\xi - [\xi]}, \qquad \xi_2 = \frac{1}{\xi_1 - [\xi_1]}, \qquad \cdots,$$

and
$$a_0 = [\xi], \qquad a_1 = [\xi_1], \qquad \ldots,$$

so that the convergents p_k/q_k of (19) are the best approximations P_k/Q_k to ξ. From this we deduce the following assertions.

THEOREM 9-4. *Every infinite regular continued fraction converges to an irrational number, the best approximations to which are afforded by the convergents of the continued fraction. Every irrational number can be expanded into an infinite regular continued fraction, and this expansion is unique. Moreover, the following identities hold, if*

$$\xi = a_0 + \frac{1}{a_1 +} \cdots :$$

$$p_{-1} = 1, \quad p_1 = a_0, \quad p_k = p_{k-1}a_k + p_{k-2},$$
$$q_{-1} = 0, \quad q_1 = 1, \quad q_k = q_{k-1}a_k + q_{k-2}, \qquad k = 1, 2, 3, \ldots, \quad (23)$$

$$q_k p_{k-1} - q_{k-1}p_k = (-1)^k, \qquad k = 0, 1, 2, \ldots, \quad (24)$$

$$q_k p_{k-2} - q_{k-2}p_k = (-1)^{k-1}a_k, \qquad k = 1, 2, 3, \ldots, \quad (25)$$

$$\xi = \frac{p_{k-1}\xi_k + p_{k-2}}{q_{k-1}\xi_k + q_{k-2}}, \qquad \text{where} \qquad \xi = a_0 + \frac{1}{a_1 +} \cdots \frac{1}{a_{k-1} +} \frac{1}{\xi_k},$$
$$k = 1, 2, 3, \ldots. \quad (26)$$

The numbers a_k are called the *partial quotients*, and the ξ_k the *complete quotients*, in the expansion.

Once the continued fraction expansion of ξ is known, the successive convergents can be computed very simply. For example, let $\xi = \sqrt{7}$. Then

$$\sqrt{7} = 2 + (\sqrt{7} - 2), \qquad a_0 = 2, \quad \xi_1 = (\sqrt{7} - 2)^{-1},$$

$$\frac{1}{\sqrt{7} - 2} = \frac{\sqrt{7} + 2}{3} = 1 + \frac{\sqrt{7} - 1}{3}, \quad a_1 = 1, \quad \xi_2 = \left(\frac{\sqrt{7} - 1}{3}\right)^{-1},$$

$$\frac{3}{\sqrt{7} - 1} = \frac{\sqrt{7} + 1}{2} = 1 + \frac{\sqrt{7} - 1}{2}, \quad a_2 = 1, \quad \xi_3 = \left(\frac{\sqrt{7} - 1}{2}\right)^{-1},$$

$$\frac{2}{\sqrt{7} - 1} = \frac{\sqrt{7} + 1}{3} = 1 + \frac{\sqrt{7} - 2}{3}, \quad a_3 = 1, \quad \xi_4 = \left(\frac{\sqrt{7} - 2}{3}\right)^{-1},$$

$$\frac{3}{\sqrt{7} - 2} = \sqrt{7} + 2 = 4 + (\sqrt{7} - 2), \quad a_4 = 4, \quad \xi_5 = (\sqrt{7} - 2)^{-1}.$$

Since $\xi_5 = \xi_1$, also $\xi_6 = \xi_2$, $\xi_7 = \xi_3$, ..., so $\{\xi_k\}$ (and therefore also $\{a_k\}$) is periodic. Thus we have the periodic expansion

$$\sqrt{7} = 2 + \frac{1}{1+} \frac{1}{1+} \frac{1}{1+} \frac{1}{4+} \frac{1}{1+} \frac{1}{1+} \frac{1}{1+} \frac{1}{4+} \cdots.$$

Using the relations (23), we construct the following table:

k	-1	0	1	2	3	4	5	6	$\cdots$
a_k		2	1	1	1	4	1	1	$\cdots$
p_k	1	2	3	5	8	37	45	82	$\cdots$
q_k	0	1	1	2	3	14	17	31	$\cdots$

Here the element $37 = p_4$, for example, is determined by multiplying $a_4 = 4$ by $p_3 = 8$ and adding $p_2 = 5$. Thus the best approximations to $\sqrt{7}$ are 3, $\frac{5}{2}$, $\frac{8}{3}$, $\frac{37}{14}$, $\frac{45}{17}$, ...

1. It can be verified, using a sufficiently good decimal approximation to π, that the beginning of the continued fraction expansion for π is

$$\pi = 3 + \frac{1}{7+} \frac{1}{15+} \frac{1}{1+} \cdots .$$

Find the first four best approximations to π.

2. Show that $\sqrt{13}$ has an expansion which is eventually periodic, and find the first few convergents.

9–4 Quadratic irrationalities. The problem of finding the best approximations to a real number x has thus been completely solved in terms of the regular continued fraction expansion of x. Of course, unless x is of a very special form, it may be impossible to give the complete expansion of x, just as one cannot give the rule of formation for the digits occurring in the decimal expansion of π. But if a decimal approximation of x is known, a corresponding part of the continued fraction expansion of x can be determined quite easily. For example, from the series expansion for e, one can easily show that

$$2.7182 < e < 2.7183.$$

By a simple computation, we find that

$$2.7182 = 2 + \frac{1}{1+} \frac{1}{2+} \frac{1}{1+} \frac{1}{1+} \frac{1}{4+} \frac{1}{1+} \frac{1}{1+} \frac{1}{1+} \frac{1}{3+} \frac{1}{1+} \frac{1}{9},$$

$$2.7183 = 2 + \frac{1}{1+} \frac{1}{2+} \frac{1}{1+} \frac{1}{1+} \frac{1}{4+} \frac{1}{1+} \frac{1}{1+} \frac{1}{19+} \frac{1}{1+} \frac{1}{1+} \frac{1}{3},$$

so that
$$e = 2 + \frac{1}{1+} \frac{1}{2+} \frac{1}{1+} \frac{1}{1+} \frac{1}{4+} \frac{1}{1+} \frac{1}{1+} \cdots .$$

(Actually, it is known that the sequence of partial quotients is

2,   1, 2, 1,   1, 4, 1,   1, 6, 1,   . . . ,   1, 2n, 1, ...)

There is, however, one simple case in which the complete expansion can be determined: that in which the partial quotients $a_0, a_1, a_2, \ldots$ constitute a sequence which is eventually periodic. Consider for example the continued fraction

$$\xi = 1 + \frac{1}{3+} \frac{1}{1+} \frac{1}{2+} \frac{1}{1+} \frac{1}{2+} \cdots ,$$

where $a_{2n} = 1$ and $a_{2n+1} = 2$ for $n \geq 1$. If ξ_2 is related to ξ as in (26), we have

$$\xi_2 = 1 + \frac{1}{2+} \frac{1}{1+} \frac{1}{2+} \cdots = 1 + \frac{1}{2+} \frac{1}{\xi_2},$$

so that

$$\xi_2 = 1 + \frac{1}{2 + \dfrac{1}{\xi_2}} = 1 + \frac{\xi_2}{2\xi_2 + 1} = \frac{3\xi_2 + 1}{2\xi_2 + 1},$$

$$2\xi_2{}^2 - 2\xi_2 - 1 = 0,$$

$$\xi_2 = \frac{-1 + \sqrt{3}}{2}.$$

(The plus sign is chosen since $\xi_2 > 0$.) Hence

$$\xi = 1 + \frac{1}{3 + \dfrac{1}{\dfrac{\sqrt{3}-1}{2}}} = \frac{4\sqrt{3} - 2}{3\sqrt{3} - 1} = \frac{17 - \sqrt{3}}{13}.$$

Conversely, if we start with a quadratic irrationality—say $\xi = 1 + \sqrt{6}$—we get

$$a_0 = [1 + \sqrt{6}] = 3,$$

$$\xi_1 = \frac{1}{\sqrt{6} - 2} = \frac{\sqrt{6} + 2}{2}, \qquad a_1 = \left[\frac{\sqrt{6} + 2}{2}\right] = 2,$$

$$\xi_2 = \frac{1}{\dfrac{\sqrt{6} + 2}{2} - 2} = \frac{2}{\sqrt{6} - 2} - \sqrt{6} + 2, \qquad a_2 = 4,$$

$$\xi_3 = \frac{1}{\sqrt{6} - 2} = \xi_1,$$

so that

$$\sqrt{6} + 1 = 3 + \frac{1}{2+} \frac{1}{4+} \frac{1}{2+} \frac{1}{4+} \cdots.$$

We can now show that these are not isolated phenomena.

THEOREM 9–5. *Every eventually periodic regular continued fraction converges to a quadratic irrationality, and every quadratic irrationality has a regular continued fraction expansion which is eventually periodic.*

Proof: The first part is quite simple. Suppose that the first period begins with a_n, and let the length of the period be h, so that $a_{k+h} = a_k$ for $k > n$. Put

$$\xi = a_0 + \frac{1}{a_1 +} \cdots, \qquad \text{and} \qquad \xi_k = a_k + \frac{1}{a_{k+1} +} \cdots,$$

so that $\xi_{k+h} = \xi_k$ for $k \geq n$. By this and equation (26),

$$\xi = \frac{p_{n-1}\xi_n + p_{n-2}}{q_{n-1}\xi_n + q_{n-2}} = \frac{p_{n+h-1}\xi_n + p_{n+h-2}}{q_{n+h-1}\xi_n + q_{n+h-2}},$$

so that ξ_n satisfies a quadratic equation with integral coefficients. Since ξ_n is obviously not rational, it is a quadratic irrationality. By (26) again, the same is true of ξ itself, since if

$$A\xi_n^2 + B\xi_n + C = 0,$$

then

$$A(-q_{n-2}\xi + p_{n-2})^2 + B(-q_{n-2}\xi + p_{n-2})(q_{n-1}\xi - p_{n-1})$$
$$+ C(q_{n-1}\xi - p_{n-1})^2 = 0,$$

and this is a quadratic equation in ξ.

The proof of the converse involves a little more computation. Suppose that

$$A\xi^2 + B\xi + C = 0,$$

where A, B, and C are integers, and ξ is irrational. Then equation (26) gives

$$A(p_{k-1}\xi_k + p_{k-2})^2 + B(p_{k-1}\xi_k + p_{k-2})(q_{k-1}\xi_k + q_{k-2})$$
$$+ C(q_{k-1}\xi_k + q_{k-2})^2 = 0,$$

or

$$A_k\xi_k^2 + B_k\xi_k + C_k = 0,$$

where the integers A_k, B_k, and C_k are given by the equations

$$A_k = Ap_{k-1}^2 + Bp_{k-1}q_{k-1} + Cq_{k-1}^2,$$

$$B_k = 2Ap_{k-1}p_{k-2} + B(p_{k-1}q_{k-2} + p_{k-2}q_{k-1}) + 2Cq_{k-1}q_{k-2},$$

$$C_k = Ap_{k-2}^2 + Bp_{k-2}q_{k-2} + Cq_{k-2}^2.$$

If $f(x) = Ax^2 + Bx + C$, then

$$A_k = q_{k-1}^2 f\left(\frac{p_{k-1}}{q_{k-1}}\right), \qquad C_k = q_{k-2}^2 f\left(\frac{p_{k-2}}{q_{k-2}}\right).$$

Using Taylor's theorem, we have

$$A_k = q_{k-1}^2 \left\{ f(\xi) + f'(\xi)\left(\frac{p_{k-1}}{q_{k-1}} - \xi\right) + \frac{1}{2}f''(\xi)\left(\frac{p_{k-1}}{q_{k-1}} - \xi\right)^2 \right\},$$

since $f'''(x)$ is identically zero. Now $f(\xi) = 0$, and

$$\xi - \frac{p_{k-1}}{q_{k-1}} = \frac{p_{k-1}\xi_k + p_{k-2}}{q_{k-1}\xi_k + q_{k-2}} - \frac{p_{k-1}}{q_{k-1}} = \frac{q_{k-1}p_{k-2} - q_{k-2}p_{k-1}}{q_{k-1}(q_{k-1}\xi_k + q_{k-2})}$$

$$= \frac{(-1)^{k-1}}{q_{k-1}(q_{k-1}\xi_k + q_{k-2})}; \tag{27}$$

since $\xi_k > 1$, it is certainly true that

$$\left| \xi - \frac{p_{k-1}}{q_{k-1}} \right| < \frac{1}{q_{k-1}^2}.$$

Hence

$$|A_k| < |f'(\xi)| + \frac{|f''(\xi)|}{2q_{k-1}^2},$$

and similarly

$$|C_k| < |f'(\xi)| + \frac{|f''(\xi)|}{2q_{k-2}^2},$$

so that $|A_k|$ and $|C_k|$ remain bounded as $k \to \infty$.

To see that $|B_k|$ is also bounded, we use the fact that all the quantities $B_k^2 - 4A_kC_k$ have the common value $B^2 - 4AC = D$. (This can be proved by a straightforward but tedious computation or, if one is acquainted with the theory of linear transformations, by noting that the expression $A_kx'^2 + B_kx'y' + C_ky'^2$ is obtained from $Ax^2 + Bxy + Cy^2$ by the unimodular substitution

$$x = p_{k-1}x' + p_{k-2}y',$$
$$y = q_{k-1}x' + q_{k-2}y',$$

and that two such forms have the same discriminant.) Since A_k and C_k are bounded and D is fixed,

$$B_k^2 = D + 4A_kC_k$$

must be bounded also.

Thus, there is a constant M such that

$$|A_k| < M, \qquad |B_k| < M, \qquad |C_k| < M$$

for all k. Since there are fewer than $(2M + 1)^3$ triples of integers numerically smaller than M, there must be three indices, say n_1, n_2, and n_3, which give the same triple:

$$A_{n_1} = A_{n_2} = A_{n_3}, \qquad B_{n_1} = B_{n_2} = B_{n_3}, \qquad C_{n_1} = C_{n_2} = C_{n_3}.$$

Since the equation $A_{n_1}x^2 + B_{n_1}x + C_{n_1} = 0$ has only two roots, two of the numbers $\xi_{n_1}, \xi_{n_2}, \xi_{n_3}$ must be equal; with proper naming, they can be taken to be ξ_{n_1} and ξ_{n_2}, where $n_1 < n_2$. If $n_2 - n_1 = h$, then $\xi_{n_1+h} = \xi_{n_1}$, and

$$\xi_{n_1+h+1} = \frac{1}{\xi_{n_1+h} - [\xi_{n_1+h}]} = \frac{1}{\xi_{n_1} - [\xi_{n_1}]} = \xi_{n_1+1},$$

$$\xi_{n_1+h+2} = \frac{1}{\xi_{n_1+h+1} - [\xi_{n_1+h+1}]} = \frac{1}{\xi_{n_1+1} - [\xi_{n_1+1}]} = \xi_{n_1+2},$$

and in general, $\xi_{k+h} = \xi_k$ for $k \geq n_1$. Thus the ξ_k's are eventually periodic. Since each a_k is determined exclusively by the corresponding ξ_k, the same is true of the a_k's, and the proof is complete.

The relation (27) is of course valid for general ξ, although it was used above only when ξ is a quadratic irrationality. It provides a proof of the following assertions.

Theorem 9–6. *If p_k/q_k is a convergent of the continued fraction expansion of ξ, then*

$$\xi - \frac{p_k}{q_k} = \frac{(-1)^k}{q_k(q_k\xi_{k+1} + q_{k-1})} . \tag{28}$$

A fortiori,

$$\frac{1}{q_k(q_k + q_{k+1})} < \left| \xi - \frac{p_k}{q_k} \right| < \frac{1}{q_k q_{k+1}}$$

and

$$\left| \xi - \frac{p_k}{q_k} \right| < \frac{1}{q_k^2} . \tag{29}$$

As a partial converse, we have

Theorem 9–7. *If* $\quad \left| \xi - \dfrac{p}{q} \right| < \dfrac{1}{2q^2},$ $\tag{30}$

then p/q is a convergent of the continued fraction expansion of ξ.

Proof: If p/q is adjacent to r/s in F_q, then the end point of $R_q(p, q)$ lying between p/q and r/s is the mediant

$$\frac{p + r}{q + s},$$

and

$$\left| \frac{p + r}{q + s} - \frac{p}{q} \right| = \left| \frac{qr - ps}{q(q + s)} \right| = \frac{1}{q(q + s)} \geq \frac{1}{2q^2}.$$

Hence, if (30) holds, either

$$\frac{p}{q} < \xi < \frac{p + r}{q + s} < \frac{r}{s} \quad \text{or} \quad \frac{r}{s} < \frac{p + r}{q + s} < \xi < \frac{p}{q},$$

and $\xi \in R_q(p, q)$, so that p/q is a best approximation, and therefore a convergent, to ξ.

PROBLEMS

1. Below is an outline of a proof that the expansion of $\sqrt{d}$ (d a positive nonsquare integer) is periodic after a_0. Fill in all details. (If $\alpha = r + s\sqrt{d}$, where r and s are rational, then $\bar{\alpha} = r - s\sqrt{d}$.)

Put $\xi = \sqrt{d} + [\sqrt{d}]$. Then $-1 < \bar{\xi} < 0$, and from the equation

$$\xi_k = a_k + \frac{1}{\xi_{k+1}}$$

it follows that $-1 < \bar{\xi}_k < 0$ for $k \geq 1$. This in turn shows that $a_k = [-1/\bar{\xi}_{k+1}]$. Now suppose that the periodicity of $\{\xi_k\}$ begins when $k = n$, and that the period is of length h, so that $\xi_n = \xi_{n+h}$. It follows that $a_{n-1} = a_{n+h-1}$, and hence that $\xi_{n-1} = \xi_{n+h-1}$, so that $\{\xi_k\}$ is periodic from the beginning.

2. It is a consequence of Theorem 8–1 that if ξ is irrational, then to each positive integer t there corresponds at least one pair of integers x, y such that

$$\left| \xi - \frac{y}{x} \right| < \frac{1}{tx}, \qquad 1 \leq x \leq t.$$

Show that this becomes false, for any irrational ξ and infinitely many t, if the second inequality above is replaced by $1 \leq x \leq t/2$. [*Hint:* Take $t = q_k + q_{k+1}$, and use Theorems 9–7 and 9–6.]

9–5 Application to Pell's equation

THEOREM 9–8. *If N and d are integers with $d > 0$ and $|N| < \sqrt{d}$, and d is not a square, then all positive solutions of the Pell equation*

$$x^2 - dy^2 = N \tag{31}$$

are such that x/y is a convergent of the continued fraction expansion of $\sqrt{d}$.

Proof: Suppose that $x + y\sqrt{d}$ is a positive solution of (31). Then, if N is positive,

$$0 < x - y\sqrt{d} = \frac{N}{x + y\sqrt{d}} < \frac{\sqrt{d}}{x + y\sqrt{d}} = \frac{1}{\dfrac{x}{\sqrt{d}} + y} = \frac{1}{y\left(\dfrac{x}{y\sqrt{d}} + 1\right)}.$$

Since $x/y > \sqrt{d}$, we have

$$\left| \sqrt{d} - \frac{x}{y} \right| < \frac{1}{2y^2}. \tag{32}$$

If N is negative, we deduce from the equation

$$y^2 - \frac{x^2}{d} = \frac{-N}{d}$$

the relations

$$0 < y - \frac{x}{\sqrt{d}} = \frac{-N/d}{y + \dfrac{x}{\sqrt{d}}} < \frac{1}{y\sqrt{d} + x} = \frac{1}{x\left(1 + \dfrac{y\sqrt{d}}{x}\right)},$$

and

$$\left| \frac{1}{\sqrt{d}} - \frac{y}{x} \right| < \frac{1}{2x^2}. \tag{33}$$

Now if

$$\sqrt{d} = a_0 + \frac{1}{a_1 +} \cdots,$$

then

$$\frac{1}{\sqrt{d}} = 0 + \frac{1}{a_0 +} \frac{1}{a_1 +} \cdots,$$

so that the convergents of the continued fraction expansion of $1/\sqrt{d}$ are $0/1$ and the reciprocals of the convergents of the continued fraction expansion of $\sqrt{d}$. Using this, the inequalities (32) and (33), and Theorem 9–7, we have the result.

This theorem provides an effective method of determining all integers N, numerically smaller than $\sqrt{d}$, for which equation (31) is solvable, for it happens that the sequence $\{p_k^2 - dq_k^2\}$ is eventually periodic, and consequently only finitely many values of k need to be examined. To see this, put $\xi = \sqrt{d}$ and

$$\sqrt{d} = \frac{p_{k-1}\xi_k + p_{k-2}}{q_{k-1}\xi_k + q_{k-2}}. \tag{34}$$

Solving for ξ_k and rationalizing the denominator, we can write

$$\xi_k = \frac{\sqrt{d} + r_k}{s_k},$$

where r_k and s_k are rational numbers. Substituting this back into (34), and replacing k by $k + 1$ throughout, we have

$$\sqrt{d} = \frac{p_k(\sqrt{d} + r_{k+1}) + p_{k-1}s_{k+1}}{q_k(\sqrt{d} + r_{k+1}) + q_{k-1}s_{k+1}},$$

or

$$(q_k r_{k+1} + q_{k-1}s_{k+1} - p_k)\sqrt{d} - (p_{k-1}s_{k+1} + p_k r_{k+1} - q_k d) = 0.$$

The rational and irrational parts must separately be zero, so

$$q_k r_{k+1} + q_{k-1}s_{k+1} = p_k, \qquad p_k r_{k+1} + p_{k-1}s_{k+1} = q_k d.$$

The determinant of this system is $q_k p_{k-1} - q_{k-1}p_k = (-1)^k$, so that

$$\begin{aligned} r_{k+1} &= (-1)^k(p_k p_{k-1} - q_k q_{k-1}d), \\ s_{k+1} &= (-1)^k(q_k^2 d - p_k^2). \end{aligned} \tag{35}$$

Now the numbers r_k and s_k are uniquely determined by ξ_k; since $\{\xi_k\}$ is eventually periodic, the same is true of $\{s_k\}$, and the eventual periodicity of $\{p_k^2 - dq_k^2\}$ follows from the second equation of (35).

The discussion of Pell's equation with $N = \pm 1$, in Chapter 8, had the serious drawback that no effective method was given for finding the fundamental solution, nor even of deciding when one exists for $N = -1$. The results obtained above entirely clarify these points: the first solution encountered, being the smallest, is the fundamental solution, and the equation $x^2 - dy^2 = -1$ is solvable if and only if a solution exists among the convergents to $\sqrt{d}$ up to the end of the second period. (For $\{s_k\}$ becomes periodic at the same point as $\{\xi_k\}$, and $\{(-1)^{k-1}s_k\}$ has period at most twice that of $\{s_k\}$.) It can be shown by the method sketched in Problem 4, below, that $s_k = 1$ for the first time at the end of the first period, so that the preceding convergent is the fundamental solution of one of the equations; if for this convergent $p_k^2 - dq_k^2 = 1$, then the equation $x^2 - dy^2 = -1$ is unsolvable, while if $p_k^2 - dq_k^2 = -1$, the convergents preceding ends of periods are alternately solutions for $N = -1$ and $N = 1$.

PROBLEMS

1. For what N with $|N| < \sqrt{7}$ is the equation $x^2 - 7y^2 = N$ solvable?

2. Show that the numbers r_k and s_k defined in this section are positive integers. [*Hint:* Use equation (28).]

3. Find the fundamental solution of $x^2 - 95y^2 = 1$; of $x^2 - 74y^2 = 1$.

4. (a) Using Problem 1, Section 9–4, show that if the length of the period in the expansion of $\sqrt{d}$ is h, then $s_h = 1$, and hence that

$$p_{h-1}^2 - dq_{h-1}^2 = (-1)^h.$$

Thus $x^2 - dy^2 = -1$ is solvable if h is odd.

(b) Using the fact that the numbers $\xi, \xi_1, \ldots, \xi_{h-1}$ are distinct, show that $s_k > 1$ if $1 \leq k \leq h - 1$. Deduce that the equation $x^2 - dy^2 = -1$ is solvable only if h is odd.

9–6 Equivalence of numbers. Because each element of the sequence $\{\xi_k\}$ depends only on the preceding one, and because the defining rule

$$\xi_k = [\xi_k] + \frac{1}{\xi_{k+1}}$$

is the same for all $k \geq 1$, it is clear that if

$$\xi = a_0 + \frac{1}{a_1 +} \cdots \frac{1}{a_{n-1} +} \frac{1}{\xi_n} = a_0 + \frac{1}{a_1 +} \cdots,$$

then

$$\xi_n = a_n + \frac{1}{a_{n+1} +} \cdots.$$

If we are interested in the possibility of finding infinitely many solutions of the inequality

$$\left| \xi - \frac{p}{q} \right| < \frac{1}{cq^2},$$

equation (28) shows that we need only examine the numbers ξ_k with large k. For this reason, we shall term two irrational numbers ξ

and ξ' *equivalent if, for some j and k, $\xi'_j = \xi_k$.* Then $\xi'_{j+m} = \xi_{k+m}$ for $m \geq 0$, and by the above remark, this means that if

$$\xi = a_0 + \cfrac{1}{a_1 +} \cdots \cfrac{1}{a_{k-1} +} \cfrac{1}{a_k +} \cfrac{1}{a_{k+1} +} \cdots$$

then

$$\xi' = b_0 + \cfrac{1}{b_1 +} \cdots \cfrac{1}{b_{j-1} +} \cfrac{1}{a_k +} \cfrac{1}{a_{k+1} +} \cdots,$$

so that

$$\xi = \frac{p_{k-1}\xi_k + p_{k-2}}{q_{k-1}\xi_k + q_{k-2}}, \qquad \xi' = \frac{p'_{j-1}\xi_k + p'_{j-2}}{q'_{j-1}\xi_k + q'_{j-2}}.$$

THEOREM 9-9. *Two irrational numbers ξ and ξ' are equivalent, in the sense that their continued fraction expansions are identical from some points on, if and only if there are integers A, B, C, and D such that*

$$\xi' = \frac{A\xi + B}{C\xi + D}, \qquad \text{where } AD - BC = \pm 1. \tag{36}$$

Proof: Eliminating ξ_k from the equations preceding the theorem gives

$$\frac{-q_{k-2}\xi + p_{k-2}}{q_{k-1}\xi - p_{k-1}} = \frac{-q'_{j-2}\xi' + p'_{j-2}}{q'_{j-1}\xi' - p'_{j-1}},$$

or

$$\xi = \frac{A\xi' + B}{C\xi' + D},$$

where

$$A = p_{k-1}q'_{j-2} - p_{k-2}q'_{j-1}, \qquad B = p_{k-2}p'_{j-1} - p_{k-1}p'_{j-2},$$
$$C = q_{k-1}q'_{j-2} - q_{k-2}q'_{j-1}, \qquad D = q_{k-2}p'_{j-1} - q_{k-1}p'_{j-2}.$$

A simple calculation shows that

$$AD - BC = (p'_{j-1}q'_{j-2} - p'_{j-2}q'_{j-1})(p_{k-1}q_{k-2} - p_{k-2}q_{k-1}) = \pm 1.$$

To complete the proof, suppose that equation (36) holds. By replacing A, B, C, and D by their negatives if necessary, we may suppose also that $C\xi + D > 0$. Substituting the value of ξ from equation (26) into (36) gives

$$\xi' = \frac{a\xi_k + b}{c\xi_k + d}, \tag{37}$$

where

$$a = Ap_{k-1} + Bq_{k-1}, \qquad b = Ap_{k-2} + Bq_{k-2},$$

$$c = Cp_{k-1} + Dq_{k-1}, \qquad d = Cp_{k-2} + Dq_{k-2},$$

and

$$ad - bc = (AD - BC)(p_{k-1}q_{k-2} - p_{k-2}q_{k-1}) = \pm 1.$$

By the inequality (29),

$$p_{k-1} = q_{k-1}\xi + \frac{\delta_{k-1}}{q_{k-1}}, \qquad p_{k-2} = q_{k-2}\xi + \frac{\delta_{k-2}}{q_{k-2}},$$

where

$$|\delta_{k-1}| < 1, \qquad |\delta_{k-2}| < 1.$$

Hence

$$c = (C\xi + D)q_{k-1} + \frac{C\delta_{k-1}}{q_{k-1}}, \qquad d = (C\xi + D)q_{k-2} + \frac{C\delta_{k-2}}{q_{k-2}}.$$

Since $C\xi + D$, q_{k-1}, and q_{k-2} are positive, and since $q_{k-1} > q_{k-2}$ and $q_k \to \infty$ with k, we have $c > d > 0$ for k sufficiently large. But by Theorem 8–14, this means that a/c and b/d are adjacent in F_c, and from (37) and the fact that $\xi_k > 1$, it is seen that ξ' lies between a/c and b/d, and is closer to a/c than is the mediant $(a + b)/(c + d)$. It follows that $\xi' \in R_{c-1}(b, d)$ and $\xi' \in R_c(a, c)$, so that b/d and a/c are successive convergents of the continued fraction expansion of ξ':

$$a = p'_{j-1}, \qquad b = p'_{j-2}, \qquad c = q'_{j-1}, \qquad d = q'_{j-2}.$$

But from

$$\xi' = \frac{p'_{j-1}\xi_k + p'_{j-2}}{q'_{j-1}\xi_k + q'_{j-2}} = \frac{p'_{j-1}\xi_j{}' + p'_{j-2}}{q'_{j-1}\xi_j{}' + q'_{j-2}}$$

it follows that $\xi_k = \xi_j{}'$, as was to be proved.

In the course of the proof, the following useful fact emerged:

THEOREM 9–10. *If a, b, c, and d are integers, and*

$$\xi = \frac{a\xi' + b}{c\xi' + d}, \qquad ad - bc = \pm 1, \qquad \xi' > 1, \quad c > d > 0,$$

then b/d and a/c are successive convergents of the continued fraction expansion of ξ, and ξ' is the corresponding complete quotient: for suitable k,

$$a = p_{k-1}, \qquad b = p_{k-2}, \qquad c = q_{k-1}, \qquad d = q_{k-2}, \qquad \xi' = \xi_k.$$

We shall use the symbol "$\cong$" to designate equivalence in the regular continued fraction sense. The notion of equivalence, together with equation (28), can be used to gain new insight concerning the Markov constant $M(\xi)$, which was defined in Section 8–4 as the upper limit of those numbers λ such that the inequality

$$\left| \xi - \frac{p}{q} \right| < \frac{1}{\lambda q^2}$$

has infinitely many solutions p, q. From (28), it is clear that $|q_k{}^2(\xi - p_k/q_k)|$ is approximately inversely proportional to ξ_k, so that $M(\xi)$ will probably have its smallest value for those ξ for which $a_k = 1$ for all large k. Now if

$$\xi = 1 + \frac{1}{1+} \frac{1}{1+} \cdots,$$

then $$\xi = 1 + \frac{1}{\xi}, \qquad \xi = \frac{1 + \sqrt{5}}{2}.$$

These remarks lead one to expect that the first part of the following theorem might be true.

THEOREM 9–11. *If* $\xi \cong \xi'$, *then* $M(\xi) = M(\xi')$. *If* $\xi \cong (1 + \sqrt{5})/2$, *then* $M(\xi) = \sqrt{5}$. *If* ξ *is irrational and not equivalent to* $(1 + \sqrt{5})/2$, *then* $M(\xi) \geq \sqrt{8}$. *If* $\xi \cong \sqrt{2}$, *then* $M(\xi) = \sqrt{8}$. *If* ξ *is not equivalent to either* $(1 + \sqrt{5})/2$ *or* $\sqrt{2}$, *and is irrational, then* $M(\xi) \geq 17/6$.

Proof: By (28), $$M(\xi) = \limsup_{k \to \infty} \left(\xi_{k+1} + \frac{q_{k-1}}{q_k} \right).$$

Now $$\xi_{k+1} = a_{k+1} + \frac{1}{a_{k+2} +} \cdots,$$

and

$$\frac{q_{k-1}}{q_k} = \frac{q_{k-1}}{q_{k-1}a_k + q_{k-2}} = \frac{1}{a_k + \dfrac{q_{k-2}}{q_{k-1}}} = \frac{1}{a_k +} \frac{1}{a_{k-1} + \dfrac{q_{k-3}}{q_{k-2}}} = \cdots$$

$$= \frac{1}{a_k +} \frac{1}{a_{k-1} +} \cdots \frac{1}{a_2 + \dfrac{q_0}{q_1}} = \frac{1}{a_k +} \cdots \frac{1}{a_1},$$

so that

$$M(\xi) = \limsup_{k \to \infty} \left\{ \left(\frac{1}{a_k +} \frac{1}{a_{k-1} +} \cdots \frac{1}{a_1} \right) + \left(a_{k+1} + \frac{1}{a_{k+2} +} \cdots \right) \right\}. \quad (38)$$

If $\xi' \cong \xi$, then $\xi_j' = \xi_k$ and $a_j' = a_k$ for all sufficiently large j and k for which $j - k$ has a suitable fixed value h. If the convergents of ξ' are p_j'/q_j', then for such j and k the continued fraction expansions of q_{k-1}/q_k and q_{j-1}'/q_j' have the same partial quotients at the beginning, and the interval of agreement can be made arbitrarily long by choosing j and k sufficiently large. Suppose that they agree in the first $l + 1$ partial quotients, that r_t/s_t $(t = 0, 1, \ldots, l)$ are the common convergents, and that

$$\frac{q_{k-1}}{q_k} = \frac{r_{l-1} \alpha_l + r_{l-2}}{s_{l-1} \alpha_l + s_{l-2}}, \qquad \frac{q_{j-1}'}{q_j'} = \frac{r_{l-1} \alpha_l' + r_{l-2}}{s_{l-1} \alpha_l' + s_{l-2}}.$$

Then using the fact that $[\alpha_l] = [\alpha_l'] \geq 1$, we have

$$\left| \frac{q_{k-1}}{q_k} - \frac{q_{j-1}'}{q_j'} \right| = \frac{|\alpha_l - \alpha_l'|}{(s_{l-1} \alpha_l + s_{l-2})(s_{l-1} \alpha_l' + s_{l-2})} \leq \frac{1}{s_{l-1}^2},$$

so that

$$\lim_{\substack{k \to \infty \\ j-k=h}} \left\{ \left(\xi_k + \frac{q_{k-1}}{q_k} \right) - \left(\xi_j' + \frac{q_{j-1}'}{q_j'} \right) \right\} = \lim_{\substack{k \to \infty \\ j-k=h}} \left(\frac{q_{k-1}}{q_k} - \frac{q_{j-1}'}{q_j'} \right) = 0,$$

and so $M(\xi) = M(\xi')$.

To prove the second assertion of Theorem 9–11, we need only notice that

$$M\left(\frac{1 + \sqrt{5}}{2} \right) = \lim_{k \to \infty} \left\{ \left(1 + \frac{1}{1 +} \cdots \right) + \left(\underbrace{\frac{1}{1 +} \frac{1}{1 +} \cdots \frac{1}{1}}_{k \text{ terms}} \right) \right\}$$

$$= \frac{1 + \sqrt{5}}{2} + \frac{1}{(1 + \sqrt{5})/2} = \sqrt{5},$$

by (38).

To prove the third part, we may suppose that $a_{k+1} \geq 2$ for infinitely many indices k. If $a_{k+1} \geq 3$ for infinitely many k, it is clear that $M(\xi) \geq 3$. Since $\sqrt{8} < 3$, we need only consider those ξ's for which a_k is either 1 or 2 for all large k. If there are infinitely many 1's and 2's, there are infinitely many values of k such that $a_k = 1$, $a_{k+1} = 2$.

But then, since the value of a continued fraction is always at least equal to its convergent with index 2,

$$a_{k+1} + \cfrac{1}{a_{k+2} +} \cdots \geq 2 + \cfrac{1}{a_{k+2} + \cfrac{1}{a_{k+3}}} \geq 2 + \cfrac{1}{2 + \cfrac{1}{1}} = \frac{7}{3}$$

and

$$\cfrac{1}{a_k +} \cfrac{1}{a_{k-1} +} \cdots \cfrac{1}{a_1} \geq \cfrac{1}{1 + \cfrac{1}{a_{k-1}}} \geq \cfrac{1}{1 + \cfrac{1}{1}} = \frac{1}{2},$$

so that

$$M(\xi) \geq \tfrac{7}{3} + \tfrac{1}{2} = \tfrac{17}{6} = 2.833 \ldots > \sqrt{8}.$$

On the other hand, if $a_k = 2$ for all large k, then

$$\xi \cong 1 + \cfrac{1}{2 +} \cfrac{1}{2 +} \cdots = \sqrt{2},$$

and

$$M(\xi) = \lim_{k \to \infty} \left\{ \left(2 + \cfrac{1}{2 +} \cdots \right) + \underbrace{\left(\cfrac{1}{2 +} \cfrac{1}{2 +} \cdots \cfrac{1}{2}\right)}_{k \text{ terms}} \right\}$$

$$= (\sqrt{2} + 1) + (\sqrt{2} - 1) = \sqrt{8}.$$

To clarify the significance of Theorem 9–11, we make use of the concept of countability, introduced by G. Cantor. Let S be an arbitrary infinite set. If it is possible to establish a one-to-one correspondence between the elements of S and the set of positive integers, then S is said to be *countable*. Another formulation of this requirement is that it should be possible to arrange the elements of S in a sequence having a first element, second element, and so on, in such a way that each element of S occurs only finitely far out in the sequence. The integers are countable, since every integer occurs in the sequence

$$0, 1, -1, 2, -2, \ldots, n, -n, \ldots.$$

The rational numbers between 0 and 1 are also countable, although they cannot be arranged by size. A suitable sequence is given by

$$\frac{1}{2}, \frac{1}{3}, \frac{2}{3}, \frac{1}{4}, \frac{3}{4}, \frac{1}{5}, \frac{2}{5}, \frac{3}{5}, \frac{4}{5}, \ldots,$$

in which the reduced fractions with denominators 2 are listed first, then those with denominators 3, etc. On the other hand, the real numbers between 0 and 1 are not countable (i.e., the set of such numbers is *uncountable*). For each such number is uniquely represented by an infinite decimal (which may consist exclusively of 0's from some point on, but not of 9's), and conversely. Suppose that the set could be arranged in a sequence, say $a_1, a_2, \ldots$, and let the decimal expansions be

$$a_1 = 0.a_{11}a_{12}a_{13} \ldots ,$$

$$a_2 = 0.a_{21}a_{22}a_{23} \ldots ,$$

$$a_3 = 0.a_{31}a_{32}a_{33} \ldots ,$$

where the a_{ij} are digits. Let $b = 0.b_1b_2b_3 \ldots$ be the real number determined according to the following rule: for $j = 1, 2, \ldots$,

$$b_j = \begin{cases} 0 & \text{if } a_{jj} \neq 0 \\ 1 & \text{if } a_{jj} = 0. \end{cases}$$

Then since $b_j \neq a_{jj}$, it is clear that $b \neq a_j$, and since this is true for every j, b is not in the sequence $a_1, a_2, \ldots$, so that the sequence does not contain every real number in the interval $0 < x < 1$.

If it can be shown that one set is countable, while another is not, then there must be some element of the second set which is not in the first. Moreover, every subset of a countable set is countable.

It is relevant to our present purpose to note that the quadruples of integers (A, B, C, D) such that $|AD - BC| = 1$ are countable. For without the restriction, we have the larger set of all quadruples of integers, and these can be arranged in a sequence by first writing $(0, 0, 0, 0)$, then all quadruples whose elements are 0 or ± 1, then those whose elements are $0, \pm 1$ or ± 2, etc. It follows from Theorem 9-9 that the set of numbers equivalent to a fixed number is countable, and it follows easily from this that the set of numbers equivalent to any of a fixed countable set of numbers is itself countable.

Theorem 9-11 contains the first two of an infinite sequence of assertions about the values less than 3 assumed by $M(\xi)$. Markov showed that there are only countably many such values, that their sole limit point is 3, and that each such value corresponds precisely to the set of numbers equivalent to a certain quadratic irrationality.

There is no really simple proof known. For later purposes, we show that these results cannot be extended to values $M(\xi) \geq 3$.

THEOREM 9–12. *There are uncountably many numbers ξ such that $M(\xi) = 3$.*

Proof: Let $r_1, r_2, \ldots$ be a strictly increasing sequence of positive integers, and let

$$\xi = 1 + \underbrace{\frac{1}{1+} \cdots \frac{1}{1+}}_{r_1} \frac{1}{2+} \frac{1}{2+} \underbrace{\frac{1}{1+} \cdots \frac{1}{1+}}_{r_2} \frac{1}{2+} \frac{1}{2+} \frac{1}{1+} \cdots,$$
$$\tag{39}$$

where there are r_1 partial quotients 1, then two 2's, then r_2 1's, then two 2's, then r_3 1's, etc. Thus two blocks consisting entirely of 1's are always separated by two 2's, and the blocks of 1's become longer as we move out in the sequence. Let

$$\beta_k = \xi_{k+1} + \frac{q_{k-1}}{q_k} = \left(a_{k+1} + \frac{1}{a_{k+2} +} \cdots\right) + \left(\frac{1}{a_k +} \frac{1}{a_{k-1} +} \cdots \frac{1}{a_1}\right).$$

If we choose k so that $a_{k+1} = 1$, then clearly $\xi_{k+1} < 2$, $q_{k-1}/q_k < 1$, and $\beta_k < 3$. If k runs through a sequence of indices such that $a_{k+1} = a_{k+2} = 2$, then

$$\beta_k = \left(2 + \frac{1}{2+} \frac{1}{1+} \frac{1}{1+} \cdots\right) + \left(\frac{1}{1+} \frac{1}{1+} \cdots \frac{1}{1}\right)$$
$$\rightarrow 2 + \frac{1}{2 + (\sqrt{5} - 1)/2} + \frac{\sqrt{5} - 1}{2} = 3,$$

while if k runs through a sequence of indices for which $a_k = a_{k+1} = 2$, then

$$\beta_k = \left(2 + \frac{1}{1+} \frac{1}{1+} \cdots\right) + \left(\frac{1}{2+} \frac{1}{1+} \frac{1}{1+} \cdots \frac{1}{1}\right)$$
$$\rightarrow 2 + \frac{1}{(1 + \sqrt{5})/2} + \frac{1}{2 + (\sqrt{5} - 1)/2} = 3.$$

Hence $M(\xi) = \lim \sup \beta_k = 3$.

To complete the proof, it is required to show that the set of inequivalent ξ's defined as in (39) is not countable. Now ξ and ξ' are equivalent if and only if the sequences $r_1, r_2, \ldots$ and $r_1', r_2', \ldots$ associated with them are identical from some points on, so that we can transfer the notion of equivalence from the numbers ξ and ξ' to the sequences $\{r_k\}$ and $\{r_k'\}$. Suppose that the inequivalent sequences among all the increasing sequences of positive integers can themselves be arranged in a sequence, say $R_1, R_2, \ldots$, where R_i stands for the sequence $r_{i1}, r_{i2}, \ldots$, with $r_{i1} < r_{i2} < \ldots$. With proper naming we can suppose that R_1 is the sequence $1, 2, 3, \ldots$ of all positive integers in order. If $i > 1$, R_i is not equivalent to R_1, and there are therefore infinitely many positive integers not included in it.

For $i > 1$ let $S_i = \{s_{ik}\}$ be the sequence complementary to R_i; that is, the positive integers, ordered by size, which do not occur in R_i. Each S_i is an infinite sequence. Now define a sequence T as follows. Pick t_1 in S_2, and then successively choose $t_2, t_3, \ldots$ so that

$$t_1 \in S_2, \qquad t_1 < t_2 \in S_3,$$

$$1 + t_2 < t_3 \in S_2, \qquad t_3 < t_4 \in S_3, \qquad t_4 < t_5 \in S_4,$$

$$1 + t_5 < t_6 \in S_2, \qquad t_6 < t_7 \in S_3, \qquad t_7 < t_8 \in S_4, \qquad t_8 < t_9 \in S_5,$$

$$\vdots$$

From this scheme it is apparent that T is an increasing sequence of integers, infinitely many of which are contained in S_k, and therefore not contained in R_k, for arbitrary $k \geq 2$. Hence T is certainly not equivalent to any of $R_2, R_3, \ldots$. Since each element $t_3, t_6, t_{10}, \ldots$ of T which lies in S_2 exceeds its predecessor in T by more than one, T is also not equivalent to R_1. Hence T is not equivalent to any R_k, contrary to the hypothesis that the sequence $\{R_k\}$ contains an element equivalent to any increasing sequence of positive integers.

PROBLEMS

1. Are the numbers $\sqrt{5}$ and $(1 + \sqrt{5})/2$ equivalent? What about $\sqrt{3}$ and $(1 + \sqrt{3})/2$?

2. Show that if ξ is irrational, then at least one of any two consecutive convergents to ξ satisfies the inequality

$$\left| \xi - \frac{p}{q} \right| < \frac{1}{2q^2}.$$

REFERENCES

Section 9–6

Markov's work appears in *Mathematische Annalen* (Leipzig) **15**, 381–407 (1879) and **17**, 379–400 (1880). A quite different treatment was given by J. W. S. Cassels, *Annals of Mathematics* **50**, 676–685 (1949).

SUPPLEMENTARY READING

Chapters 1–5

DAVENPORT, H., *The Higher Arithmetic*, London: Hutchinson & Co. (Publishers), Ltd., 1952.

DICKSON, L. E., *History of the Theory of Numbers*, Washington: Carnegie Institution of Washington, 1919. Reprinted, Chelsea Publishing Company, New York, 1950.

GRIFFIN, H., *Elementary Theory of Numbers*, New York: McGraw-Hill Book Company, Inc., 1954.

HARDY, G. H., AND E. M. WRIGHT, *An Introduction to the Theory of Numbers*, 3rd edition, New York: Oxford University Press, 1954.

JONES, B. W., *The Theory of Numbers*, New York: Rinehart & Company, Inc., 1955.

NAGELL, T., *Introduction to Number Theory*, New York: John Wiley & Sons, Inc., 1951.

ORE, Ø., *Number Theory and Its History*, New York: McGraw-Hill Book Company, Inc., 1948.

STEWART, B. M., *Theory of Numbers*, New York: The Macmillan Company, 1952.

VINOGRADOV, I. M., *Elements of Number Theory*, translation of 5th Russian edition, New York: Dover Publications, 1954.

WRIGHT, H. N., *First Course in Theory of Numbers*, John Wiley and Sons, Inc., New York, 1939.

Chapter 6

HARDY AND WRIGHT, *op. cit.*

LANDAU, E., *Handbuch der Lehre von der Verteilung der Primzahlen*, vol. 1, Leipzig: Teubner Verlagsgesellschaft, 1909. Reprinted, Chelsea Publishing Company, New York, 1953.

LANDAU, E., *Vorlesungen über Zahlentheorie*, vol. 1, part 1, Leipzig: S. Hirzel Verlag, 1927. Reprinted as *Elementare Zahlentheorie*, Chelsea Publishing Company, New York, 1950.

Chapter 7

HARDY AND WRIGHT, *op. cit.*

LANDAU, E., *Vorlesungen über Zahlentheorie.*

Chapter 8

CAHEN, E., *Théorie des Nombres*, Paris: Hermann & Cie., 1914–1924.

LANDAU, E., *Vorlesungen über Zahlentheorie.*

NAGELL, T., *op. cit.*

Chapter 9

HARDY AND WRIGHT, *op. cit.*

KOKSMA, J. F., *Diophantische Approximationen*, Berlin: Springer-Verlag OHG, 1936. (Ergebnisse der Mathematik, vol. 4, no. 4.) Reprinted, Chelsea Publishing Company, New York, 1951.

PERRON, O., *Die Lehre von den Kettenbrüchen*, 3rd edition, Stuttgart: Teubner Verlagsgesellschaft, 1954. Second edition reprinted, Chelsea Publishing Company, New York, 1950.

ZUELLIG, J., *Geometrische Deutung Unendlicher Kettenbrüche*, Zürich: Orell Füssli Verlag, 1928.

LIST OF SYMBOLS

$\pi(x)$, number of primes not exceeding x, 3

$\mid$, $\nmid$, divides, does not divide, 14

GCD, greatest common divisor, 14

(a, b), GCD of a and b, 14

LCM, least common multiple, 23

$\langle a, b \rangle$, LCM of a and b, 23

$\equiv \pmod{m}$, 24

$\varphi(m)$, Euler's function, 28

$\mathrm{ord}_m\, a$, order of $a \pmod{m}$, 43

(a/p), Legendre symbol, 45

$\parallel$, exactly divides, 52

$\lambda(m)$, 53

$\mathrm{ind}_g\, a$, 56

γ, Euler's constant, 75

(a/b), Jacobi symbol, 77

$\sigma(n)$, sum of divisors of n, 81

$\tau(n)$, number of divisors of n, 81

$\mu(n)$, Möbius function, 86

$[x]$, greatest integer not exceeding x, 89

O, o, 92

$\sim$, 92

$\zeta(s)$, Riemann's function, 119

$P_2(n)$, 128

$R[i]$, Gaussian integers, 129

$\mathbf{N}$, norm of a Gaussian integer, 129

$r_2(n)$, 132

$R[\sqrt{d}\,]$, quadratic integers, 138

$M(\xi)$, Markov's constant, 149

F_n, Farey sequence, 154

$R_N(p, q)$, 161

P_k/Q_k, best approximations, 163

p_k/q_k, convergents, 170

$\cong$, equivalent real numbers, 187

196

INDEX

Errata in **Topics in Number Theory**, Vol. I

page	line	replace	with
11	7	1–1,	1–1, there are c_n and r such that
14	−5	$0 \le r_1 < a$	$0 \le r_1 < b$
15	9	(b)	(c)
16	12	$m \ne 0$	$m > 0$
16	13	*add at end*	if $m > 0$
18	6	*insert at beginning*	positive
18	−10	of primes	of positive primes
21	−1	*add at end*	and x_0, y_0 is a particular solution.
22	−5	the integers	the nonnegative integers
67	−4	(mod m),	(mod m) and prime to m,
37	17	some $\beta < \alpha$	some β with $0 < \beta < \alpha$
40	−15	$(x - a)\mid f(x)$	$f(a) \equiv 0$
43	−13	3–16	3–17
43	−5	always	for $m > 1$
45	−14	*omit* (	
85	4	4–2	4–1
55	−15	$r \ge 1$	$r \ge 0$
58	−14	$\le \frac{1}{2}\varphi(m)$	$= \varphi(m)$

60	−9	*insert at end*	(Here, contrary to our convention, the letter e stands for the noninteger $\sum 1/n! = 2.718\ldots$)
61	12	$\mu = 1,\ 2,\ ,\ldots,\ n_1$	$\mu = 1,\ 2,\ ,\ldots,\ m_1$
62	−1	equation (12) is solvable if	the congruence (12) is solvable with $p \nmid xyz$ if
69	11	property.	property, except x itself.
69	−17	*add at end*	(Postpone this problem to Sec. 5–4.)
70	−6	$\dfrac{qx}{p} - \dfrac{1}{2} < \dfrac{q-1}{2}$	$\dfrac{qx}{p} + \dfrac{1}{2} < \dfrac{q+1}{2}$
71	5	$-2/q$	$-q/2$
72	−12	$4q' + a'$	$4qk' + a'$
77	−7	1 and odd, put	1, odd, and prime to a, put
78	18	number	number, relatively prime to the first.
85	−10	of integers	of positive integers
89	1	equal or distinct	(not necessarily distinct)
89	−13	*insert at beginning*	exists and
90	10	multiples	positive multiples
91	14	greater	smaller
92	−5	all x	all sufficiently large x
100	3	$[\log x]$	$[\log x] + 1$
103	7	sequence	sequence of real numbers
108	4	6–14	6-16

109	-12	$c_{12} = e^A$	$c_{12} = e^{2A}$
111	-5	6–13	6–17
113	-6	$p_r^{\alpha_r}$	$p_r^{\alpha_r \delta}$
115	-3	*insert at end*	$= \displaystyle\prod_{k=2}^{n} \frac{k-1}{k} \cdot \frac{k+1}{k} = \frac{1}{n} \cdot \frac{n+1}{2} > \frac{1}{2}$
115	$-2ff$	*omit remainder of proof*	
119	-11	$\displaystyle\sum_1^n \frac{1}{x}$	$\displaystyle\sum_1^{\sqrt{n}} \frac{1}{x}$
121	-11	$O(1)$	$O(\log x)$
121	-8	$n = 1$ *(3 occurrences)*	$n = 0$
124	15	given;	given by Erdös;
128	15, 16	representations *(2 occurrences)*	proper representations
129	-12	units,	units, and is not itself a unit,
132	-13	$0 \leq v \leq u$	$0 \leq v \leq 3$
134	-13	for x	for x_σ
138	14	If the leading	If $b = 0$ or the leading
141	-8	some a	some rational number a
149	11	$y^2|a| \cdot |(\xi - \xi')/2|$	$y^2|a| \cdot |3(\xi - \xi')/2|$
149	13	*replace the argument of* min *by*	$\left(\dfrac{2}{|\xi - \xi'|}, \dfrac{(3+\epsilon)|a| \cdot |\xi - \xi'|}{2} \right)$

150	-6	*replace displayed equations by*	The product of the two quantities in (16) is constant,
150	-5	the quantities in (16)	them
151	2	(18).	(18) in which $a\|y_2$.
151	-6	dy^2	$\sqrt{d}\,y^2$
164	3	$k = 0,\ 1,\ \ldots$	$k = -1,\ 0,\ 1,\ \ldots$
167	3	$Q_{k-1}x_k + P_{k-2}$	$Q_{k-1}x_k + Q_{k-2}$
174	-3	$(-1)^k$	$(-1)^{k-1}$
175	8	$1 + \dfrac{\sqrt{7}-1}{3}$ *(2 occurrences)*	$\dfrac{\sqrt{7}-2}{3}$
175	9	$\dfrac{3}{\sqrt{7}-1}$	$\dfrac{3}{\sqrt{7}-2}$
177	7	$\dfrac{-1+\sqrt{3}}{2}$	$\dfrac{1+\sqrt{3}}{2}$
177	9	*the whole line*	$\xi = 1 + \dfrac{1}{3 + \frac{1}{\frac{\sqrt{3}+1}{2}}} = 3 - \sqrt{3}.$
180	10	ξ_{1n+k}	ξ_{n_1+k}
190	-17	one set	one infinite set
190	-15	every subset	every infinite subset

TOPICS IN
NUMBER THEORY

VOLUME II

PREFACE

This book is a treatment of some advanced topics in the theory of numbers. It was written to follow the author's *"Topics in Number Theory, Volume I,"* in which elementary number theory is presented. The level of mathematical maturity required for Volume II is much higher than for Volume I. Moreover, results obtained in Volume I are used freely, and in several of the chapters a knowledge of specific topics in various other branches of mathematics is assumed. In particular, knowledge of the theory of symmetric polynomials, as well as the rule for multiplying determinants, is needed for the algebraic theory in Chapter 3, and the theory of analytic functions is used both in the theorem of Schneider in Chapter 5 and in the investigation of the distribution of primes in Chapter 7. There seemed to be no point in assuming background unnecessarily, however, so I have included brief discussions of groups and matrices, on a very elementary level, in Chapter 1.

The treatment of quadratic forms, admittedly shallow, has been based on the properties of the modular group for two reasons. In the first place, the geometric interpretation makes the usual definition of reduced forms seem quite natural, while no real insight is afforded by merely listing an unmotivated set of inequalities. In the second place, this treatment provides a simple illustration of the power of the theory associated with elliptic functions, which is of considerable importance in modern number theory. Such methods are not often taught in American universities, and I hope that this treatment may serve to stimulate interest in them.

To the best of my knowledge, the algebraic form of the Thue-Siegel-Roth theorem given in Chapter 4 has not previously appeared in print.

W. J. L.

Ann Arbor, Michigan
November 1955

CONTENTS

CHAPTER 1

BINARY QUADRATIC FORMS

1-1 Introduction. One of the subjects treated in elementary number theory is the possibility of representing a positive integer as a sum of two squares.* The expression $x^2 + y^2$ which is of interest for this problem is a special case of the general *binary quadratic form*

$$f(x, y) = ax^2 + bxy + cy^2. \tag{1}$$

(This in turn is a special case of the n-ary m-ic form, which is a homogeneous polynomial of degree m in n variables.) Systematic research in quadratic forms was begun by Gauss, and has since been extensively pursued. We shall not go very deeply into the subject, but prefer instead to develop general methods whose usefulness is not limited to the theory of quadratic forms, nor even to the theory of numbers.

Suppose that in (1) we make the linear homogeneous substitution

$$\begin{aligned} x &= \alpha x' + \beta y', \\ y &= \gamma x' + \delta y', \end{aligned} \tag{2}$$

where α, β, γ, and δ are integers and $D = \alpha\delta - \beta\gamma \neq 0$. Solving for x' and y' gives

$$\begin{aligned} x' &= \frac{\delta}{D} x - \frac{\beta}{D} y, \\ y' &= -\frac{\gamma}{D} x + \frac{\alpha}{D} y, \end{aligned} \tag{3}$$

so it is only in case $D = \pm 1$ that to each integer pair x, y corresponds an integer pair x', y' and conversely. We shall eventually suppose

---

*See, for example, LeVeque, *Topics in Number Theory*, vol. I, (Reading, Mass.: Addison-Wesley Publishing Company, Inc., 1956), Chapter 7. So much use will be made of the results obtained in this book that it will be referred to henceforth simply as Volume I.

that $D = +1$, for reasons that will appear later. Then

$$x' = \delta x - \beta y,$$
$$y' = -\gamma x + \alpha y. \tag{4}$$

Substituting (2) into (1), we have a new form in x' and y',

$$g(x', y') = Ax'^2 + Bx'y' + Cy'^2,$$

where

$$A = a\alpha^2 + b\alpha\gamma + c\gamma^2,$$
$$B = 2a\alpha\beta + b(\alpha\delta + \beta\gamma) + 2c\gamma\delta, \tag{5}$$
$$C = a\beta^2 + b\beta\delta + c\delta^2.$$

If for suitable integral values of x and y we have $f(x, y) = n$, then, for the corresponding values of x' and y' determined by equation (4), $g(x', y') = n$.

It thus appears that, as far as questions of representation are concerned, it would be senseless duplication to consider $f(x, y)$ and $g(x', y')$ separately; every integer represented by f is also represented by g, and conversely. This leads us to call f and g *equivalent*, and to write $f \sim g$, if one can be obtained from the other by a unimodular linear substitution with integral coefficients,

$$x = \alpha x' + \beta y', \qquad \alpha\delta - \beta\gamma = 1. \tag{6}$$
$$y = \gamma x' + \delta y',$$

This in turn brings up one of the principal problems of this chapter: how to decide whether two given forms are equivalent.

The substitution (2) is described quite adequately by specifying the coefficients α, β, γ, and δ; that is, by writing the *matrix*

$$\begin{pmatrix} \alpha & \beta \\ \gamma & \delta \end{pmatrix}.$$

This symbol does not represent a number, of course; it is simply a list of the coefficients of the substitution, in the order in which they occur in (2). However, we can give names to these matrices, and deduce certain of their properties from the corresponding properties of the associated substitutions. Thus if

$$M = \begin{pmatrix} \alpha & \beta \\ \gamma & \delta \end{pmatrix} \qquad \text{and} \qquad M' = \begin{pmatrix} \alpha' & \beta' \\ \gamma' & \delta' \end{pmatrix},$$

then we shall say that M and M' are equal if and only if they correspond to the same substitution, that is,

$$\alpha = \alpha', \qquad \beta = \beta', \qquad \gamma = \gamma', \qquad \delta = \delta'.$$

If for arbitrary M and M' we apply the corresponding substitutions successively, so that

$$x = \alpha x' + \beta y', \qquad x' = \alpha' x'' + \beta' y'',$$
$$y = \gamma x' + \delta y', \qquad y' = \gamma' x'' + \delta' y'',$$

we could accomplish the same thing by the single substitution

$$x = (\alpha\alpha' + \beta\gamma')x'' + (\alpha\beta' + \beta\delta')y'',$$
$$y = (\gamma\alpha' + \delta\gamma')x'' + (\gamma\beta' + \delta\delta')y''.$$

Thus, if by the *product* MM' of two matrices we mean the matrix of this latter substitution, we must define

$$\begin{pmatrix} \alpha & \beta \\ \gamma & \delta \end{pmatrix}\begin{pmatrix} \alpha' & \beta' \\ \gamma' & \delta' \end{pmatrix} = \begin{pmatrix} \alpha\alpha' + \beta\gamma' & \alpha\beta' + \beta\delta' \\ \gamma\alpha' + \delta\gamma' & \gamma\beta' + \delta\delta' \end{pmatrix}.$$

Thus the product has as element in the ith row and jth column, for each i and j, the sum of the products of the elements of the ith row of the first matrix with the corresponding elements of the jth column of the second matrix. Moreover, if the *determinant* of a matrix is defined as

$$\det \begin{pmatrix} \alpha & \beta \\ \gamma & \delta \end{pmatrix} = \begin{vmatrix} \alpha & \beta \\ \gamma & \delta \end{vmatrix} = \alpha\delta - \beta\gamma,$$

it requires only a routine calculation to show that

$$\det M \cdot \det M' = \det (MM').$$

It is to be noticed that, in general, $MM' \neq M'M$, although

$$M(M'M'') = (MM')M''.$$

Since the substitutions given by (2) and (3) are inverse to each other, it is natural to call the matrix of (3) the *inverse* of the matrix M of (2), and to designate it by M^{-1}. Then $MM^{-1} = M^{-1}M = I$, where

$$I = \begin{pmatrix} 1 & 0 \\ 0 & 1 \end{pmatrix}. \tag{7}$$

I is called the *identity matrix;* it corresponds to the trivial substitution $x = x'$, $y = y'$, and has the property that $MI = IM = M$ for every M. A square matrix has an inverse if and only if its determinant is different from zero.

Finally, we designate by $\bar{M}$ the *transpose* of M, obtained by interchanging rows and columns in M:

$$\text{if } M = \begin{pmatrix} \alpha & \beta \\ \gamma & \delta \end{pmatrix}, \quad \text{then } \bar{M} = \begin{pmatrix} \alpha & \gamma \\ \beta & \delta \end{pmatrix}.$$

The transpose of a product is the product of the transposes, in reverse order:

$$(\overline{MM'}) = \bar{M}'\bar{M}.$$

Also, the transpose of the inverse is equal to the inverse of the transpose:

$$(\bar{M})^{-1} = \overline{M^{-1}}.$$

Matrices need not be square. Thus

$$\text{if } X = (x\ y), \quad \text{then } \bar{X} = \begin{pmatrix} x \\ y \end{pmatrix};$$

note, however, that nonsquare matrices have neither determinants nor inverses.

The importance of this algebra of matrices to our present purpose lies in the fact that if

$$F = \begin{pmatrix} a & \frac{1}{2}b \\ \frac{1}{2}b & c \end{pmatrix},$$

then

$$XF\bar{X} = (x\ y) \begin{pmatrix} a & \frac{1}{2}b \\ \frac{1}{2}b & c \end{pmatrix} \begin{pmatrix} x \\ y \end{pmatrix} = (ax + \tfrac{1}{2}by \quad \tfrac{1}{2}bx + cy) \begin{pmatrix} x \\ y \end{pmatrix}$$

$$= (ax^2 + bxy + cy^2).$$

Although it is a slight abuse of language, it is convenient and in the present context harmless to identify a one-by-one matrix with the element itself, so we write

$$f(x, y) = XF\bar{X}.$$

F is called the *matrix* of the form, and $\Delta = 4 \cdot \det F = 4ac - b^2$ is called the *discriminant* of the form.

In terms of matrices, the substitution equations (2) and (3) can be written as

$$X = X'\overline{M} \qquad \text{and} \qquad X' = X\overline{M}^{-1},$$

respectively. Thus

$$f(x, y) = XF\overline{X} = (X'\overline{M})F(\overline{X'\overline{M}}) = (X'\overline{M})F(M\overline{X'})$$
$$= X'(\overline{M}FM)\overline{X'},$$

so that the matrix of g is $G = \overline{M}FM$. (The reader might test his ability to manipulate matrices by showing that the last equation is in agreement with equations (5)). Multiplying both sides of the equation $G = \overline{M}FM$ by $\overline{M}^{-1}$ on the left and M^{-1} on the right, we have

$$\overline{M}^{-1}GM^{-1} = \overline{M}^{-1}(\overline{M}FM)M^{-1} = (\overline{M}^{-1}\overline{M})F(MM^{-1}) = F.$$

If $\det M = 1$, then also $\det \overline{M} = 1$, and

$$\det G = \det (\overline{M}FM) = \det \overline{M} \cdot \det F \cdot \det M = \det F,$$

so that the discriminant of a form is not changed by a unimodular substitution.

In summary, a form with matrix F is equivalent to a form with matrix G if and only if there is a matrix M such that $G = \overline{M}FM$ and $\det M = +1$; equivalent forms have the same discriminant and represent the same integers.

The relation of "equivalence," as used here, is an equivalence relation in the technical sense.* For it is clear that

(a) $f \sim f$: $F = \overline{I}FI$;

(b) $f \sim g$ implies $g \sim f$: $G = \overline{M}FM$ implies $F = \overline{M}^{-1}GM^{-1}$;

(c) $f \sim g$ and $g \sim h$ implies $f \sim h$: $G = \overline{M}FM$ and $H = \overline{M'}GM'$
implies $H = \overline{M'}\overline{M}FMM' = (\overline{MM'})F(MM')$.

Thus all the forms equivalent to a given one are equivalent to each other, and the set of all forms splits up into equivalence classes, any two elements in one class being equivalent, and elements from different classes being inequivalent. (The equivalence classes for the relation of congruence (mod m) are simply the residue classes modulo m.) Just as we chose a system of representatives of the various residue classes modulo m, we would like to pick a system of representative forms, one from each class. It is the object of the next two sections to develop machinery by which such *reduced* forms can be obtained in a natural way.

---

*See, for example, Volume I, Section 3–1.

1. Give proofs of the following statements, for the case of two-by-two matrices:

(a) for some M and M', $\;MM' \neq M'M$,

(b) $M(M'M'') = (MM')M''$,

(c) $MM^{-1} = M^{-1}M = I, \quad IM = MI = M$,

(d) $\overline{MM'} = \overline{M}'\overline{M}, \quad (\overline{M})^{-1} = \overline{M^{-1}}$.

2. Verify directly, and also by matrix multiplication, that under the substitution

$$x = 2x' + 3y', \qquad y = x' + 2y',$$

the form $F(x, y) = 3x^2 - 7xy + 4y^2$ goes into $G(x', y') = 2x'^2 + 3x'y' + y'^2$. Compute the inverse of the matrix of the substitution, and so carry G back into F.

1–2 Groups. We say that a set G of elements $a, b, \ldots$, which need not be numbers, forms a *group* with respect to a certain operation (designated for the moment by the symbol "$\circ$"), which combines two elements to form a third, if

(a) for every a and b in G, $a \circ b$ is in G,

(b) the operation is *associative*, so that $a \circ (b \circ c) = (a \circ b) \circ c$, for every a, b, and c in G,

(c) there is an *identity* element e, such that $a \circ e = e \circ a = a$ for every a in G,

(d) every a in G has an *inverse* a^{-1} in G, such that

$$a \circ a^{-1} = a^{-1} \circ a = e.$$

Perhaps the simplest example of a group is the set of all integers, under the operation addition. In this case, the number 0 is the identity, since $a + 0 = 0 + a = a$, and the inverse of a is $-a$, since $a + (-a) = (-a) + a = 0$. This group is infinite (i.e., has infinitely many elements), but the group consisting of the four numbers $1, i, -1$, and $-i$, under the operation multiplication, is finite. If the operation consists of adding two integers and reducing the result to the least positive residue (mod m), then the numbers $1, 2, \ldots, m$ form a group, with m as the identity. Instead of a complete residue system we could consider a reduced residue system (mod m); these numbers form a group $M(m)$ in which the operation is ordinary multiplication followed by reduction (mod m). This is not quite so obvious; the identity is clearly the number $e \equiv 1 \pmod{m}$, but the existence of inverses depends on the fact that the congruence $ax \equiv 1 \pmod{m}$ is solvable if $(a, m) = 1$. Many of the results in the congruence theory which are obtained in beginning texts are simply special cases of general theorems about finite groups; it might be of interest to examine this relationship briefly before proceeding.

A subset G_1 of the elements of a group G is said to form a *subgroup* of G if it also forms a group with respect to the operation of G. Condition (b) is automatically satisfied in this case, so that one need only verify that (a) holds (which we express by saying that G_1 is closed under the operation), that the identity is in G_1, and that the inverse of each element of G_1 is in G_1.

The number of elements in a finite group is called the *order* of the group; we now show that the order of a subgroup G_1 divides the order of the group G. Suppose that a is an element of G which is not in G_1, and let aG_1 be the set of all "products" $a \circ g$, where g runs through G_1. Then no element $a \circ g$ is in G_1, for if it were, the same would be true of $a \circ g \circ g^{-1} = a$. Also, if g_1 and g_2 are distinct elements of G_1, then $a \circ g_1 \neq a \circ g_2$, since otherwise $a^{-1} \circ a \circ g_1 = a^{-1} \circ a \circ g_2$, or $g_1 = g_2$. Now suppose that b in G is not in either G_1 or aG_1; then as before, no element of bG_1 is in G_1 or aG_1, and elements of bG_1 arising from different elements of G_1 are different. This process can be continued until every element of G is in precisely one of the sets $G_1, aG_1, bG_1, \ldots$, and each set contains exactly m distinct elements, where m is the order of G_1. Clearly, if there are t such sets, then the order of G is mt, which is divisible by m.

If a is any element of G, then the powers of a (that is, a, $a^2 = a \circ a$, $a^3 = a \circ a \circ a, \ldots$) are all in G; we easily see that if G is finite, these powers form a subgroup, whose order m is such that $a^m = e$. Hence, if the order of G is mt, then $a^{mt} = e$. For the group $M(m)$ defined above, whose order is $\varphi(m)$, this statement reduces to Euler's theorem, which states that $a^{\varphi(m)} \equiv 1 \pmod{m}$ if $(a, m) = 1$. Many of the other results of Volume I follow immediately from these remarks about groups. For example, Theorem 3–19, which says that if a belongs to $t \pmod{m}$ then $t | \varphi(m)$, can be reworded in the language of groups to read, "The order of the cyclic subgroup generated by a (i.e., the group of powers of a) divides the order of the group." A primitive root of q is a generator of $M(q)$; thus Theorem 4–11 is a statement of the fact that $M(q)$ is cyclic (consists of the powers of a single element) if and only if $q = 1, 2, 4, p^\alpha, 2p^\alpha$. If $\alpha > 2$, $M(2^\alpha)$ has two generators, -1 and 5; for example, modulo 16, the powers of 5 are 5, 9, 13, and 1, and these numbers, together with their negatives, form a reduced residue system.

At present we shall do no more with finite groups, but turn our attention instead to the much more complicated multiplicative group of all two-by-two matrices with integral entries and unit determinants. This infinite group, which will be designated here by Γ, is called the *modular group*. To show that Γ is a group, we verify properties (a) through (d) above. The system is obviously closed under multiplication, since the determinant of a product is the product of the determinants of the factors. The associative property has already been verified. The identity element of Γ is I, as defined in (7). The inverse of any element

$$\begin{pmatrix} \alpha & \beta \\ \gamma & \delta \end{pmatrix}$$

is

$$\begin{pmatrix} \delta & -\beta \\ -\gamma & \alpha \end{pmatrix},$$

since

$$\begin{pmatrix} \alpha & \beta \\ \gamma & \delta \end{pmatrix}\begin{pmatrix} \delta & -\beta \\ -\gamma & \alpha \end{pmatrix} = \begin{pmatrix} \delta & -\beta \\ -\gamma & \alpha \end{pmatrix}\begin{pmatrix} \alpha & \beta \\ \gamma & \delta \end{pmatrix} = I.$$

The group Γ differs from the other examples mentioned in that it is noncommutative, since in general $MM' \neq M'M$. (Abstractly, G is said to be a *commutative* or *abelian* group if $a \circ b = b \circ a$ for every a and b in G.)

1–3 The modular group. The properties of Γ could all be developed by the use of algebra alone; we prefer instead to build up the theory with the help of a simple geometric interpretation. It is now convenient to reverse the roles of the accented and unaccented variables in the equations (2); this new notation will be used throughout the discussion of the modular group, but the original system will be reverted to when quadratic forms are again considered. To keep matters straight, (2) will be termed a *substitution*, while the modified equations will be called a *transformation*. Putting $z = x/y$ and

$z' = x'/y'$, we get

$$z' = \frac{\alpha z + \beta}{\gamma z + \delta}. \tag{8}$$

So far nothing essential has been accomplished. The crucial point lies in allowing z to range over all complex numbers, rather than the real rationals to which it was formerly restricted. Then equation (8) can be regarded as defining a transformation or mapping of the complex z-plane into the z'-plane. Somewhat more than this can be said: if

$$M = \begin{pmatrix} \alpha & \beta \\ \gamma & \delta \end{pmatrix}$$

is in Γ, so that det $M = 1$, a simple calculation shows that the imaginary parts of z and z' have the same sign. In other words, (8) maps the upper half of the z-plane (i.e., the region where the imaginary part of z is positive) into the upper half of the z'-plane, and the lower half into the lower half. Hereafter, we restrict attention to the upper half planes.

It is convenient to identify the z- and z'-planes, and to think of (8) as sending each point z of the upper half U of the complex plane into another point z' of U. We also identify the elements of Γ with the corresponding transformations (8), which has the effect of identifying the matrices

$$\begin{pmatrix} \alpha & \beta \\ \gamma & \delta \end{pmatrix} \quad \text{and} \quad \begin{pmatrix} -\alpha & -\beta \\ -\gamma & -\delta \end{pmatrix}.$$

In accordance with the earlier definition of equivalence, two points z and z' of U will be called *equivalent* if one can be mapped into the other by a transformation of Γ. As usual, this assigns each point of U to an equivalence class; two elements of the same class are equivalent, and elements from different classes are inequivalent. A region R of U is called a *fundamental region* if no two of its points are equivalent, while every point of U is equivalent to a point of R; in other words, R constitutes a complete system of representatives of the above equivalence classes. It would be more precise to refer to R as a *fundamental region of the group* Γ, since two points may be equivalent with respect to one group of transformations but not with respect to another. For example, it is clear that a fundamental region R' of a subgroup Γ' of Γ contains a fundamental region of Γ itself, if both regions exist. For any point in U, being equivalent to some point of R' under the transformations of Γ', is *a fortiori* equivalent to the same point of R' under the transformations of the larger group Γ. It may not be true, however, that any two points of R' are inequivalent with respect to Γ.

THEOREM 1–1. *The region R in U composed of all points z such that* $-\frac{1}{2} \le \operatorname{Re} z < \frac{1}{2}$ *and either* $|z| > 1$, *or else* $|z| = 1$ *and* $-\frac{1}{2} \le \operatorname{Re} z \le 0$, *is a fundamental region of Γ.* (See Fig. 1–1.)

Proof: First note that Γ has the subgroup Γ_0 of all integral translations $z' = z + \beta$. For the associated matrix

$$\begin{pmatrix} 1 & \beta \\ 0 & 1 \end{pmatrix}$$

has determinant 1, the identity transformation $z' = z$ is in Γ_0, the inverse of $z' = z + \beta$ is $z' = z - \beta$ and is in Γ_0, and the result of making two translations is again a translation. Γ_0 is cyclic, being generated by

$$z' = z + 1. \tag{9}$$

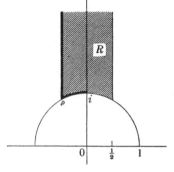

FIGURE 1–1

As a fundamental region of Γ_0 we could choose any infinite strip in U of unit width, extending parallel to the imaginary axis from the real axis. We take the following one:

$$R_0: \qquad \operatorname{Im} z > 0, \quad -\tfrac{1}{2} \le \operatorname{Re} z < \tfrac{1}{2}.$$

From the remark preceding the theorem, R_0 must contain a fundamental region of Γ if any exists. R_0 is not itself a fundamental region of Γ, however, for the point $i/2$ of R_0 is transformed into the point $2i$ of R_0 by the transformation

$$z' = -\frac{1}{z}. \tag{10}$$

With each transformation

$$T: \qquad z' = \frac{\alpha z + \beta}{\gamma z + \delta}$$

with $\gamma \neq 0$, there is associated the circle $C(T)$: $|\gamma z + \delta| = 1$, with center at $-\delta/\gamma$ and radius $1/|\gamma| \leq 1$. Now

$$\gamma z' - \alpha = \gamma \frac{\alpha z + \beta}{\gamma z + \delta} - \alpha = \frac{\beta\gamma - \alpha\delta}{\gamma z + \delta} = -\frac{1}{\gamma z + \delta},$$

so that $C(T)$ is transformed by T into $|\gamma z - \alpha| = 1$, which, by (3), is $C(T^{-1})$. More importantly, the exterior of $C(T)$ goes into the interior of $C(T^{-1})$. It is simple to deduce from this that no two points of the region R described in the theorem are equivalent. Certainly no point of R is mapped into another by an element of Γ_0. But if T is not in Γ_0, then $\gamma \neq 0$, and since the interior of R is external to all the circles $C(T)$ (inasmuch as they all have radii ≤ 1 and are centered at real points), any interior point of R is mapped by T into an interior point of one of these circles, and hence into a point outside R.

The arc A: $|z| = 1$, $-\frac{1}{2} \leq \text{Re } z \leq 0$, which forms part of the boundary of R, is also completely exterior to all the circles $C(T)$ except $|z| = 1$ and $|z + 1| = 1$. The circle $|z| = 1$ is associated with transformations

$$z' = \frac{\alpha z + \beta}{z},$$

and since the determinant must be 1, $\beta = -1$ and

$$z' = \frac{\alpha z - 1}{z} = \alpha - \frac{1}{z}, \qquad |z' - \alpha| = \frac{1}{|z|}.$$

If z is a point of A, $|z' - \alpha| = 1$, and so z' is not in R unless $\alpha = 0$ or -1. If $\alpha = 0$ we have the transformation (10), which sends A onto the arc $|z| = 1, 0 \leq \text{Re } z \leq \frac{1}{2}$; this arc has only the point i in common with R, and i goes into itself. (This means that i is equivalent to itself in two different ways: $z' = z$ and $z' = -1/z$.) If $\alpha = -1$, A goes into the arc $|z + 1| = 1$, $-1 \leq \text{Re } z \leq -\frac{1}{2}$; these two arcs have just $\rho = -\frac{1}{2} + i\sqrt{3}/2$ in common, and ρ goes into itself.

The circle $|z + 1| = 1$ is associated with transformations

$$z' = \frac{\alpha z + \beta}{z + 1} = \frac{\alpha z + (\alpha - 1)}{z + 1} = \alpha - \frac{1}{z + 1}.$$

If $|z| = 1$, then

$$\left| \frac{z' - (\alpha - 1)}{z' - \alpha} \right| = 1,$$

$$|z' - (\alpha - 1)| = |z' - \alpha|,$$

$$\operatorname{Re} z' = \alpha - \tfrac{1}{2},$$

and z' is not in R unless $\alpha = 0$. Under the transformation

$$z' = -\frac{1}{z + 1},$$

the arc A goes into the line segment $\operatorname{Re} z = -\tfrac{1}{2}$, $\tfrac{1}{2} \leq \operatorname{Im} z \leq \sqrt{3}/2$; the arc and the segment have just ρ in common, and ρ goes into itself. We have thus shown that no two points of R are equivalent, and have incidentally obtained the following result, which will be useful later.

THEOREM 1–2. *The point* $\rho = (-1 + i\sqrt{3})/2$ *is mapped into itself by the three transformations*

$$z' = z, \qquad z' = -\frac{1}{1 + z}, \qquad and \qquad z' = -\frac{z + 1}{z},$$

and by no others. The point i is mapped into itself by the two transformations

$$z' = z \qquad and \qquad z' = -\frac{1}{z},$$

and by no others. Any point of R different from ρ and i is mapped into itself only by the identity transformation $z' = z$.

To complete the proof of Theorem 1–1, we must show that any point z in U is the image of a point in R under a transformation of Γ. We do this by finding a finite sequence of transformations such that if they are successively applied to z, the final point z' is in R. Then the inverse of the product of these transformations maps z' back into z.

Designate by S the generator (9) of Γ_0, and by W the transformation (10). Let z be a point of U not in R. Then for some integer n_1, which may be positive, negative, or zero, $z_1 = S^{n_1}z = z + n_1$ is in R_0, the fundamental domain of Γ_0. If z_1 is in R, we are finished. If $|z_1| = 1$ but $0 < \operatorname{Re} z \leq \tfrac{1}{2}$, then Wz_1 is in R. If $|z_1| < 1$, then $z_2 = Wz_1$ has modulus greater than 1. In fact, if $z_1 = x_1 + iy_1$,

$$z_2 = \frac{-x_1 + iy_1}{x_1^2 + y_1^2} = x_2 + iy_2, \qquad -\tfrac{1}{2} \le x_1 < \tfrac{1}{2},$$

so that $\operatorname{Im} z_2 > \operatorname{Im} z_1$, and if $y_1 \le \tfrac{1}{2}$, then $\operatorname{Im} z_2 \ge 2 \operatorname{Im} z_1$, since then $x_1^2 + y_1^2 \le \tfrac{1}{2}$. If z_2 is in R, we are through. If not, there is a suitable exponent n_2 such that $z_3 = S^{n_2}z_2$ is in R_0, and $\operatorname{Im} z_3 = \operatorname{Im} z_2$. If z_3 is not in R, we can apply W again, and get $z_4 = WS^{n_2}WS^{n_1}z$. What we must show is that after finitely many steps, this process leads to a point in R.

As long as $y_k \le \tfrac{1}{2}$, we will have $y_{k+1} \ge 2y_k$, if $z_{k+1} = Wz_k$. Starting with a positive number (the imaginary part of z), a finite number of doublings will produce a number larger than $\tfrac{1}{2}$. So suppose that we have obtained a $z_k = x_k + iy_k$ such that

$$-\tfrac{1}{2} \le x_k < \tfrac{1}{2}, \qquad y_k > \tfrac{1}{2}, \qquad x_k^2 + y_k^2 < 1. \tag{11}$$

Then

$$z_{k+1} = -\frac{1}{z_k}, \qquad z_{k+2} = -\frac{1}{z_k} + n,$$

where n is so determined that $-\tfrac{1}{2} \le x_{k+2} < \tfrac{1}{2}$. This gives

$$z_{k+2} = \frac{nx_k - 1 + iny_k}{x_k + iy_k},$$

so that

$$|z_{k+2}|^2 = \frac{(nx_k - 1)^2 + n^2y_k^2}{x_k^2 + y_k^2}.$$

If $|n| \ge 2$,

$$|z_{k+2}|^2 \ge \frac{4 \cdot \tfrac{1}{4}}{1} = 1,$$

while if $|n| = 1$, the hypothetical inequality $|z_{k+2}|^2 < 1$ gives

$$(x_k - n)^2 + y_k^2 < x_k^2 + y_k^2,$$

which says that z_k is farther from the origin than from the point $z = n$, which is false from the first inequality of (11). Finally, if $n = 0$, then $|z_{k+2}|^2 = 1/|z_k|^2 > 1$. Hence in all cases, $|z_{k+2}| \ge 1$, and $-\tfrac{1}{2} \le \operatorname{Re} z_{k+2} < \tfrac{1}{2}$. If z_{k+2} is still not in R (which may happen if $|z_{k+2}| = 1$) then Wz_{k+2} is in R, and the proof is complete.

Moreover, the proof has shown that S and W are generators of Γ, since every transformation of Γ can be written in the form

$$S_k{}^{n_k}W \ldots WS_2{}^{n_2}WS_1{}^{n_1}.$$

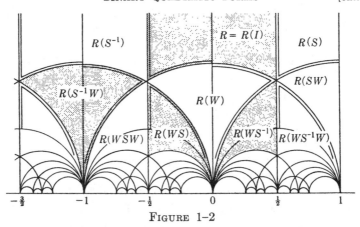

FIGURE 1–2

A geometric representation of the group Γ is given in Fig. 1–2. Here we have considered the region R as the image of itself under the identity transformation I, and have put $R = R(I)$. The congruent unshaded region to the left of R is then $R(S^{-1})$, in the sense that if a point z' of it is equivalent to a point z of R, then $z' = S^{-1}z$. To put it differently, S^{-1} maps R onto this region, just as W maps R onto the unshaded region $R(W)$ immediately below R. The semicircular arcs are portions of the circles $C(T)$; infinitely many of them terminate in each rational point on the real axis. If the drawing and shading were completed, any shaded or unshaded region could be taken as a fundamental region. Each fundamental region or "double triangle" is bounded by three arcs, with vertex angles of 0, $\pi/3$, and $\pi/3$. The heavy arc inside each region indicates the portion of the boundary which is to be included in the region.

PROBLEMS

1. Find the point in R to which the point

$$-\frac{3 + 2i}{8 + 6i}$$

is equivalent, by the method used in the proof of Theorem 1–1. Do you see an easier way, for this particular number?

2. If the term "circle" is used in the broad sense to include straight lines, show that the transformations of Γ send circles into circles. Under what circumstances are the image circles actually lines? What can be said about such a line if, for every point z on the original circle, $\text{Im } z \geq 0$?

3. Verify that, in the notation of the text, $(SW)^3 = I$.

4. Show that the transformations

$$P: \quad z' = 1 - z, \qquad Q: \quad z' = \frac{1}{z}$$

generate a group of six elements (the group of anharmonic ratios) of which a fundamental region is the set of points z such that

$$\operatorname{Im} z > 0, \qquad |z| > 1, \qquad \text{and} \qquad |z - 1| > 1,$$

together with half the boundary of this region, leading from $(1 + i\sqrt{3})/2$ to infinity in one direction. Sketch the analog of Fig. 1–2 for this group. [Note that the transformations of the group do not carry U into itself.]

1–4 Reduced definite forms. With the help of the facts now known about the modular group, we can deal with the question of how to decide whether two given binary quadratic forms are equivalent. We must consider separately the essentially different cases in which the discriminant $\Delta = 4ac - b^2$ of the form $ax^2 + bxy + cy^2$ is positive or negative. (We put aside the degenerate case in which $\Delta = 0$.) If $\Delta > 0$, the form is called *definite*, otherwise *indefinite*. The definite forms can be further classified as *positive* or *negative*, according as $a > 0$ or $a < 0$. The reason for this terminology is that the polynomial $az^2 + bz + c$ associated with a definite form has nonreal zeros, so that the form

$$y^2 \left\{ a \left(\frac{x}{y} \right)^2 + b \frac{x}{y} + c \right\}$$

has the same sign as a for every choice of x and y except $x = y = 0$, while an indefinite form can have values of either sign. We shall first consider definite forms, restricting our attention to positive forms, since the treatment of negative forms is almost identical.

Since the matrix of a form is a little cumbersome, we shall use the symbol $[a, b, c]$ to designate the form $ax^2 + bxy + cy^2$. It is to be clearly understood that this is simply an abbreviation, and cannot be combined with like symbols as matrices can.

Let us consider then a positive definite form $f(x, y) = [a, b, c]$, in which $\Delta > 0$, $a > 0$, and $c > 0$. For the time being, we do not require that a, b, and c be integers. Then the quadratic polynomial

$f(z) = az^2 + bz + c$ has zeros

$$\frac{-b \pm \sqrt{-\Delta}}{2a} \; ;$$

of these, we single out the one with positive imaginary part and call it ω. Thus to the form $[a, b, c]$ there corresponds the point ω in the upper half plane. Conversely, each point in U corresponds to exactly one form of discriminant Δ. For if z_0 is such a point, and $\bar{z}_0$ is its complex conjugate, then there is a unique number $\varkappa$ such that the quadratic expression $\varkappa(z - z_0)(z - \bar{z}_0)$ has discriminant Δ. Hence if we consider only forms of given discriminant Δ (which is all that is required in the equivalence problem, since equivalent forms have the same discriminant), there is a one-to-one correspondence between points of U and forms of that discriminant. Moreover, if the points ω_1 and ω_2 are associated with the forms f_1 and f_2 of discriminant Δ, and if a transformation T of Γ carries f_1 into f_2, then it carries ω_1 into ω_2. It therefore makes no difference whether one speaks of the form f or the point ω, as far as the operations of Γ are concerned. We call ω the *representative* of f.

It should now be clear how to decide whether or not two forms are equivalent. If they do not have the same discriminant, they are not equivalent. If they have, they are equivalent if and only if their representatives are equivalent, and this can be decided by transforming the representatives into the fundamental region R, where they must be identical to be equivalent. This leads us to define a *reduced form* as one whose representative is in R; *reduced forms are equivalent if and only if they are identical, and each class of equivalent forms contains exactly one reduced form.*

Since

$$\omega = \frac{-b + \sqrt{-\Delta}}{2a} = \frac{-b}{2a} + i\frac{\sqrt{\Delta}}{2a},$$

$$|\omega|^2 = \frac{b^2 + \Delta}{4a^2} = \frac{c}{a}, \tag{12}$$

ω is in R if and only if $-\frac{1}{2} \le -b/2a < \frac{1}{2}$, and either $c/a > 1$, or $c/a = 1$ and $-\frac{1}{2} \le -b/a \le 0$. Simplifying, we have that $[a, b, c]$ is reduced if and only if either

$$-a < b \le a < c \qquad \text{or} \qquad 0 \le b \le a = c. \tag{13}$$

Prove the assertion, made in the text, that if ω_1 and ω_2 are the representatives of the forms f_1 and f_2 with discriminant Δ, and if a T in Γ carries f_1 into f_2, it carries ω_1 into ω_2.

1-5 Reduction of definite forms. A given form can be transformed into its equivalent reduced form by exactly the process used in the proof that R is a fundamental region of Γ. That is, by a translation S^{n_1}, ω can be changed into ω', where $-\frac{1}{2} \leq \operatorname{Re} \omega' < \frac{1}{2}$; if ω' is not in R, we begin afresh with $W\omega'$, etc. The translation $z' = z + n_1$ must be such that

$$-\frac{1}{2} \leq -\frac{b}{2a} + n_1 < \frac{1}{2},$$

or

$$b = 2an_1 + r_1,$$

where $-a < r_1 \leq a$. The transformation $z' = z + n_1$ has matrix

$$S^{n_1} = \begin{pmatrix} 1 & n_1 \\ 0 & 1 \end{pmatrix},$$

but we must now revert to the inverse transformation $z = z' - n_1$ to utilize the results of Section 1-1, which were based on the equations (2). If we put

$$M = \begin{pmatrix} 1 & -n_1 \\ 0 & 1 \end{pmatrix} = S^{-n_1},$$

then, as we saw earlier, M carries a form with matrix F into one with matrix

$$G = \overline{M}FM,$$

so that in this case, if we let the result of the first translation be $f_1(x, y) = XF_1\overline{X}$, then

$$F_1 = \begin{pmatrix} 1 & 0 \\ n_1 & 1 \end{pmatrix} F \begin{pmatrix} 1 & -n_1 \\ 0 & 1 \end{pmatrix}.$$

Similarly, if F_2 is the result of applying the inversion W to F_1, then

$$F_2 = \begin{pmatrix} 0 & -1 \\ 1 & 0 \end{pmatrix} F_1 \begin{pmatrix} 0 & 1 \\ -1 & 0 \end{pmatrix}.$$

A simple calculation shows that, if $f_1 = [a, b, c]$, then $f_2 = [c, -b, a]$.

Thus we have the following algorithm for reducing $f = [a, b, c]$: find n_1 and r_1 such that

$$b = 2an_1 - r_1, \qquad -a \leq r_1 < a,$$

and compute $f_1 = [a_1, b_1, c_1]$, where

$$F_1 = \begin{pmatrix} 1 & 0 \\ -n_1 & 1 \end{pmatrix} F \begin{pmatrix} 1 & -n_1 \\ 0 & 1 \end{pmatrix},$$

so that $f_1 = [a, b - 2an_1, n_1{}^2 a - bn_1 + c]$. If f_1 is not reduced, put $f_2 = [c_1, -b_1, a_1] = [a_2, b_2, c_2]$. If f_2 is not reduced, repeat the entire procedure. For some k, f_k will be reduced.

The discussion thus far has been valid for positive definite forms with arbitrary real coefficients. For the remainder of this section and the next, we consider only *integral* forms, that is, those with integral coefficients.

THEOREM 1–3. *There are only finitely many classes of integral definite forms of given discriminant.*

Proof: To each class there belongs just one reduced form $[a, b, c]$ satisfying the conditions (13). Since

$$4a^2 \leq 4ac = \Delta + b^2 \leq \Delta + a^2,$$

the inequality $0 < a \leq \sqrt{\Delta/3}$ holds for each reduced form, so that there are only finitely many possible values of a for fixed Δ. Since $|b| \leq a$, the same is true of b, and for each pair a, b there is at most one integer c such that $4ac - b^2 = \Delta$.

If, for example, $\Delta = 3$, then $0 < a \leq 1$, so that $a = 1$ and hence $b = 0$ or 1; from this it is easily seen that the only integral reduced form of discriminant 3 is $x^2 + xy + y^2$. There is also just one class of discriminant 4, and its reduced form is $x^2 + y^2$.

PROBLEMS

1. Find all reduced integral definite forms of discriminant $\Delta \leq 20$.
2. Find the reduced form equivalent to [117, 103, 100].

1–6 Representations by definite forms. If a transformation of Γ leaves a quadratic form unchanged, it is called an *automorph* of the form. Since an automorph also leaves the representative of the form unchanged, and is the only kind of transformation which does, the following theorem is an easy consequence of Theorem 1–2.

THEOREM 1–4. *The only automorphs of* $a(x^2 + y^2)$ *are*

$$\begin{cases} x = \pm x', \\ y = \pm y', \end{cases} \quad and \quad \begin{cases} x = \pm y', \\ y = \mp x'. \end{cases}$$

The only automorphs of $a(x^2 + xy + y^2)$ *are*

$$\begin{cases} x = \pm x', \\ y = \pm y', \end{cases} \quad \begin{cases} x = \mp y', \\ y = \pm x' \pm y', \end{cases} \quad and \quad \begin{cases} x = \pm x' \pm y', \\ y = \mp x'. \end{cases}$$

Any positive reduced form distinct from these two has only the automorphs

$$\begin{cases} x = \pm x', \\ y = \pm y'. \end{cases}$$

An integer n is said to be *properly representable* by an integral form $[a, b, c]$ of discriminant Δ if there are relatively prime integers α, γ such that $a\alpha^2 + b\alpha\gamma + c\gamma^2 = n$. For such α, γ, there are β_0 and δ_0 such that $\alpha\delta_0 - \beta_0\gamma = 1$, and, in fact,

$$\alpha\delta - \beta\gamma = 1,$$

if, for some integer t,

$$\beta = \beta_0 + \alpha t,$$

$$\delta = \delta_0 + \gamma t.$$

If we make the substitution

$$\begin{aligned} x &= \alpha x' + \beta y', \\ y &= \gamma x' + \delta y', \end{aligned} \tag{14}$$

then $[a, b, c]$ goes into a form $[n, m, l]$ with first coefficient n, by equations (5). Also by (5),

$$\begin{aligned} m &= 2a\alpha(\beta_0 + \alpha t) + b(\alpha\delta_0 + \alpha\gamma t + \beta_0\gamma + \alpha\gamma t) + 2c\gamma(\delta_0 + \gamma t) \\ &= 2a\alpha\beta_0 + b(\alpha\delta_0 + \beta_0\gamma) + 2c\gamma\delta_0 + 2nt, \end{aligned}$$

so that m is determined modulo $2n$. Choose m so that $0 \leq m < 2n$; then t is fixed, β and δ are unique, and l is determined by the discriminant:

$$4ln - m^2 = \Delta.$$

THEOREM 1–5. *Let* α, γ *be a proper representation of* $n > 0$ *by the integral form* $[a, b, c]$ *of discriminant* Δ. *Then there are unique integers* β *and* δ *such that* $\alpha\delta - \beta\gamma = 1$, *and the substitution* (14) *replaces* $[a, b, c]$ *by the equivalent form* $[n, m, l]$, *where* $0 \leq m < 2n$, *m satisfies the congruence*

$$m^2 \equiv -\Delta \pmod{4n}, \tag{15}$$

and

$$l = \frac{m^2 + \Delta}{4n}. \tag{16}$$

Thus to each proper representation of n by $[a, b, c]$ there corresponds a unique form which has first coefficient n and satisfies certain auxiliary conditions. The appropriate converse, which we now consider, gives the number of such representations, and provides an effective method of finding them. If m is a solution of (15) and $0 \leq m < 2n$, then $4n - m$ is also a root, and $2n \leq 4n - m < 4n$. We shall refer to m as a *minimum root* if $0 \leq m < 2n$.

THEOREM 1–6. *Let $w(f)$ be the number of automorphs of $f = [a, b, c]$, an integral positive form of discriminant Δ. Let n be a positive integer. Corresponding to each minimum root m of the congruence (15), determine l by equation (16). Then the number of proper representations of n by f is $w(f)$ times the number of such forms $[n, m, l]$ which are equivalent to f. In particular, if there is only one class of discriminant Δ, the number of proper representations is $w(f)$ times the number of minimum roots of (15).*

Proof: Suppose that $g = [n, m, l]$ is a form of the type described in the theorem. Then if f is not equivalent to g, Theorem 1–5 shows that there is no representation of n by f corresponding to the minimum root m. If f is equivalent to g, let T be the matrix of a substitution which replaces f by g, and let A be the matrix of an automorph of f. Then

$$G = \bar{T}FT \qquad \text{and} \qquad F = \bar{A}FA,$$

so

$$(\overline{AT})F(AT) = \bar{T}\bar{A}FAT = \bar{T}FT = G,$$

so that AT is also the matrix of a substitution which carries f into g. Conversely, if for any U,

$$G = \bar{U}FU,$$

then $\bar{U}FU = \bar{T}FT$, and

$$F = \bar{T}^{-1}\bar{U}FUT^{-1},$$

so that UT^{-1} is the matrix A of an automorph of f, and $U = AT$. Hence there are exactly $w(f)$ substitutions which replace f by g.

If $$T = \begin{pmatrix} \alpha & \beta \\ \gamma & \delta \end{pmatrix},$$

and f has only two automorphs (see Theorem 1–4), then

$$AT = \begin{pmatrix} \alpha & \beta \\ \gamma & \delta \end{pmatrix} \quad \text{or} \quad \begin{pmatrix} -\alpha & -\beta \\ -\gamma & -\delta \end{pmatrix},$$

and α, γ and $-\alpha$, $-\gamma$ give two distinct proper representations, since $(\alpha, \gamma) = 1$ and therefore α and γ are not both zero. If $f \sim a(x^2 + y^2)$, then

$$AT = \begin{pmatrix} \alpha & \beta \\ \gamma & \delta \end{pmatrix} \quad \text{or} \quad \begin{pmatrix} -\alpha & -\beta \\ -\gamma & -\delta \end{pmatrix} \quad \text{or} \quad \begin{pmatrix} -\gamma & -\delta \\ \alpha & \beta \end{pmatrix}$$

$$\text{or} \quad \begin{pmatrix} \gamma & \delta \\ -\alpha & -\beta \end{pmatrix},$$

and the representations α, γ; $-\alpha$, $-\gamma$; $-\gamma$, α; and γ, $-\alpha$ are again distinct. If $f \sim a(x^2 + xy + y^2)$, then AT is one of the matrices

$$\pm \begin{pmatrix} \alpha & \beta \\ \gamma & \delta \end{pmatrix}, \quad \pm \begin{pmatrix} -\gamma & -\delta \\ \alpha + \gamma & \beta + \delta \end{pmatrix} \quad \text{or} \quad \pm \begin{pmatrix} \alpha + \gamma & \beta + \delta \\ -\alpha & -\beta \end{pmatrix},$$

and these also lead to distinct representations.

If there is only one class of discriminant Δ, then f and g are necessarily equivalent, so that all minimum roots of (15) lead to representations. The proof is complete.

In the case of *primitive* forms (those having relatively prime coefficients), $w(f)$ depends only on Δ: $w(f) = 6, 4,$ or 2 according as Δ is 3, 4, or larger than 4. If $f(x, y) = x^2 + y^2$, so that $\Delta = 4$, then m must be even to satisfy (15). Let $m = 2m_1$; then $m_1^2 \equiv -1 \pmod{n}$, and $0 \leq m < 2n$ means $0 \leq m_1 < n$, so that the number of proper representations of n as a sum of two squares is four times the number of solutions of the congruence $u^2 \equiv -1 \pmod{n}$. This result was obtained in Theorem 7–5, Volume I, by quite different methods.

PROBLEMS

1. Find β, δ, m, l of Theorem 1–5 corresponding to the proper representation 3, 5 of 118 by $[2, -5, 7]$.

2. What is the number of proper representations of 28 by [1, 1, 2]? Find them.

3. Use Theorem 1–6 to discuss the proper representability of 10 by [2, 1, 2].

4. Show that every prime congruent to 1 or 3 (mod 8) has a unique proper representation in the form $x^2 + 2y^2$ with $x > 0$, $y > 0$. More generally, show that if n is the product of powers of r such. primes, then n has 2^{r+1} proper representations in this form.

1–7 Indefinite forms. The behavior of indefinite binary forms is remarkably different from that of forms with positive discriminant. For example, any integral indefinite form whose discriminant is not the negative of a square has infinitely many automorphs, and therefore represents any integer in infinitely many ways if it represents it at all. Moreover, there seems to be no natural way to pick out a unique reduced form in each equivalence class, although we shall find a finite set of canonical forms in the case of integral forms.

Hereafter we restrict attention to integral forms [a, b, c], and put

$$D = -\Delta = b^2 - 4ac > 0.$$

If D is a square, then [a, b, c] factors into two linear factors with integral coefficients. We dismiss this degenerate case, and hereafter require that D be a nonsquare integer. Finally, for the sake of simplicity we consider only the case that [a, b, c] is primitive. We see from equations (5) (proof by contradiction) that any form [A, B, C] equivalent to a primitive form is again primitive.

As before, there is associated with [a, b, c] the quadratic equation

$$az^2 + bz + c = 0,$$

which this time has two real roots, say

$$\omega_1 = \frac{-b + \sqrt{D}}{2a}, \qquad \omega_2 = \frac{-b - \sqrt{D}}{2a}.$$

It is easily verified that a transformation of the modular group which sends [a, b, c] into [a', b', c'] sends ω_1 into ω_1' and ω_2 into ω_2', and never ω_1 into ω_2'. We call ω_1 the *first root*, and ω_2 the *second root*.

As C. Hermite noticed, there is also associated with

$$[a, b, c] = a(x - \omega_1 y)(x - \omega_2 y)$$

a family of definite forms

$$\varphi_t(x, y) = \frac{a}{2t} (x - \omega_1 y)^2 + \frac{at}{2} (x - \omega_2 y)^2,$$

where $t > 0$ is a real parameter. A simple calculation shows that the discriminant of $\varphi_t(x, y)$ is D, for every $t > 0$. Reverting to the quotient variable $z - x/y$, we find the zeros of $\varphi_t(z)$ to be those points z_t such that

$$\frac{1}{t} (z_t - \omega_1)^2 = -t(z_t - \omega_2)^2,$$

or

$$z_t - \omega_1 = \pm it(z_t - \omega_2).$$

The transformation $z' = iz$ rotates the plane about the origin through the angle $\pi/2$; it follows from the last equation that the line segment connecting z_t with ω_1 is perpendicular to the segment connecting z_t with ω_2, and hence that z_t lies on the circle having as diameter the segment which connects ω_1 and ω_2. If, as usual, we take that root z_t which has positive imaginary part as the representative of φ_t, then we have associated with $[a, b, c]$ the semicircle Σ in U connecting ω_1 and ω_2. As t varies from 0^+ to ∞, z_t describes Σ from ω_1 to ω_2; we can think of the semicircle as oriented with this sense, inasmuch as the orientation is preserved under transformations of Γ. This orientation is necessary, since otherwise there would be no way of distinguishing the (usually inequivalent) forms $[a, b, c]$ and $[-a, -b, -c]$. The form is now completely described by specifying its oriented semicircle Σ and its discriminant $-D$.

An indefinite form f will be called *reduced* if the associated semicircle intersects the fundamental region R considered earlier. Thus f is reduced if and only if the definite form φ_t is reduced for some t. The fact that any indefinite form is equivalent to a reduced form is an immediate consequence of the fact that φ_1, for example, is equivalent to a reduced definite form: the transformation which carries φ_1 into a reduced form also carries f into a reduced form. The difficulty lies in showing that each indefinite integral form is equivalent to only finitely many reduced forms. To do this, we must first examine an important subgroup of Γ which is intimately connected with f.

1–8 The automorphs of indefinite forms. A transformation of Γ which leaves $[a, b, c]$ unchanged also leaves ω_1 and ω_2 fixed. The fixed points of the transformation

$$z' = \frac{\alpha z + \beta}{\gamma z + \delta}$$

are those points ω such that

$$\omega = \frac{\alpha \omega + \beta}{\gamma \omega + \delta},$$

or

$$\gamma \omega^2 + (\delta - \alpha)\omega - \beta = 0. \tag{17}$$

Suppose that the roots of this equation are ω_1 and ω_2. These numbers are also the roots of the equation $a\omega^2 + b\omega + c = 0$; since $(a, b, c) = 1$, it follows that for some integer u,

$$\gamma = au, \tag{18}$$

$$\delta - \alpha = bu,$$

$$-\beta = cu. \tag{19}$$

Putting $\delta + \alpha = t$, we have

$$\alpha = \frac{t - bu}{2}, \qquad \delta = \frac{t + bu}{2}, \tag{20}$$

where t and u are such that

$$1 = \alpha\delta - \beta\gamma = \frac{t^2 - b^2u^2}{4} + acu^2 = \frac{t^2 - Du^2}{4},$$

or

$$t^2 - Du^2 = 4. \tag{21}$$

Conversely, if t and u are solutions of (21), and α, β, γ, and δ are determined by equations (18) through (20), then (17) reduces to $u(a\omega^2 + b\omega + c) = 0$ and $\alpha\delta - \beta\gamma = 1$. This proves

THEOREM 1–7. *The set of all automorphs of the primitive indefinite form $[a, b, c]$ is given by the set of all matrices*

$$\begin{pmatrix} \alpha & \beta \\ \gamma & \delta \end{pmatrix}$$

with α, β, γ, and δ determined by equations (18), (19), and (20), where t and u run over the integral solutions of the Pell equation (21).

Originally, automorphs were defined as substitutions giving z in terms of z', while we have here used the inverse transformation giving z' in terms of z. But if $F = \bar{A}FA$, then $F = \bar{A}^{-1}FA^{-1}$, so that the inverse of an automorph is also an automorph, and the set of all automorphs coincides with the set of all inverse automorphs. This fact has much greater significance than in its application above. For since the product of two automorphs is again an automorph, the automorphs of f form a subgroup of Γ, which we shall designate by $\Gamma_A(f)$. (The elements of $\Gamma_A(f)$ will be taken sometimes as transformations and sometimes as their matrices. The ambiguity resulting from the fact that the matrices A and $-A$ correspond to different substitutions in the form but to the same fractional transformation of Γ should cause no difficulty if the reader remains aware of it.) Using well-known properties of the solutions of Pell's equation,* $\Gamma_A(f)$ can be characterized as follows.

THEOREM 1–8. $\Gamma_A(f)$ is the infinite cyclic group generated by the matrix

$$V = \begin{pmatrix} \frac{1}{2}(t_0 - bu_0) & -cu_0 \\ au_0 & \frac{1}{2}(t_0 + bu_0) \end{pmatrix};$$

if A is any automorph of f, then $A = V^n$ for some integer n, positive, negative or zero. Here t_0, u_0 is the minimal positive solution of equation (21).

(The ambiguity mentioned above is exemplified here: every transformation $z' = (\alpha z + \beta)/(\gamma z + \delta)$ of $\Gamma_A(f)$ can be made to have matrix V^n, but the set of all substitutions which leave f fixed is given by $X = \pm X'V^n$.)

Proof: According to Theorem 1–7, $\Gamma_A(f)$ is the group of matrices

$$\begin{pmatrix} \frac{1}{2}(t - bu) & -cu \\ au & \frac{1}{2}(t + bu) \end{pmatrix}, \qquad t^2 - Du^2 = 4,$$

so it is to be shown that each of these matrices is a power of V.
If we put

$$\tfrac{1}{2}(t_0 + u_0\sqrt{D})^n = \tfrac{1}{2}(t_n + u_n\sqrt{D})$$

*Pell's equation is discussed in Volume I, Chapter 8. The minimal positive solution is described in Theorem 8–7.

for each n, then

$$\tfrac{1}{2}(t_{n+1} + u_{n+1}\sqrt{D}) = \tfrac{1}{4}(t_n + u_n\sqrt{D})(t_0 + u_0\sqrt{D})$$
$$= \tfrac{1}{4}(t_0 t_n + D u_0 u_n) + \tfrac{1}{4}(t_0 u_n + t_n u_0)\sqrt{D},$$

so that

$$t_{n+1} = \tfrac{1}{2}(t_0 t_n + D u_0 u_n), \qquad u_{n+1} = \tfrac{1}{2}(t_n u_0 + t_0 u_n).$$

Now suppose that

$$V^{n+1} = \begin{pmatrix} \tfrac{1}{2}(t_n - b u_n) & -c u_n \\ a u_n & \tfrac{1}{2}(t_n + b u_n) \end{pmatrix},$$

an assumption which is correct for $n = 0$. Then

$$V^{n+2} = V V^{n+1} = \begin{pmatrix} \tfrac{1}{2}(t_0 - b u_0) & -c u_0 \\ a u_0 & \tfrac{1}{2}(t_0 + b u_0) \end{pmatrix}$$
$$\times \begin{pmatrix} \tfrac{1}{2}(t_n - b u_n) & -c u_n \\ a u_n & \tfrac{1}{2}(t_n + b u_n) \end{pmatrix}$$
$$= \begin{pmatrix} \tfrac{1}{2}(t_{n+1} - b u_{n+1}) & -\tfrac{1}{2}c(u_0 t_n + u_n t_0) \\ \tfrac{1}{2}a(u_0 t_n + u_n t_0) & \tfrac{1}{2}(t_{n+1} + b u_{n+1}) \end{pmatrix}$$
$$= \begin{pmatrix} \tfrac{1}{2}(t_{n+1} - b u_{n+1}) & -c u_{n+1} \\ a u_{n+1} & \tfrac{1}{2}(t_{n+1} + b u_{n+1}) \end{pmatrix},$$

and by induction, V^n is of the supposed form for all $n \geq 0$. Similarly, it can be shown that

$$V^n = \begin{pmatrix} \tfrac{1}{2}(t_{n-1} - b u_{n-1}) & -c u_{n-1} \\ a u_{n-1} & \tfrac{1}{2}(t_{n-1} + b u_{n-1}) \end{pmatrix},$$

so that V^n is also of the supposed form for all $n \leq 0$. Hence the matrix corresponding to any solution of equation (21) is a power of V, and the theorem is proved.

As usual, it is useful to know a fundamental region of $\Gamma_A(f)$.

THEOREM 1–9. *Suppose that the perpendicular bisector C_0 of the segment l joining ω_1 and ω_2 is mapped by V into the circle C_1. Then C_1 does not intersect C_0, and the (infinite) region between them, together with C_1, ω_1, and ω_2, is a fundamental region of $\Gamma_A(f)$.*

Proof: If the arbitrary transformation

$$z' = T(z) = (\alpha z + \beta)/(\gamma z + \delta)$$

has the distinct fixed points z_1 and z_2, then by dividing $z' - z_1$ by $z' - z_2$ we get

$$\frac{z' - z_1}{z' - z_2} = \frac{\alpha z + \beta - z_1(\gamma z + \delta)}{\alpha z + \beta - z_2(\gamma z + \delta)} = \frac{(\alpha - \gamma z_1)z + (\beta - \delta z_1)}{(\alpha - \gamma z_2)z + (\beta - \delta z_2)}$$

$$= \frac{\alpha - \gamma z_1}{\alpha - \gamma z_2} \cdot \frac{z + (\beta - \delta z_1)/(\alpha - \gamma z_1)}{z + (\beta - \delta z_2)/(\alpha - \gamma z_2)}$$

$$= \frac{\alpha - \gamma z_1}{\alpha - \gamma z_2} \cdot \frac{z - T^{-1}(z_1)}{z - T^{-1}(z_2)} = \frac{\alpha - \gamma z_1}{\alpha - \gamma z_2} \cdot \frac{z - z_1}{z - z_2}.$$

In the case at hand, T is the transformation

$$V: \qquad z' = \frac{\frac{1}{2}(t_0 - bu_0)z - cu_0}{au_0 z + \frac{1}{2}(t_0 + bu_0)}$$

with fixed points ω_1 and ω_2, and

$$\frac{\alpha - \gamma z_1}{\alpha - \gamma z_2} = \frac{\frac{1}{2}(t_0 - bu_0) - au_0(-b + \sqrt{D})/2a}{\frac{1}{2}(t_0 - bu_0) - au_0(-b - \sqrt{D})/2a} = \frac{t_0 - \sqrt{D}\,u_0}{t_0 + \sqrt{D}\,u_0}.$$

We put

$$K = \frac{t_0 - \sqrt{D}\,u_0}{t_0 + \sqrt{D}\,u_0},$$

and have for V the representation

$$\frac{z' - \omega_1}{z' - \omega_2} = K\frac{z - \omega_1}{z - \omega_2}. \tag{22}$$

It follows that V^n is the same transformation with K replaced by K^n; this could be used to give a second proof of Theorem 1–8.

By its definition, K is a real number between 0 and 1. Since the perpendicular bisector C_0 of the segment l joining ω_1 and ω_2 has the equation $|z - \omega_1| = |z - \omega_2|$, V^n transforms it into $|z' - \omega_1| = K^n|z' - \omega_2|$, as we see by taking absolute values in (22). If we put $z = x + iy$, the last equation becomes

$$C_n: \qquad (x - \omega_1)^2 + y^2 = K^{2n}\big((x - \omega_2)^2 + y^2\big), \qquad n \gtreqless 0$$

and it is a matter of simple analytic geometry to prove the following assertions: for positive n, C_n is a circle with its center on the real axis, on the extension through ω_1 of l; it contains ω_1 in its interior;

it lies entirely on that side of C_0 on which ω_1 lies; and its radius approaches zero as n increases. For negative n, the circles C_n lie on the other side of C_0, contain ω_2, and close down on ω_2 as $|n|$ increases. Some of these circles are shown in Fig. 1-3. The lightly shaded region $R_A(1)$, which is the region described in Theorem 1-9, is the set of points z such that

$$K \le \left| \frac{z - \omega_1}{z - \omega_2} \right| < 1,$$

and it is clearly transformed by V into the set $R_A(V)$ of points z such that

$$K^2 \le \left| \frac{z - \omega_1}{z - \omega_2} \right| < K,$$

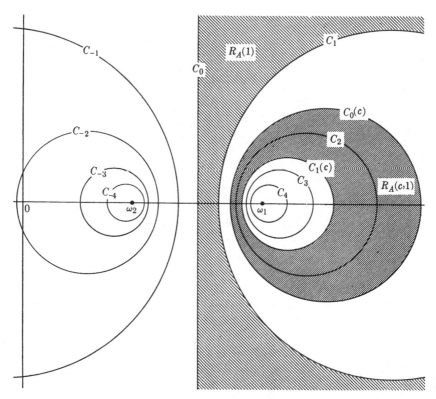

FIGURE 1-3

which is the region between C_1 and C_2, including C_2. In general, V^n transforms $R_A(1)$ into the region $R_A(V^n)$ between C_n and C_{n+1}, including C_{n+1}. Since the entire plane, excluding ω_1 and ω_2, is covered in this fashion, and no point is in two such regions, any one of them, together with ω_1 and ω_2, is a fundamental region of $\Gamma_A(f)$, and the proof is complete.

We are concerned here only with the upper half-plane U; relative to this, a fundamental region of $\Gamma_A(f)$ is that portion of any one of the above regions which lies in U.

In the next section it will be convenient to have slightly more freedom in choosing a fundamental region of $\Gamma_A(f)$. We get this by noticing that, instead of beginning with the line C_0, we could have started with any member of the family of circles

$$\left| \frac{z - \omega_1}{z - \omega_2} \right| = c. \tag{23}$$

For fixed $c > 0$, a fundamental region $R_A(c, 1)$ would then be the ring between the circle (23), which we might call $C_0(c)$, and its transform

$$C_1(c): \qquad \left| \frac{z - \omega_1}{z - \omega_2} \right| = Kc;$$

the argument given above carries through with no change except for the introduction of a factor c in certain equations. Such a region is shown heavily shaded in Fig. 1–3.

1–9 Reduction of indefinite forms. The semicircle Σ representing the form f is the upper half of the circle given parametrically by

$$\left(\frac{z - \omega_1}{z - \omega_2} \right)^2 = -t^2, \qquad 0 \le t \le \infty.$$

The generating automorph V, given by equation (22), changes Σ into the upper half of the circle

$$\left(\frac{z' - \omega_1}{z' - \omega_2} \right)^2 = -Kt^2 = -(t\sqrt{K})^2, \qquad 0 \le t \le \infty,$$

which is the same circle with a different parameter. In other words, Σ is transformed into itself by V, and hence by any element of $\Gamma_A(f)$, in the sense that each point of Σ goes into some other point of Σ, although no points of Σ remain fixed except ω_1 and ω_2. In fact, that arc of Σ which lies in a fundamental region $R_A(c, 1)$ is mapped by V^n onto the arc of Σ which lies in the region $R_A(c, V^n)$, so that these various arcs are equivalent with respect to $\Gamma_A(f)$. Hence they are also equivalent with respect to the larger group Γ.

Now imagine Σ drawn in Fig. 1–3. For suitable choice of c, the circle $C_0(c)$ defined in the last section intersects Σ at a point on the boundary of one of the transforms of R, and this is then also true of the equivalent point which is the intersection of $C_1(c)$ and Σ. The arc between these two points is thus broken up by the boundaries of the double triangles in Fig. 1–2 into a finite number, say μ, of smaller arcs. If these short arcs are transformed back into R by suitable operations of Γ, then every point of Σ is equivalent to some point on each of these new arcs; in other words, there are precisely μ elements of Γ which transform Σ into a semicircle intersecting R. Hence there are precisely μ reduced forms equivalent to f.

THEOREM 1–10. *There are only finitely many reduced forms in any equivalence class of integral primitive indefinite forms.*

Using the definition of reduced form, it is simple to characterize reduced forms in terms of their coefficients. For clearly $[a, b, c]$ is reduced if and only if one or both of the points ρ and $-\rho^2$ are inside the semicircular region bounded by Σ, or if ρ is on Σ. The points below Σ in U are the points $z = x + iy$ such that

$$a(a(x^2 + y^2) + bx + c) < 0.$$

Since ρ and $-\rho^2$ have the coordinates

$$x = \pm \frac{1}{2}, \qquad y = \frac{\sqrt{3}}{2},$$

we have that f is reduced if and only if either

$$a(2a \pm b + 2c) < 0 \qquad \text{or} \qquad 2a - b + 2c = 0. \qquad (24)$$

To find the set of reduced forms of the class containing a given form $[a, b, c]$, the procedure outlined for definite forms may first be used to reduce $\varphi_1(x, y) = a/2(x - \omega_1 y)^2 + a/2(x - \omega_2 y)^2 = ax^2 + bxy + (b^2 + D)y^2/4a$; the transformation which reduces $\varphi_1(x, y)$ also reduces $[a, b, c]$, say to $[a_1, b_1, c_1]$. Thus the semicircle Σ_1 representing $[a_1, b_1, c_1]$ intersects the fundamental region R of Γ, either in an arc or in the single point ρ. Starting from a point on Σ_1 in R, move along Σ_1 in the direction in which it is oriented. At the point at which Σ_1 leaves R, it enters one of the regions

$$R(S^{-1}),\ R(S^{-1}W),\ R(WSW),\ R(WS),\ R(W),$$

$$R(WS^{-1}),\ R(WS^{-1}W),\ R(SW),\ \text{or } R(S),$$

since these are the only regions adjacent to R (cf. Fig. 1-2). If it enters $R(T_1)$, then T_1^{-1} sends Σ_1 into a new semicircle Σ_2 (associated with $[a_2, b_2, c_2]$) which has an arc in R, and this arc is the image under T_1^{-1} of the portion of Σ_1 in $R(T_1)$. The same argument can now be applied to Σ_2, leading to a Σ_3 (associated with $[a_3, b_3, c_3]$) which has an arc in R, and this arc is the image under $T_2^{-1}T_1^{-1}$ of the arc of Σ_1 next encountered in moving along Σ_1 in the positive direction. If the process is repeated μ times, Σ_1 and $[a_1, b_1, c_1]$ will recur.

It is rather the exceptional case that Σ passes through ρ or $-\rho^2$. If it does not, the array of possible transformations listed above simplifies: the only T's to consider are then S, S^{-1}, and W. For example, consider the reduced form $[2, -4, -1]$, where $\omega_1 = (2 + \sqrt{6})/2$, $\omega_2 = (2 - \sqrt{6})/2$. Σ_1 goes from R to $R(W)$, so we make the inversion $W^{-1} = W$, or $z = -1/z'$. This replaces $[a, b, c]$ by $[c, -b, a]$, so here $[a_2, b_2, c_2] = [-1, 4, 2]$. Σ_2 goes from R to $R(S)$, so we make the translation S^{-1}, or $z = z' + 1$. In general this replaces $[a, b, c]$ by $[a, 2a + b, a + b + c]$, so here $[a_3, b_3, c_3] = [-1, 2, 5]$. Σ_3 also goes from R to $R(S)$, and we get $[a_4, b_4, c_4] = [-1, 0, 6]$. Σ_4 also goes from R to $R(S)$, and $[a_5, b_5, c_5] = [-1, -2, 5]$. A final application of S^{-1} gives $[a_6, b_6, c_6] = [-1, -4, 2]$. Since Σ_6 goes from R to $R(W)$, we invert, to get $[a_7, b_7, c_7] = [2, 4, -1]$. Σ_7 goes from R to $R(S^{-1})$, so we must make the translation $S: z = z' - 1$. In general, this replaces $[a, b, c]$ by $[a, b - 2a, a - b + c]$, so here $[a_8, b_8, c_8] = [2, 0, -3]$. A second application of S gives $[a_9, b_9, c_9] = [2, -4, -1] = [a_1, b_1, c_1]$, and we have the complete set of reduced forms for this class. If the algorithm were repeated indefinitely, a periodic sequence of forms would arise; it is therefore meaningful to speak of the *period* of reduced forms.

The following principle is useful in these calculations: If *after a translation* the inequality (24) is correct for just one choice of sign, the next step is an inversion, while if it holds for both signs, the next step is a repetition of the translation. (S is never followed by S^{-1}, nor W by W.) The reason for this should become clear upon looking back at the derivation of (24).

THEOREM 1–11. *There are only finitely many classes of integral indefinite forms of given discriminant.*

Proof: First consider the primitive forms; for them it suffices to show that there are only finitely many reduced forms of given discriminant $\Delta = -D$. From (24) we get

$$2a^2 \pm ab \leq -2ac,$$

so

$$4a^2 \pm 2ab + b^2 \leq b^2 - 4ac = D.$$

But for each choice of sign, $4a^2 \pm 2ab + b^2$ is positive definite; it therefore represents only positive integers unless $a = b = 0$, and by Theorem 1–6, each of the integers $1, 2, \ldots, D$ is represented in only finitely many ways. Hence there are only finitely many choices for a and b, and for each choice, c is fixed by the requirement $b^2 = D + 4ac$. There are therefore only finitely many reduced forms, and hence only finitely many periods, and so only finitely many classes.

If a class contains an imprimitive form, say with $(a, b, c) = d$, then every form in that class also has divisor d, so that the class consists of the elements of a class of primitive forms with smaller D, each multiplied by d. There are only finitely many such classes.

PROBLEMS

1. Find the period of reduced forms belonging to the class of

$$x^2 + 7xy + 7y^2.$$

2. Show that Theorem 1–7 remains correct if the word "indefinite" is omitted, that is, if $D < 0$ (cf. Theorem 1–4).

3. Show that there is just one class of primitive forms with $D = 20$, and one class of imprimitive forms.

1–10 Representations. The discussion occurring between Theorems 1–4 and 1–6 made no use of the definiteness of the form; it is therefore equally applicable to indefinite forms. Thus Theorem 1–6 can be recast as follows.

THEOREM 1–12. *Let $f = [a, b, c]$ be a primitive integral indefinite form of discriminant Δ, where $D = -\Delta$ is not a square. Let n be an integer. Corresponding to each minimum root m of the congruence* (15), *determine l by* (16). *If none of the forms $[n, m, l]$ is equivalent to f, there are no proper representations of n by f. If at least one of the new forms is equivalent to f, there are infinitely many proper representations of n by f; they are given by the first columns of all the matrices AT, where A can be any automorph $\pm V^n$ of f, and T is any of a set of matrices which replace f by the various equivalent forms $[n, m, l]$, each form being obtained from just one T.*

PROBLEMS

1. Discuss the proper representation of 13 by $[1, 3, -1]$.

2. Show that the odd numbers properly represented by $x^2 + 4xy - y^2$ are those of the form

$$5^\epsilon \prod_{i=1}^{r} p_i{}^{\alpha_i},$$

where $\epsilon = 0$ or 1, $r \geq 0$, and $p_i \equiv \pm 1 \pmod{10}$ for $1 \leq i \leq r$ (cf. Problem 3, Section 1–10).

CHAPTER 2

ALGEBRAIC NUMBERS

2-1 Introduction. With a few exceptions, the theory developed up to this point, both in this volume and in the preceding introductory volume, has been self-contained, in the sense that the problems, which had to do with the ordinary integers, were solved without going outside this system. When considering the distribution of primes and the theory of quadratic forms, we made use of the real and complex numbers, but not in an intrinsically arithmetic fashion. In the investigation of the representability of an integer as a sum of squares,* however, we had occasion to consider the arithmetic structure of the set of Gaussian integers, and to apply this to a problem involving ordinary integers. During the last century, it has been found that many problems in rational arithmetic are treated most naturally by introducing larger sets of "integers" and deducing, from the structure of the extended system, information about the ordinary integers. Of course, as soon as a mathematician begins to work in a new medium, to use a metaphor from art, he finds interesting questions which have little or nothing to do with the original problem. In the present case, this tendency was instrumental in the development of modern abstract algebra, a large portion of which has only a tenuous connection with number theory.

From the point of view of this text, general algebraic theory must take second place, the primary object being to give the reader an appreciation of the power afforded by the method, as well as a knowledge of some of the basic results in the subject. For this reason, the formulation will be kept as concrete as possible; there will be no striving for generality or abstractness for their own sakes. The treatment is self-contained, except for the following two theorems, whose proofs can be found, for example, in L. E. Dickson, *First Course in the Theory of Equations* (New York: John Wiley & Sons, Inc., 1921), pp. 130–131 and 124–125, respectively.

---

*See, for example, Volume I, Chapter 7.

The product $D_1 D_2$ of two determinants of the same order is another determinant of that order, whose element in row i and column j is the sum of the products of the elements of the ith row of D_1 and the corresponding elements of the jth row of D_2.

SYMMETRIC FUNCTION THEOREM.   *Any polynomial $P(x_1, \ldots, x_n)$ symmetric in $x_1, \ldots, x_n$ and of degree g in each, is equal to a polynomial of total degree q, with integral coefficients, in the elementary symmetric functions*

$$\sum x_1, \qquad \sum x_1 x_2, \qquad \ldots, \qquad x_1 x_2 \cdots x_n$$

and the coefficients of $P(x_1, \ldots, x_n)$. In particular, any symmetric polynomial with integral coefficients is equal to a polynomial in the elementary symmetric functions with integral coefficients.

If P is a polynomial in the roots of an equation $f(x) = 0$ of degree n and leading coefficient 1, and if P is symmetric in $n - 1$ of the roots, then P is equal to a polynomial, with integral coefficients, in the remaining root and the coefficients of $f(x)$ and P.

We shall also have occasion to use the so-called Fundamental Theorem of algebra; this basic assertion is proved in the remainder of the section.

FUNDAMENTAL THEOREM OF ALGEBRA.   *A polynomial $f(z) = a_0 z^n + \cdots + a_n$ having complex coefficients and positive degree, has a complex zero.   (It follows immediately that it has exactly n complex zeros, in the sense that there are complex numbers $\xi_1, \ldots, \xi_n$ such that*

$$f(z) = a_0(z - \xi_1) \cdots (z - \xi_n).)$$

Proof: Since the truth of the theorem depends on the structure of the complex numbers, it is necessary to use some properties of these numbers.   If the entire theory of functions of a complex variable is assumed, the proof is very easy indeed: an analytic function has as many zeros as poles, and a polynomial has a pole at infinity, so it must have at least one zero.   If less than this is assumed, it is reasonable to ask that as little be assumed as possible.   The proof to be given uses the fact that a real-valued continuous function of two real variables has a minimum value in any closed domain, and it assumes familiarity with the symbol $\sqrt{a}$, where a is real.   (If DeMoivre's theorem were used, to give meaning to $\sqrt[n]{a}$ for complex a, the proof would be slightly simpler.)

With the second assumption, the quadratic formula provides a proof when $n = 2$ and the coefficients are real. To solve a quadratic equation with nonreal coefficients, it may be necessary to extract the square root of a nonreal number. Let the number be $a + bi$. Then the equation $a + bi = (x + iy)^2$ gives

$$a = x^2 - y^2 \quad \text{and} \quad b = 2xy,$$

or

$$4x^4 - 4ax^2 - b^2 = 0,$$

and we can take

$$x = \sqrt{\frac{a + \sqrt{a^2 + b^2}}{2}}, \quad y = \frac{b}{2x}.$$

Before treating the general case, note first that we can write $f(x + iy) = G(x, y) + iH(x, y)$, where G and H are polynomials in the real variables x and y, with real coefficients. It follows from the continuity of G and H throughout the xy-plane that $|f(z)|$ is continuous throughout the complex z-plane, where $z = x + iy$. Moreover, for $n > 0$ and $a_0 \neq 0$ (which we henceforth assume), we have

$$\lim_{z \to \infty} |f(z)| = \infty.$$

For if max $(|a_0|, \ldots, |a_n|) = A$, then

$$|f(z)| \geq |a_0 z^n| - (|a_n| + |a_{n-1}z| + \cdots + |a_1 z^{n-1}|)$$

$$> |a_0 z^n| \left(1 - \frac{nA}{|a_0 z|}\right) \quad \text{for } |z| > 1$$

$$> \frac{|a_0 z^n|}{2} \quad\quad\quad \text{for } |z| > \max\left(\frac{2nA}{|a_0|}, 1\right).$$

Since $|f(z)|$ is continuous, it assumes a minimum value at some point in any closed circular disk with center at O, and since $|f(z)|$ becomes infinite with $|z|$, the disk can be chosen so large that this minimum occurs at an interior point ξ. We must show that $|f(\xi)| = 0$.

We now proceed by induction: suppose that every polynomial of degree less than n, with complex coefficients, has a complex zero, and that f is of degree n and $|f(z)|$ assumes its nonzero minimum at ξ.

Suppose that $f(\xi) = M$, and put

$$g(z) = \frac{f(z + \xi)}{M} = 1 + b_1 z + \cdots + b_n z^n;$$

then $|g(z)| \geq 1$ for all z. Define k as the smallest index such that $b_k \neq 0$, so that

$$g(z) = 1 + b_k z^k + \cdots + b_n z^n, \qquad k \leq n.$$

First consider the case that $k < n$. By the induction hypothesis, the equation

$$1 + b_k z^k = 0$$

has a root. Let η be this root, and put $z = \delta\eta$, where $0 < \delta < 1$. Then

$$g(\delta\eta) = 1 + b_k \delta^k \eta^k + b_{k+1} \delta^{k+1} \eta^{k+1} + \cdots + b_n \delta^n \eta^n$$
$$= 1 - \delta^k + (b_{k+1} \eta^{k+1} + \cdots + b_n \eta^n \delta^{n-k-1}) \delta^{k+1}.$$

Now if $|b_j| < B$ for $k < j \leq n$, then

$$\delta^{k+1} |b_{k+1} \eta^{k+1} + \cdots + b_n \eta^n \delta^{n-k-1}|$$
$$\leq B(1 + |\eta|)^n \delta^{k+1} (1 + \delta + \cdots + \delta^{n-k-1})$$
$$\leq Bn(1 + |\eta|)^n \delta^{k+1} = C\delta^{k+1}.$$

Thus

$$|g(\delta\eta)| < 1 - \delta^k + C\delta^{k+1} = 1 - \delta^k(1 - C\delta),$$

and for $0 < \delta < 1/C$, $|g(\delta\eta)| < 1$. This contradicts the assumption that 1 is the minimum of $|g(z)|$; hence $M = 0$.

If $k = n$, then $g(z) = 1 + b_n z^n$. If n is even, then the equation

$$z^{\frac{1}{2}n} - \sqrt{-\frac{1}{b_n}} = 0$$

is solvable, by the induction hypothesis, and any root of it is also a root of $g(z) = 0$. Hence we can suppose that n is odd. Put $b_n = c + di$. If $c \neq 0$, we put $z = -\delta \operatorname{sgn} c$ (that is, $z = \delta$ or $-\delta$, according as $c < 0$ or $c > 0$), and obtain

$$|1 + (c + di)z^n|^2 = |1 - |c|\delta^n - \delta^n di \operatorname{sgn} c|^2$$
$$= 1 - 2|c|\delta^n + (c^2 + d^2)\delta^{2n};$$

this last expression is again smaller than 1 for δ sufficiently small, and we have the same contradiction as before.

If $c = 0$, then $d \neq 0$; moreover, a sign can be chosen so that $(\pm i)^n = i$. Then if $z = \pm i\delta \operatorname{sgn} d$, we have

$$\left|1 + id(\pm i\delta \operatorname{sgn} d)^n\right| = \left|1 - |d|\delta^n\right|,$$

and this is smaller than 1 for δ sufficiently small. The proof is complete.

2–2 Polynomials and algebraic numbers. We begin by making the following definitions.

(a) R is the set of all rational numbers.

(b) $R[x]$ consists of R together with all polynomials in x with rational coefficients, the coefficient of the highest power of x being different from zero.

(c) If a polynomial $p(x)$ is in $R[x]$, deg p means the exponent of the highest power of x occurring in $p(x)$, if this is positive; if $a \neq 0$ is in R, deg $a = 0$, while if $a = 0$, deg a is not defined.

(d) A polynomial $p(x)$ in $R[x]$ is said to be *monic* if the leading coefficient is 1.

(e) If $p_1(x)$ and $p_2(x)$ are in $R[x]$, we say that $p_2(x)$ *divides* $p_1(x)$ (in symbols, $p_2(x)|p_1(x)$; the phrase *does not divide* is indicated by the symbol "$\nmid$") if there is a $q(x)$ in $R[x]$ such that $p_1(x) = p_2(x)q(x)$. Under this definition, an element of R different from zero divides every element of $R[x]$. The nonzero elements of R are therefore called *units* of $R[x]$.

(f) An element $p(x)$ is said to be *irreducible in* $R[x]$ if it cannot be written as the product of two nonunit elements of $R[x]$.

By formalizing the ordinary process of dividing one polynomial by another, it is not hard to show that if $p_1(x)$ and $p_2(x)$ are in $R[x]$, and $p_2(x)$ is not zero, then there exists a unique pair of elements $q(x)$ and $r(x)$ of $R[x]$ such that

$$p_1(x) = p_2(x)q(x) + r(x), \qquad \deg r < \deg p_2 \text{ or } r(x) = 0.$$

This analog of the division theorem for integers* forms the basis for a Euclidean algorithm, by means of which a greatest common divisor $(p_1(x), p_2(x))$ can be determined; the development is entirely parallel to that for the integers, and leads to the following theorems.

---

*See, for example, Volume I, Theorem 1–1.

THEOREM 2–1. *Given two elements* $p_1(x)$, $p_2(x)$ *of* $R[x]$, *not both zero, there is another element* $d(x)$ *which is unique to within a unit factor and which has the following properties:*

(a) $d(x)|p_1(x)$ *and* $d(x)|p_2(x)$.

(b) *If* $d_1(x)$ *is in* $R[x]$, *and divides both* $p_1(x)$ *and* $p_2(x)$, *then* $d_1(x)|d(x)$.

If $(p_1(x), p_2(x)) = d(x)$, *there are elements* $q_1(x)$ *and* $q_2(x)$ *of* $R[x]$ *such that*

$$p_1(x)q_1(x) + p_2(x)q_2(x) = d(x).$$

THEOREM 2–2. *Any nonzero element of* $R[x]$ *can be factored into a product of irreducible elements of* $R[x]$, *and this factorization is unique except for the order of factors and the presence of units.*

There is no loss in generality, and some gain in simplicity, in supposing that the various polynomials with which we deal are monic, since any polynomial can be made monic by multiplication by a unit. In this case the second part of Theorem 2–2 could be restated to read: *The factorization of a monic polynomial into irreducible monic elements is unique except for the order of factors.*

We now consider the zeros of the polynomials of $R[x]$, or, what is the same thing, the roots of equations $p(x) = 0$. If α is a root of the equation

$$p(x) \equiv x^n + r_1 x^{n-1} + r_2 x^{n-2} + \cdots + r_n = 0. \tag{1}$$

where $p(x)$ is in $R[x]$ and $n > 0$, then α is called an *algebraic number;* if $p(x)$ is irreducible in $R[x]$, α is said to be of *degree n.* (The rational numbers are algebraic numbers, since if r is in R, $x - r = 0$ has the root $x = r$. As algebraic numbers they are of degree 1, although when considered as elements of $R[x]$ the nonzero rational numbers were given degree 0.) An algebraic number α is a zero of a unique monic irreducible polynomial in $R[x]$, called the *defining polynomial* of α. For if $p(x)$ is not irreducible, it can be factored uniquely into irreducible monic factors, and α must be a zero of one of the factors. Hence α satisfies some irreducible equation, i.e., an equation in which the left side is irreducible in $R[x]$. If α satisfies two such equations, say $p(x) = 0$ and $q(x) = 0$, then it also satisfies the equation $d(x) = 0$, where $d(x) = (p(x), q(x))$. For if

$$p(x)s_1(x) + q(x)s_2(x) = d(x),$$

then

$$d(\alpha) = s_1(\alpha) \cdot 0 + s_2(\alpha) \cdot 0 = 0.$$

But since $p(x)$ and $q(x)$ are irreducible, their monic GCD is either 1 or $p(x)$. Since $1 \neq 0$, $(p(x), q(x)) = p(x)$, and $p(x) = q(x)$.

If $p(x)$ in equation (1) is the defining polynomial of α, its n zeros $\alpha_1 = \alpha, \alpha_2, \ldots, \alpha_n$ are called the *conjugates of α*. Except for an alternation in signs, the numbers $r_1, r_2, \ldots, r_n$ are simply the elementary symmetric functions of $\alpha_1 = \alpha, \alpha_2, \ldots, \alpha_n$:

$$r_1 = -\sum \alpha_1 = -(\alpha + \alpha_2 + \cdots + \alpha_n),$$

$$r_2 = \sum \alpha_1 \alpha_2 = \alpha \alpha_2 + \cdots + \alpha_{n-1} \alpha_n,$$

$$\vdots$$

$$r_n = (-1)^n \alpha \alpha_2 \cdots \alpha_n.$$

As is the case here, we shall frequently use a Greek letter, both with and without the subscript 1, to denote a single algebraic number.

THEOREM 2–3. *The sum, difference, and product of two algebraic numbers are algebraic numbers. The quotient of two algebraic numbers is an algebraic number if the denominator is not zero.*

Proof: Suppose that $\alpha = \alpha_1$ and $\beta = \beta_1$ have defining polynomials

$$p(x) = x^n + r_1 x^{n-1} + \cdots + r_n = (x - \alpha_1)(x - \alpha_2) \cdots (x - \alpha_n),$$

$$q(x) = x^m + s_1 x^{m-1} + \cdots + s_m = (x - \beta_1)(x - \beta_2) \cdots (x - \beta_m),$$

respectively. Let $\gamma_1, \gamma_2, \ldots, \gamma_{nm}$ be the numbers obtained by adding an α_i and a β_j, in all possible ways. Then the polynomial $g(x) = (x - \gamma_1)(x - \gamma_2) \cdots (x - \gamma_{nm})$ has, as coefficients, symmetric polynomials in the α_i and β_j, with integral coefficients. Let one such coefficient be $t(\alpha_1, \ldots, \alpha_n, \beta_1, \ldots, \beta_m)$. As a symmetric polynomial in the α_i it is equal to a polynomial in $r_1, \ldots, r_n$, whose coefficients are themselves polynomials in $\beta_1, \ldots, \beta_m$ with integral coefficients. These last polynomials are symmetric in $\beta_1, \ldots, \beta_m$; they are therefore integral combinations of $s_1, \ldots, s_m$, and consequently are rational numbers. Thus the coefficients of $g(x)$ are rational numbers, and $\alpha + \beta$ is an algebraic number. The same proof applies for $\alpha \cdot \beta$ and $\alpha - \beta$, with obvious changes in the definition of $\gamma_1, \ldots, \gamma_{nm}$.

If α is algebraic and different from zero, so is $1/\alpha$, for the zeros of the polynomial

$$r_n x^n + r_{n-1} x^{n-1} + \cdots + r_1 x + 1$$

are the reciprocals of those of

$$x^n + r_1 x^{n-1} + \cdots + r_n,$$

and $r_n \neq 0$. Thus the assertion that α/β is algebraic is a consequence of the fact that $\alpha \cdot \dfrac{1}{\beta}$ is algebraic.

The properties of the set of all algebraic numbers mentioned in Theorem 2–3 are shared by many sets of importance in mathematics; so many in fact that the name *field* has been reserved to describe such sets. Technically, a field F is a set of two or more elements $a, b, \ldots$, together with an equivalence relation (which we designate by an equals sign) and two operations (which we designate by the symbols "$+$" and "$\cdot$"), such that the following relations hold:

(a) For any a and b in F, either $a = b$ or $a \neq b$. If $a = b$, then $a + c = b + c$ and $a \cdot c = b \cdot c$, for every c in F.

(b) The elements form a commutative group with respect to the operation "$+$", the identity element being designated by "0". In other words, if a, b, and c are in F, then $a + b$ is in F, $a + b = b + a$, $a + (b + c) = (a + b) + c$, there is an element $-a$ in F such that $a + (-a) = (-a) + a = 0$, and $a + 0 = 0 + a = a$.

(c) The elements with 0 omitted (which we might call F^*) form a commutative group with respect to the operation "$\cdot$", the identity element being designated by "1".

(d) Multiplication is distributive with respect to addition; that is, $a \cdot (b + c) = a \cdot b + a \cdot c$ for every a, b, and c in F.

As long as one is working with a set of real or complex numbers, and ordinary multiplication, addition, and equality, one can show that the set forms a field just by showing that if a and b are in the set, so are $a \pm b$, ab, and a/b if $b \neq 0$; the other requirements are automatically fulfilled. Thus Theorem 2–3 is just the assertion that the set of all algebraic numbers is a field. Other familiar examples of fields are the set of all rational numbers, the set of all real numbers, and the set of all complex numbers. (The integers, on the other hand, do not form a field, since only the elements ± 1 have inverses, under multiplication, in the system.) In fact, every field composed of complex numbers together with the ordinary operations of addition and multiplication, contains the field R of rational numbers as a subfield. There are, however, fields with only finitely many elements. An example of such a field is the set of numbers $0, 1, \ldots, p - 1$ with the operations of addition and multiplication modulo p; in this case, $a + b$ is that element c such that $a + b \equiv c \pmod{p}$; $a \cdot b$ is that element d such that $a \cdot b \equiv d \pmod{p}$; $-a$ is 0 or $p - a$, according as a is 0 or not 0; if $a \neq 0$, a^{-1} is that element f such that $a \cdot f \equiv 1 \pmod{p}$.

The field of all algebraic numbers will play no role in the present discussion. We consider instead certain subfields of it, called *algebraic number fields*, described in the next theorem.

Let ϑ be an algebraic number, of degree $n > 1$, whose defining polynomial is $p(x)$ as given in equation (1), and whose conjugates are $\vartheta, \vartheta_2, \ldots, \vartheta_n$.

THEOREM 2–4. *The set of all numbers of the form*

$$\alpha = \frac{q_1(\vartheta)}{q_2(\vartheta)}, \tag{2}$$

where $q_1(x)$ and $q_2(x)$ are in $R[x]$ and $q_2(\vartheta) \neq 0$, is a field, which will be denoted by $R(\vartheta)$. Every element of $R(\vartheta)$ can be expressed uniquely in the form

$$\alpha = a_0 + a_1\vartheta + \cdots + a_{n-1}\vartheta^{n-1},$$

where $a_0, a_1, \ldots, a_{n-1}$ are in R.

Proof: The first part is clear, since the sum, difference, product and quotient of rational functions are again rational functions.

Since $q_2(\vartheta) \neq 0$ and $p(x)$ is irreducible, $q_2(x)$ and $p(x)$ are relatively prime, and for some $t(x)$ and $s(x)$ in $R[x]$,

$$t(x)p(x) + s(x)q_2(x) = 1.$$

This gives $s(\vartheta)q_2(\vartheta) = 1$, and

$$\alpha = \frac{q_1(\vartheta)}{q_2(\vartheta)} = s(\vartheta)q_1(\vartheta),$$

a polynomial in ϑ. Since $p(\vartheta) = 0$,

$$\vartheta^n = -r_1\vartheta^{n-1} - r_2\vartheta^{n-2} - \cdots - r_n.$$

It follows that every positive power of ϑ can be written as a polynomial in ϑ of degree $n - 1$ or less. The same is therefore true of every element α. If there were two different representations of α as polynomials in ϑ of degree $n - 1$ or less and with rational coefficients, their difference would be a polynomial of degree $n - 1$ or less which vanishes for $x = \vartheta$, which is impossible.

If α is an element of the field described in Theorem 2–4, and

$$\alpha = a_0\vartheta^{n-1} + \cdots + a_{n-1} = \varphi(\vartheta),$$

then the numbers

$$\alpha' = \alpha, \qquad \alpha'' = \varphi(\vartheta_2), \qquad \ldots, \qquad \alpha^{(n)} = \varphi(\vartheta_n)$$

are called the *field conjugates* of α. (They may not lie in the field described in Theorem 2–4.) Every field conjugate of α is also a conjugate of α in the earlier sense, for if α has the defining equation $g(x) = 0$, then $g(\varphi(x))$ vanishes for $x = \vartheta$, so that $p(x)|g(\varphi(x))$ and $g(\varphi(\vartheta_k)) = g(\alpha^{(k)}) = 0$. The converse is also true, as the following theorem shows.

THEOREM 2–5. *The set of field conjugates of an element α of $R(\vartheta)$ is either identical with the set of conjugates of α, or consists of several copies of the set of conjugates of α. (Hence $\deg \alpha | \deg \vartheta$.) The polynomial whose zeros are the field conjugates of α is a power of the defining polynomial of α; if it is equal to the defining polynomial, then $R(\alpha) = R(\vartheta)$.*

Proof: Form the *field polynomial* for α:

$$f(x) = (x - \alpha')(x - \alpha'') \cdots (x - \alpha^{(n)}).$$

Its coefficients are symmetric polynomials in the $\alpha^{(k)}$'s, and are therefore symmetric polynomials in $\vartheta_1, \ldots, \vartheta_n$, and so are rational numbers. Factor $f(x)$ into its monic irreducible factors in $R[x]$, say

$$f(x) = f_1(x) \cdot f_2(x) \cdots,$$

and let $f_1(x)$ be a factor which vanishes for $x = \alpha$. Then $f_1(\varphi(\vartheta)) = 0$, so $p(x)|f_1(\varphi(x))$, and $f_1(x)$ vanishes at $\alpha', \alpha'', \ldots, \alpha^{(n)}$. If these are distinct, $f_1(x)$ is of degree n, and $f(x)$ is irreducible. If they are not, let $\alpha, \alpha'', \ldots, \alpha^{(t)}$ be a maximal distinct set of α's. Then $f_2(x)$ vanishes for some $\alpha^{(k)}$, so $f_1(x)|f_2(x)$; since $f_2(x)$ is irreducible, $f_2(x) = cf_1(x)$, and $c = 1$ since $f_1(x)$ and $f_2(x)$ are monic. If there are other factors of $f(x)$, the argument can be repeated. Eventually, we find that

$$f(x) = (f_1(x))^{n/t}.$$

Since the zeros of $f_1(x)$, which is the defining polynomial of α, are the conjugates of α, those of $f(x)$ (that is, the field conjugates) consist of n/t copies of the set of conjugates of α.

Now suppose that $f_1(x) = f(x)$. Define

$$\varphi(x) = f(x)\left[\frac{\vartheta}{x - \alpha'} + \frac{\vartheta_2}{x - \alpha''} + \cdots + \frac{\vartheta_n}{x - \alpha^{(n)}}\right],$$

so that $\varphi(x)$ is a polynomial of degree $n - 1$ with rational coefficients. Since

$$\varphi(\alpha) = \vartheta(\alpha - \alpha'') \cdots (\alpha - \alpha^{(n)}) := \vartheta f'(\alpha),$$

we have that the number

$$\vartheta = \frac{\varphi(\alpha)}{f'(\alpha)}$$

is in $R(\alpha)$, so that $R(\vartheta)$ is a subfield of $R(\alpha)$, and $R(\alpha) = R(\vartheta)$.

The last assertion of the theorem shows that if one field $R(\alpha)$ is a proper subfield of a second field $R(\vartheta)$, then $\deg \alpha < \deg \vartheta$. For if $\deg \alpha = \deg \vartheta$, then the field polynomial of α with respect to $R(\vartheta)$ is irreducible, so that $f_1(x) = f(x)$, and $R(\alpha) = R(\vartheta)$.

The field $R(\vartheta)$ is called an *algebraic number field;* we say that $R(\vartheta)$ is obtained by *adjoining* ϑ to R, and call $R(\vartheta)$ a *simple algebraic extension* of R, of *degree n.* This same field can be obtained by adjoining various other numbers to R; for example, $R(2\vartheta) = R(\vartheta)$. If an element α of $R(\vartheta)$ is such that $R(\alpha) = R(\vartheta)$, then α is called a *primitive element* of $R(\vartheta)$. It is clear that the degrees of any two primitive elements are the same, and both are equal to the degree of the field.

There is, of course, no reason why the process of adjunction cannot be repeated; one can start from $R(\vartheta)$ and adjoin an algebraic number η to it by taking all rational functions of η whose coefficients are elements of $R(\vartheta)$. This new field is denoted by $R(\vartheta)(\eta)$, or more simply by $R(\vartheta, \eta)$.

THEOREM 2–6. *If ϑ and η are algebraic numbers, the adjunction of η to $R(\vartheta)$ gives the same field $R(\vartheta, \eta)$ as the adjunction of ϑ to $R(\eta)$.*

There exists an algebraic number ζ such that $R(\vartheta, \eta)$ is identical with $R(\zeta)$.

Proof: The first part is clear, since both $R(\vartheta, \eta)$ and $R(\eta, \vartheta)$ are identical with the field consisting of all numbers of the form

$$\frac{q_1(\vartheta, \eta)}{q_2(\vartheta, \eta)}, \qquad q_2(\vartheta, \eta) \neq 0,$$

where $q_1(x, y)$ and $q_2(x, y)$ are polynomials in two variables with rational coefficients.

If η is an element of $R(\vartheta)$, then $R(\vartheta, \eta) = R(\vartheta)$, since a rational function of a rational function is again a rational function. Assume then that ϑ and η do not lie in the fields $R(\eta)$ and $R(\vartheta)$ respectively. Let their defining polynomials be $p_1(x)$ and $p_2(x)$, and let their conjugates be $\vartheta_1, \ldots, \vartheta_n$ and $\eta_1, \ldots, \eta_m$, respectively. Let a and b be rational numbers, and let $\zeta = \zeta_1, \ldots, \zeta_{nm}$ be all expressions of the form $a\vartheta_j + b\eta_k$. Since the conjugates of ϑ are distinct, as are the conjugates of η, there is only a finite set of ratios a/b for which some two of the ζ's are equal, and we choose a and b so that a/b is not in this set. Furthermore, we order the ζi so that $\zeta = a\vartheta + b\eta$.

Now put

$$f(x) = (x - \zeta_1)(x - \zeta_2) \cdots (x - \zeta_{nm}).$$

This polynomial has no multiple zeros, and its coefficients, being symmetric in the ϑ's and η's, are rational. We show that $R(\vartheta, \eta) = R(\zeta)$. It is clear that every element of $R(\zeta)$ is in $R(\vartheta, \eta)$. Suppose on the other hand that ρ is in $R(\vartheta, \eta)$, and that

$$\rho = \frac{q_1(\vartheta, \eta)}{q_2(\vartheta, \eta)}, \qquad q_2(\vartheta, \eta) \neq 0.$$

Then we can define the numbers $\rho = \rho_1, \ldots, \rho_{nm}$ by the equation

$$\rho_i = \frac{q_1(\vartheta_j, \eta_k)}{q_2(\vartheta_j, \eta_k)},$$

where the same subscripts appear on ϑ and η in the definition of ρ_i as in the definition of ζ_i, for $i = 1, 2, \ldots, nm$. Now put

$$F(x) = f(x) \left(\frac{\rho}{x - \zeta_1} + \frac{\rho_2}{x - \zeta_2} + \cdots + \frac{\rho_{nm}}{x - \zeta_{nm}} \right);$$

by the Symmetric Function Theorem, the coefficients of $F(x)$ are rational. If $i > 1$, the polynomial

$$f(x) \frac{\rho_i}{x - \zeta_i} = \rho_i(x - \zeta) \cdots (x - \zeta_{i-1})(x - \zeta_{i+1}) \cdots (x - \zeta_{nm})$$

vanishes for $x = \zeta$, and from the representation

$$f(x) \frac{\rho}{x - \zeta} = \rho(x - \zeta_2) \cdots (x - \zeta_{nm})$$

we have

$$F(\zeta) = \rho(\zeta - \zeta_2) \cdots (\zeta - \zeta_{nm}).$$

Since

$$f'(\zeta) = (\zeta - \zeta_2) \cdots (\zeta - \zeta_{nm}) \neq 0,$$

this gives

$$\rho = \frac{F(\zeta)}{f'(\zeta)},$$

and ρ is in $R(\zeta)$.

PROBLEMS

1. Prove *Eisenstein's irreducibility criterion:* a polynomial $f(x) = a_0 + a_1x + \cdots + a_nx^n$ with integral coefficients cannot be written as a product of two or more polynomials with integral coefficients and positive degrees, if there is a prime p such that

$$p \nmid a_n, \qquad p \mid a_i \text{ if } i < n, \qquad \text{and } p^2 \nmid a_0.$$

[*Hint:* Suppose that there is such a p, but that $f(x) = g(x)h(x)$, where $g(x) = b_0 + b_1x + \cdots + b_rx^r$, $h(x) = c_0 + c_1x + \cdots + c_sx^s$. It follows that p divides exactly one of b_0 and c_0—say b_0. Let b_i be the first coefficient in $g(x)$ not divisible by p, and deduce a contradiction from the expression for a_i in terms of the b's and c's.] As we shall see later (Theorem 2–21), irreducibility over the set of polynomials with integral coefficients implies irreducibility over $R[x]$. Use this fact in Problem 2.

2. Show that the following polynomials are irreducible over $R[x]$:

(a) $x^n - p$, p a prime.

(b) $x^{p-1} + x^{p-2} + \cdots + x + 1$. [*Hint:* Replace x by $x + 1$.]

(c) $x^3 + 3x^2 + 4$.

3. Show that $R(\sqrt{2}, \sqrt{3})$ is identical with $R(\sqrt{2} + \sqrt{3})$, and find a rational function $r(x)$ with rational coefficients such that $r(\sqrt{2} + \sqrt{3}) = \sqrt{2}$.

2–3 Algebraic integers.   If the defining (monic) polynomial of an algebraic number ϑ has integral coefficients, ϑ is said to be an *algebraic integer*. This is a direct extension of the notion of ordinary or *rational* integers, which are the zeros of monic linear polynomials with integral coefficients. Hereafter we shall designate by Z the set of all rational integers.

THEOREM 2–7. *The sum, difference, and product of two algebraic integers are again algebraic integers.*

The proof follows the lines of the proof of Theorem 2–3.

THEOREM 2–8. *If α is a zero of a monic polynomial with coefficients in Z, then α is an algebraic integer.*

Proof: Suppose that $f(x) = x^N + \cdots + a_N$ is the polynomial, and that $p(x) = x^n + r_1 x^{n-1} + \cdots + r_n$ is the defining polynomial of α. Let b_0 be the LCM of the denominators of the reduced fractions $r_1, \ldots, r_n$, so that $b_0 p(x) = q(x) = b_0 x^n + b_1 x^{n-1} + \cdots + b_n$ has relatively prime rational integral coefficients. Then $q(x)$ divides $f(x)$, the coefficients in the quotient polynomial being rational, and we can write

$$\frac{f(x)}{q(x)} = \frac{cg(x)}{c'},$$

where c and c' are so chosen that $g(x)$ has relatively prime coefficients in Z. Thus $c'f(x) = cg(x)q(x)$, and the coefficients of the product $g(x)q(x)$ are relatively prime.* Since this is also true of the coefficients of $f(x)$ (for $f(x)$ is monic), we conclude that $c = c'$. Comparing the coefficients of x^N in the equation $f(x) = g(x)q(x)$, we see that $b_0|1$, and hence $b_0 = \pm 1$, which was to be proved.

THEOREM 2–9.   *If α is a root of an equation*

$$f(x) = x^n + \beta_1 x^{n-1} + \cdots + \beta_n = 0,$$

in which $\beta_1, \ldots, \beta_n$ are algebraic integers, then α is an algebraic integer.

Proof: By Theorem 2–6, $\beta_1, \ldots, \beta_n$ all lie in a simple extension field $R(\vartheta)$, of degree m, say. We can use the sets of field conjugates

---

*The reader may prove this simple fact himself, or refer to the remark preceding Theorem 3–14, Volume I.

$$(\beta_1'', \ldots, \beta_n''), \qquad \ldots, \qquad (\beta_1^{(m)}, \ldots, \beta_n^{(m)})$$

to form polynomials

$$f_2(x) = x^n + \beta_1'' x^{n-1} + \cdots + \beta_n'', \qquad \ldots,$$
$$f_m(x) = x^n + \beta_1^{(m)} x^{n-1} + \cdots + \beta_n^{(m)}.$$

The product $f(x)f_2(x) \cdots f_m(x)$ has rational integral coefficients and is monic; by Theorem 2–8, α is an algebraic integer.

The set of integers in a fixed algebraic number field $R(\vartheta)$ is also closed under addition, subtraction, and multiplication. We shall designate this set by $R[\vartheta]$, and call it the *integral domain* of the field. In particular, $R[1] = Z$ is the set of rational integers.

THEOREM 2–10. *If ϑ is an algebraic number, there exists some rational integer $a \neq 0$ such that $a\vartheta$ is an algebraic integer. If ϑ satisfies an equation $\beta_0 x^n + \cdots + \beta_n = 0$, in which $\beta_0, \ldots, \beta_n$ are algebraic integers, then $\beta_0 \vartheta$ is an algebraic integer.*

Proof: Let the defining equation of ϑ be

$$p(x) = x^n + r_1 x^{n-1} + \cdots + r_n = 0,$$

and let the LCM of the denominators of the fractions $r_1, \ldots, r_n$ be a. Then the polynomial

$$a^n p\left(\frac{x}{a}\right) = x^n + a r_1 x^{n-1} + \cdots + a^n r_n$$

has integral coefficients and is monic and irreducible; its zeros $a\vartheta$, $a\vartheta_2, \ldots, a\vartheta_n$ are therefore integers. The proof of the second part, using Theorem 2–9, is similar.

Since $R(\vartheta)$ and $R(a\vartheta)$ are identical for $a \neq 0$ in Z, any algebraic number field can be considered as the result of adjoining an algebraic integer to R.

If ϑ is an integer, so are its conjugates $\vartheta_2, \ldots, \vartheta_n$. The same is therefore true of its field conjugates.

If α is any element of the field $R(\vartheta)$ of degree n, the product $\alpha\alpha'' \cdots \alpha^{(n)}$ of all the field conjugates of α is called the *norm* of α, and denoted by $\mathbf{N}\alpha$ (a more complete notation would be $\mathbf{N}_{R(\vartheta)}\alpha$).

THEOREM 2–11. *The norm of an algebraic integer is a rational integer.*

Proof: If α has the defining equation

$$x^m + s_1 x^{m-1} + \cdots + s_m,$$

then the norm of α (in any given $R(\vartheta)$ containing α) is a power of s_m, by Theorem 2–5.

THEOREM 2–12. *If α and β are elements of $R(\vartheta)$, then*

$$\mathbf{N}(\alpha\beta) = \mathbf{N}\alpha \cdot \mathbf{N}\beta.$$

Proof: Put

$$\begin{align} \alpha &= a_0 + a_1\vartheta + \cdots + a_{n-1}\vartheta^{n-1}, \\ \beta &= b_0 + b_1\vartheta + \cdots + b_{n-1}\vartheta^{n-1}. \end{align} \tag{3}$$

Then in the product $\alpha\beta$, powers of ϑ higher than the $(n-1)$th can be reduced using the equation

$$\vartheta^{n+j} = -\vartheta^j(r_1\vartheta^{n-1} + \cdots + r_n) \tag{4}$$

derived from the defining equation of ϑ. Also $\alpha^{(k)}$ and $\beta^{(k)}$ can be obtained from (3) by replacing ϑ by ϑ_k, and in the product $\alpha^{(k)}\beta^{(k)}$, higher powers of ϑ_k can be reduced by using (4) with ϑ replaced by ϑ_k. Hence the field conjugates $(\alpha\beta)'$, $(\alpha\beta)''$, ..., $(\alpha\beta)^{(n)}$ of $\alpha\beta$ are simply $\alpha\beta$, $\alpha''\beta''$, ..., $\alpha^{(n)}\beta^{(n)}$. Thus

$$\begin{align} \mathbf{N}\alpha\beta &= (\alpha\beta)'(\alpha\beta)'' \cdots (\alpha\beta)^{(n)} \\ &= \alpha'\alpha'' \cdots \alpha^{(n)}\beta'\beta'' \cdots \beta^{(n)} = \mathbf{N}\alpha \cdot \mathbf{N}\beta. \end{align}$$

Now let α, β, ..., ν be n elements of $R(\vartheta)$, with field conjugates $\alpha^{(k)}$, $\beta^{(k)}$, ..., $\nu^{(k)}$, where $k = 1, 2, \ldots, n$. The number

$$\Delta(\alpha, \beta, \ldots, \nu) = \begin{vmatrix} \alpha & \alpha'' & \cdots & \alpha^{(n)} \\ \beta & \beta'' & \cdots & \beta^{(n)} \\ \cdot & \cdot & & \cdot \\ \cdot & \cdot & & \cdot \\ \cdot & \cdot & & \cdot \\ \nu & \nu'' & \cdots & \nu^{(n)} \end{vmatrix}^2$$

is called the *discriminant* of α, β, ..., ν. Its value is independent of the order of rows or of columns.

THEOREM 2–13. *If α, β, ..., ν are in $R[\vartheta]$, then $\Delta(\alpha, \beta, \ldots, \nu)$ is a rational integer.*

Proof: If we take the row-by-column product, we have

$$\Delta(\alpha, \ldots, \nu) = \begin{vmatrix} \alpha & \cdots & \alpha^{(n)} \\ \vdots & & \vdots \\ \nu & \cdots & \nu^{(n)} \end{vmatrix} \cdot \begin{vmatrix} \alpha & \cdots & \nu \\ \vdots & & \vdots \\ \alpha^{(n)} & \cdots & \nu^{(n)} \end{vmatrix}$$

$$= \begin{vmatrix} \alpha^2 + \cdots + (\alpha^{(n)})^2 & \cdots & \alpha\nu + \cdots + \alpha^{(n)}\nu^{(n)} \\ \vdots & & \vdots \\ \alpha\nu + \cdots + \alpha^{(n)}\nu^{(n)} & \cdots & \nu^2 + \cdots + (\nu^{(n)2}) \end{vmatrix}.$$

Just as in the proof of Theorem 2–12,

$$\alpha\beta + \alpha''\beta'' + \cdots + \alpha^{(n)}\beta^{(n)} = \alpha\beta + (\alpha\beta)'' + \cdots + (\alpha\beta)^{(n)},$$

and the sum of the field conjugates of an integer is itself a rational integer, by analogy with the proof of Theorem 2–11. Hence, the number $\Delta(\alpha, \beta, \ldots, \nu)$ can be written as a determinant with rational integral entries, and so is a rational integer.

The numbers $1, \vartheta, \ldots, \vartheta^{n-1}$ are said to form a *basis* of $R(\vartheta)$, in the sense that every element of $R(\vartheta)$ can be expressed in a unique way as a linear combination of these numbers, with coefficients in R (cf. Theorem 2–4). We now examine the possibility of finding a basis for $R[\vartheta]$; that is, a set of elements of $R[\vartheta]$ such that every element of $R[\vartheta]$ can be expressed in a unique way as a linear combination of them, the coefficients in this case being in Z. To emphasize the distinction between these two kinds of bases, the second is sometimes called an *integral basis*. Every integral basis is a basis of $R(\vartheta)$, as is immediately seen from Theorem 2–10, but the converse is false.

If $\omega_1, \ldots, \omega_n$ is to be an integral basis, then for any ρ in $R[\vartheta]$, the equation

$$\rho = x_1\omega_1 + \cdots + x_n\omega_n,$$

and therefore also the equations

$$\rho^{(k)} = x_1\omega_1^{(k)} + \cdots + x_n\omega_n^{(k)}, \qquad k = 2, \ldots, n$$

must hold for some rational integers $x_1, \ldots, x_n$. If $\Delta(\omega_1, \ldots, \omega_n) \neq 0$, this system of equations can be solved, giving each x_i as the quotient of determinants, the determinant in each denominator being a square root of $\Delta(\omega_1, \ldots, \omega_n)$. It seems plausible that the smaller $|\Delta(\omega_1, \ldots, \omega_n)|$, the better the chance of obtaining rational integral

x_i. Hence, if an integral basis always exists, the next theorem ought to be true.

THEOREM 2–14. *If $\omega_1, \omega_2, \ldots, \omega_n$ are any n integers of $R[\vartheta]$ for which $|\Delta(\omega_1, \omega_2, \ldots, \omega_n)|$ has its smallest possible value different from zero, then $\omega_1, \ldots, \omega_n$ form a basis of $R[\vartheta]$.*

Proof: Write

$$\omega_i = \sum_{j=0}^{n-1} a_{ij}\vartheta^j, \qquad i = 1, 2, \ldots, n \tag{5}$$

where the a_{ij} are in R. Then

$$\Delta(\omega_1, \ldots, \omega_n) = \begin{vmatrix} \omega_1 & \cdots & \omega_n \\ \vdots & & \vdots \\ \omega_1^{(n)} & \cdots & \omega_n^{(n)} \end{vmatrix}^2 = \begin{vmatrix} \sum_{j=0}^{n-1} a_{1j}\vartheta^j & \cdots & \sum_{j=0}^{n-1} a_{nj}\vartheta^j \\ \vdots & & \vdots \\ \sum_{j=0}^{n-1} a_{1j}\vartheta_n^j & \vdots & \sum_{j=0}^{n-1} a_{nj}\vartheta_n^j \end{vmatrix}^2,$$

and this can be factored:

$$\Delta(\omega_1, \ldots, \omega_n) = \left\{ \begin{vmatrix} 1 & \vartheta & \cdots & \vartheta^{n-1} \\ \vdots & \vdots & & \vdots \\ 1 & \vartheta_n & \cdots & \vartheta_n^{n-1} \end{vmatrix} \cdot \begin{vmatrix} a_{10} & \cdots & a_{n0} \\ \vdots & & \vdots \\ a_{1,n-1} & \cdots & a_{n,n-1} \end{vmatrix} \right\}^2$$

$$= (\det |a_{ij}|)^2 \Delta(1, \vartheta, \ldots, \vartheta^{n-1}), \tag{6}$$

where $\vartheta, \vartheta_2, \ldots, \vartheta_n$ are the conjugates of ϑ. Since $\Delta(\omega_1, \ldots, \omega_n) \neq 0$, also $\det |a_{ij}| \neq 0$, and the system of equations (5) can be solved for the numbers $1, \vartheta, \ldots, \vartheta^{n-1}$, giving linear expressions in $\omega_1, \ldots, \omega_n$. Thus every number ρ of $R[\vartheta]$ can be written in the form

$$\rho = b_1\omega_1 + \cdots + b_n\omega_n, \tag{7}$$

where $b_1, \ldots, b_n$ are rational. We must show that they are rational integers.

If this is not the case for the ρ of (7), then some b_i has a nonzero fractional part:

$$b_i = [b_i] + c,$$

where $0 < c < 1$, and the symbol $[b]$ means the largest integer not exceeding b. Put

$$\rho_i = \rho - [b_i]\omega_i = b_1\omega_1 + \cdots + c\omega_i + \cdots + b_n\omega_n.$$

In just the same way that (6) was deduced from (5), we can deduce from the system of equations

$$\omega_1 = \omega_1,$$
$$\omega_2 = \omega_2,$$
$$\vdots$$
$$\omega_{i-1} = \omega_{i-1},$$
$$\rho_i = b_1\omega_1 + b_2\omega_2 + \cdots + c\omega_i + \cdots + b_n\omega_n,$$
$$\omega_{i+1} = \omega_{i+1},$$
$$\vdots$$
$$\omega_n = \omega_n,$$

the relation

$$\Delta(\omega_1, \ldots, \rho_i, \ldots, \omega_n) = \begin{vmatrix} 1 & 0 & \cdots & & 0 \\ 0 & 1 & \cdots & & 0 \\ \vdots & \vdots & \vdots & & \vdots \\ b_1 & b_2 \ldots c \ldots & & b_n \\ \vdots & \vdots & \vdots & & \vdots \\ 0 & 0 & \cdots & & 1 \end{vmatrix}^2 \Delta(\omega_1, \omega_2, \ldots, \omega_n)$$

$$= c^2 \Delta(\omega_1, \ldots, \omega_n).$$

But this implies that the discriminant of the system $\omega_1, \ldots, \rho_i, \ldots, \omega_n$ is numerically smaller than that of $\omega_1, \ldots, \omega_n$, and is not zero, which is contrary to the hypothesis that $|\Delta(\omega_1, \ldots, \omega_n)|$ is minimal.

Any two integral bases of a single field have the same discriminant, since each is the product of the other and the square of a determinant with integral entries, as in (6). The common value is called the *discriminant of the field;* we shall designate it by Δ hereafter.

PROBLEMS

1. Let ϑ, ϑ', and ϑ'' be the roots of
   (a) $x^3 + 2x + 6 = 0$,
   (b) $x^3 - x^2 - x - 2 = 0$.
Compute the numbers $\mathbf{N}_{R(\vartheta)}(3\vartheta - 2)$.
Answer: (a) -206; (b) $4, 19$.

2. (a) Let $f(x) = a_0 x^n + \cdots + a_n$ be irreducible over R, and let $\vartheta, \vartheta'', \ldots, \vartheta^{(n)}$ be the zeros of f. Show that in $R(\vartheta)$,

$$a_0{}^n \Delta(1, \vartheta, \ldots, \vartheta^{n-1}) = (-1)^{n(n-1)/2} \prod_{i=1}^{n} f'(\vartheta^{(i)}).$$

[This depends on the well-known factorization

$$\begin{vmatrix} 1 & 1 & \cdots & 1 \\ x_1 & x_2 & \cdots & x_n \\ \cdot & \cdot & & \cdot \\ \cdot & \cdot & & \cdot \\ x_1{}^{n-1} & x_2{}^{n-1} & \cdots & x_n{}^{n-1} \end{vmatrix} = \prod_{1 \le j < i \le n} (x_i - x_j)$$

of a Vandermonde determinant.]
  (b) If in particular $f(x) = x^3 + px + q$, show that $\Delta(1, \vartheta, \vartheta^2) = -27q^2 - 4p^3$.
  3. Show that if $\alpha_1, \ldots, \alpha_n$ are elements of $R[\vartheta]$ such that $\Delta(\alpha_1, \ldots, \alpha_n)$ is square-free, then $\alpha_1, \ldots, \alpha_n$ form a basis for $R[\vartheta]$.

2-4 Units and primes in $R[\vartheta]$. If α and β are in an integral domain $R[\vartheta]$, we say that β *divides* α, and write $\beta|\alpha$, if there is another element γ of $R[\vartheta]$ such that $\alpha = \beta\gamma$. An integer ϵ such that $\epsilon|1$ is called a *unit* of $R[\vartheta]$. We say that α and β are *associates* if $\alpha = \epsilon\beta$, where ϵ is a unit.

THEOREM 2-15. *An element of $R[\vartheta]$ is a unit if and only if its norm (as an element of $R(\vartheta)$) is ± 1.*

Proof: If ϵ is a unit and

$$x^m + e_1 x^{m-1} + \cdots + e_m = 0$$

is its defining equation, then the defining equation of $1/\epsilon$ is

$$x^m + \frac{e_{m-1}}{e_m} x^{m-1} + \cdots + \frac{1}{e_m} = 0.$$

Since $1/\epsilon$ is an integer, $e_m = \pm 1$, and $\mathbf{N}(1/\epsilon)$ is a power of the constant term in the defining equation of $1/\epsilon$. (Alternatively, this result could be deduced from the multiplicativity of the norm. For if ϵ is a unit, there exists an integer ϵ_1 such that $\epsilon\epsilon_1 = 1$. Hence $1 = \mathbf{N}1 = \mathbf{N}\epsilon\epsilon_1 = \mathbf{N}\epsilon \cdot \mathbf{N}\epsilon_1$, and since the norm of an integer is a rational integer, $\mathbf{N}\epsilon = \pm 1$.)

Conversely, if the constant term in the defining equation of an element of $R[\vartheta]$ is ± 1, then the reciprocal of the element is also an element of $R[\vartheta]$, and the element is a unit.

The units of an integral domain form a multiplicative group, since the product of units is a unit, 1 is a unit, and each unit ϵ has an inverse ϵ_1 such that $\epsilon\epsilon_1 = 1$.

In the domain of rational integers, the only units are ±1; in the Gaussian domain $R[i]$, the units are ±1, $\pm i$. All these units are roots of unity, but in some domains there are units which are not roots of unity, and in fact do not have absolute value 1. This was pointed out in Chapter 8 of Volume I, but we can now go into details.

Let d be a square-free rational integer, and consider the field $R(\sqrt{d})$. As a basis for the field we can take 1, $\sqrt{d}$, so that every element of $R(\sqrt{d})$ can be uniquely expressed in the form $a + b\sqrt{d}$, where a and b are in R. If $b = 0$, then $a + b\sqrt{d}$ is an integer if and only if a is in Z. If $b \neq 0$, the defining equation of $a + b\sqrt{d}$ is

$$(x - a - b\sqrt{d})(x - a + b\sqrt{d}) = x^2 - 2ax + a^2 - db^2 = 0,$$

so that if $a + b\sqrt{d}$ is in $R[\sqrt{d}]$, both $2a$ and $a^2 - db^2$ must be rational integers. Hence $(2a)^2 - 4(a^2 - db^2) = 4db^2$ is also in Z; since d is square-free, it follows that $2b$ is in Z.

Suppose that $a = k + \frac{1}{2}$, with k in Z. Then

$$0 \equiv 4a^2 - 4db^2 \equiv 4k^2 + 4k + 1 - 4db^2 \equiv 1 - 4db^2 \pmod 4,$$

and it follows that $2b \equiv 1 \pmod 2$, and $d \equiv 1 \pmod 4$. Conversely, if a and b are halves of odd integers and $d \equiv 1 \pmod 4$, the defining equation of $a + b\sqrt{d}$ has coefficients in Z. Hence 1 and $(1 + \sqrt{d})/2$ form a basis of $R[\sqrt{d}]$, if $d \equiv 1 \pmod 4$.

If $d \equiv 2$ or $3 \pmod 4$, then a must be a rational integer. If b were of the form $k + \frac{1}{2}$, with k in Z, we should have

$$0 \equiv 4a^2 - 4db^2 \equiv -(4k^2 + 4k + 1)d \equiv -d \pmod 4,$$

and d would not be square-free. Hence in this case both a and b must be in Z, and 1, $\sqrt{d}$ form a basis of $R[\sqrt{d}]$.

THEOREM 2–16. *Let d be a square-free rational integer. Then if $d \equiv 1 \pmod 4$, the elements of $R[\sqrt{d}]$ are either of the form*

$$a + b\sqrt{d}, \qquad a \text{ and } b \text{ in } Z, \tag{8}$$

or

$$\frac{a + b\sqrt{d}}{2}, \qquad a \text{ and } b \text{ in } Z, \quad a \equiv b \equiv 1 \pmod 2,$$

and the discriminant of $R(\sqrt{d})$ is

$$\Delta = \begin{vmatrix} 1 & \frac{1}{2}(1 + \sqrt{d}) \\ 1 & \frac{1}{2}(1 - \sqrt{d}) \end{vmatrix}^2 = (-\sqrt{d})^2 = d.$$

If $d \equiv 2$ or $3 \pmod 4$, all the elements of $R[\sqrt{d}]$ are of the form (8), and the discriminant of $R(\sqrt{d})$ is

$$\Delta = \begin{vmatrix} 1 & \sqrt{d} \\ 1 & -\sqrt{d} \end{vmatrix}^2 = (-2\sqrt{d})^2 = 4d.$$

The units of $R[\sqrt{d}]$ are the integers ϵ for which $\mathbf{N}\epsilon = \pm 1$. If $d \equiv 2$ or $3 \pmod 4$, then ϵ is of the form (8), so that the units are given by the solutions of the Pell equations

$$x^2 - dy^2 = \pm 1. \tag{9}$$

If $d \equiv 1 \pmod 4$, the units are the integers of the form $(x + y\sqrt{d})/2$, where $x + y\sqrt{d}$ is a solution of one of the Pell equations

$$x^2 - dy^2 = \pm 4. \tag{10}$$

If $d < 0$, these Pell equations have only trivial solutions: (9) has solutions ± 1, 0 in all cases, and 0, ± 1 if $d = -1$, while (10) has the solution ± 2, 0 always, and ± 1, ± 1 if $d = -3$. If $d > 0$, equations (9) and (10) have infinitely many solutions.*

Returning to the general domain $R[\vartheta]$, we say that an element π is *prime* if it is not a unit and has no factors other than its associates and units.

THEOREM 2–17. *Every nonunit element of $R[\vartheta]$ can be written as a finite product of primes.*

Proof: If α in $R[\vartheta]$ is not a unit, $|\mathbf{N}\alpha| > 1$. If α is prime, we have the trivial representation $\alpha = \alpha$. If not, there is a factorization $\alpha = \beta\gamma$ into nonunits, and $\mathbf{N}\alpha = \mathbf{N}\beta \cdot \mathbf{N}\gamma$, where

$$1 < |\mathbf{N}\beta| < |\mathbf{N}\alpha|, \qquad 1 < |\mathbf{N}\gamma| < |\mathbf{N}\alpha|.$$

If either β or γ is not prime, it may be factored. The process must terminate, since the rational integer $\mathbf{N}\alpha$ has only finitely many divisors of absolute value greater than 1.

---

*This result is given in Chapter 8, Volume I. The solutions for given d can be found explicitly with the aid of Theorem 9–6 of that volume.

To see that this factorization need not be unique, consider the two representations

$$21 = 3 \cdot 7 = (4 + \sqrt{-5})(4 - \sqrt{-5})$$

of 21 in $R[\sqrt{-5}]$. Since $-5 \not\equiv 1 \pmod 4$, the integers of this domain are $a + b\sqrt{-5}$, with a and b in Z, and the units are ± 1. It is clear that no two of the numbers 3, 7, $4 + \sqrt{-5}$, $4 - \sqrt{-5}$ are associates, and we can also show that all of them are primes in $R[\sqrt{-5}]$. Suppose that

$$(a_1 + b_1\sqrt{-5})(a_2 + b_2\sqrt{-5}) = 3.$$

Then

$$\mathbf{N}(a_1 + b_1\sqrt{-5})\mathbf{N}(a_2 + b_2\sqrt{-5}) = \mathbf{N}3 = 9,$$

so that if neither factor is a unit, it must be that

$$\mathbf{N}(a_1 + b_1\sqrt{-5}) = a_1{}^2 + 5b_1{}^2 = 3. \tag{11}$$

This equation, however, has no solution in Z. By a similar argument, 7 has no proper divisors, since the equation

$$a_1{}^2 + 5b_1{}^2 = 7 \tag{12}$$

has no solution in Z. Finally, an assumed factorization of either $4 \pm \sqrt{-5}$ leads to the equation

$$\mathbf{N}(a_1 + b_1\sqrt{-5}) \cdot \mathbf{N}(a_2 + b_2\sqrt{-5}) = 21,$$

which in turn requires that either (11) or (12) hold. Hence $R[\sqrt{-5}]$ is not a unique factorization domain.

A domain $R[\vartheta]$ is called a *Euclidean domain* if for any pair of integers $\beta \neq 0$ and α of $R[\vartheta]$, there is an element γ such that

$$|\mathbf{N}(\alpha - \beta\gamma)| < |\mathbf{N}\beta|.$$

In this case, there is a Euclidean algorithm by means of which a greatest common divisor can be defined, such that if $(\alpha, \beta) = \delta$, there are integers γ_1 and γ_2 in $R[\vartheta]$ for which $\alpha\gamma_1 + \beta\gamma_2 = \delta$. It is this last property which is essential for unique factorization, since from it we get the result, equivalent to the Unique Factorization Theorem, that if $\beta|\alpha\gamma$ and $(\beta, \alpha) = 1$ then $\beta|\gamma$. For if $\gamma_1\alpha + \gamma_2\beta = 1$, then $\gamma_1\alpha\gamma + \gamma_2\beta\gamma = \gamma$; hence $\beta|\gamma$. There is no such GCD in $R[\sqrt{-5}]$.

For example, 3 and $4 + \sqrt{-5}$ must be considered as relatively prime, since they are nonassociated primes, but if we had

$$3(a + b\sqrt{-5}) + (4 + \sqrt{-5})(c + d\sqrt{-5}) = 1, \qquad a, b, c, d \text{ in } Z$$

it would follow that

$$3a + 4c - 5d = 1, \qquad 3b + c + 4d = 0.$$

Subtracting the second equation from the first, we would have

$$3(a - b + c - 3d) = 1,$$

which is palpably false.

Every Euclidean domain, then, is a unique factorization domain, although the converse is not true. The quadratic Euclidean domains are completely known: $R[\sqrt{d}]$ is Euclidean if and only if d has one of the 21 values $-11, -7, -3, -2, -1, 2, 3, 5, 6, 7, 11, 13, 17, 19, 21, 29, 33, 37, 41, 57,$ or 73.

PROBLEMS

1. Show that $R[\rho]$, where $\rho = (-1 + i\sqrt{3})/2$ is a cube root of unity, is a Euclidean domain. [Compare Theorem 7–6, Volume I.]

2. Find the GCD of $2 + \rho$ and $5 + 7\rho$ in $R[\rho]$.

3. Show that if d is square-free, and if Δ is the discriminant of $R(\sqrt{d})$, then the numbers 1 and $(\Delta + \sqrt{\Delta})/2$ form a basis of $R[\sqrt{p}]$.

2–5 Ideals. One way of restoring unique factorization consists in enlarging the set of possible divisors; we might for example try to find entities $A, B, C,$ and D of $R[\sqrt{-5}]$ which are in some sense prime, and such that

$$3 = AB, \qquad 7 = CD, \qquad 4 + \sqrt{-5} = AC, \qquad 4 - \sqrt{-5} = BD.$$

Then the two representations of 21 in $R[\sqrt{-5}]$ would no longer differ essentially; instead we would have

$$21 = (AB)(CD) = (AC)(BD) = ABCD.$$

To accomplish this without going outside the domain, we make a shift of emphasis; rather than asking for the divisors of a given number, we look for all the numbers which have a given divisor. Here two properties of the divisibility concept, in which the divisor is fixed, come to mind:

(a) If $\gamma|\alpha$, then $\gamma|\alpha\lambda$ for every integer λ.

(b) If $\gamma|\alpha$ and $\gamma|\beta$, then $\gamma|\alpha \pm \beta$.

In other words, the set of multiples of γ forms an additive group which is closed under multiplication by elements of the domain (but not necessarily in the set). If $\alpha|\beta$, then the set of multiples of α contains the set of multiples of β. The GCD (if there is one) of α and β has as multiples the set of numbers of the form $\alpha' + \beta'$, where α' and β' run independently over the multiples of α and β respectively, and this set is again an additive group closed under multiplication by elements of the domain.

Because of the repeated occurrence of this special kind of set, we give the name *ideal* to any subset (containing at least one element besides zero) of an integral domain $R[\vartheta]$ which forms a group under addition and is closed under multiplication by elements of the domain. Since there is no reason to suppose that every ideal of $R[\vartheta]$ consists of all the multiples of a single element of $R[\vartheta]$, we shall designate a general ideal by a capital letter. A *principal* ideal, consisting of all multiples of a given element α of the domain, will be designated by $[\alpha]$. (It will be clear from the context whether the brackets designate an ideal or the greatest-integer function.) But instead of a single number α, we could begin with any finite set $\alpha_1, \ldots, \alpha_m$ of elements of $R[\vartheta]$, and form all expressions

$$\lambda_1\alpha_1 + \lambda_2\alpha_2 + \cdots + \lambda_m\alpha_m,$$

where $\lambda_1, \ldots, \lambda_m$ run independently over $R[\vartheta]$; the set of such expressions again forms an ideal, which will be designated by $[\alpha_1, \ldots, \alpha_m]$. (The numbers $\alpha_1, \ldots, \alpha_m$ are called *generators* of the ideal $[\alpha_1, \ldots, \alpha_m]$.) This notation is similar to that for the GCD, if such exists, except that instead of writing $(\alpha, \beta) = \gamma$ we would now write $[\alpha, \beta] = [\gamma]$. (Two ideals are said to be equal if they consist of the same numbers.) It will be shown later that $R[\vartheta]$ is a unique factorization domain if and only if every ideal of $R[\vartheta]$ is a principal ideal. This should not be surprising, since this latter condition simply requires that any two elements of R should have a GCD in $R[\vartheta]$ which can be expressed as a linear combination of the elements.

THEOREM 2–18. *If $R(\vartheta)$ is of degree n, and A is an ideal of $R[\vartheta]$, then there exist elements $\alpha_1, \ldots, \alpha_n$ of $R[\vartheta]$ such that every element of*

A can be uniquely represented in the form

$$k_1\alpha_1 + \cdots + k_n\alpha_n, \qquad k_1, \ldots, k_n \text{ in } Z.$$

Remark: Note that the k's are rational integers, and not elements of $R[\vartheta]$. The numbers $\alpha_1, \ldots, \alpha_n$ of the theorem are called a *basis* of A; they may be taken as a set of generators of A, but may not be the smallest such set.

Proof: If the polynomial defining an element $\alpha \neq 0$ of A is $p(x)$, then for some h, the zeros of $p^h(x)$ are the field conjugates of α, so that

$$p^h(x) = x^n + a_1 x^{n-1} + \cdots \pm \mathbf{N}\alpha,$$

and $\mathbf{N}\alpha = \pm(\alpha^{n-1} + a_1\alpha^{n-2} + \cdots)\alpha$ is in A. Hence A contains a rational integer different from zero, and therefore a smallest positive integer, say a. If $\rho_1, \ldots, \rho_n$ is an integral basis of $R[\vartheta]$, then A contains $a\rho_i$ for each i. Let a_{11} be the smallest positive rational integer such that the number

$$\alpha_1 = a_{11}\rho_1$$

is in A. Since A contains $a_{11}\rho_1$ and $a\rho_2$, it contains numbers which are linear combinations of ρ_1 and ρ_2 with coefficients in Z. Of these there is one (not necessarily unique) for which the coefficient of ρ_2 is positive and minimal. Let it be

$$\alpha_2 = a_{21}\rho_1 + a_{22}\rho_2.$$

Similarly, for $\nu = 3, \ldots, n$, put

$$\alpha_\nu = a_{\nu 1}\rho_1 + a_{\nu 2}\rho_2 + \cdots + a_{\nu\nu}\rho_\nu,$$

where $a_{\nu i}$ is in Z for $1 \leq i \leq \nu$ and $a_{\nu\nu}$ is positive and minimal for α_ν in A. It is asserted that $\alpha_1, \ldots, \alpha_n$ form a basis of A.

Suppose that

$$\alpha = c_1\rho_1 + \cdots + c_n\rho_n, \qquad c_1, \ldots, c_n \text{ in } Z,$$

is in A. Then so also is $\alpha - c\alpha_n$ for every c in Z. Since

$$0 \leq c_n - a_{nn}\left[\frac{c_n}{a_{nn}}\right] < a_{nn}.$$

it follows from the minimality of a_{nn} that in the representation of the number $\alpha - [c_n/a_{nn}]\alpha_n$, the coefficient of ρ_n is 0, so that

$$\alpha - \left[\frac{c_n}{a_{nn}}\right]\alpha_n = d_1\rho_1 + \cdots + d_{n-1}\rho_{n-1}, \qquad d_1, \ldots, d_n \text{ in } Z.$$

Repeating the argument, we find that

$$\alpha - \left[\frac{c_n}{a_{nn}}\right]\alpha_n - \left[\frac{d_{n-1}}{a_{n-1,n-1}}\right]\alpha_{n-1} = e_1\rho_1 + \cdots + e_{n-2}\rho_{n-2},$$

$$e_1, \ldots, e_{n-2} \text{ in } Z.$$

After n steps, we have

$$\alpha = \left[\frac{c_n}{a_{nn}}\right]\alpha_n + \left[\frac{d_{n-1}}{a_{n-1,n-1}}\right]\alpha_{n-1} + \cdots + \left[\frac{g_1}{a_{11}}\right]\alpha_1,$$

the desired representation.

If there were two representations of the same number, their difference would be a nontrivial representation of 0:

$$k_1\alpha_1 + \cdots + k_n\alpha_n = 0, \qquad k_1^2 + \cdots + k_n^2 > 0.$$

But then also

$$k_1\alpha_1^{(m)} + \cdots + k_n\alpha_n^{(m)} = 0, \qquad m = 1, 2, \ldots, n,$$

which implies that $\Delta(\alpha_1, \ldots, \alpha_n) = 0$, contrary to the equation

$$\Delta(\alpha_1, \ldots, \alpha_n) = a_{11}^2 a_{22}^2 \cdots a_{nn}^2 \Delta(\rho_1, \ldots, \rho_n) \neq 0.$$

The proof is complete.

From their definitions, it is clear that each coefficient a_{ii} is positive and not larger than a, the smallest positive integer in A. We would like to show that bounds can also be put on the other coefficients a_{ij}, $1 \leq j < i \leq n$. We have

$$\alpha_1 = a_{11}\rho_1,$$

$$\alpha_2 = a_{21}\rho_1 + a_{22}\rho_2,$$

$$\alpha_3 = a_{31}\rho_1 + a_{32}\rho_2 + a_{33}\rho_3, \qquad (13)$$

$$\vdots$$

$$\alpha_n = a_{n1}\rho_1 + a_{n2}\rho_2 + a_{n3}\rho_3 + \cdots + a_{nn}\rho_n.$$

THEOREM 2–19. *Every ideal in $R[\vartheta]$ has a basis $\alpha_1, \ldots, \alpha_n$, given by (13), in which the numbers a_{ij} are rational integers with*

$$0 \leq a_{ij} < a_{jj} \leq a_{11}.$$

Proof: It is clear that any system of numbers $\alpha_1, \ldots, \alpha_{i-1}$, $\alpha_i - k\alpha_j, \alpha_{i+1}, \ldots, \alpha_n$, in which k is a rational integer and $j \neq i$, is also a basis of A. For if α is in A and

$$\alpha = k_1\alpha_1 + k_2\alpha_2 + \cdots + k_n\alpha_n, \qquad k_1, \ldots, k_n \text{ in } Z,$$

then

$$\alpha = k_1\alpha_1 + \cdots + (k_j + kk_i)\alpha_j + \cdots + k_i(\alpha_i - k\alpha_j) + \cdots + k_n\alpha_n.$$

In the set of equations (13), subtract a suitable multiple of α_{n-1} from α_n, so that the new coefficient of ρ_{n-1} is non-negative but smaller than $a_{n-1,n-1}$. Then subtract a suitable multiple of α_{n-2}, so that the new coefficient of ρ_{n-2} is smaller than $a_{n-2,n-2}$; this does not disturb the coefficient of ρ_{n-1}. Continuing the process, we come eventually to a basis element α_n' such that $0 \le a_{ni}' < a_{ii}$, for $i = 1, \ldots, n-1$. Then we change α_{n-1} by subtracting off suitable multiples of α_{n-2}, $\alpha_{n-3}, \ldots, \alpha_1$, etc. The result is a basis as described in the theorem.

COROLLARY. *A positive rational integer occurs in only finitely many ideals of $R[\vartheta]$.*

This follows immediately from the theorem, for if a is in A, then $a_{11} \le a$, and there are only finitely many sets of coefficients a_{ij} satisfying the conditions of the theorem.

The discriminant of the elements of a basis of an ideal is called the *discriminant of the ideal;* its value is independent of the choice of basis. For if $\alpha_1, \ldots, \alpha_n$ and $\alpha_1', \ldots, \alpha_n'$ are bases of A, then there are h_{kl} in Z such that

$$\alpha_k = \sum_{l=1}^{n} h_{kl}\alpha_l', \qquad k = 1, \ldots, n,$$

and

$$\det |h_{kl}| \ne 0.$$

Hence

$$\Delta(\alpha_1, \ldots, \alpha_n) = (\det |h_{kl}|)^2 \Delta(\alpha_1', \ldots, \alpha_n'),$$

so that the discriminants have the same sign and

$$\Delta(\alpha_1', \ldots, \alpha_n')|\Delta(\alpha_1, \ldots, \alpha_n).$$

By symmetry,

$$\Delta(\alpha_1, \ldots, \alpha_n)|\Delta(\alpha_1', \ldots, \alpha_n'),$$

and the discriminants are equal.

1. Show that every ideal in Z is principal.

2. If $A = [p, a + b\sqrt{d}]$ is an ideal of $R[\sqrt{d}]$, where p is a rational prime and d is a square-free integer not of the form $4k + 1$, show that p and $a + (b - p[b/p])\sqrt{d}$ form a basis for A.

2–6 The arithmetic of ideals.

Ideals are special kinds of sets of elements. The emphasis so far has been on the elements comprising the sets. The whole power of the theory of ideals, however, lies in considering them not as collections of elements, but as entities in their own right, which can be combined according to certain operations.

The first of these operations is multiplication. If $A = [\alpha_1, \ldots, \alpha_r]$ and $B = [\beta_1, \ldots, \beta_s]$, then the *product* AB is the ideal

$$[\alpha_1\beta_1, \ldots, \alpha_1\beta_s, \alpha_2\beta_1, \ldots, \alpha_r\beta_s].$$

The product ideal does not depend on the representation chosen for A and B. To show this, let $AB = C$, and suppose that also

$$A = [\alpha_1', \ldots, \alpha_{r'}'], \qquad B = [\beta_1', \ldots, \beta_{s'}'].$$

To keep matters straight, designate these last ideals by A' and B', even though they are equal to A and B. We must show that every element of C is also an element of $A'B' = C'$, and conversely.

First of all, α_i' is in A and β_j' is in B, so that we can write

$$\alpha_i' = \lambda_1\alpha_1 + \cdots + \lambda_r\alpha_r, \qquad \beta_j' = \mu_1\beta_1 + \cdots + \mu_s\beta_s.$$

Hence the number

$$\alpha_i'\beta_j' = \sum \lambda_k\mu_l\alpha_k\beta_l = \sum \nu_{kl}\alpha_k\beta_l$$

is in C for $1 \leq i \leq r'$, $1 \leq j \leq s'$. Since C is an ideal, every linear combination of the numbers $\alpha_i'\beta_j'$ is in C; thus C' is a subset of C. Hence $C = C'$, by symmetry.

THEOREM 2–20. *If A is an ideal of $R[\vartheta]$, there exists an ideal B of $R[\vartheta]$ such that AB is a principal ideal $[a]$, where a is in Z.*

Remark: It is this theorem which is the crux of the whole matter. As indicated in the discussion at the beginning of Section 2–5, we are trying to enlarge the set of possible divisors of an integer by introducing ideal elements. Given any such divisor, there should certainly be a second divisor whose product with the first is the original integer.

Since we have taken divisors as sets, we must identify the original integer with the set of all its multiples. It should be noted that all the associates of a given integer generate the same principal ideal.

Proof: Suppose $A = [\alpha_1, \ldots, \alpha_r]$, and put

$$f(x) = \alpha_1 + \alpha_2 x + \cdots + \alpha_r x^{r-1}.$$

By representing $\alpha_1, \ldots, \alpha_r$ as polynomials in ϑ, and replacing ϑ in all the polynomials by $\vartheta_2, \vartheta_3, \ldots, \vartheta_n$ in turn, we get sets $\alpha_1^{(\nu)}, \ldots, \alpha_r^{(\nu)}$, where $\nu = 2, 3, \ldots, n$. We define

$$g(x) = \prod_{\nu=2}^{n} (\alpha_1^{(\nu)} + \alpha_2^{(\nu)} x + \cdots + \alpha_r^{(\nu)} x^{r-1})$$

$$= \beta_1 + \beta_2 x + \cdots + \beta_s x^{s-1}.$$

The β's are symmetric polynomials, with rational integral coefficients, in all the conjugates of $\alpha_1, \ldots, \alpha_r$ except $\alpha_1, \ldots, \alpha_r$ themselves. Hence they are polynomials in $\alpha_1, \ldots, \alpha_r$, with coefficients in Z, and therefore are in $R[\vartheta]$. It is asserted that the ideal $B = [\beta_1, \ldots, \beta_s]$ satisfies the conditions of the theorem.

Put

$$f(x)g(x) = \gamma_1 + \gamma_2 x + \cdots + \gamma_{r+s-1} x^{r+s-2}.$$

Since each γ is a symmetric polynomial, with rational integral coefficients, in each α_i and its conjugates, the γ's are themselves rational integers. Let their GCD be a. Then a can be represented as a linear combination of $\gamma_1, \ldots, \gamma_{r+s-1}$, with coefficients in Z; since $\gamma_1, \ldots, \gamma_{r+s-1}$ are obviously in AB, a is in AB, and so $[a]$ is a subset of AB.

If we knew that a divides every product $\alpha_i \beta_j$, then we would know that every element of AB is contained in $[a]$. The proof will therefore be complete when we prove Theorem 2–21, which is A. Hurwitz' extension of a theorem due to Gauss.

THEOREM 2–21. *Let*

$$A(x) = \alpha_0 x^r + \cdots + \alpha_r, \qquad B(x) = \beta_0 x^s + \cdots + \beta_s,$$

where $\alpha_0 \beta_0 \neq 0$, be polynomials with integral algebraic coefficients. If an algebraic integer δ divides every coefficient of

$$C(x) = A(x)B(x) = c_0 x^t + \cdots + c_t,$$

in the sense that each quotient c_i/δ is an algebraic integer, then δ also divides every product $\alpha_k \cdot \beta_l$.

Proof: First we prove a lemma: *if*

$$f(x) = \delta_0 x^u + \cdots + \delta_u, \qquad \delta_0 \neq 0,$$

is any polynomial with integral algebraic coefficients and a zero ρ, then $f(x)/(x - \rho)$ has integral coefficients. The proof is by induction on u. If $u = 1$, then $f(x) = \delta_0 x + \delta_1$, and

$$\frac{f(x)}{x - \rho} = \frac{\delta_0 x + \delta_1}{x + \delta_1/\delta_0} = \delta_0$$

is an integer. Suppose the lemma true for all polynomials of degree less than u. Then the polynomial

$$Q(x) = f(x) - \delta_0 x^{u-1}(x - \rho)$$

has integral algebraic coefficients (by the second part of Theorem 2–10), and has degree less than u and vanishes for $x = \rho$. By the induction hypothesis,

$$\frac{Q(x)}{x - \rho} = \frac{f(x)}{x - \rho} - \delta_0 x^{u-1}$$

has integral algebraic coefficients, and the same is therefore true of $f(x)/(x - \rho)$. The lemma follows by the induction principle.

By repeated application of the lemma, we deduce that if $f(x) = \delta_0(x - \rho_1) \cdots (x - \rho_u)$, then any product $\delta_0 \rho_1 \cdots \rho_k$ is an integer.

Returning to Theorem 2–21, suppose that

$$A(x) = \alpha_0(x - \rho_1) \cdots (x - \rho_r),$$

$$B(x) = \beta_0(x - \sigma_1) \cdots (x - \sigma_s).$$

By assumption, the polynomial

$$\frac{C(x)}{\delta} = \frac{\alpha_0 \beta_0}{\delta}(x - \rho_1) \cdots (x - \sigma_s)$$

has integral coefficients, and it follows that any product

$$\frac{\alpha_0 \beta_0}{\delta} \rho_{n_1} \cdots \rho_{n_i} \sigma_{m_1} \cdots \sigma_{m_j}, \qquad \begin{cases} 1 \leq n_1 < n_2 < \cdots < n_i \leq r, \\ 1 \leq m_1 < m_2 < \cdots < m_j \leq s, \end{cases} \tag{14}$$

is an integer. Since α_k/α_0 and β_l/β_0 are elementary symmetric functions in the ρ's and σ's, respectively, the number

$$\frac{\alpha_k\beta_l}{\delta} = \frac{\alpha_0\beta_0}{\delta} \cdot \frac{\alpha_k}{\alpha_0} \cdot \frac{\beta_l}{\beta_0}$$

is a sum of terms of the form (14), and is therefore an integer. The proof is complete.

THEOREM 2–22. *If $AC - BC$, then $A - B$.*

Remark: Note that there is no zero ideal.

Proof: Let D be an ideal such that $CD = [e]$, a principal ideal. Then $ACD = BCD$, so $A[e] = B[e]$. Thus e times any element of A is equal to e times some element of B, and so $A = B$.

If $A = BC$, then we say that C *divides* A, and write $C|A$.

THEOREM 2–23. *$A|C$ if and only if every element of C is in A.*

Proof: If $A = [\alpha_1, \ldots, \alpha_r]$ and $B = [\beta_1, \ldots, \beta_s]$, then $AB = C = [\ldots, \alpha_i\beta_j, \ldots]$, so every element of C is in A, and also in B.

Conversely, suppose that every element of C is in A. Then every element of CD is in AD, for every D. Choose D so that $AD = [e]$ is principal, and let $CD = [\sigma_1, \ldots, \sigma_t]$. Then for each i with $1 \le i \le t$, $\sigma_i = e\lambda_i$ for a suitable integer λ_i. Hence $CD = [e][\lambda_1, \ldots, \lambda_t] = AD[\lambda_1, \ldots, \lambda_t]$, and by Theorem 2–22, $C = A[\lambda_1, \ldots, \lambda_t]$, so that $A|C$.

THEOREM 2–24. *An ideal is divisible by only a finite number of ideals.*

Proof: If the ideal is A, choose B so that $AB = [c]$, where c is a positive integer. Then c is in A and in every divisor of A, and by the corollary to Theorem 2–19, there are only finitely many such ideals.

A common divisor of A and B which is divisible by every common divisor is called a *greatest common divisor* (GCD) of A and B.

THEOREM 2–25. *Every pair of ideals A and B has a unique GCD, (A, B). It is composed of the numbers $\alpha + \beta$, where α runs over A and β over B.*

Proof: Let D be the set described in the theorem; it is clearly an ideal. Since 0 is in A and B, D contains every element of A and of B, and so is a divisor of A and of B. Any common divisor of A and B

contains all the elements of A and of B, and since it is closed under addition, it contains all numbers $\alpha + \beta$, and so divides D.

If D' is also a GCD of A and B, then D and D' are divisors of each other, and so each contains the other. Thus $D = D'$.

If the GCD of A and B is $[1]$, we say that A and B are *relatively prime*. As an immediate consequence of this definition and Theorem 2–25, we have

THEOREM 2–26. *If A and B are relatively prime, there exist α in A and β in B such that $\alpha + \beta = 1$.*

THEOREM 2–27. *If $A|BC$ and A is prime to B, it divides C.*

Proof: Choose α in A and β in B so that $\alpha + \beta = 1$. Then if γ is in C, $\alpha\gamma + \beta\gamma = \gamma$, and $\beta\gamma$ and $\alpha\gamma$ are in A, so that γ is in A. Hence $A|C$.

If A has no divisors except itself and $[1]$, then A is said to be *prime*.

THEOREM 2–28. *Every ideal can be represented as a finite product of prime ideals, and the representation is unique except for the order of factors.*

The finiteness of the representation follows from Theorem 2–24, and the uniqueness from Theorem 2–27.

In particular, it follows that the principal ideal generated by any element of $R[\vartheta]$ has a unique factorization into prime ideals of $R[\vartheta]$. If these prime factors are themselves always principal ideals, we might expect that ideals can be dispensed with entirely, and that there is then unique factorization of the numbers themselves.

THEOREM 2–29. *A necessary and sufficient condition that $R[\vartheta]$ be a unique factorization domain is that every ideal of $R[\vartheta]$ be a principal ideal.*

Proof: Uniqueness of factorization in $R[\vartheta]$ is equivalent to the property:

$$\text{if } \alpha|\beta\gamma \text{ and } \alpha \text{ and } \beta \text{ are relatively prime, then } \alpha|\gamma. \tag{15}$$

For if the domain has this property, unique factorization can be proved in the usual way, while if factorization is unique and $\alpha|\beta\gamma$, then every prime π dividing α must occur in the factorization of $\beta\gamma$; since this factorization is the product of the factorizations of β and γ, if π does not occur in β it must occur in γ.

Suppose that factorization is unique in $R[\vartheta]$, so that (15) holds. Then if π is a prime number, $[\pi]$ is a prime ideal. For if $[\pi] = AB$, where neither A nor B is $[\pi]$, there would exist an α in A and a β in B, neither of which is divisible by π, while their product is.

Let P be any prime ideal, and $\alpha = \pi_1{}^{n_1} \ldots \pi_r{}^{n_r}$ any element of P. Then

$$[\alpha] = [\pi_1]^{n_1} \ldots [\pi_r]^{n_r},$$

and since α is in P, so is every element of $[\alpha]$, whence $P|[\alpha]$ and P is one of the principal ideals $[\pi_k]$. Since every prime ideal is principal, every ideal is principal.

Now suppose that every ideal in $R[\vartheta]$ is principal, and that α and β are relatively prime. Then $[\alpha, \beta] = [\gamma]$, for some γ, and every linear combination $\lambda\alpha + \mu\beta$ is a multiple of γ. Taking $\lambda = 1$ and $\mu = 0$, we have $\gamma|\alpha$; for $\lambda = 0$ and $\mu = 1$, we obtain $\gamma|\beta$. Hence γ is a unit, $[\alpha, \beta] = [1]$, and we can take $\gamma = 1$. Thus there are λ and μ such that $\lambda\alpha + \mu\beta = 1$, so that if $\alpha|\beta\gamma$, then α divides $\lambda\alpha\gamma + \mu\beta\gamma = \gamma$, and (15) holds. Hence factorization is unique.

PROBLEMS

1. Using Theorem 2–21, reformulate and prove the new version of Eisenstein's irreducibility criterion, as given in Problem 1, Section 2–2.

2. Show that if $A = [a_1 + b_1\sqrt{d}, a_2 + b_2\sqrt{d}]$ is an ideal of $R[\sqrt{d}]$, then the product of A with its *conjugate ideal* $A' = [a_1 - b_1\sqrt{d}, a_2 - b_2\sqrt{d}]$ is principal.

2–7 Congruences. The norm of an ideal.

Two elements α and β of $R[\vartheta]$ will be said to be *congruent modulo an ideal A* if their difference lies in A, that is, if A divides the ideal $[\alpha - \beta]$. This is a natural extension of the earlier notion of congruence of rational integers, if the modulus m is identified with the principal ideal $[m]$. The familiar properties of congruences are easily seen to hold.

For fixed α, the set of all elements of $R[\vartheta]$ which are congruent to α modulo A is called a *residue class modulo A*.

THEOREM 2–30. *There are only finitely many residue classes modulo A.*

Proof: Choose B so that $AB = [c]$, where c is a rational integer. Then $\alpha_1 \not\equiv \alpha_2 \pmod{A}$ implies that $\alpha_1 \not\equiv \alpha_2 \pmod{[c]}$, since $A | [c]$ and therefore A contains $[c]$. So if we can show that there are only finitely many elements, no two of which are congruent modulo $[c]$, the theorem follows. But this is an immediate consequence of the fact that in the basis representation

$$\alpha = r_1\omega_1 + \cdots + r_n\omega_n,$$

where $\omega_1, \ldots, \omega_n$ form an integral basis of $R[\vartheta]$, each of the rational integral coefficients $r_1, \ldots, r_n$ has only c possible values modulo c, and that if

$$r_i \equiv r_i' \pmod{c}, \qquad i = 1, \ldots, n,$$

then

$$r_1\omega_1 + \cdots + r_n\omega_n \equiv r_1'\omega_1 + \cdots + r_n'\omega_n \pmod{[c]}.$$

The number of residue classes modulo A is called the *norm* of A, written $\mathbf{N}A$. For the time being, it is necessary to distinguish between $\mathbf{N}\alpha$ and $\mathbf{N}[\alpha]$, the norms of the number α and the ideal $[\alpha]$, respectively. However, we shall soon see that the two quantities are essentially the same.

THEOREM 2–31. *If $R(\vartheta)$ has discriminant Δ, and A is an ideal of $R[\vartheta]$ having discriminant $\Delta(A)$, then*

$$\Delta(A) = (\mathbf{N}A)^2\Delta.$$

Proof: Let $\alpha_1, \cdots, \alpha_n$ be the basis of A described in Theorem 2–19, and let $\rho_1, \cdots, \rho_n$ be a basis of $R[\vartheta]$. Then

$$\Delta(A) = (a_{11} \cdots a_{nn})^2\Delta,$$

and we must show that $\mathbf{N}A = a_{11} \cdots a_{nn}$; that is, that there are $a_{11} \cdots a_{nn}$ numbers of $R[\vartheta]$, no two of which are congruent modulo A and such that every element of $R[\vartheta]$ is congruent to one of them. We show that this is true of the numbers

$$r_1\rho_1 + \cdots + r_n\rho_n,$$

where $0 \le r_k < a_{kk}$ for $k = 1, \ldots, n$. If two of these numbers are congruent, say

$$r_1\rho_1 + \cdots + r_n\rho_n \equiv r_1'\rho_1 + \cdots + r_n'\rho_n \pmod{A},$$

and $r_n \ge r_n'$, then

$$(r_1 - r_1')\rho_1 + \cdots + (r_n - r_n')\rho_n \equiv 0 \pmod{A}.$$

But a_{nn} is the smallest positive rational integer for which any number of the form

$$s_1\rho_1 + \cdots + s_{n-1}\rho_{n-1} + a_{nn}\rho_n$$

is in A; since $0 \leq r_n - r_n' < a_{nn}$, it follows that $r_n - r_n' = 0$. Similarly, $r_{n-1} = r_{n-1}', \ldots, r_1 = r_1'$.

If

$$\beta = s_1\rho_1 + \cdots + s_n\rho_n,$$

then

$$\beta - \left[\frac{s_n}{a_{nn}}\right]\alpha_n = s_1'\rho_1 + \cdots + s_{n-1}'\rho_{n-1} + b_n\rho_n,$$

where $0 \leq b_n < a_{nn}$. By iteration,

$$\beta - \left[\frac{s_n}{a_{nn}}\right]\alpha_n - \left[\frac{s_{n-1}'}{a_{n-1,n-1}}\right]\alpha_{n-1} - \cdots = b_1\rho_1 + \cdots + b_n\rho_n,$$

where $0 \leq b_k < a_{kk}$ for $k = 1, \ldots, n$, and

$$\beta \equiv b_1\rho_1 + \cdots + b_n\rho_n \pmod{A}.$$

COROLLARY. $\mathbf{N}[\alpha] = |\mathbf{N}\alpha|$.

For $\alpha\rho_1, \ldots, \alpha\rho_n$ is clearly a basis for $[\alpha]$, and

$$\Delta(\alpha\rho_1, \ldots, \alpha\rho_n) = (\mathbf{N}\alpha)^2\Delta,$$

so that $(\mathbf{N}\alpha)^2 = (\mathbf{N}[\alpha])^2$. But $\mathbf{N}[\alpha]$, being the number of residue classes, is positive.

THEOREM 2–32. *If A and B are ideals, then there is an α in A such that $([\alpha], AB) = A$.*

If such an α exists, then clearly $[\alpha] = AC$, where $(B, C) = [1]$. If we rephrase the theorem, its close relation to Theorem 2–20 becomes clear: given two ideals A and B, there is a C such that AC is principal and $(B, C) = [1]$.

Proof: Let $P_1, \ldots, P_r$ be the distinct primes dividing AB, and let

$$A = P_1^{e_1} \cdots P_r^{e_r}, \qquad e_i \geq 0.$$

Put

$$D_i = \prod_{\substack{1 \leq j \leq r \\ j \neq i}} P_j^{e_j+1}, \qquad i = 1, \ldots, r.$$

Since $(D_1, \ldots, D_r) = [1]$, there are numbers δ_i in D_i, for $i = 1, \ldots, r$, such that

$$\delta_1 + \cdots + \delta_r = 1.$$

Then $[\delta_i]$ is divisible by D_i, and therefore by P_k for $k \neq i$, and therefore not by P_i, since 1 is not. Now let α_i be an element of $P_i^{e_i}$ which does not occur in $P_i^{e_i+1}$, for $i = 1, \ldots, r$, and put

$$\alpha = \alpha_1 \delta_1 + \cdots + \alpha_r \delta_r.$$

Then for each i, every term but one in this representation of α occurs in $P_i^{e_i+1}$, while the remaining term occurs in $P_i^{e_i}$ but not in $P_i^{e_i+1}$. Hence $A|[\alpha]$, but

$$\left(\frac{[\alpha]}{A}, B \right) = [1].$$

THEOREM 2–33. *The congruence*

$$\alpha \xi \equiv \beta \pmod{A}$$

is solvable if and only if $D|[\beta]$, where $D = ([\alpha], A)$. The solution, if it exists, is unique modulo A/D.

Proof: If ξ is a solution, then $\alpha \xi - \beta = \gamma$ is in A, and therefore in D. Since also $\alpha \xi$ is in D, it follows that β is in D, so $D|[\beta]$.

If β is in D, then it is the sum of an element of $[\alpha]$ and an element of A; that is, $\beta = \alpha \xi + \delta$. Since $\delta \equiv 0 \pmod{A}$, $\beta \equiv \alpha \xi \pmod{A}$.

If $\alpha \xi \equiv \alpha \xi' \equiv \beta \pmod{A}$, then $\alpha(\xi - \xi') \equiv 0 \pmod{A}$. Hence if $[\alpha] = DA_1$ and $A = DA_2$, then $(A_1, A_2) = [1]$ and

$$DA_2|DA_1[\xi - \xi'],$$

$$A_2|A_1[\xi - \xi'],$$

$$\xi \equiv \xi' \pmod{A_2}.$$

THEOREM 2–34. $N(AB) = NA \cdot NB$.

Proof: By Theorem 2–32, there is a γ such that

$$([\gamma], AB) = A.$$

Let $NA = n_1$, $NB = n_2$, and let $\alpha_1, \ldots, \alpha_{n_1}$ and $\beta_1, \ldots, \beta_{n_2}$ be complete residue systems modulo A and B, respectively. We shall show that the $n_1 n_2$ numbers $\alpha_i + \gamma \beta_j$ form a complete residue system modulo AB.

If
$$\alpha_i + \gamma\beta_j \equiv \alpha_k + \gamma\beta_l \pmod{AB},$$
then
$$\gamma(\beta_j - \beta_l) \equiv \alpha_k - \alpha_i \pmod{AB},$$

and by Theorem 2–33, $([\gamma], AB)|[\alpha_k - \alpha_i]$, so that $A|[\alpha_k - \alpha_i]$. But this gives $\alpha_k \equiv \alpha_i \pmod A$, so $k = i$. Moreover,

$$\gamma(\beta_j - \beta_l) = 0 \pmod{AB},$$

$$\beta_j - \beta_l \equiv 0 \pmod B,$$

$$j = l.$$

To show that every integer δ is congruent to one of the above numbers, choose α_i so that $\delta \equiv \alpha_i \pmod A$. Then the congruence

$$\gamma\xi \equiv \delta - \alpha_i \pmod{AB}$$

is solvable, since $([\gamma], AB) = A$ is a divisor of $[\delta - \alpha_i]$. Finally, ξ is unique modulo B, and can therefore be taken to be one of the numbers β_j.

THEOREM 2–35. **NA is an element of A.**

Proof: If $\alpha_1, \ldots, \alpha_{NA}$ is a complete residue system modulo A, then so is $\alpha_1 + 1, \ldots, \alpha_{NA} + 1$. Hence

$$\alpha_1 + \cdots + \alpha_{NA} \equiv (\alpha_1 + 1) + \cdots + (\alpha_{NA} + 1) \pmod A,$$

$$0 \equiv NA \pmod A.$$

COROLLARY. *There are only finitely many ideals of given norm.*

For by the corollary to Theorem 2–19, a positive rational integer occurs in only finitely many ideals.

PROBLEMS

1. Show that if P is a prime ideal of $R[\vartheta]$, the congruence

$$x^m + \alpha_1 x^{m-1} + \cdots + \alpha_m \equiv 0 \pmod P$$

with coefficients in $R[\vartheta]$ has at most m incongruent solutions modulo P.

2. Show that if P is a prime ideal of $R[\vartheta]$, α is an element of $R[\vartheta]$, and $P\!\nmid\![\alpha]$, then

$$\alpha^{NP-1} \equiv 1 \pmod P.$$

2–8 Prime ideals

THEOREM 2–36. *If* $\mathbf{N}A$ *is prime, so is* A.

This follows immediately from Theorem 2–34.

THEOREM 2–37. *There are infinitely many prime ideals* P *in any domain* $R[\vartheta]$. *Each such* P *divides exactly one rational prime* p, *and* $\mathbf{N}P = p^f$, *where* f, *called the* degree *of* P, *is a positive integer not exceeding the degree of* $R(\vartheta)$.

Proof: Let p be a rational prime, and let P be one of the factors of $[p]$ in $R[\vartheta]$. Then if P also divided the ideal defined by another rational prime p', it would divide their GCD, which is $[1]$. Hence each P divides at most one p, and each of the infinitely many rational primes p is divisible by at least one P, so that there must be infinitely many P's.

Now let a be a rational integer such that $P|[a]$; by Theorem 2–35, we could take $a = \mathbf{N}P$. If $a = p_1 \cdots p_r$, then

$$P|[p_1] \cdots [p_r],$$

and so $P|[p_i]$ for some i.

Finally, if $P|[p]$ then $[p] = PA$ for some A. By the corollary to Theorem 2–31,

$$\mathbf{N}[p] = |\mathbf{N}p| = p^n,$$

and so $\mathbf{N}(PA) = \mathbf{N}P \cdot \mathbf{N}A = p^n$. Hence $\mathbf{N}P|p^n$, and the proof is complete.

Theorem 2–37 shows that the primes of $R[\vartheta]$ are to be found among the factors of the principal ideals $[p]$. Only partial information is available about the way these ideals decompose, and the derivation of most of what is known is too intricate for inclusion here, but we can prove the simpler half of a famous theorem due to Dedekind, which states that $[p]$ *is divisible by the square of a prime ideal in* $R[\vartheta]$ *if and only if* p *divides* Δ, *the discriminant of* $R(\vartheta)$.

THEOREM 2–38. *If* p *does not divide* Δ, *then* $[p]$ *factors as a product of* (one or more) *distinct prime ideals.*

Proof: Suppose that $P^2|[p]$, so that $[p] = P^2M$. Choose an element α of PM which does not belong to P^2M, so that $p|\alpha^2$ but $p\nmid\alpha$. Since $p \geq 2$, $p|(\alpha\beta)^p$ for every β in $R[\vartheta]$.

For an arbitrary element γ of $R(\vartheta)$, define $\mathbf{S}\gamma$, the *trace* of γ, by the equation

$$\mathbf{S}\gamma = \gamma' + \cdots + \gamma^{(n)},$$

where $\gamma', \ldots, \gamma^{(n)}$ are the field conjugates of γ. By the Symmetric Function Theorem, $\mathbf{S}\gamma$ is in Z if γ is an integer, and it is clear that $\mathbf{S}(r\gamma) = r\mathbf{S}\gamma$ if r is rational. In particular,

$$\mathbf{S}\frac{(\alpha\beta)^p}{p} = \frac{\mathbf{S}(\alpha\beta)^p}{p}$$

is in Z, so that $\mathbf{S}(\alpha\beta)^p$ is in $[p]$. By the multinomial theorem, if $\beta', \ldots, \beta^{(n)}$ are the field conjugates of β, then

$$\begin{aligned}
(\mathbf{S}(\alpha\beta))^p &= (\alpha'\beta' + \cdots + \alpha^{(n)}\beta^{(n)})^p \\
&\equiv (\alpha'\beta')^p + \cdots + (\alpha^{(n)}\beta^{(n)})^p = \mathbf{S}((\alpha\beta)^p) \\
&\equiv 0 \pmod{p},
\end{aligned}$$

and since $\mathbf{S}(\alpha\beta)$ is a rational integer, $p|\mathbf{S}(\alpha\beta)$.

Now let $\rho_1, \ldots, \rho_n$ be an integral basis for $R[\vartheta]$. Then

$$\alpha = h_1\rho_1 + \cdots + h_n\rho_n,$$

where the h's are rational integers not all divisible by p, since $p \nmid \alpha$. For $1 \leq i \leq n$ we have

$$\mathbf{S}(\alpha\rho_i) = \mathbf{S}\left(\sum_{j=1}^{n} h_j\rho_j\rho_i\right) = \sum_{j=1}^{n} h_j\mathbf{S}(\rho_i\rho_j).$$

Let

$$d = \det |\mathbf{S}(\rho_i\rho_j)|,$$

and let A_{ij} be the cofactor of $\mathbf{S}(\rho_i\rho_j)$ in d. Then for $k = 1, 2, \ldots, n$,

$$\sum_{i=1}^{n} A_{ik} \sum_{j=1}^{n} h_j\mathbf{S}(\rho_i\rho_j) = \sum_{j=1}^{n} h_j \sum_{i=1}^{n} A_{ik}\mathbf{S}(\rho_i\rho_j) = dh_k.$$

Since

$$p|\sum_{j} h_j\mathbf{S}(\rho_i\rho_j)$$

for each i, it is also true that $p|dh_k$ for each k; p therefore divides d. Finally,

$$d = \det |\mathbf{S}(\rho_i\rho_j)| = \det |\sum_{k} \rho_i^{(k)}\rho_j^{(k)}| = \det |\rho_i^{(k)}|^2 = \Delta;$$

hence $p|\Delta$.

As an illustration of the present theorem, note that in the field $R(i)$, of discriminant -4, we have

$$[2] = [1 + i]^2,$$
$$[p] = [a + bi][a - bi], \qquad \text{if } a^2 + b^2 = p \equiv 1 \;(\text{mod } 4),$$
$$[q] = P, \text{ a prime ideal}, \qquad \text{if } q \equiv 3 \;(\text{mod } 4).$$

Here $[1 + i]$, $[a + bi]$, and $[a - bi]$ are prime ideals of degree 1, while each P is of degree 2.*

THEOREM 2–39. *Each ideal $[p]$, where p is a rational prime, splits into at most n ideal factors in the integral domain of a field of degree n.*

Proof: If
$$[p] = P_1 \cdots P_r,$$
then
$$p^n = \mathbf{N}[p] = \mathbf{N}P_1 \cdots \mathbf{N}P_r,$$
and for each i, $\mathbf{N}P_i > 1$. Hence $r \leq n$.

PROBLEMS

1. In the domain $R[\sqrt{-5}]$, put

$$A = [3, 4 + \sqrt{-5}], \qquad B = [3, 4 - \sqrt{-5}], \qquad C = [7, 4 + \sqrt{-5}],$$
$$D = [7, 4 - \sqrt{-5}].$$

Show that $AB = [3]$, $CD = [7]$, $AC = [4 + \sqrt{-5}]$, $BD = [4 - \sqrt{-5}]$, and that A, B, C, and D are prime ideals. Factor $[1 + 2\sqrt{-5}]$.

2. Let $R(\sqrt{d})$, where d is square-free, have discriminant Δ. If q is an odd prime dividing Δ, show that the ideals

$$\left[q, \frac{\Delta + \sqrt{\Delta}}{2} \right] \qquad \text{and} \qquad \left[q, \frac{\Delta - \sqrt{\Delta}}{2} \right]$$

are equal, and that their product is q. Show also that if Δ is even, then $[2] = [2, \sqrt{d}]^2$ for $d \equiv 2 \;(\text{mod } 4)$ and that $[2] = [2, 1 + \sqrt{d}]^2$ if $d \equiv 3$ (mod 4). This completes the proof of Dedekind's theorem, stated just before Theorem 2–38, in the case of a quadratic field.

2–9. Units of algebraic number fields. We saw in Section 2–4 that the units of a quadratic field $R(\sqrt{d})$ are determined by the solutions of the Pell equation with $N = \pm 1$ or ± 4, and it is an easy conse-

---

*Compare the remark following Theorem 7–7, Volume I.

quence of this relationship and standard properties of Pell's equation*
that the group of units of $R(\sqrt{d})$ has a *basis*, consisting of -1 and
the fundamental solution ϵ of the appropriate Pell equation. That is,
every unit of $R(\sqrt{d})$ can be written in the form $(-1)^{\alpha}\epsilon^{\beta}$, where α is
0 or 1 and β ranges over Z. We shall now show that this property is
not peculiar to quadratic fields, but that in fact the group of units in
each algebraic number field has a finite basis. (In general, if G is a
commutative multiplicative group and $b_1, \ldots, b_m$ are elements of G,
they are said to form a basis for G if every element of G can be represented in the form $b_1{}^{n_1} \cdots b_m{}^{n_m}$, and in every such representation of
the unit element e of G, the factor $b_i{}^{n_i} = e$ for $1 \leq i \leq m$.) This
theorem, which is due to Dirichlet, can be sharpened by giving
the exact number of basis elements, but for many purposes, including
the application to be made in the next chapter, the finiteness of the
basis suffices. The upper bound which we shall obtain is actually the
correct number.

We introduce the symbol $\lceil \alpha \rceil$ to designate the maximum of the
absolute values of the conjugates of the algebraic number α, and
denote by K a fixed algebraic number field.

THEOREM 2–40. *If a is a fixed positive number, there are only
finitely many integers α of K such that*

$$\lceil \alpha \rceil \leq a.$$

If all conjugates of α have absolute value 1, then α is a root of unity.

Proof: If $\lceil \alpha \rceil \leq a$ and deg $\alpha = n$, then each of the elementary
symmetric functions in α and its conjugates is numerically smaller
than some bound depending only on a and n. If α is an integer in K,
then n cannot exceed the degree of K, so that there are available only
finitely many coefficients for the defining polynomial of α, and there
are, therefore, only finitely many such α's in K.
If $|\alpha^{(i)}| = 1$ for $i = 1, \ldots, n$, then $\lceil \alpha^m \rceil = 1$ for all m in Z, so that
by what we have just proved, $\alpha^{m_1} = \alpha^{m_2}$ for some distinct exponents
m_1 and m_2. Hence $\alpha^{m_1 - m_2} = 1$, so that α is a root of unity.

THEOREM 2–41. *The group U of roots of unity in K is a finite cyclic
group.*

---

*See, for example, Theorems 8–5, 8–6, and 8–7, Volume I.

Proof: If ζ is a root of unity, then $|\zeta| = 1$, and the finiteness of the group follows from the preceding theorem. Let the various elements u_i of U be primitive w_i-th roots of unity, for $i = 1, \ldots, t$, and put

$$w = \max (w_1, \ldots, w_t).$$

For fixed i, the numbers $e^{2\pi i a/w_i}$ and $e^{2\pi i b/w}$ are in U, for every a and b in Z. If $(w_i, w) = d$, choose a and b so that $aw + bw_i = d$; then the product

$$e^{2\pi i(a/w_i + b/w)} = e^{2\pi i d/w_i w} = e^{2\pi i/\langle w_i, w\rangle}$$

is in U. It follows that the LCM of w_i and w does not exceed w, so that $w_i | w$ for $i = 1, \ldots, t$. Since the powers of

$$\zeta_0 = e^{2\pi i/w}$$

include all dth roots of unity if $d|w$, it is clear that ζ_0 generates U.

Now let ϑ, of degree n, be a primitive element of K, so that $K = R(\vartheta)$, and arrange the conjugates of ϑ in such an order that $\vartheta^{(1)}, \ldots, \vartheta^{(r_1)}$ are real, while $\vartheta^{(r_1+1)}, \ldots, \vartheta^{(n)}$ are not real. (Note that it is not necessarily true that $\vartheta^{(1)} = \vartheta$.) Then $n - r_1$ is an even number, say $2r_2$, and we can further order the nonreal conjugates so that $\vartheta^{(r_1+j)}$ and $\vartheta^{(r_1+r_2+j)}$ are complex-conjugate, for $j = 1, \ldots, r_2$. If α is any number of K, the field conjugates of α are such that $\alpha^{(1)}, \ldots, \alpha^{(r_1)}$ are real, while $\alpha^{(r_1+j)}$ and $\alpha^{(r_1+r_2+j)}$ are complex-conjugate for $j = 1, \ldots, r_2$. Of course some of these latter numbers may also be real, but in any case

$$|\alpha^{(r_1+j)}| = |\alpha^{(r_1+r_2+j)}|, \qquad \text{for } j = 1, \ldots, r_2. \tag{16}$$

If $\epsilon_1, \ldots, \epsilon_k$ are units of K, they are said to be *independent* if the relation

$$\epsilon_1^{a_1} \cdots \epsilon_k^{a_k} = 1, \qquad a_1, \ldots, a_k \text{ in } Z, \tag{17}$$

holds only for $a_1 = \cdots = a_k = 0$.

THEOREM 2–42. *Units $\epsilon_1, \ldots, \epsilon_k$ in K are independent if and only if the sole solution of the system*

$$\sum_{m=1}^{k} x_m \log |\epsilon_m^{(i)}| = 0, \qquad i = 1, 2, \ldots, r, \tag{18}$$

in rational integers is $x_1 = \cdots = x_k = 0$. Here $r = r_1 + r_2 - 1$.

Proof: Suppose that (17) has a solution in which not all the a's are zero. Then the analogous equation with each ϵ_m replaced by $\epsilon_m^{(i)}$ also holds, so that

$$|\epsilon_1^{(i)}|^{a_1} \cdots |\epsilon_k^{(i)}|^{a_k} = 1, \qquad i = 1, \ldots, n,$$

and

$$\sum_{m=1}^{k} a_m \log |\epsilon_m^{(i)}| = 0, \qquad i = 1, \ldots, n. \tag{19}$$

Conversely, if (19) holds with not all the rational integers $a_1, \ldots, a_k$ equal to zero, then $\epsilon_1^{a_1} \cdots \epsilon_k^{a_k}$ is an integer of K all of whose conjugates have absolute value 1; it is therefore a root of unity whose wth power is 1, and (17) holds with $a_1, \ldots, a_k$ replaced by $wa_1, \ldots, wa_k$. Hence the nontrivial solvability in Z of (19) is equivalent to the dependence of $\epsilon_1, \ldots, \epsilon_k$.

The truth of the theorem will now follow if we can prove that if the equations (19) hold with $i = 1, \ldots, r_1 + r_2 - 1$, then the remaining $n - r_1 - r_2 + 1$ equations are also correct. To show this, suppose that the first $r_1 + r_2 - 1$ equations are true, and define

$$e_i = \begin{cases} 1 & \text{for } 1 \le i \le r_1, \\ 2 & \text{for } r_1 + 1 \le i \le n. \end{cases}$$

Since each ϵ_m is a unit, its norm $\epsilon_m^{(1)} \cdots \epsilon_m^{(n)}$ has absolute value 1; by (16),

$$\sum_{i=1}^{n} \log |\epsilon_m^{(i)}| = \sum_{i=1}^{r_1+r_2} e_i \log |\epsilon_m^{(i)}| = 0.$$

Hence

$$\sum_{m=1}^{k} a_m \sum_{i=1}^{r_1+r_2} e_i \log |\epsilon_m^{(i)}| = \sum_{i=1}^{r_1+r_2} e_i \sum_{m=1}^{k} a_m \log |\epsilon_m^{(i)}| = 0,$$

so that

$$e_{r_1+r_2} \cdot \sum_{m=1}^{k} a_m \log |\epsilon_m^{(r_1+r_2)}| = -\sum_{i=1}^{r_1+r_2-1} e_i \sum_{m=1}^{k} a_m \log |\epsilon_m^{(i)}| = 0.$$

Thus (19) also holds for $i = r_1 + r_2$, and so, by (16), for

$$i = 1, 2, \ldots, n.$$

THEOREM 2–43. *If the relation* (18) *holds for some set of real numbers* $x_1, \ldots, x_k$ *which are not all zero, it also holds for rational integers* $x_1, \ldots, x_k$ *which are not all zero.*

Proof: Suppose the hypothesis fulfilled. Since the system (18) is certainly nontrivially solvable in rational integers if some ϵ_m is in U, it suffices to consider the case that all the units are of infinite order. Then each unit separately is independent. Now suppose that the units $\epsilon_1, \ldots, \epsilon_q$ are such that the equations

$$\sum_{m=1}^{q-1} \alpha_m \log |\epsilon_m^{(i)}| = 0, \qquad i = 1, \ldots, r, \tag{20}$$

have the single real solution $\alpha_1 = \cdots = \alpha_{q-1} = 0$, while the system

$$\sum_{m=1}^{q} \alpha_m \log |\epsilon_m^{(i)}| = 0, \qquad i = 1, \ldots, r, \tag{21}$$

has a nontrivial real solution $\alpha_1, \ldots, \alpha_q$. Then $2 \leq q \leq k$, $\alpha_q \neq 0$, and the ratios $\alpha_1/\alpha_q, \ldots, \alpha_{q-1}/\alpha_q$ are uniquely determined, since otherwise the differences of the respective ratios would provide a nontrivial solution of (20). If we can show that these ratios are rational numbers, the theorem will result by taking a suitable common integral multiple of the numbers

$$x_m = \begin{cases} \dfrac{\alpha_m}{\alpha_q} & \text{for } 1 \leq m \leq q, \\ 0 & \text{for } q < m \leq k. \end{cases}$$

If we put $\alpha_m/\alpha_q = -\beta_m$ for $m = 1, \ldots, q-1$, equations (21) imply that

$$\log |\epsilon_q^{(i)}| = \sum_{m=1}^{q-1} \beta_m \log |\epsilon_m^{(i)}|, \qquad i = 1, \ldots, n. \tag{22}$$

Now consider the set of all units η with the property that

$$\log |\eta^{(i)}| = \sum_{m=1}^{q-1} \gamma_m \log |\epsilon_m^{(i)}|, \qquad i = 1, \ldots, n, \tag{23}$$

for suitable real numbers $\gamma_1, \ldots, \gamma_{q-1}$. For such an η the coefficients γ_m are unique. We call the set $\gamma_1, \ldots, \gamma_{q-1}$ of real numbers *proper* if η, as defined in (23) with these γ's, actually is a unit, and if in addition $|\gamma_1| < 1, \ldots, |\gamma_{q-1}| < 1$. If $\gamma_1, \ldots, \gamma_{q-1}$ is a proper set, then

$$\left|\log|\eta^{(i)}|\right| < \sum_{m=1}^{q-1} \left|\log|\epsilon_m^{(i)}|\right|,$$

and, by Theorem 2–40, there are only finitely many (say H) proper sets. On the other hand, if $\gamma_1, \ldots, \gamma_{q-1}$ is proper, so also is

$$N\gamma_1 - [N\gamma_1], \ldots, N\gamma_{q-1} - [N\gamma_{q-1}],$$

if N is a rational integer. For

$$\sum_{m=1}^{q-1} (N\gamma_m - [N\gamma_m]) \log |\epsilon_m^{(i)}| = \log |\eta^{(i)}|^N - \sum_{m=1}^{q-1} \log |\epsilon_m^{(i)}|^{[N\gamma_m]},$$

which is the logarithm of a product of powers of units, and is therefore the logarithm of a unit. Now if any β_m were irrational, then no two of the numbers $N\beta_m - [N\beta_m]$, where N runs over Z, would be equal, and we should have infinitely many proper sets. This contradiction establishes the theorem.

THEOREM 2–44. *If* $\epsilon_1, \ldots, \epsilon_k$ *are units such that the only real solution of* (18) *is the trivial solution, then there is a rational integer* M *with the following property: in order that a number* η *such that*

$$\log |\eta^{(i)}| = \sum_{m=1}^{k} \gamma_m \log |\epsilon_m^{(i)}|, \qquad i = 1, \ldots, n,$$

be a unit of K, *it is necessary that all the numbers* $M\gamma_m$ *be rational integers.*

Proof: The hypothesis is that which was used in the preceding proof, except that we have replaced $q - 1$ by k. Suppose that $\gamma_m = a/b$, where a and b are rational integers with $b > 0$ and $(a, b) = 1$, and m is one of the integers $1, \ldots, k$. Then $N\gamma_m - [N\gamma_m]$ assumes the b values $0/b, 1/b, \ldots, (b - 1)/b$, so that $b \leq H$, where H is the number of proper sets. Hence, $b | H!$, and we can take $M = H!$.

THEOREM 2–45. *The group* E *of all units of* K *has a finite basis, the number of basis elements of infinite order being at most* r.

Proof: The system (18) of r linear homogeneous equations in k unknowns is certainly nontrivially solvable in reals if $k > r$, and it follows from Theorem 2–43 that there are at most r independent units in K. Let k be the exact maximal number of independent units, and let $\epsilon_1, \ldots, \epsilon_k$ be such a set. Then by Theorem 2–44, for every unit η of K there are $g_1, \ldots, g_k$ in Z such that

$$\log |\eta^{(i)}| = \sum_{m=1}^{k} \frac{g_m}{M} \log |\epsilon_m^{(i)}|, \qquad i = 1, \ldots n.$$

By the second part of Theorem 2–40, and Theorem 2–41, it follows that

$$\eta^M = \epsilon_1{}^{g_1} \cdots \epsilon_k{}^{g_k}\zeta_0{}^g,$$

so that $\epsilon_1, \ldots, \epsilon_k, \zeta_0$ form a basis for the group of Mth powers of units. Now define the numbers $\xi_0, \ldots, \xi_k$ by the equations

$$\xi_0 = \zeta_0{}^{1/M}, \qquad \xi_1 = \epsilon_1{}^{1/M}, \qquad \ldots, \qquad \xi_k = \epsilon_k{}^{1/M},$$

where an arbitrary but fixed Mth root is taken in each case. The numbers ξ_m may not lie in K, but they form a basis for a group E_M of complex numbers, and E_M clearly contains E as a subgroup. The theorem is therefore a consequence of the following general principle.

THEOREM 2–46. *If G is a commutative group having a basis of n elements, every subgroup of G also has a basis, of at most n elements.*

Proof: Suppose that $\lambda_1, \ldots, \lambda_n$ is a basis for G, that S is a subgroup of G, and that some λ_i actually occurs in the representation of some element s of S. Let I_i be the set of all exponents which occur on λ_i in the representations of the various elements of S. If a is in I_i, so is ka for k in Z, and if a and a' are in I_i, so is $a - a'$. Hence I_i is an ideal in Z, and is therefore a principal ideal, say $I_i = [a_i{}^*]$.

We now proceed by induction on n. If $n = 1$, then $\lambda_1{}^{a_1*}$ is a basis for S, by what we have just proved. Suppose that the theorem is true for every commutative group with $n - 1$ basis elements, and suppose that G has n basis elements, say $\lambda_1, \ldots, \lambda_n$. Let S be a subgroup of G. If every element of S can be written in the form

$$\lambda_1{}^{a_1} \cdots \lambda_{n-1}{}^{a_{n-1}},$$

the result follows from the induction hypothesis. Otherwise, suppose that $I_n = [a]$, and let λ be an element of S in whose basis representation λ_n occurs with exponent a. Then for every s in S there exists a b_s in Z such that $s\lambda^{b_s}$ has a representation

$$s\lambda^{b_s} = \lambda_1{}^{a_1} \cdots \lambda_{n-1}{}^{a_{n-1}}.$$

The set of numbers of the form $s\lambda^{b_s}$ is therefore a subgroup of the group G' which has $\lambda_1, \ldots, \lambda_{n-1}$ as a basis, and by the induction hypothesis this subgroup also has a basis, of at most $n - 1$ elements. This latter basis, together with λ, clearly constitutes a basis for S.

REFERENCES

Section 2-4

The complete tabulation of Euclidean domains is the work of many writers. K. Inkeri (*Annales Academiae Scientiarum Fennicae, Series A* (Helsinki) I, Mathematics-Physics, **41**, 35pp. (1947)) supplied the last link in a chain of theorems which together show that if $d > 100$, then $R(\sqrt{d})$ is not Euclidean. E. S. Barnes and H. P. F. Swinnerton-Dyer (*Acta Mathematica* (Stockholm) **87**, 259–323 (1952)) showed that, contrary to what had been believed, $R(\sqrt{97})$ is not Euclidean. P. Varnavides (*Proceedings Konink. Nederlandsche Akademie van Wetenschappen, Series A* (Amsterdam) **55**, 111–122 (1952) or *Indagationes Mathematicae* (Amsterdam) **14**, 111–122 (1952)) showed that the values of d listed in the text yield Euclidean domains.

Section 2-9

The material of this section is adapted from E. Hecke, *Vorlesungen über die Theorie der Algebraischen Zahlen*, Leipzig: Akademische Verlagsgesellschaft m.b.H., 1923; reprinted by Chelsea Publishing Company, New York, 1948; pp. 116–131. It is proved there that the upper bound obtained in the text is exact.

CHAPTER 3

APPLICATIONS TO RATIONAL NUMBER THEORY

3-1 Introduction. As was suggested in the preceding chapter, there are many problems in rational number theory which are most naturally treated in the more extensive framework of an algebraic number field. Chief among these are various Diophantine equations; indeed, it was the study of Fermat's equation, $x^n + y^n = z^n$, $n \geq 3$, which was originally responsible for the development of ideal theory. While this approach has not led to a complete verification of Fermat's conjecture in all cases, it has produced results which would probably never have been obtained using rational methods alone. In the first part of this chapter we will discuss some results of this kind due to E. Kummer. Here heavy use will be made of ideal theory.

The latter portion of the chapter is primarily concerned with a theorem due to B. Delauney and T. Nagell, which asserts that the cubic analog of Pell's equation,

$$x^3 + dy^3 = 1,$$

has at most one solution in nonzero rational integers x, y, and completely characterizes this possible solution. (In the next chapter we shall prove a less precise result about the general equation $x^n + dy^n = 1$, $n \geq 3$.) Use is made here of the insolvability in Z of

$$x^3 + y^3 = z^3,$$

but otherwise the two parts are mutually independent.

3-2 Equivalence and class number. We say that the ideals A and B of $R[\vartheta]$ are *equivalent*, and write $A \sim B$, if there are nonzero elements α and β of $R[\vartheta]$ such that

$$[\alpha]A = [\beta]B.$$

It is easily seen that "$\sim$" is an equivalence relation. Moreover, if

$A \sim B$ and $C \sim D$, then $AC \sim BD$, and if $AC \sim BC$ then $A \sim B$.

THEOREM 3-1. *All principal ideals are equivalent. Any ideal equivalent to a principal ideal is principal.*

Proof: The first statement is trivial, since

$$[\alpha][\beta] = [\beta][\alpha].$$

If $A \sim [\alpha]$, then for some β and γ,

$$[\beta]A = [\alpha][\gamma] = [\alpha\gamma],$$

and hence

$$[\beta]|[\alpha\gamma],$$

$$\beta|\alpha\gamma,$$

$$\alpha\gamma = \beta\delta,$$

$$[\beta]A = [\alpha\gamma] = [\beta][\delta],$$

$$A = [\delta].$$

Since equivalence is an equivalence relation, the ideals of $R[\vartheta]$ can be separated into equivalence classes in the usual way. The number h of such classes is called the *class number* of the field; according to Theorem 3-1, $h = 1$ if and only if every ideal is principal, i.e., if and only if $R[\vartheta]$ is a unique factorization domain. We shall now show that h is always finite.

THEOREM 3-2. *There is a positive constant c, which depends only on the field, such that each ideal A divides a principal ideal AB for which*

$$\mathbf{N}AB \le c\mathbf{N}A.$$

Proof: Let $\rho_1, \ldots, \rho_n$ be a field basis, and let $\rho_1^{(s)}, \ldots, \rho_n^{(s)}$ $(s = 1, \ldots, n)$ be the field conjugates of these numbers. We shall show that the theorem is true with

$$c = \prod_{s=1}^{n} (|\rho_1^{(s)}| + \cdots + |\rho_n^{(s)}|).$$

Let A be an arbitrary ideal, and let k be the greatest rational integer not exceeding $(\mathbf{N}A)^{1/n}$, so that $k^n \le \mathbf{N}A < (k+1)^n$. Then if $t_1, \ldots, t_n$ range independently over the integers $0, 1, \ldots, k$, there are determined $(k + 1)^n$ different integers

$$t_1\rho_1 + \cdots + t_n\rho_n,$$

and two of them must be congruent modulo A :

$$u_1\rho_1 + \cdots + u_n\rho_n \equiv v_1\rho_1 + \cdots + v_n\rho_n \pmod{A}.$$

Thus

$$\alpha = (u_1 - v_1)\rho_1 + \cdots + (u_n - v_n)\rho_n$$

is in A, so that $A|[\alpha]$, and

$$\mathbf{N}[\alpha] = |\mathbf{N}\alpha| = \left| \prod_{s=1}^{n} \left(\sum_{i=1}^{n} (u_i - v_i)\rho_i^{(s)} \right) \right| \leq \prod_{s=1}^{n} \sum_{i=1}^{n} |u_i - v_i| \cdot |\rho_i^{(s)}|$$

$$\leq \prod_{s=1}^{n} \sum_{i=1}^{n} k|\rho_i^{(s)}| = ck^n \leq c\mathbf{N}A.$$

THEOREM 3–3.   *The class number of any algebraic number field is finite.*

Proof: It suffices to show that in each class there is an ideal B such that $\mathbf{N}B \leq c$, by the corollary to Theorem 2–35. Let C be an arbitrary ideal of a given class, and determine A so that AC is principal. Then by Theorem 3–2, there is an ideal B such that AB is principal and $\mathbf{N}AB \leq c\mathbf{N}A$. Then $AB \sim AC$, $B \sim C$, and

$$\mathbf{N}B = \frac{\mathbf{N}AB}{\mathbf{N}A} \leq c.$$

THEOREM 3–4.   *If h is the class number, the hth power of any ideal is principal.*

Proof: If $A_1, \ldots, A_h$ is a complete system of representatives of the various classes, and A is arbitrary, then $AA_1, \ldots, AA_h$ is another such system. Hence

$$A_1 \cdots A_h \sim AA_1 \cdots AA_h = A^h A_1 \cdots A_h,$$

so $A^h \sim [1]$ and A^h is principal.

THEOREM 3–5.   *If p is a rational prime and $p \nmid h$, then $A^p \sim B^p$ implies $A \sim B$.*

Proof: Since $p \nmid h$, there are positive x and y in Z such that

$$px - hy = 1.$$

From the fact that $A^p \sim B^p$ we have

$$[\alpha]A^p = [\beta]B^p,$$
$$[\alpha]^x A^{px} = [\beta]^x B^{px},$$
$$[\alpha]^x A^{hy} A = [\beta]^x B^{hy} B,$$

and by Theorem 3–4, $A \sim B$.

Theorem 3–5 shows that the primes which do not divide h enjoy a property not shared by other primes. This is of great importance in the investigation of Fermat's equation.

PROBLEMS

1. Let α and β be algebraic integers, not both zero. Show that there is an integer δ such that, first, $\delta|\alpha$ and $\delta|\beta$ (in the sense that α/δ and β/δ are again integers), and, second, for suitable integers ξ and η, $\delta = \alpha\xi + \beta\eta$. Show that this GCD is unique up to an algebraic unit (i.e., an integer which divides 1). [*Hint:* First settle the case $\alpha\beta = 0$. In the other case, let K be an algebraic number field of class number h, containing both α and β. Then $[\alpha, \beta]^h = [\gamma]$, for some γ in K. Let δ be an integer such that $\delta^h = \gamma$, and show that the equation $[\alpha, \beta]^h = [\gamma]$ still holds when $[\alpha, \beta]$ and $[\gamma]$ are interpreted as ideals in $K(\delta)$. Deduce that $[\alpha, \beta] = [\delta]$ in $K(\delta)$.] Does the Unique Factorization Theorem hold in the domain of all algebraic integers?

2. Let K be an algebraic number field. Show that to each ideal A of K there corresponds an integer α (not necessarily in K) such that the elements of A are exactly those integers of K which are divisible by α.

3-3 The cyclotomic field K_p.

Let p be an odd prime, let

$$\Phi(x) = \frac{x^p - 1}{x - 1} = x^{p-1} + x^{p-2} + \cdots + 1,$$

and let $\zeta = e^{2\pi i/p}$, so that the zeros of Φ are $\zeta, \zeta^2, \ldots, \zeta^{p-1}$, the primitive pth roots of unity. The field $R(\zeta) = R(\zeta^2) = \cdots = R(\zeta^{p-1}) = K_p$ is called a *cyclotomic field*. It is clearly of degree $p - 1$ at most. We put $1 - \zeta = \pi$. (The fact that the symbol π is used for two different numbers should occasion no confusion; the number $\pi = 3.14159 \ldots$ will occur only in the argument of the exponential function.)

THEOREM 3–6. *In K_p, the ideal $[p]$ has the factorization*

$$[p] = [\pi]^{p-1};$$

$[\pi]$ *is prime, and* $\mathbf{N}[\pi] = p$; Φ *is irreducible, and* K_p *is of degree* $p - 1$.

Proof: Since ζ is an integer of K_p, so is

$$\epsilon_r = \frac{1 - \zeta^r}{1 - \zeta} = 1 + \zeta + \cdots + \zeta^{r-1}, \qquad 1 \leq r \leq p - 1;$$

if now an r' is chosen so that $rr' \equiv 1 \pmod{p}$, then $\zeta^{rr'} = \zeta$, so that

$$\epsilon_r^{-1} = \frac{1 - \zeta^{rr'}}{1 - \zeta^r} = 1 + \zeta^r + \cdots + \zeta^{r(r'-1)}$$

is also an integer. Hence ϵ_r is a unit of K_p, and

$$p = \Phi(1) = \prod_{r=1}^{p-1} (1 - \zeta^r) = (1 - \zeta)^{p-1} \prod_{r=1}^{p-1} \epsilon_r = \epsilon(1 - \zeta)^{p-1},$$

where ϵ is a unit. It follows from this equation that $[p] = [\pi]^{p-1}$, and also that $\mathbf{N}\pi = p$. By Theorem 2–39, deg $K_p \geq p - 1$, so that $[\pi]$ is prime, deg $K_p = p - 1$, and Φ is irreducible. (For a different proof of the irreducibility of Φ, see Problem 1, Section 2–2.)

Hereafter, we designate $[\pi]$ by P.

THEOREM 3–7. *Writing* $\Delta(1, \zeta, \ldots, \zeta^{p-2}) = \Delta(\zeta)$, *we have*

$$\Delta(\zeta) = (-1)^{(p-1)/2} p^{p-2}.$$

Proof: From the representations

$$\Phi(x) = \prod_{r=1}^{p-1} (x - \zeta^r) = \frac{x^p - 1}{x - 1},$$

we obtain

$$\Phi'(\zeta^s) = \prod_{\substack{1 \leq r \leq p-1 \\ r \neq s}} (\zeta^s - \zeta^r) = \frac{p(\zeta^s - 1)\zeta^{s(p-1)} - (\zeta^{ps} - 1)}{(\zeta^s - 1)^2}$$

$$= \frac{p\zeta^{s(p-1)}}{\zeta^s - 1} = -\frac{p}{\zeta^s(1 - \zeta^s)}.$$

Since

$$\Delta(\zeta) = \begin{vmatrix} 1 & \zeta & \cdots & \zeta^{p-2} \\ 1 & \zeta^2 & \cdots & \zeta^{2(p-2)} \\ \vdots & \vdots & & \vdots \\ 1 & \zeta^{p-1} & \cdots & \zeta^{(p-2)(p-1)} \end{vmatrix}^2 = \prod_{1 \leq r < s \leq p-1} (\zeta^r - \zeta^s)^2,$$

we have

$$\Delta(\zeta) = (-1)^{\frac{1}{2}(p-1)(p-2)} \prod_{s=1}^{p-1} \prod_{\substack{1 \le r \le p-1 \\ r \ne s}} (\zeta^s - \zeta^r)$$

$$= (-1)^{\frac{1}{2}(p-1)} \prod_{s=1}^{p-1} \Phi'(\zeta^s) = (-1)^{\frac{1}{2}(p-1)} \frac{(-1)^{p-1} p^{p-1}}{\mathbf{N}\zeta \mathbf{N}\pi}$$

$$= (-1)^{\frac{1}{2}(p-1)} p^{p-2}.$$

THEOREM 3–8. *The numbers* $1, \zeta, \ldots, \zeta^{p-2}$ *form an integral basis for* K_p, *so that*

$$\Delta = \Delta(\zeta) = (-1)^{\frac{1}{2}(p-1)} p^{p-2}.$$

Proof: Suppose that α is an integer of K_p, and that

$$\alpha = r_0 + r_1\zeta + \cdots + r_{p-2}\zeta^{p-2},$$

where the r's are rational. Then for $k = 0, 1, \ldots, p-1$,

$$\zeta^k \alpha = \sum_{j=0}^{p-2} r_j \zeta^{j+k},$$

and since the trace function is clearly additive,

$$\mathbf{S}(\zeta^k \alpha) = \sum_{j=0}^{p-2} \mathbf{S}(r_j \zeta^{j+k}) = \sum_{j=0}^{p-2} r_j \mathbf{S}(\zeta^j \zeta^k).$$

Solving this system of equations for the numbers r_j, we obtain

$$r_j = \frac{\text{a determinant in } \alpha \text{ and } \zeta}{\det |\mathbf{S}(\zeta^j \zeta^k)|}.$$

But as we saw in the proof of Theorem 2–38, $\det |\mathbf{S}(\zeta^j\zeta^k)| = \Delta(\zeta)$; since the determinant in the numerator has the rational value $r_j\Delta(\zeta)$, and is clearly an integer of K_p, it is a rational integer. Thus α can be written in the form

$$\alpha = \frac{c_0 + c_1\zeta + \cdots + c_{p-2}\zeta^{p-2}}{p^{p-2}} = \frac{d_0 + d_1\pi + \cdots + d_{p-2}\pi^{p-2}}{p^{p-2}},$$

where the c's, and therefore also the d's, are in Z. Since α is an integer,

$$p | (d_0 + d_1\pi + \cdots + d_{p-2}\pi^{p-2}),$$

and since $P|[p]$,

$$P | [d_0 + d_1\pi + \cdots + d_{p-2}\pi^{p-2}],$$

so $P|[d_0]$. It follows that $\mathbf{N}P|\mathbf{N}[d_0]$, $p|d_0{}^{p-1}$, and finally $p|d_0$. This argument may be repeated $p-2$ times, to show that $p|d_k$ for $k = 1, \ldots, p-2$, so that

$$\alpha = \frac{e_0 + e_1\pi + \cdots + e_{p-2}\pi^{p-2}}{p^{p-3}},$$

where the e's are rational integers. Repeating the entire argument $p-3$ times, we see that

$$\alpha = f_0 + f_1\pi + \cdots + f_{p-2}\pi^{p-2},$$

where the f's are in Z. Hence $1, \pi, \ldots, \pi^{p-2}$ form an integral basis for K_p. But from the equations

$$\pi = 1 - \zeta, \qquad\qquad \zeta = 1 - \pi,$$
$$\pi^2 = 1 - 2\zeta + \zeta^2, \quad \text{and} \quad \zeta^2 = 1 - 2\pi + \pi^2,$$
$$\vdots \qquad\qquad\qquad\qquad \vdots$$

we see that $\Delta(\pi) = a^2 \Delta(\zeta)$ and $\Delta(\zeta) = a^2 \Delta(\pi)$, where a is a certain determinant with binomial coefficients as entries. Hence $a^2 = 1$, $\Delta(\zeta) = \Delta(\pi)$, and $1, \zeta, \ldots, \zeta^{p-2}$ also form an integral basis, by Theorem 2–14.

THEOREM 3–9. *If α is an integer of K_p, there is a rational integer, a, such that*

$$\alpha^p \equiv a \pmod{P^p}.$$

Proof: Since $\mathbf{N}P = p$, the incongruent numbers $0, 1, \ldots, p-1$ form a complete residue system modulo P, so that for suitable b in Z,

$$\alpha \equiv b \pmod{P}.$$

But

$$\alpha^p - b^p = \prod_{r=0}^{p-1} (\alpha - \zeta^r b),$$

and since $\zeta \equiv 1 \pmod{P}$,

$$\alpha^p - b^p \equiv \prod_{r=0}^{p-1} (\alpha - b) \equiv 0 \pmod{P^p},$$

so that we can take $a = b^p$.

If $P \nmid [\alpha]$ and $\alpha \equiv a \pmod{P^2}$ for some a in Z, then α is said to be *primary*.

THEOREM 3-10. *If $\pi \nmid \alpha$, then for some positive rational integer f, $\zeta^f \alpha$ is primary.*

Proof: For suitable a and b in Z,

$$\alpha \equiv a + b\pi \pmod{P^2},$$

and $\pi \nmid \alpha$, so that $p \nmid a$. Choose f so that

$$af \equiv b \pmod{p}.$$

Then since

$$\zeta^f = (1 - \pi)^f \equiv 1 - f\pi \pmod{P^2},$$

we have

$$\zeta^f \alpha \equiv (1 - \pi f)(a + b\pi) \equiv a + \pi(-af + b) \equiv a \pmod{P^2}.$$

We now investigate the units of K_p.

THEOREM 3-11. *The only roots of unity in K_p are the numbers $\pm \zeta^r$, $0 \le r < p$.*

Proof: The roots of unity are the numbers

$$e^{2\pi i t/m},$$

where t and m are rational integers, and $(t, m) = 1$. If such a number is in K_p, and if $tt' \equiv 1 \pmod{m}$, then also

$$e^{2\pi i tt'/m} = e^{2\pi i/m}$$

is in K_p. The numbers mentioned in the theorem are the $(2p)$th roots of unity, so we need only show that $e^{2\pi i/m}$ is not in K_p if $m \nmid 2p$. If $m \nmid 2p$, then either $4 | m$, or some odd prime $q \neq p$ divides m, or $p^2 | m$. Suppose that $e^{2\pi i/m}$ is in K_p.

If $4 | m$, then

$$e^{2\pi i/4} = i$$

is in K_p. But then so are $1 + i$ and $1 - i$, and

$$[1 + i] = [1 - i] \qquad \text{and} \qquad [2] = [1 + i]^2,$$

contrary to Theorem 2-38.

If $q | m$, then

$$\sigma = e^{2\pi i/q}$$

is in K_p. But then the reasoning used in the proof of Theorem 3-6 shows that

$$[q] = [1 - \sigma]^{q-1},$$

again contradicting Theorem 2–38.

If $p^2 | m$, then

$$\xi = e^{2\pi i/p^2}$$

is in K_p. But ξ is a zero of

$$\frac{x^{p^2} - 1}{x^p - 1} = x^{p(p-1)} + \cdots + x^p + 1 = \prod_{\substack{1 \le m \le p^2 \\ p \nmid m}} (x - \xi^m),$$

and

$$p = \prod_{\substack{1 \le m \le p^2 \\ p \nmid m}} (1 - \xi^m).$$

As before, the factors in this product are associated, and we get

$$[p] = [1 - \xi]^{p(p-1)},$$

contradicting Theorem 2–39.

THEOREM 3–12.　*Each unit ϵ of K_p can be written in the form*

$$\epsilon = \zeta^g \eta,$$

where g is a positive rational integer and η is real.

Proof: Express ϵ in terms of the integral basis $1, \zeta, \ldots, \zeta^{p-2}$:

$$\epsilon = f(\zeta),$$

where f is a polynomial with rational integral coefficients. Then clearly $\epsilon_s = f(\zeta^s)$ is also a unit, since $\mathbf{N}\epsilon = \epsilon_1 \cdots \epsilon_{p-1} = \pm 1$. Also,

$$\epsilon_{p-s} = f(\zeta^{p-s}) = f(\zeta^{-s}) = \overline{f(\zeta^s)} = \overline{\epsilon_s},$$

where the bar denotes the complex conjugate, so that $\epsilon_s \epsilon_{p-s} = |\epsilon_s|^2 > 0$, and

$$\mathbf{N}\epsilon = \prod_{s=1}^{\frac{1}{2}(p-1)} \epsilon_s \epsilon_{p-s} > 0,$$

so that $\mathbf{N}\epsilon = 1$.

Since $\epsilon_s = \overline{\epsilon_{p-s}}$,

$$\left| \frac{\epsilon_s}{\epsilon_{p-s}} \right| = 1.$$

The polynomial

$$\prod_{s=1}^{p-1}\left(x - \frac{\epsilon_s}{\epsilon_{p-s}}\right) = \prod_{s=1}^{p-1}(\epsilon_{p-s}x - \epsilon_s)$$

has coefficients in Z, so, by Theorem 2–40, $\epsilon_1/\epsilon_{p-1}$ is a root of unity, and by Theorem 3–11,

$$\epsilon_1 = \pm\, \zeta^m \epsilon_{p-1}.$$

Since either m or $p + m$ is even, and since

$$\zeta^m = \zeta^{p+m},$$

we can write

$$\epsilon_1 = \pm \zeta^{2g}\epsilon_{p-1}.$$

The proof will be complete if it can be shown that the plus sign is appropriate here, since then the quantities $\epsilon_1\zeta^{-g}$ and $\epsilon_{p-1}\zeta^g$ are simultaneously equal and complex-conjugate, and are therefore real, so that $\epsilon = \epsilon_1 = \zeta^g(\epsilon_{p-1}\zeta^g)$.

To show this, choose a from among $0, 1, \ldots, p - 1$ so that

$$\zeta^{-g}\epsilon \equiv a \ (\mathrm{mod}\ P).$$

Then

$$\mu = \frac{\zeta^{-g}\epsilon - a}{\pi}$$

is an integer in K_p, as is

$$\bar{\mu} = \frac{\bar{\zeta}^{-g}\bar{\epsilon} - a}{\bar{\pi}} = \frac{\zeta^g\epsilon_{p-1} - a}{\bar{\pi}}.$$

Since $\bar{\pi} = 1 - \zeta^{p-1}$ is an associate of π, it follows that

$$\frac{\zeta^g\epsilon_{p-1} - a}{\pi}$$

is an integer of K_p, so that

$$\zeta^g\epsilon_{p-1} \equiv a \equiv \zeta^{-g}\epsilon \ (\mathrm{mod}\ P).$$

Thus

$$\frac{\epsilon}{\epsilon_{p-1}} = \pm\zeta^{2g} \qquad \text{and} \qquad \frac{\epsilon}{\epsilon_{p-1}} \equiv \zeta^{2g} \ (\mathrm{mod}\ P).$$

If the minus sign obtains in the equation, we have

$$-\zeta^{2g} \equiv \zeta^{2g} \pmod{P},$$

$$2\zeta^{2g} \equiv 0 \pmod{P},$$

$$P|[2\zeta^{2g}],$$

$$NP|2^{p-1},$$

contrary to the fact that $NP = p$. The proof is complete.

PROBLEMS

1. Let p and q be distinct odd primes, and let ζ be a primitive pth root of unity.

   (a) Show that

   $$p = \prod_{a=1}^{p-1} (1 - \zeta^{2a}) = (-1)^{(p-1)/2} \prod_{a=1}^{\frac{1}{2}(p-1)} (\zeta^a - \zeta^{-a})^2.$$

   (b) Show that

   $$(\zeta^a - \zeta^{-a})^q \equiv \zeta^{qa} - \zeta^{-qa} \pmod{q}.$$

   (c) Deduce that

   $$p^{\frac{1}{2}(q-1)} \equiv (-1)^{\frac{1}{2}(p-1)\cdot\frac{1}{2}(q-1)} \prod_{a=1}^{\frac{1}{2}(p-1)} \frac{\zeta^{aq} - \zeta^{-aq}}{\zeta^a - \zeta^{-a}} \pmod{q}.$$

   (d) Show that the second factor on the right side of the last congruence above is $(-1)^\mu$, where μ is the number of numerically smallest residues (mod p) among $q, 2q, \ldots, \frac{1}{2}(p-1)q$ which are negative, and so obtain a proof of the law of quadratic reciprocity.

2. For an odd prime p and a positive integer h, put

   $$\Phi_h(x) = \frac{x^{p^h} - 1}{x^{p^{h-1}} - 1} = x^{p^{h-1}(p-1)} + x^{p^{h-1}(p-2)} + \cdots + x^{p^{h-1}} + 1,$$

   and let ζ be a zero of Φ_h and $K_{p^h} = R(\zeta)$. Then the degree of K_{p^h} is at most $\varphi(p^h) = t$. Put $1 - \zeta = \pi$, and $[\pi] = P$.

   (a) Show that in K_{p^h}, the ideal $[p]$ has the factorization P^t, P is prime, $NP = p$, $\Phi_h(x)$ is irreducible and K_{p^h} is of degree t.

   (b) Show that $\Delta(\zeta) = (-1)^{\frac{1}{2}t(t-1)} \cdot p^{p^{h-1}(hp-h-1)}$. [*Hint:* Notice that $1 - \zeta^{p^{h-1}}$ is a zero of $\Phi_1(1 - x)$, an irreducible polynomial of degree $p - 1$ with leading coefficient $(-1)^{p-1}$ and constant term p, and deduce that $N(1 - \zeta^{p^{h-1}}) = ((-1)^{p-1}p)^{p^{h-1}}$.]

   (c) Show that if L is any prime ideal in K_{p^h} different from P, and if $\zeta^a \equiv \zeta^b \pmod{L}$, then $a \equiv b \pmod{p^h}$.

3–4 Fermat's equation.   For the sake of completeness, we consider first the equation

$$x^n + y^n = z^n \qquad (1)$$

for the cases $n = 2, 4$, and 3.   When these have been disposed of, Fermat's assertion would be proved if it could be shown that (1) has no solutions in rational integers x, y, z, with $xyz \neq 0$, if n is a prime larger than 3.

The proof that (1) is impossible when $n = 4$ depends on the following theorem, which characterizes the solutions of (1) when $n = 2$.

THEOREM 3–13.   *A general primitive solution* (*i.e., a solution in which* $(x, y, z) = 1$) *of*

$$x^2 + y^2 = z^2, \qquad y \text{ even}, \quad x > 0, \quad y > 0, \quad z > 0$$

is given by

$$x = a^2 - b^2, \qquad y = 2ab, \qquad z = a^2 + b^2,$$

where a *and* b *are prime to each other and not both odd, and* $a > b > 0$.

Remark: It is clear that one of x and y must be even, since otherwise $x^2 + y^2 \equiv 2 \not\equiv z^2 \pmod 4$.   There is no loss in generality in assuming that it is y which is even.

Proof: Suppose that $x^2 + y^2 = z^2$.   Since $(x, y, z) = 1$, also $(y, z) = 1$, so that $(z - y, z + y) = 1$ or 2.   But z is odd and y is even, so that $(z - y, z + y) = 1$.   Hence, from the equation

$$x^2 = (z - y)(z + y),$$

we deduce that $z - y$ and $z + y$ must be squares, since they are positive.   Now if t and u are fixed integers of the same parity (both odd or both even), there are integers a and b such that $t = a + b$ and $u = a - b$.   Hence we can put

$$z - y = (a - b)^2, \qquad z + y = (a + b)^2,$$

which gives

$$z = \frac{(a - b)^2 + (a + b)^2}{2} = a^2 + b^2,$$

$$y = \frac{(a + b)^2 - (a - b)^2}{2} = 2ab,$$

$$x = (a - b)(a + b) = a^2 - b^2.$$

Since $(z - x, z + x) = (2a^2, 2b^2) = 2$, we must choose a and b so that $(a, b) = 1$. Since x is odd, $a + b$ must be odd. Since $y > 0$, a and b must have the same sign, and since $x > 0$, $|a| > |b|$. Since the pairs a, b and $-a, -b$ give the same solution, we can suppose that $a > b > 0$.

THEOREM 3–14.   *The equation $x^4 + y^4 = z^4$ is not solvable in non-zero rational integers.*

Proof: It suffices to show that there is no primitive solution of the equation

$$x^4 + y^4 = z^2.$$

Suppose that x, y, and z constitute such a solution; with no loss in generality we can take $x > 0, y > 0, z > 0$, and y even. Writing the supposed relation in the form

$$(x^2)^2 + (y^2)^2 = z^2,$$

we have from the preceding theorem that

$$x^2 = a^2 - b^2, \qquad y^2 = 2ab, \qquad z = a^2 + b^2,$$

where $(a, b) = 1$ and exactly one of a and b is odd. If a were even, we would have

$$1 \equiv x^2 = a^2 - b^2 \equiv -1 \ (\text{mod } 4),$$

so $2|b$. We apply Theorem 3–13 again, this time to the equation $x^2 + b^2 = a^2$, and obtain

$$x = p^2 - q^2, \qquad b = 2pq, \quad a = p^2 + q^2,$$

where $(p, q) = 1, p > q > 0$, and not both of p and q are odd. From

$$y^2 = 2ab$$

we have

$$y^2 = 4pq(p^2 + q^2).$$

Here p, q and $p^2 + q^2$ are relatively prime in pairs, so each must be a square:

$$p = r^2, \qquad q = s^2, \qquad p^2 + q^2 = t^2,$$

from which

$$r^4 + s^4 = t^2.$$

Now

$$x = r^4 - s^4, \qquad y = 2rst, \qquad z = a^2 + b^2 = r^8 + 6r^4s^4 + s^8,$$

so that $$z > (r^4 + s^4)^2 = t^4,$$

or $t < z^{\frac{1}{4}}$. It follows that if one solution of $x^4 + y^4 = z^2$ were known, another solution r, s, t could be found for which $rst \neq 0$ and $0 < t < z^{\frac{1}{4}}$. But this would give an infinite decreasing sequence of positive integers.

The case $n = 3$ is rather more difficult, since it is necessary to work in the quadratic field $K_3 = R(\zeta)$, where $\zeta = (-1 + i\sqrt{3})/2$ is a primitive cube root of unity. Not all the complications of the general case are present, however, since there is unique factorization of the integers of K_3, as the following theorem shows.

THEOREM 3–15. *Given any two integers α and γ of K_3, of which $\gamma \neq 0$, there are integers $\varkappa$ and ρ such that*

$$\alpha = \varkappa\gamma + \rho, \qquad 0 \le \mathbf{N}\rho < \mathbf{N}\gamma.$$

The integers of K_3 therefore form a Euclidean domain.

Proof: Since 1 and ζ form an integral basis for K_3, we can write

$$\frac{\alpha}{\gamma} = \frac{a + b\zeta}{c + d\zeta} = \frac{(a + b\zeta)(c + d\zeta^2)}{c^2 - cd + d^2} = R + S\zeta,$$

where a, b, c, and d are rational integers, and R and S are rational. Choose x and y in Z such that

$$|R - x| \le \tfrac{1}{2}, \qquad |S - y| \le \tfrac{1}{2};$$

then

$$\left|\frac{\alpha}{\gamma} - (x + y\zeta)\right|^2 = (R - x)^2 - (R - x)(S - y) + (S - y)^2 \le \frac{3}{4}.$$

Hence, if $\varkappa = x + y\zeta$ and $\rho = \alpha - \varkappa\gamma$, then

$$\mathbf{N}\rho \le \tfrac{3}{4}\mathbf{N}\gamma < \mathbf{N}\gamma, \qquad \text{and} \qquad \mathbf{N}\rho = \rho\bar{\rho} = |\rho|^2 \ge 0.$$

THEOREM 3–16. *The equation*

$$\xi^3 + \eta^3 + \vartheta^3 = 0 \tag{2}$$

has no solution in nonzero integers of K_3. It therefore has no solution in nonzero rational integers.

Proof: We first note that one of ξ, η, and ϑ must be divisible by the prime $\pi = 1 - \zeta$, if (2) holds. For put

$$\xi + \eta = \rho, \qquad \eta + \vartheta = \sigma, \qquad \vartheta + \xi = \tau.$$

Then a simple calculation, using (2), shows that

$$(\rho + \sigma + \tau)^3 = 24\rho\sigma\tau.$$

Since the expression on the right side of this equation is divisible by

$$3 = -\zeta^2\pi^2,$$

the left side must be divisible by π, and therefore by π^3. Returning to the right side, it follows that one of ρ, σ, or τ must be divisible by π. If $\pi|\rho$, then $\pi|(\xi^3 + \eta^3)$, so $\pi|\vartheta^3$, and finally $\pi|\vartheta$.

If there were a common factor in two of ξ, η, and ϑ, it would also occur in the third, and could be divided out; so suppose that (2) holds, that ξ, η, and ϑ are relatively prime in pairs, and that $\pi|\vartheta$.

By Theorem 3–10, we may suppose that an appropriate power of ζ has been introduced into ξ and η so that

$$\xi \equiv 1, \qquad \eta \equiv -1 \pmod{3},$$

which we express by putting

$$\xi = 1 + 3\alpha, \qquad \eta = -1 + 3\beta,$$

where α and β are integers of K_3. Put

$$A = \frac{\xi + \zeta\eta}{\pi}, \qquad B = \frac{\zeta\xi + \eta}{\pi}, \qquad C = \frac{\zeta^2(\xi + \eta)}{\pi};$$

these numbers are integers of K_3, since

$$A = 1 + \frac{3}{\pi}(\alpha + \beta\zeta),$$

$$B = -1 + \frac{3}{\pi}(\zeta\alpha + \beta),$$

$$C = \frac{3}{\pi}(\alpha + \beta)\zeta^2.$$

Moreover,

$$A + B + C = 0, \tag{3}$$

$$ABC = \frac{\xi^3 + \eta^3}{\pi^3} = \left(\frac{-\vartheta}{\pi}\right)^3, \tag{4}$$

$$\xi = -\zeta A + \zeta^2 B, \qquad \eta = \zeta^2 A - \zeta B. \tag{5}$$

From (5) we see that $(A, B) = 1$, since otherwise ξ and η would have a common factor. From (3), also $(A, C) = (B, C) = 1$.

It follows from (4) that A, B, and C must all be cubes, say $A = \varphi^3, B = \chi^3, C = \psi^3$, and

$$\varphi^3 + \chi^3 + \psi^3 = 0.$$

Now $A \equiv 1, \qquad B \equiv -1, \qquad C \equiv 0 \pmod{\pi},$

so that from (4), ψ contains a smaller power of π than does ϑ.

Repeating the argument a sufficient number of times, we would arrive eventually at a solution of (2) in which no variable is divisible by π, which is impossible.

3–5 Kummer's theorem. If p is an odd rational prime, and its associated cyclotomic field K_p has class number h, then p is said to be *regular* if $p{\nmid}h$. According to Theorem 3–5, if p is a regular prime and A and B are ideals in K_p such that $A^p \sim B^p$, then $A \sim B$. It was this essential property of the regular primes which enabled Kummer to prove that Fermat's conjecture is correct for all regular primes. (Unfortunately, there are infinitely many irregular ones.) We shall not be able to prove Kummer's theorem in its entirety, but shall have to assume without proof a difficult preliminary result. We can, however, prove the following theorem.

THEOREM 3–17. *If p is regular, the equation*

$$x^p + y^p + z^p = 0 \tag{6}$$

has no solution in rational integers x, y, z for which $p{\nmid}xyz$.

Proof: Suppose that the theorem is false, and that x, y, and z satisfy all the requirements. We can assume that $(x, y) = 1$ and $p > 3$; as usual, ζ is a primitive pth root of unity, and $P = [1 - \zeta]$. From (6) we obtain

$$\prod_{m=0}^{p-1} (x + \zeta^m y) = -z^p,$$

so that

$$\prod_{m=0}^{p-1} [x + \zeta^m y] = [z]^p. \tag{7}$$

Now no two of the factors on the left have a common factor. For, if Q is a prime ideal such that $Q|[x + \zeta^{m_1}y]$ and $Q|[x + \zeta^{m_2}y]$ for $m_1 < m_2$, then

$$Q|[\zeta^{m_1}(1 - \zeta^{m_2-m_1})y],$$

and hence $Q|P[y]$. But from (7), $Q|[z]$, so $Q \neq P$ (since $p \nmid z$); hence $Q|[y]$. But then also $Q|[x]$, and we deduce that $NQ|y^{p-1}$ and $NQ|x^{p-1}$, which is contrary to the assumption that $(x, y) = 1$.

It follows that each factor on the left side of (7) is the pth power of an ideal. If

$$[x + \zeta y] = A^p,$$

then $A^p \sim [1] = [1]^p$, so that by Theorem 3–5, A itself is principal, say $A = [\alpha]$. Then

$$[x + \zeta y] = [\alpha]^p = [\alpha^p].$$

Hence

$$x + \zeta y = \epsilon \alpha^p,$$

where ϵ is a unit of K_p. Using the canonical form for units in K_p obtained in Theorem 3–12, we have

$$x + \zeta y = \zeta^g \eta \alpha^p, \qquad 0 \le g \le p - 1,$$

where η is real. By Theorem 3–9, since $[p]|P^p$,

$$\alpha^p \equiv a \pmod{[p]}$$

for some a in Z, so that

$$x + \zeta y \equiv \zeta^g \sigma \pmod{[p]},$$

where σ is a real integer of K_p. The complex conjugate of the integer

$$\frac{\zeta^{-g}(x + \zeta y) - \sigma}{p}$$

is also a field conjugate, and is therefore also an integer. Since $\bar{p} = p$ and $\bar{\sigma} = \sigma$, we have

$$\sigma \equiv (x + \zeta y)\zeta^{-g} \pmod{[p]},$$

and

$$\sigma \equiv (x + \zeta^{-1}y)\zeta^{g} \pmod{[p]},$$

so that

$$x\zeta^{-g} + y\zeta^{1-g} - x\zeta^{g} - y\zeta^{g-1} \equiv 0 \pmod{[p]}. \tag{8}$$

Two of these exponents must be congruent modulo p. For suppose that they are all distinct, and put

$$\beta = \frac{x}{p}\zeta^{-g} + \frac{y}{p}\zeta^{1-g} - \frac{x}{p}\zeta^{g} - \frac{y}{p}\zeta^{g-1}.$$

Then $p\beta$ has a representation in terms of distinct elements of an integral basis, the coefficients not being divisible by p. But since β is an integer, $p\beta$ also has a representation in which the coefficients are divisible by p, and this is contrary to the definition of a basis. We conclude that g must have one of the values 0, 1, or $(p + 1)/2$ (that is, $2g \equiv 1 \pmod{p}$).

If $g = 0$, the congruence (8) gives

$$y\zeta - y\zeta^{-1} \equiv 0 \pmod{[p]},$$

whence, since $\zeta^2 - 1$ is an associate of π,

$$P^2|[y]P, \qquad P|[y], \qquad p|y,$$

which is false. If $g = 1$, then (8) yields

$$x(1 - \zeta^2) \equiv 0 \pmod{[p]},$$

which implies that $p|x$, which is also false. Finally, if $g = (p + 1)/2$, then from (8) we get

$$(x - y)\pi \equiv 0 \pmod{[p]},$$

which gives

$$x \equiv y \pmod{p}.$$

Interchanging y and z in equation (7), we deduce that also

$$x \equiv z \pmod{p}.$$

But then equation (6) implies that

$$x^p + y^p + z^p \equiv 3x^p \equiv 0 \pmod{p},$$

which is false since $p > 3$ and $p \nmid x$. Hence the theorem is not false.

Because of its methodological interest, we deduce the general Kummer theorem from the following lemma, whose proof is too long for inclusion here:

KUMMER'S LEMMA. *Let p be a regular prime. Then if ϵ is a unit of K_p and a is a rational integer such that*

$$\epsilon \equiv a \pmod{P^p},$$

then ϵ is the pth power of another unit of K_p.

This is a partial converse of Theorem 3–9. Using it, we can generalize Theorem 3–17 in two ways: by allowing x, y, and z to be integers of K_p instead of rational integers, or by dropping the restriction that $p \nmid xyz$.

THEOREM 3–18. *If p is a regular prime, the equation*

$$x^p + y^p + z^p = 0$$

has no solution in nonzero integers x, y, z of K_p for which $\pi | xyz$. It therefore has no solution in nonzero rational integers x, y, z for which $p | xyz$, and therefore (by Theorem 3–17) no nonzero rational integral solutions.

Proof: We first show that the equation

$$x^p + y^p = \epsilon \pi^{up} z'^p, \qquad \pi \nmid xyz', \quad \epsilon \text{ a unit of } K_p, \tag{9}$$

has no nontrivial solution if $u = 1$. Equation (9) is a generalization of the equation obtained from (6) by supposing that $z = z' \pi^u$, where $\pi \nmid z'$.

We may suppose that x and y have no common numerical factor, since it would also occur in z and could be canceled out. (Notice that it cannot be assumed that the ideals $[x]$ and $[y]$ are relatively prime, since $[x, y]$ may not be principal.) We may also suppose that x and y are primary, since they may be multiplied by appropriate powers of ζ without affecting (9). If (9) is written in the form

$$\prod_{m=0}^{p-1} (x + \zeta^m y) = \epsilon \pi^{up} z'^p, \tag{9'}$$

it is clear that at least one of the factors on the left, say $x + \zeta^i y$, is divisible by π. Since, however, the differences

$$(x + \zeta^k y) - (x + \zeta^i y) = (\zeta^k - \zeta^i)y$$

and

$$\zeta^i(x + \zeta^k y) - \zeta^k(x + \zeta^i y) = -(\zeta^k - \zeta^i)x$$

are also divisible by π, each factor on the left in (9') must be divisible by π. If two factors were divisible by π^2, we would have

$$\pi^2 | (\zeta^k - \zeta^i)y,$$

$$\pi^2 | \epsilon' \pi y \qquad (\epsilon' \text{ a unit}),$$

$$\pi | y,$$

and similarly $\pi | x$, contrary to assumption. On the other hand, since

x and y are primary, there is an a in Z such that

$$x + y \equiv a \pmod{P^2}.$$

But then

$$a \equiv x + y \equiv 0 \pmod{P},$$

$$p \mid a,$$

$$P^2 \mid [a],$$

$$x + y \equiv 0 \pmod{P^2}.$$

Thus the total number of factors of π on the left side of (9') is at least $p + 1$, so that $u > 1$.

Now rewrite (9') as

$$\prod_{m=0}^{p-1} [x + \zeta^m y] = P^{up}[z']^p. \tag{9''}$$

Any common factor different from P of two ideals in the product on the left side of (9'') must be a factor of both $[x]$ and $[y]$, and therefore of $[z']$. After dividing out every such common factor, as well as one factor of P from each ideal, the ideals remaining on the left are pairwise prime, and their product is a pth power; therefore each factor separately is a pth power.

Combining all these results, we can write

$$[x + y] = P^{p(u-1)+1} J_0{}^p \, D,$$

$$[x + \zeta^m y] = P \, J_m{}^p \, D, \qquad m = 1, \ldots, p - 1,$$

where $D = [x, y]$, and $J_0, J_1, \ldots, J_{p-1}$ are certain ideals not divisible by P. If we put $t_m = p(u - 1) + 1$ or 1, according as $m = 0$ or $m > 0$, we have, for $m \neq 1$,

$$[x + \zeta y] P^{t_m} J_m{}^p \, D = [x + \zeta y][x + \zeta^m y] = P J_1{}^p \, D[x + \zeta^m y];$$

since P is a principal ideal, it follows that

$$J_m{}^p \, D \sim J_1{}^p \, D,$$

so that $J_m{}^p \sim J_1{}^p$, and by Theorem 3–5, $J_m \sim J_1$. Thus integers γ_m and δ_m (which are not divisible by π) exist such that

$$[\gamma_m] J_m = [\delta_m] J_1, \qquad m = 0, 2, 3, \ldots, p - 1. \tag{10}$$

Raising both sides of (10) (with $m = 0$) to the pth power, and then multiplying through by $D \, P^{p(u-1)+1}$, we have

$$[\gamma_0]^p D \, P^{p(u-1)+1} \, J_0{}^p = D \, P \, J_1{}^p \cdot P^{p(u-1)}[\delta_0]^p,$$

or

$$[\gamma_0{}^p][x + y] = [x + \zeta y] \, P^{p(u-1)}[\delta_0{}^p].$$

Similarly,

$$D \, P[\gamma_2]^p J_2{}^p = D \, P[\delta_2]^p \, J_1{}^p,$$

$$[x + \zeta^2 y][\gamma_2{}^p] = [x + \zeta y][\delta_2{}^p],$$

so that

$$\gamma_0{}^p (x + y) = \epsilon_1 (x + \zeta y)\pi^{p(u-1)}\delta_0{}^p,$$

$$\gamma_2{}^p (x + \zeta^2 y) = \epsilon_2 (x + \zeta y)\delta_2{}^p, \tag{11}$$

where ϵ_1 and ϵ_2 are units.

We now use the identity

$$(x + \zeta^2 y) + (x + y)\zeta = (x + \zeta y)(1 + \zeta).$$

We multiply through by $\gamma_0{}^p\gamma_2{}^p$, and in the resulting equation replace the left sides of equations (11) by the right sides. After canceling the common factor $x + \zeta y$, there results

$$\epsilon_2 (\gamma_0\delta_2)^p + \epsilon_1\zeta\pi^{p(u-1)} (\gamma_2\delta_0)^p = (1 + \zeta)(\gamma_0\gamma_2)^p.$$

Since ϵ_1, ϵ_2, ζ, and $1 + \zeta = (1 - \zeta^2)/(1 - \zeta)$ are units, this equation is of the form

$$\xi^p + \epsilon_3\eta^p = \epsilon_4\pi^{p(u-1)}\vartheta^p, \tag{12}$$

where ϵ_3 and ϵ_4 are units and $\pi\nmid\xi\eta\vartheta$. By Theorem 3–9,

$$\xi^p \equiv a_1, \qquad \eta^p \equiv a_2 \ (\text{mod } P^p),$$

where a_1 and a_2 are rational integers; since $u > 1$, (12) gives

$$a_1 + \epsilon_3 a_2 \equiv 0 \ (\text{mod } P^p).$$

Since $\pi\nmid\eta$, also $\pi\nmid a_2$, so that $p\nmid a_2$. Choose a_3 so that $a_2 a_3 \equiv 1 \ (\text{mod } p^2)$; then

$$a_2 a_3 \equiv 1 \ (\text{mod } P^p),$$

$$a_1 a_3 + \epsilon_3 \equiv 0 \ (\text{mod } P^p).$$

By Kummer's lemma, $\epsilon_3 = \epsilon_5{}^p$, and (12) becomes

$$\xi^p + (\epsilon_5\eta)^p = \epsilon_4\pi^{p(u-1)}\vartheta^p,$$

which is an equation of the form (9) with u replaced by $u - 1$. Repeating the argument $u - 2$ times, we would have a solution of (9) with $u = 1$, which is impossible.

Before leaving the subject of Fermat's conjecture, it might be of some interest to mention certain other facts known about it. We consider only the solvability of equation (6),

$$x^p + y^p + z^p = 0,$$

in Z.

It was proved by Wieferich in 1909 that if (6) holds in integers x, y, and z such that $p \nmid xyz$ (the so-called Case I), then

$$2^{p-1} \equiv 1 \pmod{p^2}.$$

Later investigators have shown that in Case I,

$$q^{p-1} \equiv 1 \pmod{p^2}$$

for every prime $q \leq 43$; J. B. Rosser used this fact to show that there are no solutions in Case I for $p < 41{,}000{,}000$. D. H. and Emma Lehmer later extended Rosser's method to prove Fermat's conjecture in Case I for $p < 253{,}747{,}889$. This in turn implies that if there is a solution in Case I, it must be that $\log \log z > 23$.

Without the restriction to Case I, Theorem 3–18 disposes of the regular primes. Kummer also found criteria to handle the irregular primes less than 164; this was pushed on to all $p < 619$ by H. S. Vandiver and his collaborators, and quite recently D. H. and E. Lehmer and Vandiver have used high-speed computing techniques to settle the problem for all $p < 2000$. It turns out that of the 302 primes less than 2000, 118 are irregular; while it is not known that there are infinitely many regular primes, there is nothing in the limited data available to indicate that there are only finitely many.

3–6 The equation $x^2 + 2 = y^3$. For the remainder of this chapter we shall be primarily concerned with the cubic analog of Pell's equation. At one point in the argument, however, we shall need the following auxiliary result.

THEOREM 3–19. *The only solutions in Z of the equation*

$$x^2 + 2 = y^3 \tag{13}$$

are $x = \pm 5$, $y = 3$.

Proof: Following Euler's idea, we make use of the arithmetic of the quadratic field $R(\sqrt{-2})$. By Theorem 2–16, the integers of this field are of the form $a + b\sqrt{-2}$, where a and b are rational integers. By a proof exactly paralleling that of Theorem 3–15, it can be shown that they form a Euclidean domain: given a, b, c, d in Z, with $cd \neq 0$, there are e, f, g, h in Z such that

$$a + b\sqrt{-2} = (c + d\sqrt{-2})(e + f\sqrt{-2}) + (g + h\sqrt{-2}),$$
$$g^2 + 2h^2 < c^2 + 2d^2.$$

It follows that $R[\sqrt{-2}]$ is a unique factorization domain.

We first show that if x and y satisfy (13), then $x + \sqrt{-2}$ and $x - \sqrt{-2}$ are relatively prime. It is clear that

$$(x + \sqrt{-2}, x - \sqrt{-2}) \mid -2\sqrt{-2},$$

and since $-2\sqrt{-2} = (\sqrt{-2})^3$ and $\sqrt{-2}$ is prime in the domain (by Theorem 2–39), it must be that $(x + \sqrt{-2}, x - \sqrt{-2}) = (\sqrt{-2})^m$, $0 \leq m \leq 3$. But if $x + \sqrt{-2} = (a + b\sqrt{-2})\sqrt{-2}$, then $x = -2b$, whence, by (13),

$$4b^2 + 2 = y^3,$$

$$y^3 \equiv 2 \pmod{4},$$

which is impossible.

Since the only units of $R(\sqrt{-2})$ are ± 1, it follows from (13) that

$$x + \sqrt{-2} = (a + b\sqrt{-2})^3,$$

where a and b are rational integers, and equating real and imaginary parts gives

$$a^3 - 6ab^2 = x,$$

$$3a^2b - 2b^3 = 1.$$

From the second of these equations it follows that $b = \pm 1$, and hence that $3a^2 - 2 = \pm 1$, or $a = \pm 1$. From the first, $x = \pm 1 \mp 6 = \pm 5$.

3–7 Pure cubic fields. The field $L = R(\sqrt[3]{d})$, in which $d > 1$ is a cube-free rational integer and $\sqrt[3]{d}$ is real, is called a *pure cubic field*. In this section we determine an integral basis for L and note certain other properties.

Since d is cube-free, we can write

$$d = ab^2,$$

where ab is square-free. Since $\sqrt[3]{d^2} = b\sqrt[3]{a^2b}$, the numbers 1, $\sqrt[3]{ab^2}$, $\sqrt[3]{a^2b}$ form a basis for L. Following Dedekind, we say that L is of the *first* or *second kind*, according as 9 does not, or does, divide $a^2 - b^2$. The reason for the distinction is made clear in the following theorem.

THEOREM 3–20. *The numbers*

$$1, \; \sqrt[3]{ab^2}, \; \sqrt[3]{a^2b}$$

form an integral basis for L if it is of the first kind. The numbers

$$\tfrac{1}{3}(1 + a\sqrt[3]{ab^2} + b\sqrt[3]{a^2b}), \; \sqrt[3]{ab^2}, \; \sqrt[3]{a^2b}$$

form an integral basis for L if it is of the second kind.

Remark: Note that the second basis represents every integer represented by the first, since

$$3z_1 \frac{1 + a\sqrt[3]{ab^2} + b\sqrt[3]{a^2b}}{3} + (z_2 - az_1)\sqrt[3]{ab^2} + (z_3 - bz_1)\sqrt[3]{a^2b}$$

$$= z_1 + z_2\sqrt[3]{ab^2} + z_3\sqrt[3]{a^2b}.$$

Proof: Suppose that ω is an integer in L, and that

$$\omega = x_1 + x_2\sqrt[3]{ab^2} + x_3\sqrt[3]{a^2b}, \qquad x_1, x_2, x_3 \text{ in } R.$$

Then the conjugates of ω are

$$\omega' = x_1 + \rho x_2\sqrt[3]{ab^2} + \rho^2 x_3\sqrt[3]{a^2b},$$

$$\omega'' = x_1 + \rho^2 x_2\sqrt[3]{ab^2} + \rho x_3\sqrt[3]{a^2b},$$

where ρ is a primitive cube root of unity. We see that

$$\omega + \omega' + \omega'' = 3x_1,$$

$$\sqrt[3]{a^2b}(\omega + \rho^2\omega' + \rho\omega'') = 3abx_2,$$

$$\sqrt[3]{ab^2}(\omega + \rho\omega' + \rho^2\omega'') = 3abx_3,$$

and since the left sides of these equations are algebraic integers and the right sides are rational, it follows that the numbers $3x_1$, $3abx_2$, $3abx_3$ are rational integers. Hence for any integer ω in L, there are y_1, y_2, y_3 in Z such that

$$3ab\omega = y_1 + y_2\sqrt[3]{ab^2} + y_3\sqrt[3]{a^2b}. \tag{14}$$

We show first that ab is a divisor of y_1, y_2, and y_3, and so can be omitted in (14).

Let p be a rational prime dividing a, and let P be a prime ideal of L which divides $[p]$. It was supposed that ab is square-free; *a fortiori*, $(a, b) = 1$, and $P \nmid [b]$. If we put

$$\alpha = \sqrt[3]{ab^2}, \qquad \beta = \sqrt[3]{a^2b},$$

then $P|[\alpha]^3$, so $P^3|[\alpha]^3$; since L is of degree 3, it follows from Theorem 2-39 that $[p] = P^3$. Hence $P||[\alpha]$ and $P^2||[\beta]$.

Now suppose, in accordance with (14), that

$$y_1 + y_2\alpha + y_3\beta \equiv 0 \pmod{3ab}.$$

Then

$$y_1 + y_2\alpha + y_3\beta \equiv 0 \pmod{P^3},$$

$$y_1 \equiv 0 \pmod{P},$$

$$y_1 \equiv 0 \pmod{p}, \tag{15}$$

$$y_1 \equiv 0 \pmod{P^3},$$

$$y_2\alpha + y_3\beta \equiv 0 \pmod{P^3},$$

$$y_2\alpha \equiv 0 \pmod{P^2},$$

$$y_2 \equiv 0 \pmod{p}, \tag{16}$$

$$y_3\beta \equiv 0 \pmod{P^3},$$

$$y_3 \equiv 0 \pmod{p}. \tag{17}$$

By equations (15), (16), and (17), and the fact that p was an arbitrary prime divisor of a, we see that a divides y_1, y_2, and y_3. Similarly, b divides y_1, y_2, and y_3. It follows that there are z_1, z_2, z_3 in Z such that

$$3\omega = z_1 + z_2\alpha + z_3\beta. \tag{18}$$

Let the defining equation of ω be

$$x^3 + c_1x^2 + c_2x + c_3 = 0, \qquad c_1, c_2, c_3 \text{ in } Z.$$

Then by (18) and the analogous equations for $3\omega'$ and $3\omega''$,

$$c_1 = -(\omega + \omega' + \omega'') = -z_1,$$

$$c_2 = \omega\omega' + \omega\omega'' + \omega'\omega'' = \tfrac{1}{3}(z_1{}^2 - abz_2z_3), \tag{19}$$

$$c_3 = -\omega\omega'\omega'' = -\tfrac{1}{27}(z_1{}^3 + ab^2z_2{}^3 + a^2bz_3{}^3 - 3abz_1z_2z_3). \tag{20}$$

Suppose that $3|a$; then $3\nmid b$, and L is of the first kind. Since c_2 is in Z, $3|z_1$. Since c_3 is in Z,

$$0 \equiv -27c_3 \equiv 3\frac{a}{3}b^2z_2{}^3 \pmod 9,$$

whence $3|z_2$, and by (20) again, $3|z_3$. In this case, then, the numbers 1, $\sqrt[3]{ab^2}$, $\sqrt[3]{a^2b}$ constitute an integral basis for L. A similar argument applies in the case that $3|b$.

Suppose now that $3\nmid ab$, so that

$$a^2 \equiv b^2 \equiv 1 \pmod 3. \tag{21}$$

If $3|z_1$, then by (19), $3|z_2z_3$; if $3|z_2$, say, then it follows from (20) that also $3|z_3$. Similarly, if $3|z_2$, then also $3|z_1$ and $3|z_3$. Hence 3 divides all or none of z_1, z_2, z_3; in the first case ω is of the form specified in the theorem.

We now examine the possibility that ω in (19) is an integer, but that $3\nmid z_1z_2z_3$. Then by (20), (21), and Fermat's theorem,

$$z_1{}^3 + ab^2z_2{}^3 + a^2bz_3{}^3 \equiv 0 \pmod 3,$$

$$z_1 + az_2 + bz_3 \equiv 0 \pmod 3,$$

$$z_1 \equiv az_2 \equiv bz_3 \pmod 3,$$

$$z_2 \equiv az_1, \qquad z_3 \equiv bz_1 \pmod 3,$$

$$z_2 = az_1 + 3t_2, \qquad z_3 = bz_1 + 3t_3.$$

Substituting these expressions for z_2 and z_3 into (20), we obtain

$$
\begin{aligned}
-27c_3 &= z_1{}^3 + ab^2(az_1 + 3t_2)^3 + a^2b(bz_1 + 3t_3)^3 \\
&\quad - 3abz_1(az_1 + 3t_2)(bz_1 + 3t_3) \\
&= z_1{}^3(1 + a^4b^2 + a^2b^4 - 3a^2b^2) \\
&\quad + 9z_1{}^2(a^3b^2t_2 + a^2b^3t_3 - a^2bt_3 - ab^2t_2) \\
&\quad + 27z_1(a^2b^2t_2{}^2 + a^2b^2t_3{}^2 - abt_2t_3) + 27(ab^2t_2{}^3 + a^2bt_3{}^3) \\
&\equiv z_1{}^3(1 + a^4b^2 + a^2b^4 - 3a^2b^2) \\
&\quad + 9z_1{}^2(ab^2t(a^2 - 1) + a^2bt_3(b^2 - 1)) \pmod{27}.
\end{aligned}
$$

By (21),

$$0 \equiv -27c_3 \equiv z_1{}^3(1 + a^4b^2 + a^2b^4 - 3a^2b^2) \pmod{27},$$

and it follows that

$$1 + a^4b^2 + a^2b^4 - 3a^2b^2 \equiv 0 \pmod{27}. \tag{22}$$

Using (21), we can put

$$b^2 = a^2 + 3f + 9g,$$

where f and g are rational integers and $0 \leq f \leq 2$. Then the congruence (22) reduces to

$$\varphi(f, a) = 2a^6 + (9f - 3)a^4 + 9(f^2 - f)a^2 + 1 \equiv 0 \pmod{27}.$$

For $f = 0$, this becomes

$$(a^2 - 1)^2(2a^2 + 1) \equiv 0 \pmod{27},$$

which is true for every a not divisible by 3, since

$$2a^2 + 1 \equiv a^2 - 1 \equiv 0 \pmod 3.$$

Moreover, for every a such that $3 \nmid a$,

$$\varphi(1, a) \equiv \varphi(0, a) + 9a^4 \equiv 9a^4 \not\equiv 0 \pmod{27},$$

$$\varphi(2, a) \equiv \varphi(0, a) + 18a^4 + 18a^2 \equiv 18a^2(a^2 + 1) \not\equiv 0 \pmod{27}.$$

Thus we find that if $3 \nmid z_1 z_2 z_3$, than c_3 is in Z if and only if

$$a^2 \equiv b^2 \not\equiv 0 \pmod 9,$$

(i.e., if and only if L is of the second kind) and $az_2 \equiv bz_3 \pmod 3$. If this is the case, then c_1 and c_2 are also rational integers, and

$$\omega = \tfrac{1}{3}(z_1 + (az_1 + 3t_2)\alpha + (bz_1 + 3t_3)\beta)$$

$$= z_1 \frac{1 + a\alpha + b\beta}{3} + t_2\alpha + t_3\beta$$

is an integer in L. The proof is complete.

In the course of the proof, it appeared that if $\omega = (x + y\alpha + z\beta)/3$ is an integer, and if one of x, y, z is divisible by 3, all of them are. In particular, if $x + y\alpha$ is an integer and x and y are rational, they are also integers.

We now consider the units of L. If L is of the first kind, then

$$\eta = x + y\alpha + z\beta$$

is a unit if and only if $\mathbf{N}\eta = \eta\eta'\eta'' = \pm 1$, or

$$x^3 + ab^2y^3 + a^2bz^3 - 3abxyz = \pm 1. \tag{23}$$

If L is of the second kind, then

$$\eta = \frac{u}{3}(1 + a\alpha + b\beta) + v\alpha + w\beta$$

is a unit if and only if

$$x^3 + ab^2y^3 + a^2bz^3 - 3abxyz = \pm 27, \qquad (24)$$

where $u = x$, $au + 3v = y$, and $bu + 3w = z$. If η is positive, the plus sign must be chosen in (23) and (24), since η' and η'' are complex conjugates.

The field L has the property that each of its elements is either rational or of degree three. For if there were an element of degree two, L would be an extension of the field generated by that element, and so would be of even degree. It follows that ± 1 are the only roots of unity in L. Since α has one real and two nonreal conjugates, we see by Theorem 2–45 that either L has only the units ± 1, or else there is a fundamental unit ξ, which may be chosen between 0 and 1, such that every unit η of L can be expressed in the form

$$\eta = \pm \xi^n,$$

where n is a rational integer, positive, negative, or zero.

A positive unit of the form $\eta = x + y\alpha$ is always smaller than 1. For since $x^3 + dy^3 = 1$, we have

$$\eta^{-1} = \eta'\eta'' = x^2 - xy\alpha + y^2\alpha^2 \geq 1 + \alpha + \alpha^2 > 3,$$

since xy is negative. Consequently, for such a unit we have

$$\eta = \xi^n, \qquad n > 0.$$

The same remarks apply to a positive unit of the form $x + z\beta$.

3–8 Two lemmas. For simplicity in notation, we define the binomial coefficient $\binom{m}{k}$ to be zero for $k > m$. Here and hereafter in this chapter, lower-case Latin letters stand for rational integers, unless otherwise specified.

THEOREM 3–21. *Let m be a positive integer. Then*

$$\binom{m}{0} + \binom{m}{3} + \binom{m}{6} + \cdots \not\equiv 0 \pmod{3}.$$

Proof: Put

$$S_0 = \binom{m}{0} + \binom{m}{3} + \binom{m}{6} + \cdots,$$

$$S_1 = \binom{m}{1} + \binom{m}{4} + \binom{m}{7} + \cdots,$$

$$S_2 = \binom{m}{2} + \binom{m}{5} + \binom{m}{8} + \cdots.$$

Then

$$S_0 + S_1 + S_2 = 2^m \equiv (-1)^m \pmod{3},$$

and

$$S_2 = \binom{m}{1}\frac{m-1}{2} + \binom{m}{4}\frac{m-4}{5} + \cdots \equiv -mS_1 + S_1 \pmod{3},$$

$$S_1 = \binom{m}{0}\frac{m}{1} + \binom{m}{3}\frac{m-3}{4} + \cdots \equiv mS_0 \pmod{3},$$

so that

$$(1 + 2m - m^2)S_0 \equiv (-1)^m \pmod{3}.$$

THEOREM 3–22. *Suppose that x and y are integers such that $(x, dy) = 1$, and suppose that*

$$(x + y\sqrt[3]{d})^n = X + Y\sqrt[3]{d} + Z(\sqrt[3]{d})^2,$$

where X, Y, and Z are rational and $n > 1$. Then $XYZ \neq 0$ except in the following cases:

$$(\sqrt[3]{10} - 1)^5 = 99 - 45\sqrt[3]{10},$$

$$(\sqrt[3]{4} - 1)^4 = -15 + 12\sqrt[3]{2}.$$

Proof: Since $(x, d) = 1$, it is clear that $X \neq 0$. Suppose that $Z = 0$, so that

$$\binom{n}{2}x^{n-2}y^2 + \binom{n}{5}x^{n-5}y^5 d + \binom{n}{8}x^{n-8}y^8 d^2 + \cdots = 0. \quad (25)$$

Dividing by $\binom{n}{2}y^2$, this becomes

$$-x^{n-2} = \sum_{k \geq 1} \binom{n-2}{3k}\frac{2x^{n-3k-2}y^{3k}d^k}{(3k+1)(3k+2)}. \quad (26)$$

Let q be a prime divisor of y. Then since $q^{3k} \geq 2^{3k} > 3k + 2$ for $k \geq 1$, each term in the last sum is divisible by q, which is impossible since $(x, y) = 1$. Hence $y = \pm 1$.

When $n \equiv 0 \pmod 3$, equation (26) can be written in the form

$$-y^{n-3}d^{\frac{1}{3}(n-3)} = \sum_{k \geq 1} \binom{n-1}{3k} \frac{x^{3k}y^{n-3k-3}d^{\frac{1}{3}(n-3)-k}}{3k+1} ;$$

when $n \equiv 1 \pmod 3$,

$$-y^{n-4}d^{\frac{1}{3}(n-4)} = \sum_{k \geq 1} \binom{n-2}{3k} \frac{2x^{3k}y^{n-3k-4}d^{\frac{1}{3}(n-4)-k}}{(3k+1)(3k+2)} ;$$

and when $n \equiv 2 \pmod 3$,

$$-y^{n-2}d^{\frac{1}{3}(n-2)} = \sum_{k \geq 1} \binom{n}{3k} x^{3k}y^{n-3k-2}d^{\frac{1}{3}(n-2)-k}.$$

The same argument now shows that $x = \pm 1$, and since it is clear from (25) that $xy < 0$, we have $x = -y$.

Now let q be a prime divisor of d, and suppose that $q^\alpha \| d$ (that is, $q^\alpha | d$ but $q^{\alpha+1} \nmid d$). If $q^\alpha > 5$, then $q^{\alpha k} > 5^k \geq 3k + 2$ for $k \geq 1$, so that each term in the sum in (26) is divisible by q, which is impossible since $(x, d) = 1$. If $q = 3$, then since $3 \nmid (3k + 1)(3k + 2)$ we reach the same contradiction. Hence $q^\alpha = 2$ or 5, and $d = 2, 5,$ or 10.

The information obtained so far shows that

$$1 - \binom{n-2}{3}\frac{2d}{45} + \binom{n-2}{6}\frac{2d^2}{7 \cdot 8} - \cdots = 0. \qquad (27)$$

If $d = 10$, this becomes

$$\binom{n-2}{3} - 1 = \frac{1}{6}(n-5)(n^2 - 4n + 6)$$

$$= \sum_{k \geq 2} (-1)^k \binom{n-2}{3k} \frac{2 \cdot 10^k}{(3k+1)(3k+2)}.$$

This equation is true for $n = 5$, and leads to the first of the exceptions mentioned in the theorem. For other values of n, we may divide through by $(n-5)/6$ and obtain

$$n^2 - 4n + 6$$
$$= -\sum_{k \geq 1} (-1)^k \binom{n-6}{3k-1} \frac{(n-2)(n-3)(n-4) \cdot 12 \cdot 10^{k+1}}{3k(3k+1)(3k+2)(3k+3)(3k+4)(3k+5)}.$$

The highest power of 5 which divides the denominator of a term in the sum is clearly at most $5(3k + 5)$, and since $5^{k+1} > 5(3k + 5)$ for $k \geq 2$, we have

$$n^2 - 4n + 6$$
$$= (n-2)^2 + 2 \equiv \binom{n-6}{2} \frac{(n-2)(n-3)(n-4) \cdot 12 \cdot 10^2}{3 \cdot 4 \cdot 5 \cdot 6 \cdot 7 \cdot 8} \equiv 0 \pmod{5},$$

which is false since -2 is a quadratic nonresidue of 5.

When $d = 2$ or 5, equation (27) leads to the congruence

$$1 + \binom{n-2}{3} + \binom{n-2}{6} + \cdots \equiv 0 \pmod{3},$$

which is false by Theorem 3–21.

There remains only the possibility that $Y = 0$. The proof that this happens only in the case of the second exception mentioned in the theorem is completely similar to what has just been done for the case $Z = 0$, and we leave the details to the reader. (The only variation lies in the fact that d may now have the sole prime divisor 2, so that $d = 2$ or 4.)

3–9 The Delaunay-Nagell theorem. As we shall see in the next chapter, there is a general theorem which implies that the equation

$$ax^3 + by^3 = c \tag{28}$$

has only finitely many solutions in integers x, y if a, b, and c are nonzero integers. In certain special cases, however, it is possible to make more precise statements about the number and nature of possible solutions. We shall concern ourselves here with the equation

$$x^3 + dy^3 = 1, \tag{29}$$

which was first considered in detail by B. Delaunay. His method was later refined by T. Nagell, who also applied it to (28) in the case that $c = 1$ or 3. Nagell's result concerning (29) is as follows.

THEOREM 3–23. *Equation (29) has at most one solution in integers x, y different from zero. If x_1, y_1 is a solution, the number $x_1 + y_1 \sqrt[3]{d}$ is either the fundamental unit of $L = R(\sqrt[3]{d})$ or its square; the latter can happen for only finitely many values of d.*

If $d = \pm 1$, (29) has only trivial solutions. If d contains a cube larger than 1, it can be absorbed into the factor y^3. Hence we can assume that d is cube-free and larger than 1.

The idea of the proof is quite simple. If

$$\mathbf{N}(x_1 + y_1\sqrt[3]{d}) = x_1{}^3 + dy_1{}^3 = 1, \qquad y_1 \neq 0,$$

then $x_1 + y_1\sqrt[3]{d}$ is a positive unit of L, and as such is a positive power of the fundamental unit ξ mentioned at the end of Section 3-7. It therefore suffices to show that no power of a positive unit smaller than 1, with exponent larger than 2, is of the special form $x + y\sqrt[3]{d}$, and to show that the square of a unit is of this form in only finitely many cases. We divide the proof into four parts, summarized in the next four theorems.

THEOREM 3-24.   *The square of an irrational unit of L of the form*

$$\eta = x + y\alpha + z\beta, \qquad x, y, z \text{ in } Z$$

is itself of the form $X + Y\alpha$ *only if*

$$\eta = 1 + \sqrt[3]{20} - \sqrt[3]{50}.$$

The square of a unit of L of the form

$$\eta = \tfrac{1}{3}(x + y\alpha + z\beta), \qquad 3 \nmid xyz,$$

(if such exists) is itself of the form $X + Y\alpha$ *for only finitely many values of d.*

Proof: Let $\qquad \eta = x + y\alpha + z\beta$

be a positive unit of L, so that, by (23),

$$x^3 + ab^2y^3 + a^2bz^3 - 3abxyz = 1 \tag{30}$$

and

$$\eta^2 = (x^2 + 2abyz) + (2xy + az^2)\alpha + (2xz + by^2)\beta.$$

If the coefficient of β in this last expression is 0, then

$$z = -\frac{by^2}{2x},$$

and substituting this into (30) we obtain

$$x^3 + dy^3 - d^2\frac{y^6}{8x^3} + 3d\frac{y^3}{2} = 1,$$

or
$$d^2y^6 - 20x^3dy^3 - 8(x^6 - x^3) = 0,$$

whence
$$dy^3 = 10x^3 \pm 2x\sqrt{27x^4 - 2x}. \tag{31}$$

Thus the number $27x^4 - 2x$ must be a square:
$$(27x^3 - 2)x = t^2. \tag{32}$$

If x is even, then $(27x^3 - 2, x) = 2$, so that
$$27x^3 - 2 = \pm 2u^2, \qquad x = \pm 2v^2.$$

Since -1 is a quadratic nonresidue of 3, we must choose the lower sign, and eliminating x we obtain
$$108v^6 + 1 = u^2,$$
$$(u - 1)(u + 1) = 108v^6.$$

Since $(u - 1, u + 1) = 2$, this implies that
$$u \pm 1 = 54r^6, \qquad u \mp 1 = 2s^6,$$

whence
$$27r^6 - s^6 = (3r^2)^3 - (s^2)^3 = \pm 1.$$

From the truth of Fermat's conjecture for $n = 3$, it follows that $r = 0$, which gives $v = 0$ and $x = 0$. But then also $y = 0$, by (31), which is impossible since $z\beta$ is not a unit.

If x is odd, (32) yields
$$27x^3 - 2 = \pm u^2, \qquad x = \pm v^2.$$

Here the upper sign must be chosen, and we have
$$(3x)^3 = u^2 + 2,$$

which by Theorem 3–19 has the sole solution $x = 1$, $u = \pm 5$. By (31), $dy^3 = 10 \pm 10$, so that $d = 20$, $y = 1$. (If $y = 0$, then $z = 0$, and η is rational.) The sole solution is therefore
$$(1 + \sqrt[3]{20} - \sqrt[3]{50})^2 = -19 + 7\sqrt[3]{20}.$$

Now let η be a positive unit of the form
$$\eta = \tfrac{1}{3}(x + y\alpha + z\beta).$$

Then by (24),
$$x^3 + ab^2y^3 + a^2bz^3 - 3abxyz = 27, \tag{33}$$

and
$$9\eta^2 = (x^2 + 2abyz) + (2xy + az^2)\alpha + (2xz + by^2)\beta. \tag{34}$$

If $3|x$, also $3|y$ and $3|z$, and we have already treated this case. Suppose that $3 \nmid x$. If the coefficient of β in the expression for η^2 is 0, we again have

$$z = -\frac{by^2}{2x}.$$

Substituting this into (33), it follows that

$$dy^3 = 10x^3 \pm 6x\sqrt{3x^4 - 6x}, \tag{35}$$

so that

$$3x^4 - 6x = t^2. \tag{36}$$

If x is even, the fact that $3 \nmid x$ implies that

$$x^3 - 2 = \pm 6u^2, \qquad x = \pm 2v^2,$$

whence

$$\pm 4v^6 - 1 = \pm 3u^2.$$

Since $3 \nmid (4v^6 + 1)$, we must choose the upper sign; the last equation can then be written as

$$(u + 1)^3 - (u - 1)^3 = (2v^2)^3,$$

so that $|u| = 1$. Hence $x = 2$, and by (35), $dy^3 = 80 \pm 72$. The lower sign yields $d = 1$ or 8, both of which are excluded. Hence $d = 19$, $y = 2$, and $z = -1$. The only solution in this case is

$$\left(\frac{2 + 2\sqrt[3]{19} - (\sqrt[3]{19})^2}{3}\right)^2 = -8 + 3\sqrt[3]{19}.$$

If x is odd, (36) implies that

$$x^3 - 2 = \pm 3u^2, \qquad x = \pm v^2,$$

so that

$$\pm v^6 - 2 = \pm 3u^2.$$

The lower sign must be chosen: $x = -v^2$ and

$$3u^2 - 2 = v^6. \tag{37}$$

But it is an immediate consequence of Theorem 4–17, to be proved in the next chapter, that (37) has only finitely many solutions, and the proof is complete.

We note for future use that if u, v satisfy (37), then v must be odd.

THEOREM 3–25.   *The fourth power of a positive irrational unit of L is never of the form $X + Y\alpha$.*

Proof: Let ϵ be such a unit,

$$\epsilon = \tfrac{1}{3}(x_1 + y_1\alpha + z_1\beta),$$

and suppose that

$$\epsilon^4 = X + Y\alpha.$$

Then since the coefficient of β in ϵ^4 is 0, we have

$$6bx_1^2 y_1^2 + 4x_1^3 z_1 + 4ab^2 y_1^3 z_1 + 12abx_1 y_1 z_1^2 + a^2 b z_1^4 = 0. \quad (38)$$

If we put

$$\eta = \epsilon^2 = \tfrac{1}{3}(x + y\alpha + z\beta),$$

then

$$x = \tfrac{1}{3}(x_1^2 + 2aby_1 z_1),$$

$$y = \tfrac{1}{3}(2x_1 y_1 + az_1^2),$$

$$z = \tfrac{1}{3}(2x_1 z_1 + by_1^2).$$

Since $\eta^2 = X + Y\alpha$, we can apply Theorem 3–24. The cases

$$d = 20, \qquad x = y = -z = 3,$$

$$d = 19, \qquad x = y = 2, \quad z = -1$$

are impossible, since in the first the above equation for z becomes $-9 = 2x_1 z_1 + 2y_1^2$, while in the second the system is easily seen to be inconsistent for all choices of signs of x_1, y_1, z_1. Hence it must be that $x = -v^2$, where v is odd, so that

$$3v^2 + x_1^2 = -2aby_1 z_1.$$

Since v is odd, so is x_1, so that $3v^2 + x_1^2 \equiv 4 \pmod 8$. Hence three of the numbers a, b, y_1, z_1 are odd, and the fourth is even. By (38), $a^2 b z_1^4$ is even, so y_1 is odd. If either a or z_1 is even, (38) implies that $6bx_1^2 y_1^2 \equiv 0 \pmod 4$, which is false. If b is even, (38) implies that $a^2 b z_1^4 \equiv 0 \pmod 4$, which is false since b is square-free. The proof is complete.

THEOREM 3–26. *The cube of a positive irrational unit of L is never of the form $X + Y\alpha$.*

Proof: If

$$\eta = \tfrac{1}{3}(x + y\alpha + z\beta)$$

is a positive unit, the coefficient of β in η^3 is

$$\tfrac{1}{9}(bxy^2 + x^2 z + abyz^2).$$

We see from the equation

$$x^3 + ab^2y^3 + a^2bz^3 - 3abxyz = 27 \tag{39}$$

that $(x, b) = 1$, and deduce from the equation

$$bxy^2 + x^2z + abyz^2 = 0 \tag{40}$$

that $b|z$. From (39) again, $\delta = (x, y, z) = 1$ or 3. Since $\eta \neq \pm 1$, y and z are not both zero, and we can write

$$x = \delta d_1 d_2 x_1, \qquad y = \delta d_2 d_3 y_1, \qquad z = \delta b d_1 d_3 z_1, \tag{41}$$

where

$$\left(\frac{x}{\delta}, \frac{z}{\delta b}\right) = |d_1|, \qquad \left(\frac{x}{\delta}, \frac{y}{\delta}\right) = |d_2|, \qquad \left(\frac{y}{\delta}, \frac{z}{\delta b}\right) = |d_3|,$$

and $x_1 > 0$, $y_1 > 0$, $z_1 > 0$. The numbers $d_1 x_1$, $d_2 y_1$, $d_3 z_1$ are relatively prime in pairs. Substituting the values from (41) into (39) and dividing by $\delta^3 b d_1 d_2 d_3$, we obtain

$$d_2{}^2 d_3 x_1 y_1{}^2 + d_1{}^2 d_2 x_1{}^2 z_1 + ab^2 d_1 d_3{}^2 y_1 z_1{}^2 = 0.$$

It follows from this that $d_1|x_1$, $d_2|y_1$, and $d_3|z_1$. Putting

$$x_1 = d_1 x_2, \qquad y_1 = d_2 y_2, \qquad z_1 = d_3 z_2,$$

substituting, and dividing by $d_1 d_2 d_3$, we obtain

$$d_2{}^3 x_2 y_2{}^2 + d_1{}^3 x_2{}^2 z_2 + ab^2 d_3{}^3 y_2 z_2{}^2 = 0.$$

A consequence of this is that $x_2|ab^2 d_3{}^3 y_2 z_2{}^2$, which in turn implies that $x_2 = 1$. Similarly, $y_2 = z_2 = 1$, so that

$$d_1{}^3 + d_2{}^3 + ab^2 d_3{}^3 = 0 \tag{42}$$

and

$$x = \delta d_1{}^2 d_2, \qquad y = \delta d_2{}^2 d_3, \qquad z = \delta b d_1 d_3{}^2.$$

Substituting these values into (39), we obtain

$$d_1{}^6 d_2{}^3 + ab^2 d_2{}^6 d_3{}^2 + a^2 b^4 d_1{}^3 d_3{}^6 - 3ab^2 d_1{}^3 d_2{}^3 d_3{}^3 = \frac{27}{\delta^3}.$$

Eliminating $ab^2 d_3{}^3$ between this equation and (42), we have

$$d_1{}^9 + 6d_1{}^6 d_2{}^3 + 3d_1{}^3 d_2{}^6 - d_2{}^9 = \left(\frac{3}{\delta}\right)^3, \tag{43}$$

and putting $d_1{}^3 = u$, $d_2{}^3 = v$, $3/\delta = w$, this becomes

$$u^3 + 6u^2v + 3uv^2 - v^3 = w^3. \tag{44}$$

But it is easily verified that

$$(u^3 + 6u^2v + 3uv^2 - v^3)U^3 = V^3 + W^3,$$

where

$$U = u^2 + uv + v^2, \qquad V = u^3 + 3u^2v - v^3, \qquad W = 3u^2v + 3uv^2.$$

Since neither U nor V is zero for relatively prime u and v, (44) can hold with $w \neq 0$ only if $W = 0$, that is, if $u = -v$. In this case $w = v$. Since $(d_1, d_2) = 1$, it follows that $d_1 = -1$, $d_2 = 1$, $\delta = 3$. This, however, leads to the values $x = 3$, $y = -3$, $z = 0$, for which the coefficient of β in η^3 is not zero.

THEOREM 3–27.   *If $p > 3$ is prime and*

$$\eta = \tfrac{1}{3}(x + y\alpha + z\beta)$$

is a positive unit smaller than 1, then η^p is not of the form $X + Y\alpha$.

Proof: Suppose that $z = 0$. Then $3|x$ and $3|y$, and

$$\mathbf{N}\eta = \left(\frac{x}{3}\right)^3 + d\left(\frac{y}{3}\right)^3 = 1,$$

so that $\left(\dfrac{x}{3}, d\dfrac{y}{3}\right) = 1$; by Theorem 3–22, the coefficient of β in η^p is not zero. Thus $z \neq 0$. By the same reasoning (applied in the field $L' = R(\beta) = R(\alpha) = L$), y cannot be zero.

As we saw in the proof of Theorem 3–20, it follows from the representation

$$\omega = x_1 + x_2\alpha + x_3\beta$$

of an arbitrary integer ω of L that

$$\alpha(\omega + \rho\omega' + \rho^2\omega'') = 3abx_3.$$

Taking $\omega = \eta^p$, we see that if the coefficient of β in η^p is zero, it must be that

$$\left(\frac{x + y\alpha + z\beta}{3}\right)^p + \rho\left(\frac{x + y\rho\alpha + z\rho^2\beta}{3}\right)^p$$

$$+ \rho^2\left(\frac{x + y\rho^2\alpha + z\rho\beta}{3}\right)^p = 0. \qquad (45)$$

Suppose first that $p \equiv 1 \pmod 3$. Then since $\rho^p = \rho$, (45) can be written in the form

$$\left(\frac{x\rho + y\rho^2\alpha + z\beta}{3}\right)^p + \left(\frac{x\rho^2 + y\rho\alpha + z\beta}{3}\right)^p = -\left(\frac{x + y\alpha + z\beta}{3}\right)^p.$$

Since p is odd, the left side is divisible by

$$\frac{x\rho + y\rho^2\alpha + z\beta}{3} + \frac{x\rho^2 + y\rho\alpha + z\beta}{3} = \frac{-x - y\alpha + 2z\beta}{3};$$

this number is an integer, and since it divides η^p, it is a unit. Consequently,

$$-x^3 - ab^2y^3 + 8a^2bz^3 - 6abxyz = \pm 27.$$

Since η is a positive unit, also

$$x^3 + ab^2y^3 + a^2bz^3 - 3abxyz = 27,$$

and by addition,

$$9a^2bz^3 - 9abxyz = 0 \text{ or } 54.$$

In the first case $az^2 - xy$ must be zero. But this number is the coefficient of α in $3/\eta$, and as we saw at the end of Section 3–7, it is not zero, since $1/\eta > 1$.

In the second case we have

$$abz(az^2 - xy) = 6.$$

But then x, y, z are not all divisible by 3, so that L is of the second kind. This is impossible, since if $ab|6$ then $a^2 - b^2 \not\equiv 0 \pmod 9$.

The case in which $p \equiv 2 \pmod 3$ proceeds similarly. Equation (45) can be written in the form

$$\left(\frac{x\rho^2 + y\alpha + z\rho\beta}{3}\right)^p + \left(\frac{x\rho + y\alpha + z\rho^2\beta}{3}\right)^p = -\left(\frac{x + y\alpha + z\beta}{3}\right)^p,$$

from which it follows that the number

$$\frac{x\rho^2 + y\alpha + z\rho\beta}{3} + \frac{x\rho + y\alpha + z\rho^2\beta}{3} = \frac{-x + 2y\alpha - z\beta}{3}$$

is a unit. As before,

$$9ab^2y^3 - 9abxyz = 0 \text{ or } 54.$$

Since $by^2 - xz$ is the coefficient of β in $3/\eta$, it is not zero. But it is also impossible that $(by^2 - xz)|6$ and $ab|6$, since then L must be of both the first and second kinds. The proof is complete.

Theorems 3–25, 3–26, and 3–27 show that any nonzero solution of $x^3 + dy^3 = 1$ must correspond either to the fundamental unit of L, or to its square. Not both of these numbers can lead to solutions, by Theorem 3–22 with $n = 1$. This completes the proof of Theorem 3–23.

REFERENCES

Section 3–5

For a complete exposition of what is known concerning Fermat's conjecture, see H. S. Vandiver, "Fermat's last theorem: the history and the nature of the results concerning it," *American Mathematical Monthly* **53**, 555–578 (1946). The result of Lehmer, Lehmer, and Vandiver was announced in *Proceedings of the National Academy of Sciences* **40**, 25–33 (1954). Landau gives a proof of Kummer's lemma; see his *Vorlesungen über Zahlentheorie*, vol. 3, Leipzig: S. Hirzel Verlag, 1927.

Section 3–6

The equation $y^2 = x^3 + k$ was the subject of L. J. Mordell's inaugural address, *A Chapter in the Theory of Numbers*, New York: Cambridge University Press, 1947. Also see Dickson's *History of the Theory of Numbers*, Washington: Carnegie Institution of Washington, 1919; reprinted, Chelsea Publishing Company, New York, 1950; vol. 2, pp. 531–539.

Section 3–7

Dedekind's fundamental paper on pure cubic fields is in *Journal für die Reine und Angewandte Mathematik* (Berlin) **121**, 40–123 (1899).

Sections 3–8, 3–9

We have followed the treatment by Nagell, *Journal des Mathématiques Pures et Appliquées* (Paris) **4**, 209–270 (1925). Delaunay (*Comptes Rendus Hebdomadaires des Séances de l'Académie des Sciences* (Paris) **171**, 336 (1920) and **172**, 434 (1921)) announced that equation (28) has at most five solutions in case $c = 1$. His work on (29) was announced in *Comptes Rendus* **162**, 150–151 (1916).

CHAPTER 4

THE THUE-SIEGEL-ROTH THEOREM

4–1 Introduction. It is shown in introductory texts in number theory* that if α is a quadratic irrationality (that is, an algebraic number of degree two), then there is a positive constant c such that

$$\left| \alpha - \frac{p}{q} \right| > \frac{c}{q^2}$$

for every pair of rational integers p, q with $q > 0$. The idea used there suffices to prove the following generalization, which is due to J. Liouville.

THEOREM 4–1. *If α is an algebraic number of degree $n \geq 2$, then there exists a positive constant c such that*

$$\left| \alpha - \frac{p}{q} \right| > \frac{c}{q^n} \tag{1}$$

for every pair of rational integers p, q with $q > 0$.

Proof: Let α be a zero of the irreducible polynomial

$$f(x) = a_0 x^n + \cdots + a_n, \qquad a_0 > 0,$$

with coefficients in Z, and let $\alpha_1 = \alpha, \alpha_2, \ldots, \alpha_n$ be its conjugates, so that

$$f(x) = a_0(x - \alpha)(x - \alpha_2) \cdots (x - \alpha_n).$$

Then the number

$$q^n f\left(\frac{p}{q}\right) = a_0 p^n + a_1 p^{n-1} q + \cdots + a_n q^n$$

is a rational integer different from zero, and it therefore has absolute

---

* See for example, Volume I, Section 8–4. In Section 8–5 Hurwitz' theorem is stated and proved, and in Chapter 9 the problem of approximating real numbers by rationals is considered; all this material is assumed in the present section.

value at least 1. Hence

$$\left| \alpha - \frac{p}{q} \right| = \frac{\left| q^n f\left(\frac{p}{q}\right) \right|}{a_0 q^n \prod_{k=2}^{n} \left| \alpha_k - \frac{p}{q} \right|} \geq \frac{1}{a_0 q^n \prod_{k=2}^{n} \left| \alpha_k - \frac{p}{q} \right|}.$$

Put

$$\beta = \lceil \alpha \rceil = \max\,(|\alpha|, \ldots, |\alpha_n|).$$

We consider two cases, according as $|p/q|$ is greater than 2β or not. In the first case we have the trivial lower bound

$$\left| \alpha - \frac{p}{q} \right| > \beta \geq \frac{\beta}{q^n}.$$

In the second case the inequality $|\alpha_k - p/q| \leq 3\beta$ holds for $k = 2, \ldots, n$, and, by the inequality of the preceding paragraph,

$$\left| \alpha - \frac{p}{q} \right| \geq \frac{1}{a_0 q^n (3\beta)^{n-1}}.$$

Thus, the theorem holds with

$$c = \min\left(\beta, \frac{1}{a_0 (3\beta)^{n-1}} \right).$$

Liouville used this theorem to show the existence of nonalgebraic numbers; this will be discussed in detail in the next chapter. At the moment, let us consider a hypothetical improvement of Theorem 4–1, in which the inequality (1) is replaced by

$$\left| \alpha - \frac{p}{q} \right| > \frac{c}{q^\nu}, \tag{2}$$

where ν is any number smaller than n. A. Thue noticed that if such a theorem could be proved, it would have the important consequence that the Diophantine equation

$$q^n f\left(\frac{p}{q}\right) = a_0 p^n + a_1 p^{n-1} q + \cdots + a_n q^n = A \tag{3}$$

can have only finitely many solutions for any fixed rational integer A different from zero, if $f(x)$ has distinct zeros. To see this, let the zeros of $f(x)$ again be $\alpha_1 = \alpha, \ldots, \alpha_n$, and put

$$\gamma = \min_{i \neq j}\,(|\alpha_i - \alpha_j|).$$

Suppose that (3) has infinitely many solutions p, q. Then there must be at least one α_i, which by suitable naming we can take to be α, which is a limit point of the numbers p/q, since otherwise the quantity

$$q^n f\left(\frac{p}{q}\right) = a_0 q^n \prod_{k=1}^{n}\left(\frac{p}{q} - \alpha_k\right)$$

is certainly not bounded as q increases indefinitely. There must therefore be infinitely many solutions of (3) for which $|\alpha - p/q| < \gamma/2$. But for all such solutions,

$$\left|\alpha - \frac{p}{q}\right| = \frac{A}{a_0 q^n \prod_{k=2}^{n}\left|\dfrac{p}{q} - \alpha_k\right|} \leq \frac{A}{a_0\left(\dfrac{\gamma}{2}\right)^{n-1}} \cdot \frac{1}{q^n},$$

and this is at variance with (2) if ν is a constant smaller than n and q is sufficiently large.

Thue showed that (2) holds with

$$\nu = \frac{n}{2} + 1.$$

Later C. L. Siegel improved Thue's result, showing that (2) holds with

$$\nu > \min_{\substack{1 \leq s \leq n-1 \\ s \in Z}}\left(\frac{n}{s+1} + s\right),$$

and in particular with $\nu = 2\sqrt{n}$. In 1947 F. J. Dyson made the further improvement $\nu > \sqrt{2n}$, and finally in 1955 K. F. Roth proved that (2) holds with $\nu = 2 + \epsilon$, for each $\epsilon > 0$, for all but a finite number of fractions p/q. This is the best theorem possible if ν is to be independent of q, since Hurwitz' theorem shows that the corresponding statement is false for every irrational algebraic number, for $\nu = 2$ and suitable c. Roth's work is similar in some respects to a simplification of Dyson's proof, published by T. Schneider in 1948.

In addition to the problem of sharpening Theorem 4–1 by decreasing the exponent of q, we may also consider the question of extending the methods so as to analyze the approximability of an algebraic number by other algebraic numbers. This is not mere generalization for its own sake: as we saw in the preceding chapter, it is natural to consider the solvability of $x^p + y^p = z^p$ in a larger set of integers than Z, and the same is true of many other Diophantine equations. But if the variables in an equation range over the integers of an algebraic number field, then to the extent that approximation theorems are useful at all they must be formulated in terms of algebraic rather than rational numbers.

While Siegel gave many algebraic variants of his basic result, Roth presented a detailed proof only in the rational case. In this chapter we give a complete proof of a useful algebraic version of Roth's theorem. Unfortunately, the proof is complicated; the student might profit by first examining Schneider's work mentioned above.

We shall proceed as follows. In the next three sections we shall make some definitions, and obtain some preliminary results, which are needed for the proof of the main theorem: in Section 4–2 some properties of polynomials will be treated, in Section 4–3 the concept of the generalized Wronskian will be introduced, and in Section 4–4 the index of a polynomial will be defined and discussed. Then we shall proceed to prove, in Sections 4–5 and 4–6, several lemmas on which the proof of the main theorem depends, and finally, in Section 4–7, we shall state and prove the Thue-Siegel-Roth theorem itself. In the remainder of the chapter, some applications of the theorem will be taken up.

4–2 Polynomials. If $P(z)$ is a polynomial with arbitrary complex coefficients, we denote by $||P||$ the maximum of the absolute values of its coefficients. If α is an algebraic number and $P(z) = 0$ is its defining equation, so that P is irreducible and has relatively prime coefficients in Z, we define the *height* $H(\alpha)$ to be $||P||$. Finally, if P has algebraic coefficients, we designate by $\overline{|P|}$ the maximum of the absolute values of their conjugates. Clearly $||P|| = \overline{|P|}$ if P has coefficients in Z, and for a nonzero constant polynomial $P(z) = \alpha$ the new definition of $\overline{|\alpha|}$ agrees with the old one.

Except when a polynomial is written as a determinant, it will be supposed that no two terms have the same exponents on the variable, or sets of exponents on the variables.

THEOREM 4–2. *Let $l, \lambda_1, \ldots, \lambda_h$ be complex numbers, and put*

$$L(z) = l \prod_{k=1}^{h} (z - \lambda_k).$$

Then
$$|l| \prod_{k=1}^{h} (1 + |\lambda_k|) \leq 6^h \|L\|.$$

Proof: There is no loss in generality in supposing that $l = 1$, since a change in l affects in the same way the two sides of the inequality to be proved. Let $\lambda_1, \ldots, \lambda_t$ be those of the λ's such that $|\lambda_k| \leq 2$. If $f(z) = \prod_{k=1}^{t} (z - \lambda_k)$, then there is a complex number z_0 with $|z_0| = 1$ for which $|f(z_0)| \geq 1$. To see this, let ϵ be a $(t + 1)$th root of unity, and suppose that

$$f(z) = \sum_{r=0}^{t} \mu_r z^r, \qquad \mu_t = 1.$$

Then

$$\sum_{\nu=0}^{t} \epsilon^\nu f(\epsilon^\nu) = \sum_{\nu=0}^{t} \epsilon^\nu \sum_{r=0}^{t} \mu_r \epsilon^{\nu r} = \sum_{r=0}^{t} \mu_r \sum_{\nu=0}^{t} \epsilon^{\nu(r+1)}.$$

But

$$\sum_{\nu=0}^{t} \epsilon^{\nu(r+1)} = \begin{cases} 0 & \text{if } (t+1) \nmid (r+1), \\ t+1 & \text{if } (t+1) \mid (r+1), \end{cases} \tag{4}$$

and since $r \leq t$, $(t + 1) \mid (r + 1)$ if and only if $r = t$. Hence

$$\sum_{\nu=0}^{t} \epsilon^\nu f(\epsilon^\nu) = (t+1)\mu_t = t + 1,$$

so that one of the $t + 1$ numbers $|f(\epsilon^\nu)|$ is at least 1. Thus

$$\prod_{k=1}^{t} (1 + |\lambda_k|) \leq (1 + 2)^t = 3^t \leq 3^t \left| \prod_{k=1}^{t} (z_0 - \lambda_k) \right|. \tag{5}$$

If $t < h$, then for $k = t + 1, \ldots, h$ we have $|\lambda_k| > 2$ and

$$\frac{1 + |\lambda_k|}{|z_0 - \lambda_k|} \leq \frac{1 + |\lambda_k|}{|\lambda_k| - |z_0|} = \frac{|\lambda_k| + 1}{|\lambda_k| - 1} = 1 + \frac{2}{|\lambda_k| - 1} < 1 + \frac{2}{2 - 1} = 3,$$

so that

$$\prod_{k=t+1}^{h} (1 + |\lambda_k|) < 3^{h-t} \left| \prod_{k=t+1}^{h} (z_0 - \lambda_k) \right|.$$

Combining this with (5), we have

$$\prod_{k=1}^{h} (1 + |\lambda_k|) < 3^h \left| \prod_{k=1}^{h} (z_0 - \lambda_k) \right| \leq 3^h \|L\| (|z_0|^h + \cdots + 1)$$

$$= 3^h (h + 1) \|L\| \leq 6^h \|L\|.$$

THEOREM 4–3. *Suppose that $f(z)$ and $g(z)$ are polynomials with complex coefficients, of degrees n and m respectively. Suppose further that the coefficient of z^m in $g(z)$ has absolute value at least 1. Then*

$$||f|| \leq 6^{n+m}||fg||.$$

Proof: Let

$$f(z) = a_0(z - \lambda_1) \cdots (z - \lambda_n),$$

$$g(z) = b_0(z - \lambda_{n+1}) \cdots (z - \lambda_{n+m}).$$

Then

$$||f|| \leq \left\| a_0 \prod_{k=1}^{n} (z + |\lambda_k|) \right\| \leq |a_0 b_0| \cdot \prod_{k=1}^{n} (1 + |\lambda_k|)$$

$$\leq |a_0 b_0| \prod_{k=1}^{n+m} (1 + |\lambda_k|),$$

and the desired result follows from Theorem 4–2.

THEOREM 4–4. *If $f(z)$ is an arbitrary polynomial of degree n, with real coefficients, then*

$$||f||^m \leq (mn + 1)||f^m||. \tag{6}$$

Proof: Let $f(z) = a_0 + a_1 z + \cdots + a_n z^n$, and let $||f|| = a$. The theorem is certainly true if either $|a_0| = a$ or $|a_n| = a$, since the first and last coefficients in $f^m(z)$ are the mth powers of a_0 and a_n, respectively, so that in this case $||f^m|| \geq ||f||^m$. If we put

$$f^*(z) = z^n f\left(\frac{1}{z}\right),$$

then clearly

$$||f^*|| = ||f|| \qquad \text{and} \qquad ||(f^*)^m|| = ||(f^m)^*||,$$

so that we can suppose, with no loss in generality, that the numerically largest of all the coefficients in $f(z)$ is a_t, where $\frac{1}{2}n \leq t < n$.

Put $$g(z, \theta) = f(z) - ae^{i\theta}z^n,$$

and let $\alpha = \alpha(\theta)$ be the numerically largest of the zeros of $g(z, \theta)$ for each θ. *The inequality* (6) *holds if, for some* θ, $|\alpha(\theta)| \geq 1$. For

$$|f^m(\alpha)| = |ae^{i\theta}\alpha^n|^m = a^m|\alpha|^{mn},$$

while for $|\alpha| \geq 1$,

$$|f^m(\alpha)| \leq ||f^m||(1 + |\alpha| + \cdots + |\alpha|^{mn}) \leq ||f^m||(mn + 1)|\alpha|^{mn},$$

so that

$$a^m = ||f||^m \leq ||f^m||(mn + 1).$$

We know that $|a_n| < a$. Hence if $f(1) > a$, then

$$g(1, 0) > a - a = 0, \qquad g(\infty, 0) = -\infty,$$

and $1 < |\alpha(0)| < \infty$. Similiarly, if $f(1) < -a$, then

$$g(1, \pi) < -a + a = 0, \qquad g(\infty, \pi) = \infty,$$

and $1 < |\alpha(\pi)| < \infty$. This proves the theorem unless $|f(1)| \leq a$, which we henceforth assume.

Now put $z = e^{i\varphi}$, so that

$$g(e^{i\varphi}, \theta) = f(e^{i\varphi}) - ae^{i(\theta + n\varphi)}.$$

If we find a φ_0 such that $|f(e^{i\varphi_0})| = a$, then θ_0 can be determined so that $g(e^{i\varphi_0}, \theta_0) = 0$; this gives $|\alpha(\theta_0)| \geq 1$ and proves the theorem. Since $|f(e^{i\varphi})|$ is a continuous function of φ, and since $|f(1)| \leq a$, it suffices to prove the existence of a φ_0 such that $|f(e^{i\varphi_0})| \geq a$.

Let ϵ be a primitive $(t + 1)$th root of unity, where $|a_t| = a$ and $\frac{1}{2}n \leq t < n$. Then

$$\sum_{\nu=0}^{t} \epsilon^{\nu} f(\epsilon^{\nu}) = \sum_{\nu=0}^{t} \epsilon^{\nu} \sum_{k=0}^{n} a_k \epsilon^{\nu k} = \sum_{k=0}^{n} a_k \sum_{\nu=0}^{t} \epsilon^{\nu(k+1)}.$$

Since $k \leq n$ and $t \geq \frac{1}{2}n$, we have that $(t + 1)|(k + 1)$ if and only if $k = t$. Hence, by (4),

$$\sum_{\nu=0}^{t} \epsilon^{\nu} f(\epsilon^{\nu}) = a_t(t + 1),$$

so that for some ν,

$$|\epsilon^{\nu} f(\epsilon^{\nu})| = |f(\epsilon^{\nu})| \geq |a_t| = a.$$

The proof is complete.

THEOREM 4-5. *If $f_1(z), \ldots, f_t(z)$ are polynomials with algebraic coefficients, then*

$$\left| \overline{\prod_{\nu=1}^{t} f_\nu} \right| \leq \prod_{\nu=1}^{t} (1 + \deg f_\nu) \prod_{\nu=1}^{t} \boxed{f_\nu}.$$

Proof: There is no loss in generality in supposing that

$$\deg f_1 \geq \deg f_2 \geq \cdots \geq \deg f_t.$$

The product $f_1 f_2$ is a polynomial each of whose coefficients is a sum of products of a coefficient of f_1 and a coefficient of f_2, the number of summands being at most $1 + \deg f_2$. Hence

$$\lceil f_1 f_2 \rceil = (1 + \deg f_2) \lceil f_1 \rceil \lceil f_2 \rceil.$$

Similarly,

$$\lceil f_1 f_2 f_3 \rceil \leq (1 + \deg f_3) \lceil f_1 f_2 \rceil \lceil f_3 \rceil \leq (1 + \deg f_3)(1 + \deg f_2) \lceil f_1 \rceil \lceil f_2 \rceil \lceil f_3 \rceil,$$

and so on.

THEOREM 4-6. *Let p and r be positive integers, with $1 \leq r < p$. Suppose that $F(z_1, \ldots, z_p)$, $G(z_1, \ldots, z_r)$, and $H(z_{r+1}, \ldots, z_p)$ are polynomials with coefficients in an algebraic number field K, those of F being integers, and suppose that*

$$F(z_1, \ldots, z_p) = G(z_1, \ldots, z_r) H(z_{r+1}, \ldots, z_p).$$

Then if γ is any coefficient in F, there is a factorization $\gamma = \alpha\beta$ in K such that the coefficients in αH and βG are integers in K.

Proof: Let the coefficients in G be $\alpha_1, \ldots, \alpha_s$, and those in H be $\beta_1, \ldots, \beta_t$, in some order. Then, since the variables in G and H are disjoint, the coefficients in F are simply the products $\alpha_i \beta_j$. Since the coefficients in F are integers, all the products $\alpha_i \beta_1, \ldots, \alpha_i \beta_t$ are integers, as are all the products $\beta_j \alpha_1, \ldots, \beta_j \alpha_s$. But these two sets of numbers are just the coefficients in $\alpha_1 H$ and $\beta_1 G$.

4-3 Generalized Wronskians. Polynomials $f_0(z_1, \ldots, z_p), \ldots,$ $f_{l-1}(z_1, \ldots, z_p)$ with coefficients in an algebraic number field K are said to be *linearly dependent* if some linear combination of them, with constant coefficients in K which are not all zero, vanishes identically, and are otherwise said to be *independent*. In the case of a single independent variable, it is well known that the question of independence of a set of functions can sometimes be settled by reference to their Wronskian. For our purposes it is convenient to define this as the determinant

$$W(z) = \det\left(\frac{1}{\mu!} \frac{d^\mu}{dz^\mu} f_\nu(z)\right), \qquad \mu, \nu = 0, 1, \ldots, l-1,$$

which differs from the usual definition only in the presence of the nonzero constant factor

$$\frac{1}{0! 1! \cdots (l-1)!}.$$

The exact relation of the behavior of the Wronskian to independence, as applied to polynomials, is indicated in the first part of the next theorem.

For functions of several variables, the situation is not quite so simple, since there are then several partial derivatives to consider. We proceed as follows. Let $\Delta_0, \Delta_1, \ldots, \Delta_\mu, \ldots, \Delta_{l-1}$ be differential operators of the form

$$\frac{1}{j_1! \cdots j_p!} \left(\frac{\partial}{\partial z_1}\right)^{j_1} \cdots \left(\frac{\partial}{\partial z_p}\right)^{j_p},$$

such that the *order* $j_1 + \cdots + j_p$ of Δ_μ does not exceed μ, for $0 \le \mu \le l - 1$. Then the function

$$G(z_1, \ldots, z_p) = \begin{vmatrix} \Delta_0 f_0 & \Delta_0 f_1 & \cdots & \Delta_0 f_{l-1} \\ \Delta_1 f_0 & \Delta_1 f_1 & \cdots & \Delta_1 f_{l-1} \\ \vdots & \vdots & & \vdots \\ \Delta_{l-1} f_0 & \Delta_{l-1} f_1 & \cdots & \Delta_{l-1} f_{l-1} \end{vmatrix}$$

is called a *generalized Wronskian* of $f_0, \ldots, f_{l-1}$. Except in the trivial case $p = l = 1$, there are several Δ_μ's for each μ, and hence more than one generalized Wronskian. In the case of functions of one variable, the ordinary Wronskian is that generalized Wronskian for which the order of Δ_μ is exactly μ, for $0 \le \mu \le l - 1$.

THEOREM 4-7. (a) *If $f_0, \ldots, f_{l-1}$ are l polynomials over K in the single variable z, whose Wronskian $W(z)$ vanishes identically, then they are dependent over K.*

(b) *If $f_0, \ldots, f_{l-1}$ are l polynomials over K in the variables $z_1, \ldots, z_p$, for which every generalized Wronskian $G_l(z_1, \ldots, z_p)$ vanishes identically, then they are dependent over K.*

Proof: (a) The proof in this case is by induction. If $l = 1$, then $W(z) = f_0(z)$, and the truth of the theorem is obvious.

Take $l > 1$, and suppose that the theorem is true for every set of $l - 1$ polynomials, $f_0, f_1, \ldots, f_{l-2}$, over K; suppose also that the Wronskian W_l of $f_0, \ldots, f_{l-1}$ vanishes identically. If $f_0, \ldots, f_{l-2}$ are dependent, so are $f_0, \ldots, f_{l-1}$, and the assertion is proved. Suppose then that $f_0, \ldots, f_{l-2}$ are independent, so that their Wronskian W_{l-1} is not identically zero. Now W_{l-1}, being a polynomial, has only finitely many zeros; let I be an interval in which it does not vanish, and take z in I. For such z, the system of equations

$$\sum_{k=0}^{l-2} f_k^{(j)}(z)y_k = f_{l-1}^{(j)}(z), \qquad j = 0, 1, \ldots, l-2, \qquad (7)$$

can be solved for the y's as rational functions of z. But then, by subtracting appropriate multiples of each column of W_l from its last column, we obtain

$$0 = 1! \cdots (l-1)! \, W_l$$

$$= \begin{vmatrix} f_0(z) & f_1(z) & \cdots & 0 \\ f_0'(z) & f_1'(z) & \cdots & 0 \\ \vdots & \vdots & & \vdots \\ f_0^{(l-1)}(z) & f_1^{(l-1)}(z) & \cdots & f_{l-1}^{(l-1)}(z) - \sum_{k=0}^{l-2} f_k^{(l-1)}(z)y_k \end{vmatrix}$$

$$= 1! \cdots (l-1)! \left(f_{l-1}^{(l-1)}(z) - \sum_{k=0}^{l-2} f_k^{(l-1)}(z)y_k \right) W_{l-1},$$

so that also

$$\sum_{k=0}^{l-2} f_k^{(l-1)}(z)y_k = f_{l-1}^{(l-1)}(z). \qquad (8)$$

Differentiating (7) gives

$$\sum_{k=0}^{l-2} f_k^{(j+1)}(z)y_k + \sum_{k=0}^{l-2} f_k^{(j)}(z)y_k' = f_{l-1}^{(j+1)}(z), \qquad j = 0, \ldots, l-2,$$

and comparison of this for $j = l-2$ with (8), and for $j = 0, \ldots, l-3$ with (7), shows that

$$\sum_{k=0}^{l-2} f_k^{(j)}(z)y_k' = 0, \qquad j = 0, \ldots, l-2.$$

Since $W_{l-1} \neq 0$, it must be that

$$y_0' = \cdots = y_{l-2}' = 0,$$

so that the y's are constants, say $y_k = c_k$, and they are clearly in K. But then the polynomial

$$\sum_{k=0}^{l-2} c_k f_k(z) - f_{l-1}(z)$$

vanishes throughout I, and therefore identically, so that the l polynomials $f_0, f_1, \ldots, f_{l-1}$ are dependent.

(b) This case is proved by contradiction. Suppose that the l polynomials $f_0(z_1, \ldots, z_p), \ldots, f_{l-1}(z_1, \ldots, z_p)$ are independent, and suppose further that for each ν, f_ν is of degree less than k in each of its arguments, so that we can write

$$f_\nu(z_1, \ldots, z_p) = \sum_{k_1=0}^{k-1} \cdots \sum_{k_p=0}^{k-1} b_\nu(k_1, \ldots, k_p)z_1{}^{k_1} \cdots z_p{}^{k_p},$$

$$0 \leq \nu \leq l - 1.$$

Then the polynomials $f_\nu(t, t^k, t^{k^2}, \ldots, t^{k^{p-1}})$ are linearly independent. For otherwise there would be an identity in t of the form

$$\sum_{\nu=0}^{l-1} c_\nu \sum_{k_1=0}^{k-1} \cdots \sum_{k_p=0}^{k-1} b_\nu(k_1, \ldots, k_p)t^{k_1+k_2k+\cdots+k_pk^{p-1}} = 0,$$

or

$$\sum_{k_1=0}^{k-1} \cdots \sum_{k_p=0}^{k-1} \left(\sum_{\nu=0}^{l-1} c_\nu b_\nu(k_1, \ldots, k_p)\right) t^{k_1+k_2k+\cdots+k_pk^{p-1}} = 0,$$

and it would follow from the uniqueness of the representation of an integer to the base k that for each set of exponents $k_1, \ldots, k_p$,

$$\sum_{\nu=0}^{l-1} c_\nu b_\nu(k_1, \ldots, k_p) = 0,$$

whence

$$\sum_{\nu=0}^{l-1} c_\nu f_\nu(z_1, \ldots, z_p) = 0,$$

contrary to assumption.

We know therefore that the Wronskian

$$W(t) = \det\left(\frac{1}{\mu!}\left(\frac{d}{dt}\right)^\mu f_\nu(t, t^k, \ldots, t^{k^{p-1}})\right), \qquad \mu, \nu = 0, \ldots, l-1,$$

does not vanish identically. By a standard differentiation formula,

$$\frac{d}{dt} f_\nu(t, \ldots, t^{k^{p-1}}) = \sum_{j=1}^{p} \frac{\partial}{\partial z_j} f_\nu(z_1, \ldots, z_p)\Big|_{(t,\ldots,t^{k^{p-1}})} \frac{dt^{k^{j-1}}}{dt}$$

and it follows easily by induction on μ that an operator identity

$$\left(\frac{d}{dt}\right)^\mu = \varphi_1(t)\Delta^{(1)} + \cdots + \varphi_r(t)\Delta^{(r)}$$

holds, where $\Delta^{(1)}, \ldots, \Delta^{(r)}$ are differential operators of orders not exceeding μ, r depends only on μ and p, and $\varphi_1, \ldots, \varphi_r$ are polynomials with rational coefficients. Using this in the above expression for $W(t)$, and writing the resulting determinant as a sum of other determinants, an expression for $W(t)$ of the form

$$W(t) = \psi_1(t)G_1(t, \ldots, t^{k^{p-1}}) + \cdots + \psi_s(t)G_s(t, \ldots, t^{k^{p-1}})$$

results, in which $\psi_1, \ldots, \psi_s$ are polynomials and $G_1, \ldots, G_s$ are generalized Wronskians of $f_1, \ldots, f_{l-1}$. Since $W(t)$ does not vanish identically, there is an i for which $G_i(t, \ldots, t^{k^{p-1}})$ is not identically zero, and *a fortiori* $G_i(z_1, \ldots, z_p)$ is not identically zero.

THEOREM 4–8. *Let $R(z_1, \ldots, z_p)$ be a polynomial in $p \geq 2$ variables, with integral coefficients in K such that*

$$0 < \lceil R \rceil \leq B.$$

Let R be of degree at most r_j in z_j, for $j = 1, \ldots, p$. Then there is an l in Z with

$$1 \leq l \leq r_p + 1, \tag{9}$$

there is an integer β in K, and there are differential operators $\Delta_0, \ldots, \Delta_{l-1}$ on the variables $z_1, \ldots, z_{p-1}$, of orders at most $0, \ldots, l - 1$, respectively, such that if

$$F(z_1, \ldots, z_p) = \beta \det\left(\Delta_\mu \frac{1}{\nu!}\left(\frac{\partial}{\partial z_p}\right)^\nu R\right), \quad \mu, \nu = 0, \ldots, l - 1, \tag{10}$$

then

   (a) *F has integral coefficients in K and is not identically zero;*
   (b) *a decomposition*

$$F(z_1, \ldots, z_p) = U(z_1, \ldots, z_{p-1})V(z_p) \tag{11}$$

holds, where U and V have integral coefficients in K, U is of degree at most lr_j in z_j for $j = 1, \ldots, p - 1$, and V is of degree at most lr_p in z_p;

   (c) *the following bound holds:*

$$\lceil F \rceil \leq \{(r_1 + 1) \cdots (r_p + 1)\}^{2l} 2^{2(r_1 + \cdots + r_p)l} l!^2 B^{2l}.$$

Proof: Write R as a polynomial in z_p:

$$R(z_1, \ldots, z_p) = \sum_{\varkappa=0}^{r_p} S_\varkappa(z_1, \ldots, z_{p-1}) z_p^\varkappa.$$

The polynomials $S_\varkappa$ need not be independent; let $\psi_\nu(z_1, \ldots, z_{p-1})$, for $\nu = 0, \ldots, l - 1$, be a maximal set of independent polynomials among the S_x, so that $1 \leq l \leq r_p + 1$. Then there are constants $\beta_{\nu x}$ in K such that for $x = 0, \ldots, r_p$,

$$S_x(z_1, \ldots, z_{p-1}) = \sum_{\nu=0}^{l-1} \beta_{\nu x} \psi_\nu(z_1, \ldots, z_{p-1}). \tag{12}$$

If we put

$$\varphi_\nu(z_p) = \sum_{x=0}^{r_p} \beta_{\nu x} z_p^x, \qquad \nu = 0, \ldots, l-1,$$

then

$$R(z_1, \ldots, z_p) = \sum_{\nu=0}^{l-1} \psi_\nu(z_1, \ldots, z_{p-1}) \, \varphi_\nu(z_p), \tag{13}$$

and $\varphi_0, \ldots, \varphi_{l-1}$ are independent. For if $\delta_0, \ldots, \delta_{l-1}$ are constants such that

$$\delta_0 \varphi_0(z_p) + \cdots + \delta_{l-1} \varphi_{l-1}(z_p) = 0,$$

the coefficient of each power of z_p must be zero, so that

$$\delta_0 \beta_{0x} + \cdots + \delta_{l-1} \beta_{l-1,x} = 0 \tag{14}$$

for $x = 0, \ldots, r_p$. For fixed ν_0 with $0 \leq \nu_0 \leq l - 1$, choose x_0 so that $S_{x_0}(z_1, \ldots, z_{p-1}) = \psi_{\nu_0}(z_1, \ldots, z_{p-1})$; this is possible since the ψ's are a subset of the S's. Then (12) shows that

$$\beta_{\nu x_0} = \begin{cases} 1 & \text{if } \nu = \nu_0, \\ 0 & \text{if } \nu \neq \nu_0. \end{cases}$$

Choosing $x = x_0$ in (14), we obtain $\delta_{\nu_0} = 0$. Since ν_0 is arbitrary, every $\delta_i = 0$.

Let $W(z_p)$ be the Wronskian of $\varphi_0, \ldots, \varphi_{l-1}$; it is a polynomial with coefficients in K, and it does not vanish identically. Let $G(z_1, \ldots, z_{p-1})$ be some generalized Wronskian of $\psi_0, \ldots, \psi_{l-1}$ which is not identically zero. Then

$$W(z_p) = \det\left(\frac{1}{\mu!} \left(\frac{d}{dz_p} \right)^\mu \varphi_\nu(z_p) \right), \qquad \mu, \nu = 0, \ldots, l-1,$$

$$G(z_1, \ldots, z_{p-1}) = \det\left(\Delta_\mu \psi_\nu(z_1, \ldots, z_{p-1}) \right),$$

where $\Delta_0, \ldots, \Delta_{l-1}$ are differential operators on $z_1, \ldots, z_{p-1}$, of orders at most $0, \ldots, l - 1$ respectively. Taking the row-by-row product of G and W, we obtain

$$GW = \det\left(\sum_{\rho=0}^{l-1} \Delta_\mu \frac{1}{\nu!} \left(\frac{\partial}{\partial z_p} \right)^\nu \varphi_\rho(z_p) \, \psi_\rho(z_1, \ldots, z_{p-1}) \right),$$

or
$$GW = \det\left(\Delta_\mu \frac{1}{\nu!}\left(\frac{\partial}{\partial z_p}\right)^\nu R\right). \tag{15}$$

Since W is a determinant of order l whose elements are polynomials in z_p of degrees at most r_p, it is clear that deg $W \leq lr_p$. Similarly, G is of degree at most lr_j in z_j, for $j = 1, \ldots, p-1$.

In the expression (15) for GW, we can write R as the sum of $(r_1 + 1) \cdots (r_p + 1)$ terms of the form

$$\alpha_{s_1 \cdots s_p} z_1{}^{s_1} \cdots z_p{}^{s_p}.$$

The determinant can then be written as a sum of

$$((r_1 + 1) \cdots (r_p + 1))^l$$

new determinants, each having entries of the form

$$\alpha_{s_1 \cdots s_p} \Delta_\mu \frac{1}{\nu!}\left(\frac{\partial}{\partial z_p}\right)^\nu z_1{}^{s_1} \cdots z_p{}^{s_p} = a\alpha_{s_1 \cdots s_p} z_1{}^{t_1} \cdots z_p{}^{t_p},$$

in which $t_j \leq s_j$ for $j = 1, \ldots, p$. Here

$$a = \binom{s_1}{t_1} \cdots \binom{s_p}{t_p} \leq 2^{s_1 + \cdots + s_p} \leq 2^{r_1 + \cdots + r_p}.$$

Thus the entries of each new determinant are such that the maxima of the absolute values of their conjugates do not exceed

$$2^{r_1 + \cdots + r_p}B,$$

and hence

$$\overline{|GW|} \leq ((r_1 + 1) \cdots (r_p + 1))^l l! 2^{(r_1 + \cdots + r_p)l} B^l.$$

The coefficients in GW are integers in K. It follows from Theorem 4–6 that if β is any one of them which is not zero, there is a factorization $\beta = \beta_1\beta_2$ in K such that $\beta_1 G = U$ and $\beta_2 W = V$ have integral coefficients in K, and

$$\beta GW = F = UV.$$

By the bound just obtained for $\overline{|GW|}$, we have

$$0 < \overline{|F|} \leq \overline{|GW|}^2 \leq ((r_1 + 1) \cdots (r_p + 1))^{2l} l!^2 2^{2(r_1 + \cdots + r_p)l} B^{2l}.$$

4–4 The index. Let $P(z_1, \ldots, z_p)$ be any polynomial in p variables which does not vanish identically. Let $\alpha_1, \ldots, \alpha_p$ be any complex numbers, and let $r_1, \ldots, r_p$ be any positive numbers. We define the *index θ of P at the point* $(\alpha_1, \ldots, \alpha_p)$ *relative to* $r_1, \ldots, r_p$ as follows. Expand $P(\alpha_1 + y_1, \ldots, \alpha_p + y_p)$ as a polynomial in $y_1, \ldots, y_p$, say

$$P(\alpha_1 + y_1, \ldots, \alpha_p + y_p) = \sum_{j_1=0}^{\infty} \cdots \sum_{j_p=0}^{\infty} c(j_1, \ldots, j_p) y_1^{j_1} \cdots y_p^{j_p}.$$

Then

$$\theta = \min \left(\frac{j_1}{r_1} + \cdots + \frac{j_p}{r_p} \right),$$

the minimum being extended over all sets of non-negative integers $j_1, \ldots, j_p$ for which $c(j_1, \ldots, j_p) \neq 0$, or, equivalently, for which

$$\left(\frac{\partial}{\partial z_1} \right)^{j_1} \cdots \left(\frac{\partial}{\partial z_p} \right)^{j_p} P(\alpha_1, \ldots, \alpha_p) \neq 0.$$

Note that $\theta \geq 0$ always, and that $\theta = 0$ if and only if

$$P(\alpha_1, \ldots, \alpha_p) \neq 0.$$

Moreover, if any derived polynomial

$$\left(\frac{\partial}{\partial z_1} \right)^{k_1} \cdots \left(\frac{\partial}{\partial z_p} \right)^{k_p} P(z_1, \ldots, z_p)$$

is not identically zero, it is clear that its index at $(\alpha_1, \ldots, \alpha_p)$ relative to $r_1, \ldots, r_p$ is at least

$$\theta - \frac{k_1}{r_1} - \cdots - \frac{k_p}{r_p}.$$

The following properties, which we list in a theorem for later reference, are also immediate consequences of the definition.

THEOREM 4–9. *Let* $P(z_1, \ldots, z_p)$ *and* $Q(z_1, \ldots, z_p)$ *be polynomials, neither of which vanishes identically. Then if we consider indices formed at the same point* $(\alpha_1, \ldots, \alpha_p)$ *relative to the same numbers* $r_1, \ldots, r_p$, *the following relations hold:*

$$\text{index } (P + Q) \geq \min (\text{index } P, \text{index } Q), \qquad (16)$$

$$\text{index } PQ = \text{index } P + \text{index } Q. \qquad (17)$$

Equation (17) *remains true if* P *is a polynomial in* $z_1, \ldots, z_{p-1}$ *only, and* Q *is a polynomial in* z_p *only, provided that the index of* P *is taken at* $(\alpha_1, \ldots, \alpha_{p-1})$ *relative to* $r_1, \ldots, r_{p-1}$, *and that of* Q *at* α_p *relative to* r_p.

Now let $r_1, \ldots, r_m$ be positive integers, and suppose that $B \geq 1$. We consider the set $\mathcal{R}_m = \mathcal{R}_m(B; r_1, \ldots, r_m)$ of polynomials $R(z_1, \ldots, z_m)$ which satisfy the following conditions:

    (a) R has integral coefficients in K, and is not identically zero.

    (b) R is of degree at most r_j in z_j, for $j = 1, \ldots, m$.

    (c) $\overline{|R|} \leq B$.

Let $\zeta_1, \ldots, \zeta_m$ be algebraic numbers (not necessarily in K) of heights $H(\zeta_1) = q_1, \ldots, H(\zeta_m) = q_m$. Let $\theta(R)$ denote the index of $R(z_1, \ldots, z_m)$ at the point $(\zeta_1, \ldots, \zeta_m)$ relative to $r_1, \ldots, r_m$. Our object in the present section is to obtain, under certain conditions, an upper bound for $\theta(R)$ in terms of $B, q_1, \ldots, q_m, r_1, \ldots, r_m$. We therefore define

$$\Theta_m(B; q_1, \ldots, q_m; r_1, \ldots, r_m) = \sup \theta(R), \qquad (18)$$

the supremum, or least upper bound, being taken over all R in $\mathcal{R}$ and all integers $\zeta_1, \ldots, \zeta_m$ of heights $q_1, \ldots, q_m$, respectively.

The double significance of $r_1, \ldots, r_m$ in the definition (18) should be noted; these numbers occur both in the definition of the index and in condition (b) above.

We proceed by induction on m. In Theorem 4–10 the case $m = 1$ is treated, in Theorem 4–11 there is given a recurrence relation between Θ_{m-1} and Θ_m, and in Theorem 4–12 an explicit bound is obtained.

THEOREM 4–10.

$$\Theta_1(B; q_1; r_1) \leq \frac{3N(N + 1)}{\log q_1} + \frac{N \log B}{r_1 \log q_1}.$$

Proof: Let the defining polynomial of ζ_1 be

$$\chi(z_1) = d_0 z_1^h + \cdots + d_h, \qquad d_0 \neq 0,$$

where $d_0, \ldots, d_h$ are relatively prime rational integers, so that

$$\|\chi\| = H(\zeta_1) = q_1 = \max(|d_0|, \ldots, |d_h|).$$

Each polynomial R in $\mathcal{R}_1$ has integral coefficients in K; regarding these coefficients as polynomials in a single primitive element, we can obtain other polynomials from R by successively replacing this primitive element throughout by its various conjugates. Let R^* be the product of these N polynomials. By the Symmetric Function

Theorem, R^* has coefficients in Z. Also, $\deg R^* = Nr_1$, and by Theorem 4–5,

$$\|R^*\| \leq (1 + r_1)^N B^N.$$

By the definition of the index, $R(z_1)$ is divisible by $(z_1 - \zeta_1)^{r_1\theta}$, and the same is therefore true of $R^*(z_1)$. Since $R^*(z_1)$ has coefficients in Z, it is divisible by $\chi^{r_1\theta}$. One consequence of this fact is that $hr_1\theta \leq Nr_1$. Also, it follows from Theorem 4–3 that

$$\|\chi^{r_1\theta}\| \leq 6^{Nr_1}\|R^*\|,$$

and, by Theorem 4–4,

$$\begin{aligned}
q_1{}^{r_1\theta} = \|\chi\|^{r_1\theta} &\leq (hr_1\theta + 1)6^{Nr_1}\|R^*\| \\
&\leq (Nr_1 + 1)^{N+1}6^{Nr_1}B^N < 2^{Nr_1(N+1)}6^{Nr_1}B^N \\
&< 12^{N(N+1)r_1}B^N.
\end{aligned}$$

Hence

$$\theta < \frac{N(N + 1)\log 12}{\log q_1} + \frac{N\log B}{r_1 \log q_1},$$

and the theorem follows from the fact that $\log 12 < 3$.

THEOREM 4–11. *Let $p \geq 2$ be a positive integer, let $r_1, \ldots, r_p$ be positive integers such that*

$$r_p > 10\delta^{-1}, \qquad \frac{r_{j-1}}{r_j} > \delta^{-1}, \qquad \text{for } j = 2, \ldots, p, \qquad (19)$$

where $0 < \delta < 1$, and let $q_1, \ldots, q_p$ be positive integers. Then

$$\Theta_p(B; q_1, \ldots, q_p; r_1, \ldots, r_p) \leq 2\,\max\,(\Phi + \Phi^{\frac{1}{2}} + \delta^{\frac{1}{2}}), \qquad (20)$$

where the maximum is taken over integers l satisfying

$$1 \leq l \leq r_p + 1, \qquad (21)$$

and where

$$\Phi = \Theta_1(M; q_p; lr_p) + \Theta_{p-1}(M; q_1, \ldots, q_{p-1}; lr_1, \ldots, lr_{p-1}) \qquad (22)$$

and

$$M = (r_1 + 1)^{2pl}2^{2r_1pl}l!^2 B^{2l}. \qquad (23)$$

Proof: Let $R(z_1, \ldots, z_p)$ be any polynomial of the class $\mathcal{R}_p(B; r_1, \ldots, r_p)$ and let $\zeta_1, \ldots, \zeta_p$ be algebraic numbers of heights $q_1, \ldots, q_p$ respectively. Then R satisfies the hypotheses of Theorem 4–8, so that there are numbers l and β and a polynomial $F(z_1, \ldots, z_p)$

having the properties listed there. By Theorem 4–8,

$$\overline{|F|} \le ((r_1 + 1) \cdots (r_p + 1))^{2l} 2^{2(r_1 + \cdots + r_p)l} l!^2 B^{2l}$$

and hence

$$\overline{|F|} < (r_1 + 1)^{2pl} 2^{2r_1 pl} l!^2 B^{2l} = M,$$

since $r_1 > r_2 > \cdots > r_p$ by (19).   From the factorization

$$F(z_1, \ldots, z_p) = U(z_1, \ldots, z_{p-1}) V(z_p)$$

and the fact that the arguments of U and V are disjoint, it follows that also

$$\overline{|U|} < M, \qquad \overline{|V|} < M.$$

The polynomial $U(z_1, \ldots, z_{p-1})$ has degree at most lr_j in z_j, for $j = 1, \ldots, p - 1$.   It is therefore an element of the class

$$\mathcal{R}_{p-1}(M; lr_1, \ldots, lr_{p-1}).$$

Hence, its index at $(\zeta_1, \ldots, \zeta_{p-1})$ relative to $lr_1, \ldots, lr_{p-1}$ is at most

$$\Theta_{p-1}(M; q_1, \ldots, q_{p-1}; lr_1, \ldots, lr_{p-1}).$$

It follows from the definition of the index that the index of U at that point relative to $r_1, \ldots, r_{p-1}$ is at most

$$l\Theta_{p-1}(M; q_1, \ldots, q_{p-1}; lr_1, \ldots, lr_{p-1}).$$

Similarly, $V(z_p)$ is an element of the class $\mathcal{R}_1(M; lr_p)$, and its index at ζ_p relative to r_p is at most

$$l\Theta_1(M; q_p; lr_p).$$

By the last sentence of Theorem 4–9, the index of $F = UV$ at $(\zeta_1, \ldots, \zeta_p)$ relative to $r_1, \ldots, r_p$ is the sum of the indices of U and V, so that

$$\text{index } F \le l\Phi, \tag{24}$$

where Φ is defined in (22).

We now deduce from the determinantal representation of F in equation (10) a lower bound for the index of F in terms of the index θ of R.   Consider first any differential operator of the form

$$\Delta = \frac{1}{i_1! \cdots i_{p-1}!} \left(\frac{\partial}{\partial z_1}\right)^{i_1} \cdots \left(\frac{\partial}{\partial z_{p-1}}\right)^{i_{p-1}},$$

of order

$$w = i_1 + \cdots + i_{p-1} \le l - 1.$$

If the polynomial

$$\Delta \frac{1}{\nu!} \left(\frac{\partial}{\partial z_p} \right)^\nu R(z_1, \ldots, z_p)$$

does not vanish identically, its index at $(\zeta_1, \ldots, \zeta_p)$ relative to $r_1, \ldots, r_p$ is at least

$$\theta - \frac{i_1}{r_1} - \cdots - \frac{i_{p-1}}{r_{p-1}} - \frac{\nu}{r_p} \geq \theta - \frac{w}{r_{p-1}} - \frac{\nu}{r_p}.$$

Now

$$\frac{w}{r_{p-1}} \leq \frac{l-1}{r_{p-1}} \leq \frac{r_p}{r_{p-1}} < \delta,$$

by the inequalities (21) and (19). Hence, since the index is non-negative, it must be at least

$$\max \left(0, \theta - \delta - \frac{\nu}{r_p} \right) \geq \max \left(0, \theta - \frac{\nu}{r_p} \right) - \delta.$$

If we expand the determinant on the right side of (10), we obtain for F a sum of $l!$ terms, a typical term being

$$\pm \beta (\Delta_{\mu_0} R) \left(\Delta_{\mu_1} \frac{1}{1!} \frac{\partial}{\partial z_p} R \right) \cdots \left(\Delta_{\mu_{l-1}} \frac{1}{(l-1)!} \left(\frac{\partial}{\partial z_p} \right)^{l-1} R \right),$$

where $\Delta_{\mu_0}, \ldots, \Delta_{\mu_{l-1}}$ are differential operators on $z_1, \ldots, z_{p-1}$ whose orders are at most $l - 1$. By Theorem 4–9, the index of such a term, if it does not vanish identically, is at least

$$\sum_{\nu=0}^{l-1} \max \left(0, \theta - \frac{\nu}{r_p} \right) - l\delta.$$

Since F is a sum of such terms, it follows from Theorem 4–9 again that

$$\text{index } F \geq \sum_{\nu=0}^{l-1} \max \left(0, \theta - \frac{\nu}{r_p} \right) - l\delta.$$

We may suppose that $\theta r_p > 10$, since otherwise

$$\theta < 10 r_p^{-1} < \delta < 2\delta^{\frac{1}{2}}$$

and the desired inequality for θ then holds. Under this supposition, $[\theta r_p]^2 > 2\theta^2 r_p{}^2 / 3$. Hence if $\theta r_p < l$, we have

$$\sum_{\nu=0}^{l-1} \max\left(0, \theta - \frac{\nu}{r_p}\right) = r_p^{-1} \sum_{\nu=0}^{[\theta r_p]} (\theta r_p - \nu)$$

$$\geq \tfrac{1}{2} r_p^{-1} [\theta r_p]^2$$

$$\geq \tfrac{1}{3} \theta^2 r_p,$$

while if $\theta r_p \geq l$, then

$$\sum_{\nu=0}^{l-1} \max\left(0, \theta - \frac{\nu}{r_p}\right) = \sum_{\nu=0}^{l-1} \left(\theta - \frac{\nu}{r_p}\right) \geq \frac{1}{2} l\theta.$$

Hence

$$\text{index } F \geq \min\left(\tfrac{1}{2} l\theta, \tfrac{1}{3} r_p \theta^2\right) - l\delta. \tag{25}$$

Combining (24) and (25), we obtain

$$\min\left(\tfrac{1}{2} l\theta, \tfrac{1}{3} r_p \theta^2\right) \leq l(\Phi + \delta).$$

Thus either $\theta < 2(\Phi + \delta)$, in which case θ satisfies the desired inequality, or

$$\tfrac{1}{3} r_p \theta^2 \leq l(\Phi + \delta) \leq (r_p + 1)(\Phi + \delta).$$

Since $r_p + 1 < 4r_p/3$ by (19), this gives

$$\theta \leq 2(\Phi + \delta)^{\frac{1}{2}} \leq 2(\Phi^{\frac{1}{2}} + \delta^{\frac{1}{2}}),$$

and the proof is complete.

THEOREM 4–12.   *Let m be a positive integer, and suppose that*

$$0 < \delta < \frac{1}{m2^m(N+1)^2}. \tag{26}$$

Let $r_1, \ldots, r_m$ be positive integers such that

$$r_m > 10\delta^{-1}, \qquad \frac{r_{j-1}}{r_j} > \delta^{-1}, \qquad for \ j = 2, \ldots, m. \tag{27}$$

Let $q_1, \ldots, q_m$ be positive integers such that

$$log \ q_1 > 2\delta^{-1} \, m(2m+1), \tag{28}$$

$$r_j \, log \ q_j \geq r_1 \, log \ q_1, \qquad for \ j = 2, \ldots, m, \tag{29}$$

$$log \ q_1 > 3\delta^{-1} N(N+1). \tag{30}$$

Then

$$\Theta_m(q_1{}^{\delta r_1}; q_1, \ldots, q_m; r_1, \ldots, r_m) < 10^m \delta^{(\frac{1}{2})^m}. \tag{31}$$

Proof: The proof is by induction on m. For $m = 1$, we apply Theorem 4–10, together with the inequalities (30) and (26), and obtain

$$\Theta_1(q_1{}^{\delta r_1}; q_1; r_1) < \frac{3N(N+1)}{\log q_1} + \frac{N \log (q_1{}^{\delta r_1})}{r_1 \log q_1} < (N+1)\delta < 10\delta^{\frac{1}{2}},$$

which is the desired inequality.

Now suppose that $p \geq 2$ is an integer, and that the theorem holds when $m = p - 1$. When $m = p$, the hypotheses of the present theorem are more stringent than those of Theorem 4–11, so that the latter is applicable here. We must estimate M and Φ.

We have

$$M = (r_1 + 1)^{2pl} 2^{2r_1pl} l!^2 B^{2l} \leq \big((r_1 + 1)^{2p} 2^{2r_1p} l^2 q_1{}^{2\delta r_1}\big)^l.$$

Since $l \leq r_p + 1 < r_1 + 1 \leq 2^{r_1}$, it follows that

$$M < (2^{(4p+2)r_1} q_1{}^{2\delta r_1})^l < (e^{(4p+2)r_1} q_1{}^{2\delta r_1})^l.$$

By (28) with $m = p$, we have $4p + 2 < \delta p^{-1} \log q_1$, so that

$$M < q_1{}^{\delta_1 l r_1},$$

where
$$\delta_1 = 2\delta(1 + p^{-1}). \tag{32}$$

Thus
$$\Theta_1(M; q_p; l r_p) \leq \Theta_1(q^{\delta_1 l r_1}; q_p; l r_p) \tag{33}$$
and

$$\Theta_{p-1}(M; q_1, \ldots, q_{p-1}; l r_1, \ldots, l r_{p-1})$$
$$\leq \Theta_{p-1}(q_1{}^{\delta_1 l r_1}; q_1, \ldots, q_{p-1}; l r_1, \ldots, l r_{p-1}). \tag{34}$$

Moreover, (32), together with the inequality (26) with $m = p$, implies that

$$\delta_1 < \frac{1 + p^{-1}}{p 2^{p-1}(N+1)^2} < \frac{1}{(p-1)2^{p-1}(N+1)^2}. \tag{35}$$

In particular, $(N+1)\delta_1 < \delta_1^{\frac{1}{2}}$.

It follows from (30), and the fact that $q_p > q_1$, that

$$\log q_p > 3\delta^{-1} N(N+1).$$

Hence by Theorem 4–10, the right side of (33) does not exceed

$$\delta + \frac{N\delta_1 l r_1 \log q_1}{l r_p \log q_p} \leq \delta + N\delta_1 < (N+1)\delta_1 < \delta_1^{\frac{1}{2}};$$

here we have used (29).

To estimate the right side of (34), we use the induction hypothesis, that the theorem holds when $m = p - 1$. The conditions of the theorem are satisfied for $m = p - 1$, if we replace δ by δ_1 and $r_1, \ldots, r_{p-1}$ by $lr_1, \ldots, lr_{p-1}$; since $\delta_1 > \delta$, this is obvious for all the relations but (26), which has already been verified in (35). It follows that

$$\Theta_{p-1}(q_1{}^{\delta_1 lr_1}; q_1, \ldots, q_{p-1}; lr_1, \ldots, lr_{p-1}) < 10^{p-1}\delta_1{}^{(\frac{1}{2})^{p-1}}.$$

Hence, since $\delta_1 < 4\delta$, the two results just proved imply that

$$\Phi < 2\delta^{\frac{1}{2}} + 2(10^{p-1}\delta^{(\frac{1}{2})^{p-1}}) < 3(10^{p-1}\delta^{(\frac{1}{2})^{p-1}}).$$

Finally, (20) gives

$$\Theta_p(q_1{}^{\delta r_1}; q_1, \ldots, q_p; r_1, \ldots, r_p)$$
$$< 2\{3(10^{p-1}\delta^{(\frac{1}{2})^{p-1}}) + 3^{\frac{1}{2}}10^{\frac{1}{2}(p-1)}\delta^{(\frac{1}{2})^p} + \delta^{\frac{1}{2}}\}$$
$$< 2\left(\frac{3}{10} + \frac{3^{\frac{1}{2}}}{10^{\frac{1}{2}}} + \frac{1}{10^2}\right)10^p\delta^{(\frac{1}{2})^p} < 10^p\delta^{(\frac{1}{2})^p}.$$

4–5 A combinatorial lemma

THEOREM 4–13. *If $r_1, \ldots, r_m$ are any positive integers, and $\lambda > 0$, then the number $A_m(\lambda)$ of sets of integers $j_1, \ldots, j_m$ which satisfy the inequalities*

$$0 \leq j_1 \leq r_1, \qquad \ldots, \qquad 0 \leq j_m \leq r_m,$$
$$\frac{j_1}{r_1} + \cdots + \frac{j_m}{r_m} \leq \frac{1}{2}(m - \lambda)$$

does not exceed

$$2m^{\frac{1}{2}}\lambda^{-1}(r_1 + 1) \cdots (r_m + 1).$$

Proof: We proceed by induction on m. The theorem holds for $m = 1$, since the number of integers j_1 such that

$$0 \leq j_1 \leq r_1, \qquad j_1 \leq \tfrac{1}{2}(1 - \lambda)r_1$$

is at most $r_1 + 1$, and is 0 if $\lambda > 1$.

Now suppose $m > 1$. The result is trivial if $\lambda \leq 2m^{\frac{1}{2}}$, since then the conditions on the individual j's give an improvement of the desired upper bound. Hence we may suppose that $\lambda > 2m^{\frac{1}{2}}$. If we fix j_m, we must count the sets of integers $j_1, \ldots, j_{m-1}$ such that

$$0 \leq j_1 \leq r_1, \qquad \ldots, \qquad 0 \leq j_{m-1} \leq r_{m-1},$$

$$\frac{j_1}{r_1} + \cdots + \frac{j_{m-1}}{r_{m-1}} \leq \frac{1}{2}\left(m - \lambda - \frac{2j_m}{r_m}\right).$$

Putting

$$m - \lambda - \frac{2j_m}{r_m} = (m - 1) - \lambda',$$

or, what is the same thing,

$$\lambda' = \lambda'(j_m) = \lambda - 1 + \frac{2j_m}{r_m},$$

we see that

$$A_m(\lambda) = \sum_{j_m=0}^{r_m} A_{m-1}(\lambda'(j_m)).$$

By the induction hypothesis,

$$A_m(\lambda) \leq 2(m - 1)^{\frac{1}{2}}(r_1 + 1) \cdots (r_{m-1} + 1) \sum_{j=0}^{r_m} \left(\lambda - 1 + \frac{2j}{r_m}\right)^{-1},$$

and it suffices to prove that

$$\sum_{j=0}^{r} \left(\lambda - 1 + \frac{2j}{r}\right)^{-1} \leq \lambda^{-1}(m - 1)^{-\frac{1}{2}} m^{\frac{1}{2}}(r + 1)$$

for all positive integers r and m, if $\lambda > 2m^{\frac{1}{2}}$.

If r is even, we put $j = \frac{1}{2}r + k$ and obtain the sum

$$\sum_{k=-\frac{1}{2}r}^{\frac{1}{2}r} \left(\lambda + \frac{2k}{r}\right)^{-1} = \lambda^{-1} + \sum_{k=1}^{\frac{1}{2}r} \left\{\left(\lambda + \frac{2k}{r}\right)^{-1} + \left(\lambda - \frac{2k}{r}\right)^{-1}\right\}$$

$$= \lambda^{-1} + \sum_{k=1}^{\frac{1}{2}r} 2\lambda \left(\lambda^2 - \frac{4k^2}{r^2}\right)^{-1}$$

$$\leq \lambda^{-1} + 2\lambda \sum_{k=1}^{\frac{1}{2}r} (\lambda^2 - 1)^{-1}$$

$$= \lambda^{-1} + 2\lambda^{-1} \sum_{k=1}^{\frac{1}{2}r} (1 - \lambda^{-2})^{-1}$$

$$\leq \lambda^{-1}(r + 1)(1 - \lambda^{-2})^{-1}.$$

Since $1 - \lambda^{-2} > 1 - m^{-1}/4 > (1 - m^{-1})^{\frac{1}{2}}$, we have the desired inequality.

If r is odd, we put $j = (r - 1)/2 + k$ and obtain the sum

$$\sum_{k=-\frac{1}{2}(r-1)}^{\frac{1}{2}(r+1)} \left(\lambda + \frac{2k - 1}{r} \right)^{-1}$$

$$= \sum_{k=1}^{\frac{1}{2}(r+1)} \left\{ \left(\lambda + \frac{2k - 1}{r} \right)^{-1} + \left(\lambda - \frac{2k - 1}{r} \right)^{-1} \right\}$$

$$= 2\lambda \sum_{k=1}^{\frac{1}{2}(r+1)} \left(\lambda^2 - \frac{(2k - 1)^2}{r^2} \right)^{-1}$$

$$\leq \lambda(\lambda^2 - 1)^{-1}(r + 1),$$

and the result is as before.

4–6 The approximation polynomial. Let α be an algebraic integer of degree $n \geq 2$ over K, so that α is a zero of a polynomial which has integral coefficients in K and which cannot be factored into a product of such polynomials of positive degrees. Let $L = K(\alpha)$ be the field obtained by adjoining α to K. Finally, let $\omega_1, \ldots, \omega_N$ be an integral basis for K, and put

$$\lceil \alpha \rceil = b_1, \qquad \max\left(\lceil \omega_1 \rceil, \ldots, \lceil \omega_N \rceil \right) = b_2. \tag{36}$$

In the remainder of the proof we shall be concerned with a single set of values of m, δ, q_1, ζ_1, $\ldots$, q_m, ζ_m, r_1, $\ldots$, r_m, which will be chosen later in the order just specified. The choice will be made so as to satisfy the following conditions:

$$0 < \delta < m^{-1}2^{-m}(N + 1)^{-2}, \tag{37}$$

$$10^m \delta^{(\frac{1}{2})^m} + 2(1 + 3\delta)nm^{\frac{1}{2}} < \frac{m}{2}, \tag{38}$$

$$r_m > 10\delta^{-1}, \qquad \frac{r_{j-1}}{r_j} > \delta^{-1}, \qquad \text{for } j = 2, \ldots, m, \tag{39}$$

$$\delta^2 \log q_1 > 2m + 1 + m \log (b_1 + 1) + 4b_2 N, \tag{40}$$

$$r_j \log q_j \geq r_1 \log q_1, \qquad \text{for } j = 2, \ldots, m, \tag{41}$$

$$\log q_1 > 3\delta^{-1}N(N + 1). \tag{42}$$

Notice that these conditions imply those of Theorem 4–12, since (37) and (40) together imply that $\delta \log q_1 > 2m(2m + 1)$.

Define λ, μ, η, B_1 by the equations

$$\lambda = 4(1 + 3\delta)nm^{\frac{1}{2}}, \tag{43}$$

$$\mu = \tfrac{1}{2}(m - \lambda), \tag{44}$$

$$\eta = 10^m \delta^{(\frac{1}{2})^m}, \tag{45}$$

$$B_1 = [q_1{}^{\delta r_1}]. \tag{46}$$

Then (38) is equivalent to

$$\eta < \mu. \tag{47}$$

Also,

$$q_1{}^{\frac{1}{2}\delta r_1} < B_1,$$

since $\sqrt{x} < x - 1 < [x]$ for all $x > (3 + \sqrt{5})/2$, and

$$q^{\delta r_1} > q^{\delta^2 r_1} > e^{(2m+1)r_1} > e^{3rm} > e^{30}.$$

We come now to the main lemma, which will be the only one to which reference is made in the eventual proof of the Thue-Siegel-Roth theorem.

THEOREM 4–14.   *Suppose that the conditions (37) through (42) are satisfied, and suppose that $\zeta_1, \ldots, \zeta_m$ are algebraic numbers of heights $q_1, \ldots, q_m$, respectively. Then there exists a polynomial $Q(z_1, \ldots, z_m)$ with integral coefficients in K and of degree at most r_j in z_j, for $j = 1, \ldots, m$, such that*
   (a) *the index of Q at the point $(\alpha, \ldots, \alpha)$ relative to $r_1, \ldots, r_m$ is at least $\mu - \eta$;*
   (b) $Q(\zeta_1, \ldots, \zeta_m) \neq 0$;
   (c) *for all derivatives*

$$Q_{i_1 \cdots i_m}(z_1, \ldots, z_m) = \frac{1}{i_1! \cdots i_m!} \left(\frac{\partial}{\partial z_1}\right)^{i_1} \cdots \left(\frac{\partial}{\partial z_m}\right)^{i_m} Q,$$

where $i_1, \ldots, i_m$ are non-negative integers, the inequality

$$|Q_{i_1 \cdots i_m}(z_1, \ldots, z_m)| < B_1^{1+3\delta}(1 + |z_1|)^{r_1} \cdots (1 + |z_m|)^{r_m}$$

holds, and the corresponding inequality also holds if the coefficients in Q are replaced by their respective field conjugates.

Proof: Let $c_1, \ldots, c_N$ range independently over the non-negative rational integers not exceeding B_1, and let C be the set of integers of K of the form

$$c_1\omega_1 + \cdots + c_N\omega_N.$$

The number of elements of C is $(1 + B_1)^N$, and if we put

$$(1 + r_1) \cdots (1 + r_m) = r,$$

there are

$$(1 + B_1)^{Nr} \tag{48}$$

distinct polynomials

$$P(z_1, \ldots, z_m) = \sum_{s_1=0}^{r_1} \cdots \sum_{s_m=0}^{r_m} \gamma(s_1, \ldots, s_m) z_1^{s_1} \cdots z_m^{s_m}$$

whose coefficients $\gamma(s_1, \ldots, s_m)$ belong to C. For $\gamma(s_1, \cdots, s_m)$ in C,

$$\overline{|\gamma(s_1, \ldots, s_m)|} \leq b_2 B_1 N, \tag{49}$$

and if we put

$$P_{j_1 \cdots j_m}(z_1, \ldots, z_m)$$

$$= \frac{1}{j_1! \cdots j_m!} \left(\frac{\partial}{\partial z_1}\right)^{j_1} \cdots \left(\frac{\partial}{\partial z_m}\right)^{j_m} P(z_1, \ldots, z_m)$$

$$= \sum_{s_1=0}^{r_1} \cdots \sum_{s_m=0}^{r_m} \gamma(s_1, \ldots, s_m) \binom{s_1}{j_1} \cdots \binom{s_m}{j_m} z_1^{s_1-j_1} \cdots z_m^{s_m-j_m},$$

then

$$\overline{|P_{j_1 \cdots j_m}|} \leq 2^{r_1+\cdots+r_m} b_2 B_1 N \leq b_2 N 2^{mr_1} B_1 < b_2 N B_1^{1+\delta},$$

since $mr_1 \log 2 < \frac{1}{2} \delta^2 r_1 \log q_1$ by (40). Now replace all of $z_1, \ldots, z_m$ by α. Since the total number of terms is at most r, and since, by (40),

$$r = (r_1 + 1) \cdots (r_m + 1) \leq 2^{r_1+\cdots+r_m} \leq (b_1 + 1)^{mr_1} < B_1^{\delta}, \tag{50}$$

we obtain the bound

$$\overline{|P_{j_1 \cdots j_m}(\alpha, \ldots, \alpha)|} \leq b_2 N B_1^{1+\delta} r b_1^{r_1+\cdots+r_m}$$

$$\leq b_2 N B_1^{1+3\delta}.$$

Let ϑ be a primitive element of L, so that $L = R(\vartheta)$. Order the conjugates of ϑ so that $\vartheta_1, \ldots, \vartheta_{\rho_1}$ are real and $\vartheta_{\rho_1+\nu}$ and $\vartheta_{\rho_1+\rho_2+\nu}$ are complex-conjugate for $\nu = 1, \ldots, \rho_2$, so that $\rho_1 + 2\rho_2 = nN$. Let ξ be a fixed one of the numbers $P_{j_1 \cdots j_m}(\alpha, \ldots, \alpha)$, where $j_1, \ldots, j_m$ satisfy the inequalities

$$0 \leq j_1 \leq r_1, \quad \cdots, \quad 0 \leq j_m \leq r_m, \quad \frac{j_1}{r_1} + \cdots + \frac{j_m}{r_m} \leq \mu. \tag{51}$$

Then ξ can be written as a polynomial in ϑ, with rational coefficients, and as such has field conjugates $\xi^{(\nu)}$, $\nu = 1, \ldots, nN$. Hence we can define nN real numbers $\xi_1, \ldots, \xi_{nN}$ by the equations

$$\xi_\nu = \xi^{(\nu)}, \qquad \text{for } \nu = 1, \ldots, \rho_1,$$

$$\xi_\nu + i\xi_{\nu+\rho_2} = \xi^{(\nu)}, \qquad \text{for } \rho_1 + 1 \leq \nu \leq \rho_1 + \rho_2.$$

Collecting them in a fixed order for fixed coefficients $\gamma(s_1, \ldots, s_m)$ and for all $j_1, \ldots, j_m$ satisfying the inequalities (51), we have a set of numbers which can be considered as coordinates of a point; by Theorem 4–13 there are

$$M \leq 2nNm^{\frac{1}{2}}\lambda^{-1}r$$

coordinates, and each is numerically smaller than $[b_2NB_1{}^{1+3\delta}]+1 = t$. Thus all the points, for the various sets of coefficients in C, lie in a cube of edge $2t$ in M-dimensional space. If each edge is divided into $3t$ equal parts, we get $(3t)^M$ subcubes of edge $\frac{2}{3}$. By (48), if

$$(1 + B_1)^{Nr} > (3t)^M, \tag{52}$$

there are more points than subcubes, and the points corresponding to two different polynomials $P^*(z_1, \ldots, z_m)$ and $P^{**}(z_1, \ldots, z_m)$ lie in the same subcube. If we put

$$\bar{P}(z_1, \ldots, z_m) = P^*(z_1, \ldots, z_m) - P^{**}(z_1, \ldots, z_m),$$

then

$$\overline{\left|\bar{P}_{j_1\cdots j_m}(\alpha, \ldots, \alpha)\right|} \leq \sqrt{2} \cdot \frac{2}{3} < 1$$

for $j_1, \ldots, j_m$ as in (51). Since $\bar{P}_{j_1\cdots j_m}(\alpha, \ldots, \alpha)$ is an algebraic integer whose norm is numerically smaller than 1, it must be zero. Hence the index of $\bar{P}$ at the point $(\alpha, \ldots, \alpha)$ relative to $r_1, \ldots, r_m$ is at least μ. Also the coefficients $\bar{\gamma}(s_1, \ldots, s_m)$ in $\bar{P}$ are integers of K, not all zero, such that the relation (49) holds.

To verify (52), notice that by the inequality (40),

$$q_1{}^{\delta r_1} > 4b_2N,$$

and hence

$$B_1 > 4b_2N,$$

$$B_1{}^{Nr} > (4b_2NB_1)^{\frac{1}{2}Nr},$$

$$B_1{}^{Nr} > (3b_2NB_1{}^{1+3\delta} + 3)^{\frac{1}{2}Nr(1+3\delta)^{-1}},$$

$$(1 + B_1)^{Nr} > (3t)^M.$$

We now apply Theorem 4–12, the hypotheses of which are satisfied, as was noted earlier. Since $\bar{P}$ belongs to the class $\mathscr{R}_m(q_1{}^{\delta r_1}; r_1, \ldots, r_m)$, its index at $(\zeta_1, \ldots, \zeta_m)$ relative to $r_1, \ldots, r_m$ is less than η, defined in (45). Hence $\bar{P}$ possesses some derivative

$$Q(z_1, \ldots, z_m) = \frac{1}{k_1! \cdots k_m!} \left(\frac{\partial}{\partial z_1}\right)^{k_1} \cdots \left(\frac{\partial}{\partial z_m}\right)^{k_m} \bar{P},$$

with

$$\frac{k_1}{r_1} + \cdots + \frac{k_m}{r_m} < \eta,$$

such that

$$Q(\zeta_1, \ldots, \zeta_m) \neq 0.$$

The index of Q at the point $(\alpha, \ldots, \alpha)$ relative to $r_1, \ldots, r_m$ is at least $\mu - \eta$. Thus Q has the properties (a) and (b) of Theorem 4–14.

From the relations (49) and (50),

$$\overline{|Q|} \leq 2^{r_1 + \cdots + r_m} b_2 N B_1 < 2^{m r_1} b_2 N B_1 < b_2 N B_1{}^{1+\delta}.$$

Hence for an arbitrary derivative,

$$\overline{|Q_{i_1 \cdots i_m}|} \leq 2^{r_1 + \cdots + r_m} b_2 N B_1{}^{1+\delta} < b_2 N B_1{}^{1+2\delta}.$$

Finally,

$$|Q_{i_1 \cdots i_m}(z_1, \ldots, z_m)| < b_2 N B_1{}^{1+2\delta} \prod_{\nu=1}^{m} (1 + |z_\nu| + \cdots + |z_\nu|^{r_\nu})$$

$$< b_2 N B_1{}^{1+2\delta} \prod_{\nu=1}^{m} (1 + |z_\nu|)^{r_\nu}$$

$$< B_1{}^{1+3\delta} \prod_{\nu=1}^{m} (1 + |z_\nu|)^{r_\nu},$$

since $b_2 N < B_1{}^{\delta}$ by (40). The same inequality holds for the conjugate polynomials, and the proof is complete.

4–7 The Thue-Siegel-Roth theorem

THEOREM 4–15. *Let K be an algebraic number field of degree N, and let α be algebraic. Then for each $\varkappa > 2$, the inequality*

$$|\alpha - \zeta| < \frac{1}{(H(\zeta))^{\varkappa}} \tag{53}$$

has only finitely many solutions ζ in K.

Proof: We shall suppose that the theorem is false, so that (53) has infinitely many solutions, and produce a contradiction. We may suppose also that α is an integer. For if not there is a positive rational integer a such that $a\alpha$ is an algebraic integer, and for each solution ζ of (53) we have

$$|a\alpha - a\zeta| < \frac{a}{(H(\zeta))^\varkappa} \leq \frac{a^{\varkappa N+1}}{(H(a\zeta))^\varkappa}.$$

Hence for arbitrary $\epsilon > 0$, and for all solutions ζ with $H(\zeta)$ sufficiently large,

$$|a\alpha - a\zeta| < \frac{1}{(H(a\zeta))^{\varkappa-\epsilon}},$$

and ϵ can be chosen so small that $\varkappa - \epsilon > 2$.

Finally, it suffices to prove that (53) has only finitely many solutions in primitive elements ζ of K. For an algebraic number field has only finitely many subfields, and every element of K is a primitive element of some one of its subfields; moreover, the inequality in question does not depend on the degree of α over K.

We first choose m so large that $m > 4nm^{\frac{1}{2}}$ and

$$\frac{2m}{m - 4nm^{\frac{1}{2}}} < \varkappa, \tag{54}$$

which is possible since $\varkappa > 2$. For sufficiently small δ we have

$$m - 4(1 + 3\delta)nm^{\frac{1}{2}} - 2\eta > 0,$$

where η, given by (45), becomes arbitrarily small with δ. This condition is the same as that of (38). We choose δ to satisfy this and the inequality (37), and finally the inequality

$$\frac{2m(1 + \delta) + 2\delta N(2 + 5\delta)}{m - 4(1 + 3\delta)nm^{\frac{1}{2}} - 2\eta} < \varkappa, \tag{55}$$

which is possible in view of (54). The inequality (55) is equivalent to

$$\frac{m(1 + \delta) + \delta N(2 + 5\delta)}{\mu - \eta} < \varkappa, \tag{56}$$

by equations (43) and (44).

Having chosen m and δ, we now choose a solution ζ_1 of (53) (a primitive element of K) with $H(\zeta_1) = q_1$ and with q_1 so large as to satisfy (40) and (42). We then choose further primitive solutions $\zeta_2, \ldots, \zeta_m$ of heights $q_2, \ldots, q_m$, such that for $j = 2, \ldots, m$,

$$\frac{\log q_j}{\log q_{j-1}} > \frac{2}{\delta}. \tag{57}$$

We now take r_1 to be any integer such that

$$r_1 > \frac{10 \log q_m}{\delta \log q_1}, \tag{58}$$

and define r_j, for $j = 2, \ldots, m$, by

$$\frac{r_1 \log q_1}{\log q_j} \leq r_j < \frac{r_1 \log q_1}{\log q_j} + 1. \tag{59}$$

Then the inequality (41) is satisfied. Also,

$$\frac{r_j \log q_j}{r_1 \log q_1} < 1 + \frac{\log q_j}{r_1 \log q_1} \leq 1 + \frac{\log q_m}{r_1 \log q_1} < 1 + \frac{\delta}{10}, \tag{60}$$

by (58). The conditions (39) are satisfied, since

$$r_m \geq \frac{r_1 \log q_1}{\log q_m} > 10\delta^{-1},$$

and

$$\frac{r_{j-1}}{r_j} > \frac{\log q_j}{\log q_{j-1}} \left(1 + \frac{\delta}{10}\right)^{-1} > \delta^{-1},$$

by (59), (60), and (57).

We know from Theorem 4–14 that there exists a polynomial $Q(z_1, \ldots, z_m)$, whose properties are listed in that theorem. Let $\zeta_1, \ldots, \zeta_m$ in K be zeros of irreducible polynomials of degree N with relatively prime coefficients in Z, the coefficients of z^N being $k_1, \ldots, k_m$, respectively. Then the number

$$\varphi = Q(\zeta_1, \ldots, \zeta_m)$$

is an element of K. If the field conjugates of ζ_i are $\zeta_i', \zeta_i'', \ldots$, for $i = 1, \ldots, m$, then $N\varphi$ is a sum of products of powers of the $\zeta_i^{(j)}$ with integral coefficients from K, and in each such product a factor $\zeta_i^{(j)}$ occurs to the power r_i at most. In the proof of Theorem 2–21, it was shown that the product of k_i and any set of distinct conjugates of ζ_i is an algebraic integer. For each i, the field conjugates of ζ_i

are distinct, because ζ_i is a primitive element of K. It follows that $k_1{}^{r_1} \cdots k_m{}^{r_m} \mathbf{N}\varphi$ is an algebraic integer, and since it is also rational it is a rational integer, so that

$$|k_1{}^{r_1} \cdots k_m{}^{r_m} \mathbf{N}\varphi| \geq 1. \tag{61}$$

On the other hand, we have

$$Q(\zeta_1, \ldots, \zeta_m)$$
$$= \sum_{i_1=0}^{r_1} \cdots \sum_{i_m=0}^{r_m} Q_{i_1 \cdots i_m}(\alpha, \ldots, \alpha)(\zeta_1 - \alpha)^{i_1} \cdots (\zeta_m - \alpha)^{i_m},$$

and, by part (a) of Theorem 4–14, the terms with

$$\frac{i_1}{r_1} + \cdots + \frac{i_m}{r_m} < \mu - \eta$$

all vanish. In all other terms we have

$$|(\zeta_1 - \alpha)^{i_1} \cdots (\zeta_m - \alpha)^{i_m}| < (q_1{}^{i_1} \cdots q_m{}^{i_m})^{-\varkappa}$$
$$= \{q_1{}^{i_1/r_1}(q_2{}^{r_2/r_1})^{i_2/r_2} \cdots (q_m{}^{r_m/r_1})^{i_m/r_m}\}^{-r_1\varkappa}$$
$$\leq (q_1{}^{i_1/r_1} \cdots q_1{}^{i_m/r_m})^{-r_1\varkappa}$$
$$< q_1{}^{-r_1(\mu-\eta)\varkappa},$$

since $q_j{}^{r_j/r_1} > q_1$ by (41). Hence, using part (c) of Theorem 4–14, we have

$$|\varphi| < (r_1 + 1) \cdots (r_m + 1)B_1{}^{1+3\delta}(1 + b_1)^{mr_1}q_1{}^{-r_1(\mu-\eta)\varkappa}$$
$$< B_1{}^{1+5\delta}q_1{}^{-r_1(\mu-\eta)\varkappa},$$

and by using part (c) again, together with Theorem 4–2, we obtain

$$|k_1{}^{r_1} \cdots k_m{}^{r_m}\mathbf{N}\varphi| < B_1{}^{1+5\delta}q_1{}^{-r_1(\mu-\eta)\varkappa}B_1{}^{(N-1)(1+3\delta)}$$
$$\times \prod_{i=1}^{m} \left\{ k_i \prod_{j=1}^{N} (1 + |\zeta_i{}^{(j)}|) \right\}^{r_i}$$
$$< B_1{}^{N(1+5\delta)}q_1{}^{-r_1(\mu-\eta)\varkappa} \prod_{i=1}^{m} (6^N q_i)^{r_i}.$$

Now, by (50),

$$6^{N(r_1+\cdots+r_m)} < 2^{3N(r_1+\cdots+r_m)} < B_1{}^{3\delta N} < q_1{}^{3\delta^2 N r_1} < q_1{}^{\delta N r_1},$$

so that

$$|k_1{}^{r_1} \cdots k_m{}^{r_m}\mathbf{N}\varphi| < q_1{}^{\delta N r_1(1+5\delta)+\delta N r_1+(r_1+\cdots+r_m)-r_1(\mu-\eta)\varkappa}$$
$$< q_1{}^{\delta N r_1(2+5\delta)+mr_1(1+\delta)-r_1(\mu-\eta)\varkappa}.$$

This, together with (61), implies that

$$\delta N(2 + 5\delta) + m(1 + \delta) > (\mu - \eta)\varkappa,$$

or

$$\varkappa < \frac{m(1 + \delta) + \delta N(2 + 5\delta)}{\mu - \eta},$$

which contradicts (56). This completes the proof.

4–8 Applications to Diophantine equations. The Thue-Siegel-Roth theorem will now be applied to show that a rather large variety of Diophantine equations have only finitely many solutions.

THEOREM 4–16. *Let $U(x, y)$ be a binary form of degree n, without multiple linear factors, whose coefficients belong to an algebraic number field K_0 of degree h. Let x and y be integral variables of K_0. Suppose that*

$$n > 2h.$$

Let $V(x, y)$ be any polynomial of total degree $\nu < n - 2h$ which has coefficients in K_0 and has no common factor with $U(x, y)$. Then the equation

$$U(x, y) = V(x, y) \tag{62}$$

has only finitely many solutions.

Proof: Just as in the representation theory for binary quadratic forms, it makes no difference whether we consider (62) or an equation obtained from it by a substitution $x = ax' + by'$, $y = cx' + dy'$, where a, b, c, d are in Z, and $|ad - bc| = 1$. If $U(x, y) = a_0 x^n + \cdots + a_n y^n$, then

$$U(x, ax + y) = U(1, a)x^n + \cdots + a_n y^n,$$

$$U(x + by, y) = a_0 x^n + \cdots + U(b, 1)y^n.$$

Choose a in K_0 so that $U(1, a) \neq 0$, and put $U(x, ax + y) = U_1(x, y)$. Then choose b in K_0 so that $U_1(b, 1) \neq 0$, and put $U_1(x + by, y) = U_2(x, y)$. Dropping the subscript, we see that there is no loss in generality in supposing that the coefficients of x^n and y^n in $U(x, y)$ are different from zero, and we can write

$$U(x, y) = \alpha y^n \prod_{k=1}^{n} \left(\frac{x}{y} - \xi_k\right), \tag{63}$$

where neither α nor any ξ_k is zero. By assumption, the numbers ξ_k are distinct, so if we put

$$c_1 = \min_{j \neq k} (|\xi_j - \xi_k|),$$

then $c_1 > 0$, and for every x and y, at least $n - 1$ of the factors in the product occurring in (63) have absolute values not less than $\frac{1}{2}c_1$.

Let $x = \eta$ and $y = \zeta \neq 0$ be integers of K_0, with field conjugates $\eta^{(1)}, \ldots, \eta^{(h)}, \zeta^{(1)}, \ldots, \zeta^{(h)}$. Then as we saw in Theorem 2–5,

$$\prod_{j=1}^{h} (\zeta^{(j)}t - \eta^{(j)}) = (Q(t))^f,$$

where $Q(t)$ is an irreducible polynomial with coefficients in Z, and $1 \leq f \leq h$. Let $M = \max(\lceil \zeta \rceil, \lceil \eta \rceil)$, and name the conjugates so that

$$Q(t) = \prod_{j=1}^{h/f} (\zeta^{(j)}t - \eta^{(j)}).$$

Then the coefficients of $Q(t)$ are numerically smaller then the corresponding coefficients of

$$\prod_{j=1}^{h/f} (Mt + M),$$

so that $\|Q\| \leq (2M)^{h/f}$. A fortiori, $H(\eta/\zeta) \leq (2M)^{h/f}$.

Now by Theorem 4–15, there are only finitely many solutions of the inequality

$$\left| \xi - \frac{\eta}{\zeta} \right| < \frac{1}{H(\eta/\zeta)^{2+\epsilon'}}$$

for fixed $\epsilon' > 0$. Hence for M sufficiently large, and $\epsilon = \epsilon'h$,

$$\left| \xi - \frac{\eta}{\zeta} \right| \geq \frac{1}{(2M)^{h(2+\epsilon')/f}} \geq \frac{1}{(2M)^{2h+\epsilon}},$$

at least if the left side is not zero. This is certainly true of the solutions of (62), since $U(x, y)$ and $V(x, y)$ have no common factor. The same argument applies to the numbers $\eta^{(j)}/\zeta^{(j)}$ and ξ_k, for $1 \leq j \leq h, 1 \leq k \leq n$; we see that for $\epsilon > 0$ and M sufficiently large, the inequality

$$\left| \xi_k - \frac{\eta^{(j)}}{\zeta^{(j)}} \right| > \frac{1}{(2M)^{2h+\epsilon}}, \qquad j = 1, \ldots, h; \quad k = 1, \ldots, n,$$

holds for every solution of (62). There is no loss in generality in supposing that $M = |\zeta^{(1)}| = |\zeta|$, since (62) remains correct after replacing all quantities by their conjugates and if necessary interchanging x and y. Hence, for large M,

$$|U(\eta, \zeta)| > |\alpha| M^n \left(\frac{c_1}{2}\right)^{n-1} \frac{1}{(2M)^{2h+\epsilon}}.$$

On the other hand, there is a constant c_2, depending only on the coefficients of V, such that

$$|V(\eta, \zeta)| < c_2 M^\nu.$$

If we choose $\epsilon < n - 2h - \nu$, then for sufficiently large M,

$$|U(\eta, \zeta)| > |V(\eta, \zeta)|.$$

But a bound on M implies a bound on the integral coefficients of the polynomials defining η and ζ, so that there are only finitely many solutions of (62).

COROLLARY. *If $U(x, y)$ is a binary form of degree $n > 2$, with coefficients in Z and without repeated linear factors, and if $a \neq 0$ is a rational integer, there are only finitely many rational integral solutions of the equation $U(x, y) = a$. In particular, the equation*

$$ax^n + by^n = c$$

has only finitely many solutions in Z if a, b, and c are in Z, $abc \neq 0$, and $n \geq 3$.

This follows immediately from the theorem, with $K_0 = R$, $h = 1$, and $n - 2h > 0$. The special case mentioned includes the higher-degree analog of Pell's equation, $x^n - dy^n = N$.

In the above considerations, strong use was made of the homogeneity of $U(x, y)$. If a Diophantine equation is not of the form specified in Theorem 4–16, it may still be possible to relate its solvability to that of one of this form. We now consider such a case.

4–9 A special equation. It was conjectured by E. Catalan in 1842 that 8 and 9 are the only two consecutive integers larger than 1 which are powers of other integers. This has never been proved; it has not even been shown that no three consecutive integers are powers, although it is trivial that no four can be, since one must be of the form $4k + 2$. In slightly different terms, the problem is to show that the Diophantine equation

$$x^w - y^z = 1 \tag{64}$$

has no solutions with w and z larger than 1, except for that mentioned. Various special cases arise by fixing, or specializing in some other way, one or more of the variables in (64). The case we are now going to examine is that in which the exponents are fixed, so that we consider the equation

$$x^m - y^n = 1. \tag{65}$$

Catalan's conjecture would be proved if it could be shown that for each pair of integers m and n larger than 1, (65) has no positive solutions except that mentioned. Since this seems to be unfeasible, we consider the more modest question of whether (65) can have infinitely many solutions. This, at last, is a question that can be answered. It is a very weak consequence of the following theorem, due to Mahler, that (65) has only finitely many solutions if $m \geq 2$, $n \geq 3$.

THEOREM 4–17.   *Suppose that $m \geq 2$, $n \geq 3$, $ab \neq 0$, $(x, y) = 1$. Then as* max $(|x|, |y|) \to \infty$, *the greatest prime factor of*

$$ax^m + by^n$$

tends to infinity.

Since $x^2 - y^2 = 1$ has only the obvious solutions $x = \pm 1$, $y = 0$, the new problem is completely solved. Unfortunately Mahler's proof, which depends on a p-adic version of the Thue-Siegel theorem, cannot be included here. We can, however, obtain partial results of some interest.

If mn is even, the fact that (65) has only finitely many solutions is a consequence of the next theorem, which is a special case of a theorem proved anonymously and published by L. J. Mordell.

THEOREM 4–18.   *Let $f(x)$ be a polynomial of degree $n \geq 3$, with coefficients in Z and with distinct zeros, and let a be any nonzero rational integer. Then the equation*

$$ay^2 = f(x) \tag{66}$$

has only finitely many solutions x, y in Z.

Proof: Suppose that

$$f(x) = a_0(x - \xi_1) \cdots (x - \xi_n),$$

and that (66) has infinitely many solutions. The numbers $\alpha_j = a_0\xi_j$, for $j = 1, \ldots, n$, are algebraic integers, and if (66) holds, then

$$aa_0^{n-1}y^2 = (a_0x - \alpha_1) \cdots (a_0x - \alpha_n).$$

Let $K = R(\xi_1, \ldots, \xi_n)$ be the splitting field of f. Any ideal in K dividing $[a_0x - \alpha_i]$ and $[a_0x - \alpha_j]$ also divides $[\alpha_i - \alpha_j]$, so that the norm of such a common divisor is a divisor of the discriminant d of f. Hence, if P is a prime ideal divisor of y and $\mathbf{N}P > d$, then for some i, $P^2|[a_0x - \alpha_i]$. Since there are only finitely many ideals with norms smaller than d, and only finitely many divisors of $a_0^{n-1}a$, it follows that for each i,

$$[a_0x - \alpha_i] = B_iC_i^2, \tag{67}$$

where B_i and C_i are ideals, and B_i runs over a finite set of ideals.

Let D run over a fixed system of representatives of the various ideal classes in K; the number of D's is finite. Then for each i and some D, $C_i \sim D$, so that

$$[\beta]C_i = [\delta]D,$$

for some β and δ. We shall show that β can be chosen from a finite set of integers of K. Let $([\beta], [\delta]) = E$, and put $[\beta] = EF$, $[\delta] = EG$. Then $EFC_i = EDG$, whence $FC_i = DG$; thus $F|D$, and F is one of a finite set of ideals. By Theorem 3–2, there is an H with norm less than c (so that H is one of a finite set) such that $FH = [\gamma]$ is principal. Thus

$$[\gamma]C_i = (GH)D.$$

Since $C_i \sim D$, also $[\gamma] \sim GH$; hence $GH = [\varsigma_i]$, and

$$[\gamma]C_i = [\varsigma_i]D,$$

where γ is one of a finite set of integers.

By (67),

$$[\gamma^2][a_0x - \alpha_i] = B_i[\varsigma_i^2]D^2,$$

from which it follows that B_iD^2 is principal, say $B_iD^2 = [\eta_i]$. Thus for some unit ϵ_i,

$$\gamma_i^2(a_0x - \alpha_i) = \epsilon_i\eta_i\varsigma_i^2.$$

By Dirichlet's theorem on units, ϵ_i can be written as $\epsilon_i'\epsilon_i''^2$, where ϵ_i' is one of a finite number of units. Finally, for $i = 1, \ldots, n$,

$$a_0x - \alpha_i = \varkappa_i\lambda_i^2,$$

where $\lambda_1, \ldots, \lambda_n$ are integers of K, and $\varkappa_1, \ldots, \varkappa_n$ are certain ones of finitely many numbers of K. Hence

$$\varkappa_1 \lambda_1{}^2 - \varkappa_2 \lambda_2{}^2 = \alpha_2 - \alpha_1 \neq 0,$$

$$\varkappa_2 \lambda_2{}^2 - \varkappa_3 \lambda_3{}^2 = \alpha_3 - \alpha_2 \neq 0,$$

$$\varkappa_3 \lambda_3{}^2 - \varkappa_1 \lambda_1{}^2 = \alpha_1 - \alpha_3 \neq 0.$$

Now let $L = K(\sqrt{\varkappa_1}, \sqrt{\varkappa_2}, \sqrt{\varkappa_3})$. Then, in L,

$$(\lambda_1 \sqrt{\varkappa_1} - \lambda_2 \sqrt{\varkappa_2})(\lambda_1 \sqrt{\varkappa_1} + \lambda_2 \sqrt{\varkappa_2}) = \alpha_2 - \alpha_1,$$

and since the denominators of $\varkappa_1$ and $\varkappa_2$ can be taken to be bounded, it follows that

$$\lambda_1 \sqrt{\varkappa_1} - \lambda_2 \sqrt{\varkappa_2} = \beta_3 \epsilon_3{}^l,$$

where β_3 is one of finitely many elements of L, ϵ_3 is a unit of L, and $l > 1$ is an arbitrary positive integer. Similarly,

$$\lambda_2 \sqrt{\varkappa_2} - \lambda_3 \sqrt{\varkappa_3} = \beta_1 \epsilon_1{}^l,$$

$$\lambda_3 \sqrt{\varkappa_3} - \lambda_1 \sqrt{\varkappa_1} = \beta_2 \epsilon_2{}^l.$$

But then

$$\frac{\beta_1}{\beta_3}\left(\frac{\epsilon_1}{\epsilon_3}\right)^l + \frac{\beta_2}{\beta_3}\left(\frac{\epsilon_2}{\epsilon_3}\right)^l = -1. \tag{68}$$

If there were only finitely many distinct ratios ϵ_1/ϵ_3, there would be a finite set of coefficients φ such that

$$\sqrt{a_0 x - \alpha_2} - \sqrt{a_0 x - \alpha_3} = \varphi(\sqrt{a_0 x - \alpha_1} - \sqrt{a_0 x - \alpha_2})$$

for every solution x of (66) and for suitable determination of the radicals. This is clearly impossible, so (68) must have infinitely many solutions in integers ϵ_1/ϵ_3, ϵ_2/ϵ_3 of L. But for l sufficiently large, this is in contradiction with Theorem 4–16. Hence the supposition that (66) has infinitely many solutions is not tenable, and the proof is complete.

Returning to equation (65), we see that the only possible solutions have $x = 0$ or ± 1, if $(m, n) > 1$. For the problem that remains, it suffices to consider the case in which $m = p$ and $n = q$ are distinct odd primes. This was treated by M. Newman, whose work was not published. A slightly strengthened version of his result, obtained by applying Theorem 4–16 rather than the analogous consequence of the Thue-Siegel theorem, follows.

THEOREM 4–19. *If p and q are distinct odd primes such that $q > 2(p - 1)$ and q does not divide the class number of the cyclotomic field $K_p = R(\zeta)$, where $\zeta = exp(2\pi i/p)$, then the equations*

$$x^p - y^q = \pm 1 \qquad (69)$$

have only finitely many solutions x, y in Z.

Proof: We carry out the proof only for the equation $x^p - y^q = 1$; the alternate case requires only trivial modifications. Put $1 - \zeta = \pi$ and $[\pi] = P$, so that P is a prime ideal of K_p, by Theorem 3–6. Let h be the class number of K_p.

If x and y satisfy (69) with the plus sign, then

$$[x - 1][x - \zeta] \cdots [x - \zeta^{p-1}] = [y]^q. \qquad (70)$$

Put

$$D_{rs} = [x - \zeta^r, x - \zeta^s] \qquad \text{for } 0 \leq r \leq p - 1,$$
$$0 \leq s \leq p - 1, \quad r \neq s.$$

Then

$$D_{rs} = [x - \zeta^r, \zeta^r - \zeta^s] = [x - 1 + 1 - \zeta^r, \zeta^r - \zeta^s]$$

$$= \left[x - 1 + \frac{1 - \zeta^r}{1 - \zeta} \pi, \frac{\zeta^r - \zeta^s}{1 - \zeta} \pi \right] = \left[x - 1 + \frac{1 - \zeta^r}{1 - \zeta} \pi, \pi \right]$$

$$= [x - 1, \pi],$$

since $(\zeta^k - \zeta^l)/(1 - \zeta)$ is a unit if $p \nmid (k - l)$. Thus D_{rs} is the same for all r and s, and, since $D_{rs} | P$ and P is prime, either $D_{rs} = [1]$ or $D_{rs} = P$. We consider the two cases separately.

If $D_{rs} = [1]$, then the ideals $[x - \zeta^r]$ are pairwise relatively prime; since their product is a qth power, there are ideals $A_0, \ldots, A_{p-1}$ such that

$$[x - \zeta^r] = A_r{}^q, \qquad r = 0, \ldots, p - 1. \qquad (71)$$

Suppose that e_r is the smallest positive integer such that $A_r{}^{e_r}$ is principal; by Theorem 3–4, $e_r | h$, and by (71), $e_r | q$. But q is prime and $q \nmid h$, so $e_r = 1$ and A_r is principal. Hence there are integers α and β and units ϵ and ϵ' of K_p such that $x - 1 = \epsilon\alpha^q$ and $x - \zeta = \epsilon'\beta^q$, whence

$$\epsilon'\beta^q - \epsilon\alpha^q = \pi. \tag{72}$$

By Theorem 2–45, the units of K_p have a finite basis, so that each unit has a representation $\epsilon_1 \cdot \epsilon_2{}^q$, where ϵ_1 is one of the finite number of units obtained by taking products of powers of the basis elements, with exponents non-negative and smaller than q. Thus (72) implies that one of the finitely many equations

$$\epsilon_1{}'(\epsilon_2{}'\beta)^q - \epsilon_1(\epsilon_2\alpha)^q = \pi \tag{73}$$

must hold. But for each choice of ϵ_1 and $\epsilon_1{}'$, (73) has only finitely many integral solutions $\epsilon_2\alpha$, $\epsilon_2{}'\beta$ in K_p; this is evident from Theorem 4–16 with $K_0 = K_p$, $h = p - 1$, $n = q > 2(p - 1)$, $\nu = 0$. Hence x, and therefore also y, has only finitely many possible values.

The proof for the case $D_{rs} = P$ proceeds similarly. We put $x - 1 = \pi w$ and $y = \pi^m z$, where w and z are integers of K_p with $[\pi, z] = [1]$. Then (70) becomes

$$[w]\left[w + \frac{1 - \zeta}{1 - \zeta}\right] \cdots \left[w + \frac{1 - \zeta^{p-1}}{1 - \zeta}\right] = P^{mq-p}[z]^q,$$

and since the ideals on the left are pairwise relatively prime, there is a t with $0 \le t \le p - 1$ such that

$$P^{mq-p}\left|\left[w + \frac{1 - \zeta^t}{1 - \zeta}\right]\right.$$

Thus there are ideals $A_0, \ldots, A_{p-1}$ such that

$$\left[w + \frac{1 - \zeta^t}{1 - \zeta}\right] = P^{mq-p}A_t{}^q,$$

$$\left[w + \frac{1 - \zeta^r}{1 - \zeta}\right] = A_r{}^q, \qquad \text{for } 0 \le r \le p - 1, \quad r \ne t.$$

As before, it follows that all the ideals A_r are principal (for $r = t$, use the fact that an ideal equivalent to a principal ideal is principal). Since $p > 2$, there are distinct rational integers r and s different from t such that $0 \le r \le p - 1$, $0 \le s \le p - 1$. Then for integers α and β and units ϵ and ϵ' of K_p,

$$w + \frac{1 - \zeta^r}{1 - \zeta} = \epsilon\alpha^q, \qquad w + \frac{1 - \zeta^s}{1 - \zeta} = \epsilon'\beta^q,$$

so that

$$\epsilon'\beta^q - \epsilon\alpha^q = \frac{\zeta^r - \zeta^s}{1 - \zeta},$$

and the expression on the right is not zero. The earlier reasoning shows that the theorem is also true in this case.

PROBLEMS

1. Extend Theorem 4–18 to the case that f may have multiple zeros, but has at least three distinct zeros of odd orders.

2. Deduce from the finiteness of the number of solutions of (66) that as the integral variable x tends to infinity, the greatest prime divisor of $f(x)$ does also. [*Hint:* Assume that for infinitely many x, $f(x)$ is a product of powers of a fixed finite set of primes, and obtain a contradiction.]

REFERENCES

Section 4–1

See the following papers: J. Liouville, *Journal des Mathématiques Pures et Appliquées* (Paris) **16**, 133–142 (1851); A. Thue, *Journal für die Reine und Angewandte Mathematik* (Berlin) **135**, 284–305 (1909); C. L. Siegel, *Mathematische Zeitschrift* (Berlin) **10**, 173–213 (1921); F. J. Dyson, *Acta Mathematica* (Stockholm) **79**, 225–240 (1947); T. Schneider, *Archiv der Mathematik* (Karlsruhe) **1**, 288–295 (1948–1949); K. F. Roth, *Mathematika* (London) **2**, 1–20 (1955); Corrigendum, *Mathematika* **2**, 168 (1955).

The paper by Siegel contains many variants and applications of the Thue-Siegel theorem.

Section 4–7

The literature concerning Catalan's conjecture is reviewed by R. Obláth, *Revista Matemática Hispano-Americana* (Madrid) **1**, 122–140 (1941). Mahler's theorem appeared in *Nieuw Archief voor Wiskunde* (Amsterdam) **1**, 113–122 (1953). Theorem 4–17 appeared in *Journal of the London Mathematical Society* **2**, 66–68 (1926).

CHAPTER 5

IRRATIONALITY AND TRANSCENDENCE

5-1 Irrational numbers. One of the oldest results in the theory of numbers is that $\sqrt{2}$ is irrational; this was known to the Pythagoreans in the fifth century B.C. The proof, when suitably generalized with the help of the Unique Factorization Theorem, leads to the well-known rule for determining the possible rational zeros of a polynomial with rational integral coefficients; this in turn makes it possible to show, if such is the case, that a given polynomial has only irrational zeros. Thus the numbers given implicitly as zeros of polynomials can be trivially classified as rational or irrational.

If a number is given by its decimal expansion, one has only to determine whether its digits eventually recur periodically to know whether or not it is irrational. For example, the number

$$0.1234567891011\ldots,$$

whose successive digits are formed in an obvious fashion, is clearly irrational, since arbitrarily long blocks of a single digit occur, precluding periodicity. Similarly, using the regular continued fraction expansion of a real number, one can identify not only the rational numbers but also the quadratic irrationalities. (Unfortunately, there is no simple algorithm known which singles out the algebraic numbers of fixed degree $n \geq 3$ in a distinctive way.)

If a real number x is not given in one of these convenient forms, the problem of deciding whether or not it is rational may be decidedly nontrivial. It is, for example, not known whether Euler's constant, defined as

$$\lim_{n \to \infty} \left(1 + \frac{1}{2} + \frac{1}{3} + \cdots + \frac{1}{n} - \log n \right),$$

is rational. Aside from properties of special algorithms, the only method available for investigating such questions depends on the following observation. If $x = a/b$ is rational, then for every pair of integers p and q, the number $qx - p$ is some integral multiple of $1/b$,

161

so that it is impossible to find an infinite sequence of pairs p_n and q_n such that

$$|q_1 x - p_1| > |q_2 x - p_2| > |q_3 x - p_3| > \cdots. \qquad (1)$$

More generally, no such sequence can be found for which

$$|q_n x - p_n| \neq 0 \text{ for every } n, \qquad \text{and} \qquad \lim_{n \to \infty} |q_n x - p_n| = 0. \qquad (2)$$

On the other hand, when x is irrational there are infinitely many solutions of the inequality

$$0 < |qx - p| < \frac{1}{q}.$$

We therefore have

THEOREM 5–1. *Each of the following is a necessary and sufficient condition for the irrationality of a real number x:*

(a) *there are integers $p_1, q_1, p_2, q_2, \ldots$, such that the inequalities (1) hold;*

(b) *there are integers $p_1, q_1, p_2, q_2, \ldots$, such that the conditions (2) hold.*

As a simple application of this principle, we prove

THEOREM 5–2. *The number e is irrational.*

Proof: We recall the expansion

$$\frac{1}{e} = 1 - \frac{1}{1!} + \frac{1}{2!} - \cdots + \frac{(-1)^n}{n!} + \cdots.$$

It is well known that if $a_0, a_1, \ldots$ is an unbounded increasing sequence of positive numbers, then the series

$$\sum_{k=0}^{\infty} \frac{(-1)^k}{a_k} \qquad (3)$$

converges to its sum S in such a way that

$$0 < \left| S - \sum_{k=0}^{n} \frac{(-1)^k}{a_k} \right| < \frac{1}{a_{n+1}}$$

for $n \geq 0$. Hence if we put $q_n = n!$ and

$$p_n = n! \sum_{k=0}^{n} \frac{(-1)^k}{k!},$$

then p_n and q_n are integers and

$$0 < \left| q_n \cdot \frac{1}{e} - p_n \right| = n! \left| \frac{1}{e} - \sum_{k=0}^{n} \frac{(-1)^k}{k!} \right| < \frac{n!}{(n+1)!} = \frac{1}{n+1}.$$

It follows that $1/e$, and hence e itself, is irrational. (This is a variant of the original proof due to Fourier.) More generally, the same argument shows that if the LCM of the integers $a_1, \ldots, a_n$ is $o(a_{n+1})$ as $n \to \infty$, then the series (3) converges to an irrational number.

For completeness, we give a proof due to I. Niven that π is irrational. It is short and simple to follow, but to one unfamiliar with older work it must appear completely unmotivated.

THEOREM 5–3. *The number π is irrational.*

Proof: Suppose on the contrary that $\pi = a/b$, where a and b are integers. Put

$$f(x) = \frac{x^n(a - bx)^n}{n!},$$

and

$$F(x) = f(x) - f''(x) + f^{(iv)}(x) - \cdots + (-1)^n f^{(2n)}(x),$$

where the positive integer n will be specified later. Now $f(0) = f'(0) = \cdots = f^{(n-1)}(0) = 0$, and if we write

$$f(x) = \frac{a_0 x^n + a_1 x^{n+1} + \cdots + a_n x^{2n}}{n!},$$

we see that for $n \le k \le 2n$,

$$f^{(k)}(x) = \frac{1}{n!} \sum_{l=0}^{n} (n+l)(n+l-1) \cdots (n+l-k+1) a_l x^{n+l-k}$$

$$= \frac{1}{n!} \sum_{l=0}^{n} \frac{(n+l)!}{(n+l-k)!} a_l x^{n+l-k},$$

so that

$$f^{(k)}(0) = \frac{1}{n!} \cdot k! a_{k-n}.$$

Hence $f^{(j)}(0) \in Z$, and since $f(x) = f(\pi - x)$, also $f^{(j)}(\pi) \in Z$, for $0 \le j \le 2n$. Finally, $F(0)$ and $F(\pi)$ must be integers.

On the other hand,

$$\frac{d}{dx}\left(F'(x)\sin x - F(x)\cos x\right) = F''(x)\sin x + F(x)\sin x$$
$$= f(x)\sin x,$$

so that

$$\int_0^\pi f(x)\sin x\,dx = [F'(x)\sin x - F(x)\cos x]_0^\pi = F(\pi) + F(0).$$

But for $0 < x < \pi$,

$$0 < f(x)\sin x < \frac{\pi^n a^n}{n!},$$

so that the above integral is positive but arbitrarily small for n sufficiently large. But this is impossible, since $F(0) + F(\pi)$ is an integer. The contradiction establishes the theorem.

PROBLEM

Given a real number x, define the sequence $\{x_k\}$ of real numbers and the sequence $\{a_k\}$ of integers by the conditions

$$[x] = a_0, \qquad x_1 = x - [x],$$

$$x_1 = \frac{1}{a_1} + x_2, \qquad \text{where } \frac{1}{a_1} \le x_1 < \frac{1}{a_1 - 1},$$

$$x_2 = \frac{1}{a_2} + x_3, \qquad \text{where } \frac{1}{a_2} \le x_2 < \frac{1}{a_2 - 1},$$

$$\vdots$$

$$x_k = \frac{1}{a_k} + x_{k+1}, \qquad \text{where } \frac{1}{a_k} \le x_k < \frac{1}{a_k - 1},$$

$$\vdots$$

Thus

$$x = a_0 + \frac{1}{a_1} + \frac{1}{a_2} + \cdots.$$

Show that this expansion terminates if and only if x is rational. Show also that if x has an infinite series expansion

$$x = b_0 + \frac{1}{b_1} + \frac{1}{b_2} + \cdots.$$

where the numbers b_k are integers with $b_{k+1} > b_k^2$, then $b_k = a_k$ for all k, and x is irrational.

5–2 The existence of transcendental numbers. One class of irrationals, the algebraic numbers, has been treated in some detail in the preceding chapters. We now consider the complementary set of *transcendental* numbers: those complex numbers which do not satisfy any rational algebraic equation with coefficients in Z. It is by no means obvious that this set is nonvacuous; the first proof, given by Liouville in 1844, depends on the fact (see Theorem 4–1) that if α is algebraic of degree $n \geq 2$, then there is a constant C such that the inequality

$$\left| \alpha - \frac{p}{q} \right| < \frac{C}{q^n}$$

has no solution p, q in Z. If a number ξ can be found such that for every $\omega > 0$ the inequality

$$0 < |q\xi - p| < \frac{1}{q^\omega}, \qquad q > 1, \tag{4}$$

has a solution, then ξ cannot be algebraic of any degree, and must therefore be transcendental.

An example of a *Liouville number,* for which (4) always has a solution, is given by

$$\xi = \sum_{k=1}^{\infty} (-1)^k a^{-b_k},$$

where $a > 1$ is a fixed integer and $b_1, b_2, \ldots$ is an increasing sequence of positive integers such that

$$\limsup_{k \to \infty} \frac{b_{k+1}}{b_k} = \infty.$$

For, given ω, there is an $n = n(\omega)$ for which $b_{n+1}/b_n > \omega + 1$, and if we put

$$q = a^{b_n}, \qquad p = q \sum_{k=1}^{n} (-1)^k a^{-b_k},$$

then p and q are integers, and

$$0 < |q\xi - p| < q \cdot \frac{1}{a^{b_{n+1}}} = \frac{1}{q^{-1+b_{n+1}/b_n}} < \frac{1}{q^\omega}.$$

It should be emphasized that the condition (4), while sufficient for transcendence, is by no means necessary, even for real numbers.

For, using a modification of an argument due to Cantor, we can give a second proof of the existence of transcendental numbers, and in particular of numbers of this kind for which the inequality

$$\left| \xi - \frac{p}{q} \right| < \frac{1 - \epsilon}{3q^2}$$

has only finitely many solutions for fixed $\epsilon > 0$. It is known[*] that there are uncountably many irrational numbers ξ for which $M(\xi) = 3$, where $M(\xi)$ is the supremum of the numbers λ for which the inequality

$$\left| \xi - \frac{p}{q} \right| < \frac{1}{\lambda q^2}$$

has infinitely many solutions. Hence if the algebraic numbers are countable, it follows that there are nonalgebraic numbers for which $M(\xi) = 3$.

To order the algebraic numbers, we associate with each non-constant polynomial $P(x) = a_0 x^n + \cdots + a_n$ with integral coefficients the number $h(P) = n + |a_0| + \cdots + |a_n|$. There are no polynomials with $h(P) = 1$. If $h(P) = 2$, then $P(x) = x$ or $-x$. If $h(P) = 3$, then $P(x)$ is one of $\pm x \pm 1$, $\pm 2x$, $\pm x^2$, all combinations of signs being allowed. In general, it is clear that if $k \geq 2$, there are only finitely many polynomials such that $h(P) = k$. Hence all polynomials with integral coefficients can be arranged in a sequence: first those with $h(P) = 2$, in some order, then those with $h(P) = 3$, in some order, etc. Suppose that $P_1(x)$, $P_2(x)$, ... is such a sequence. Each $P_k(x)$ has finitely many zeros; write down all the zeros of $P_1(x)$ in some order, then all those of $P_2(x)$ in some order, etc. Let this sequence be $\beta_1, \beta_2, \ldots$. Now if $\beta_2 = \beta_1$, delete β_2; if $\beta_3 = \beta_2$ or β_1, delete β_3; and in general, if β_k is equal to some β with smaller subscript, delete β_k. Then the resulting sequence $\alpha_1, \alpha_2, \ldots$ contains all algebraic numbers, each just once.

To summarize, if a number can be approximated sufficiently well by rational numbers, it is transcendental, but there are transcendental numbers which cannot be approximated even as well as some quadratic irrationalities.

---

[*] See, for example, Volume I, Theorem 9–12.

1. Show that ξ is a Liouville number if the partial quotients in its continued fraction expansion,

$$\xi = a_0 + \cfrac{1}{a_1 + } \cdots,$$

have the property that

$$\lim_{k \to \infty} \sup \frac{\log a_{k+1}}{\log ((a_1 + 1) \cdots (a_k + 1))} = \infty,$$

[*Hint:* Show from the recursion relation for the successive convergents that $q_k < (a_1 + 1) \cdots (a_k + 1)$, and then use Theorem 2–6.]

2. Investigate the implications of Theorem 4–15 as regards transcendental numbers.

5–3 A criterion for transcendence. In order to obtain an approximability condition which is equivalent to transcendence, we must replace the linear expresion $q\xi - p$ occurring in the inequality (4) by a polynomial in ξ.

THEOREM 5–4. *A real or complex number ξ is transcendental if and only if there corresponds to each $\omega > 0$ a positive integer n, such that the inequality*

$$0 < |x_0 + x_1\xi + \cdots + x_n\xi^n| < X^{-\omega} \tag{5}$$

has infinitely many integral solutions $x_0, \ldots, x_n$, where

$$X = \max (|x_0|, \ldots, |x_n|).$$

It is to be noticed that the Liouville numbers (those for which (4) has a solution for each ω) are precisely the numbers for which we can take $n = 1$ for every ω. In general, however, n increases with ω.

Proof: We first prove that the condition is sufficient. Let $\alpha = \alpha_1$ be algebraic of degree g, let $f(x) = a_0 + a_1x + \cdots + a_gx^g$ be that multiple of its defining polynomial which has relatively prime coefficients in Z, with $a_g > 0$, and let $\alpha_1, \ldots, \alpha_g$ be its conjugates. Let $h(x) = x_0 + x_1x + \cdots + x_nx^n$ $(x_n > 0)$ be any polynomial with integral coefficients, and with zeros $\beta_1, \ldots, \beta_n$ distinct from $\alpha_1, \ldots, \alpha_g$. Then

$$0 < \left| \frac{1}{a_g{}^n} \prod_{i=1}^{n} f(\beta_i) \right| = \left| \prod_{i=1}^{n} \prod_{j=1}^{g} (\beta_i - \alpha_j) \right| = \left| \prod_{j=1}^{g} \prod_{i=1}^{n} (\beta_i - \alpha_j) \right|$$

$$= \left| \frac{1}{x_n{}^g} \prod_{j=1}^{g} h(\alpha_j) \right|,$$

so that if $X = \max (|x_0|, \ldots, |x_n|)$, then

$$0 < |h(\alpha)| = \frac{\left| x_n{}^g \prod_{i=1}^{n} f(\beta_i) \right|}{a_g{}^n \prod_{j=2}^{g} |h(\alpha_j)|} = \frac{\left| x_n{}^g \prod_{i=1}^{n} f(\beta_i) \right|}{a_g{}^n \prod_{j=2}^{g} \left(\dfrac{|h(\alpha_j)|}{X} \right) \cdot X^{g-1}}. \qquad (6)$$

But

$$\prod_{i=1}^{n} f(\beta_i)$$

is a symmetric polynomial with integral coefficients in the β's, of degree g in each β, and is therefore, by the Symmetric Function Theorem, a polynomial of total degree g, with integral coefficients, in the elementary symmetric functions x_{n-1}/x_n, $-x_{n-2}/x_n, \ldots,$ $\pm x_0/x_n$. Hence the numerator in the expression (6) is a positive integer, and we have

$$|h(\alpha)| \geq \frac{1}{a_g{}^n \prod_{j=2}^{n} \left| \dfrac{h(\alpha_j)}{X} \right| \cdot X^{g-1}}.$$

Now if $r = \lceil \alpha \rceil$, then

$$\left| \frac{h(\alpha_j)}{X} \right| \leq 1 + |\alpha_j| + |\alpha_j|^2 + \cdots + |\alpha_j|^n \leq 1 + r + r^2 + \cdots + r^n,$$

so that the quantity

$$\frac{1}{a_g{}^n \prod_{j=2}^{n} \left| \dfrac{h(\alpha_j)}{X} \right|}$$

has a positive lower bound $A(n, \alpha)$ depending only on α and n. Thus

$$|h(\alpha)| \geq \frac{A(n, \alpha)}{X^{g-1}}. \qquad (7)$$

It follows that if (5) has infinitely many solutions with fixed n, ξ cannot be algebraic of degree less than $\omega + 1$. Since ω can be arbitrarily large, ξ cannot be algebraic.

The necessity of the condition of Theorem 5–4 is a consequence of the following more general theorem.

THEOREM 5–5. *If $\vartheta_1, \ldots, \vartheta_n$ are complex numbers, then for a suitable c which depends only on n and $\vartheta_1, \ldots, \vartheta_n$, the inequality*

$$|x_0 + x_1\vartheta_1 + \cdots + x_n\vartheta_n| < \frac{c}{X^{\frac{1}{2}(n-1)}} \tag{8}$$

has infinitely many integral solutions $x_0, \ldots, x_n$.

If ξ is transcendental, we can take $\vartheta_k = \xi^k$; since no polynomial in ξ vanishes, it follows that (5) has infinitely many solutions if $n = [2\omega + 2]$.

Proof: The theorem is trivial if $n = 1$. For $n > 1$, put

$$c' = c'(\vartheta_1, \ldots, \vartheta_n) = 1 + |\vartheta_1| + \cdots + |\vartheta_n|,$$

let $h \geq 2$ be a positive integer, and let $x_0', x_1', \ldots, x_n'$ range independently over the integers from $-h$ to h inclusive. Since each of the $n + 1$ numbers x_k' can assume any of $2h + 1$ values, there are $(2h + 1)^{n+1} = t$ expressions

$$L(\vartheta_1, \ldots, \vartheta_n) = x_0' + x_1'\vartheta_1 + \cdots + x_n'\vartheta_n, \qquad |x_k'| \leq h.$$

Let these be, in some order, $L_1, \ldots, L_t$. Clearly

$$|L_i(\vartheta_1, \ldots, \vartheta_n)| \leq c'h,$$

so that all the points $L_i(\vartheta_1, \ldots, \vartheta_n)$ lie in the square of side $2c'h$ with its center at the origin of the complex plane. Subdivide this square into m^2 subsquares of side $2c'h/m$ each; then if $m^2 < t$, there must be at least one subsquare containing more than one point $L(\vartheta_1, \ldots, \vartheta_n)$. We can fulfill the condition $m^2 < t$ by taking

$$m = [(2h + 1)^{\frac{1}{2}(n+1)}] - 1.$$

For this m, suppose that the points

$$L_1(\vartheta_1, \ldots, \vartheta_n) = x_0' + x_1'\vartheta_1 + \cdots + x_n'\vartheta_n$$

and $\qquad L_2(\vartheta_1, \ldots, \vartheta_n) = \bar{x}_0' + \bar{x}_1'\vartheta_1 + \cdots + \bar{x}_n'\vartheta_n$

lie in a common subsquare; the distance between them does not exceed the length of the diagonal of the subsquare, which is $2\sqrt{2}\,c'h/m$. So if we put $x_0 = x_0' - \bar{x}_0', \ldots, x_n = x_n' - \bar{x}_n'$ (so that $X \leq h - (-h) = 2h$), and

$$\begin{aligned} L(\vartheta_1, \ldots, \vartheta_n) &= L_1(\vartheta_1, \ldots, \vartheta_n) - L_2(\vartheta_1, \ldots, \vartheta_n) \\ &= x_0 + x_1\vartheta_1 + \cdots + x_n\vartheta_n, \end{aligned}$$

then

$$|L(\vartheta_1, \ldots, \vartheta_n)| \leq \frac{2\sqrt{2}\, c'h}{[(2h+1)^{\frac{1}{2}(n+1)}] - 1} \leq \frac{4c'h}{(2h)^{\frac{1}{2}(n+1)}}$$

$$= \frac{2c'}{(2h)^{\frac{1}{2}(n-1)}} \leq \frac{2c'}{X^{\frac{1}{2}(n-1)}} . \qquad (9)$$

Hence (8) has at least one solution, with $c = 2c'$.

If $L(\vartheta_1, \ldots, \vartheta_n) = 0$, then $xL(\vartheta_1, \ldots, \vartheta_n) = L(x\vartheta_1, \ldots, x\vartheta_n) = 0$ for every integer x, and (8) has infinitely many solutions. In the contrary case, choose h_1 so large that

$$|L(\vartheta_1, \ldots, \vartheta_n)| > \frac{2c'}{(2h_1)^{\frac{1}{2}(n-1)}} ,$$

and repeat the entire argument with h replaced by h_1. Calling the new form thus produced $L^{(1)}$, we have, by the analog of (9) and the definition of h_1, that

$$|L^{(1)}(\vartheta_1, \ldots, \vartheta_n)| < |L(\vartheta_1, \ldots, \vartheta_n)|,$$

so that we have a second solution of (7). Continuing the process, we can obtain arbitrarily many solutions.

PROBLEM

Show that if the numbers $\vartheta_1, \cdots, \vartheta_n$ are real, then Theorem 4–5 remains correct if the inequality (8) is replaced by

$$|x_0 + x_1\vartheta_1 + \cdots + x_n\vartheta_n| < \frac{c}{X^{n-1}} .$$

5–4 Measure of transcendence. Mahler's classification. In light of Theorem 5–4, we make the following definition: a function $\varphi(n, t)$ is called a *transcendence measure* for the transcendental number ξ if for each n there is a constant c_n such that for every $X \geq 1$,

$$|x_0 + x_1\xi + \cdots + x_n\xi^n| > c_n\varphi(n, X)$$

for each set of integers $x_0, \ldots, x_n$ of height $X = \max(|x_0|, \ldots, |x_n|)$. By Theorem 5–5, any such $\varphi(n, t)$ is no larger than $t^{-\frac{1}{2}(n-1)}$. A theorem giving a measure of transcendence of a number ξ represents a refinement of the assertion that ξ is transcendental; such measures have been given for certain numbers. In Section 5–5 we shall determine a measure of transcendence for e.

Mahler has elaborated on the theory of transcendence measure in the following way. Let z be a complex number, and put

$$\omega_n(X, z) = \omega_n(X) = \min \left(\left| \sum_{k=0}^{n} x_k z^k \right| \right), \tag{10}$$

where the minimum is extended over all those sets of rational integral coefficients $x_0, \ldots, x_n$ of heights at most X for which

$$\sum_{k=0}^{n} x_k z^k \neq 0.$$

Then $\omega_n(X)$ is at most 1, and is a nonincreasing function of both X and n. Put

$$\omega_n(X) = X^{-\rho_n(X)}, \tag{11}$$

so that

$$\rho_n(X) = \frac{\log \left(1/\omega_n(X) \right)}{\log X},$$

and let

$$\omega_n(z) = \omega_n = \limsup_{X \to \infty} \rho_n(X),$$

$$\omega(z) = \omega = \limsup_{n \to \infty} \frac{\omega_n}{n}.$$

Each of ω_n and ω is either $+\infty$ or a non-negative number. If ω_n is infinite and $n' > n$, then $\omega_{n'}$ is also infinite; hence there is an index $\mu(z) = \mu$, which may be finite or infinite, such that ω_n is finite for $n < \mu$ and infinite for $n \geq \mu$. The two quantities ω, μ are never finite simultaneously, for the finiteness of μ implies that there is an $n < \infty$ such that $\omega_n = \infty$, whence $\omega = \infty$. The number z is called

an A-number,   if $\omega = 0$, $\mu = \infty$,
an S-number,   if $0 < \omega < \infty$, $\mu = \infty$,
a T-number,   if $\omega = \infty$, $\mu = \infty$,
a U-number,   if $\omega = \infty$, $\mu < \infty$.

If μ is finite, then there is a fixed integer n such that for every $\sigma > 0$ there are integers $x_0, \ldots, x_n$ such that

$$|x_0 + x_1 z + \cdots + x_n z^n| < X^{-\sigma}.$$

For the case $n = 1$, this is exactly the definition of the Liouville numbers, so that the U-numbers may be regarded as higher degree analogs of Liouville numbers. The author has shown that there are U-numbers of every degree.

If z is algebraic, the inequality (7) shows that $\rho_n(X)$, and hence also ω_n, remains bounded as $n \to \infty$, so that $\omega = 0$ and z is an A-number. If, on the other hand, z is transcendental, it follows from Theorem 5-5 that $\rho_n \geq \frac{1}{2}(n-1)$, whence $\omega \geq \frac{1}{2}$. Thus the A-numbers are precisely the algebraic numbers.

The existence of T-numbers has never been proved.

THEOREM 5-6. *If the complex numbers z and w are algebraically dependent, that is, if there is a polynomial $F(x, y)$ with coefficients in Z such that $F(z, w) = 0$, then they belong to the same class.*

Proof: If z is algebraic and w is algebraically dependent on z, then w is clearly also algebraic. We may therefore suppose that z and w are transcendental.

Let
$$F(x, y) = \sum_{h=0}^{M} \sum_{k=0}^{N} a_{hk} x^h y^k,$$

and suppose that F is irreducible. (One consequence of this assumption is that no polynomial in x alone is a factor of F.) Write

$$F(x, y) = \sum_{h=0}^{M} A_h(y) x^h,$$

where
$$A_h(y) = \sum_{k=0}^{N} a_{hk} y^k.$$

We may suppose that $A_M(y)$ is not identically zero.

Let $A(x) = a_0 + \cdots + a_n x^n$ be a polynomial for which the minimum is achieved in the definition (10) of $\omega_n(X, z)$, so that in particular $\max(|a_k|) \leq X$. We shall obtain inequalities relating $\omega(z)$ and $\omega(w)$; since in the definition of these quantities the first limit is taken on X, we temporarily regard n as fixed and X as a parameter.

Since it is not the case that for each fixed y the polynomials $F(x, y)$ and $A(x)$ have a common zero, we know by a standard theorem* that

---

* See, for example, B. L. van der Waerden, *Modern Algebra* (English edition, translated by Fred Blum from the second revised German edition), New York: Frederick Ungar Publishing Co., 1949, Vol. 1, pp. 83–85.

the resultant

$$R(y) = \begin{vmatrix} a_0 & \cdots & \cdots & \cdots & a_n & 0 & \cdots & 0 \\ 0 & a_0 & \cdots & \cdots \cdots & \cdots & a_n & 0 & \cdots \\ & & \cdot & & & \cdot & & \\ & & \cdot & & & & \cdot & \\ & & & \cdot & & & & \\ 0 & \cdots & 0 & a_0 & \cdots & \cdots & \cdots & a_n \\ A_0(y) & \cdots & \cdots & \cdots \cdots & \cdots & A_M(y) & 0 & \cdots \\ & & \cdot & & & & \cdot & \\ & & & \cdot & & & & \cdot \\ \cdots & 0 & A_0(y) & \cdots & \cdots & \cdots & \cdots & A_M(y) \end{vmatrix} \begin{matrix} \\ \left.\vphantom{\begin{matrix}a\\a\\a\end{matrix}}\right\} M \text{ rows} \\ \\ \left.\vphantom{\begin{matrix}a\\a\\a\end{matrix}}\right\} n \text{ rows} \end{matrix}$$

is not identically zero. $R(y)$ is a polynomial in y of degree nN at most, with coefficients in Z. Since F is a fixed polynomial throughout, the coefficients in $R(y)$ do not exceed $c_1 X^M$, where c_1 is a constant depending only on n and F.

If for each l with $2 \le l \le M + n$, the lth column in the determinant for $R(y)$ is multiplied by x^{l-1} and added to the first column, the new first column is

$$A(x), \ xA(x), \ \ldots, \ x^{M-1}A(x), \ F(x,y), \ xF(x,y), \ \ldots, \ x^{n-1}F(x,y).$$

Expanding by minors of the new first column, we obtain an identity

$$R(y) = A(x)g(x,y) + F(x,y)h(x,y),$$

from which

$$R(w) = A(z)g(z,w).$$

Regarding $g(x,y)$ as a sum of minors, we see that its coefficients are rational integers not exceeding $c_2 X^{M-1}$ in absolute value, so that

$$|g(z,w)| < c_3 X^{M-1}.$$

Hence

$$|A(z)| > c_3^{-1} X^{-M+1} |R(w)|.$$

But

$$|R(w)| \ge \omega_{nN}(c_1 X^M, w),$$

so

$$|A(z)| > c_3^{-1} X^{-M+1} \omega_{nN}(c_1 X^M, w).$$

It follows from the definition of $A(x)$ that

$$\omega_n(X, z) \geq c_3^{-1} X^{-M+1} \omega_{nN}(c_1 X^M, w),$$

and so we obtain

$$\omega_n(z) = \limsup_X \frac{\log\left(1/\omega_n(X, z)\right)}{\log X} \leq M - 1 + M\omega_{nN}(w),$$

$$\omega(z) = \limsup_n \frac{\omega_n(z)}{n}$$

$$\leq \limsup_n \frac{(M - 1)N + MN\omega_{nN}(w)}{nN} \leq MN\omega(w),$$

and

$$\mu(w) \leq N\mu(z).$$

By symmetry,

$$\omega(w) \leq MN\omega(z) \quad \text{and} \quad \mu(z) \leq M\mu(w).$$

Thus $\omega(z)$ and $\omega(w)$ are simultaneously finite or infinite, as are $\mu(z)$ and $\mu(w)$; hence z and w are in the same class.

5–5 Arithmetic properties of the exponential function. In this section we shall prove a theorem due to Mahler which simultaneously shows that e is an S-number (and therefore transcendental), gives a transcendence measure for e, and shows that π is transcendental. The transcendence measure is not the most precise one known, but more exact results are more difficult to prove.

We begin with an algebraic analog of Theorem 5–5. Let $\omega_1, \ldots, \omega_m$ be distinct complex numbers (having no connection with the function $\omega_n(z)$ of the preceding section), and let $r_1, \ldots, r_m$ be positive integers. Instead of asking for rational integers $x_0, \ldots, x_m$ for which the quantity $x_0 + x_1\omega_1 + \cdots + x_m\omega_m$ is numerically small, we shall investigate the polynomials

$$A_k(z) = A_k(z; r_1, \ldots, r_m; \omega_1, \ldots, \omega_m), \qquad k = 1, \ldots, m,$$

of respective degrees $r_1 - 1, \ldots, r_m - 1$ at most, for which the function

$$R(z) = R(z; r_1, \ldots, r_m; \omega_1, \ldots, \omega_m)$$
$$= A_1(z)e^{\omega_1 z} + \cdots + A_m(z)e^{\omega_m z} \tag{12}$$

is *algebraically* small, i.e., has a Maclaurin expansion beginning with a large power of z. The total number of coefficients among the polynomials $A_k(z)$ is $r = r_1 + \cdots + r_m$; if they are taken as un-determined constants, then the conditions

$$R(0) = 0, \qquad R'(0) = 0, \qquad \ldots, \qquad R^{(r-2)}(0) = 0$$

yield a system of $r - 1$ linear homogeneous equations in these r unknowns. Such a system always has solutions distinct from $(0, 0, \ldots, 0)$. Let $R(z)$ temporarily designate any of the functions obtained in this manner; thus $R(z)$, which is not identically zero, certainly has a zero of order $r - 1$ at $z = 0$, and could conceivably have one of higher order there. Suppose that the actual order is $r - 1 + E$, so that $R(z)$ has an expansion

$$R(z) = \sum_{h=r+E-1}^{\infty} a_h z^h, \qquad a_{r+E-1} \neq 0.$$

The non-negative integer E is called the *excess*, and m is called the *order*, of $R(z)$. We first show that the excess is always equal to zero.

At least one of the polynomials $A_k(z)$ does not vanish identically, and with no loss in generality we may suppose it to be $A_1(z)$. It is easily proved by induction that if $D = d/dz$,

$$D^\alpha e^{\omega z} A(z) = e^{\omega z} (D + \omega)^\alpha A(z) \tag{13}$$

for every positive integer α and every function $A(z)$ with sufficiently many derivatives. Moreover, if $A(z)$ is a polynomial which is not identically zero, and $\omega \neq 0$, then $(D + \omega)^\alpha A(z)$ is a polynomial of the same degree as $A(z)$. Hence

$$D^{r_m} e^{-\omega_m z} R_m(z)$$
$$= D^{r_m} (A_1(z) e^{(\omega_1 - \omega_m)z} + \cdots + A_{m-1}(z) e^{(\omega_{m-1} - \omega_m)z} + A_m(z))$$
$$= A_1{}^*(z) e^{(\omega_1 - \omega_m)z} + \cdots + A{}^*_{m-1}(z) e^{(\omega_{m-1} - \omega_m)z}$$
$$= R(z; r_1, \ldots, r_{m-1}; \omega_1 - \omega_m, \ldots, \omega_{m-1} - \omega_m),$$

where $A_1{}^*$ is not identically zero and, as implied by the notation, $\deg A_k{}^* \leq r_k - 1$ for $k = 1, \ldots, m - 1$. Clearly

$$R^{(\rho)}(0; r_1, \ldots, r_{m-1}; \omega_1 - \omega_m, \ldots, \omega_{m-1} - \omega_m) = 0$$

for $\rho = 0, 1, \ldots, r + E - r_m - 1$, so that from an R-function of order m and excess E we have obtained another of order $m - 1$ and excess E. Repeating the process, we come finally to a function $R_1(z) = R(z; r_1; \omega) = \tilde{A}(z) e^{\omega z}$ of order 1 and excess E. But if $\tilde{A}(z) = \tilde{a}_0 + \tilde{a}_1 z + \cdots + \tilde{a}_{r_1-1} z^{r_1-1}$, the conditions $R_1(0) = \cdots = R_1{}^{(r_1-2)}(0) = 0$ give $\tilde{a}_0 = \cdots = \tilde{a}_{r_1-2} = 0$, so that there is certainly no such function which does not vanish identically, if $E > 0$. Hence $R^{(r-1)}(0) \neq 0$, or equivalently, the coefficient of z^{r-1} in the Maclaurin expansion of $R(z)$ is not zero, while all preceding coefficients are zero. Introducing an appropriate numerical factor, we can put

$$R(z) = \frac{z^{r-1}}{(r-1)!} + b_r z^r + \cdots .$$

The function R and the coefficients $b_r, b_{r+1}, \ldots$ are now uniquely determined, since if there were two such functions for given $\omega_1, \ldots, \omega_m, r_1, \ldots, r_m$, their difference would have positive excess. Moreover, while we have so far known only that not all of the polynomials $A_1(z), \ldots, A_m(z)$ are identically zero, we now see that in fact they are of exact degrees $r_1 - 1, \ldots, r_m - 1$, respectively, since otherwise we could have begun with lower degree polynomials and arrived at a function of positive excess. Finally, we see that $R(z)$ is symmetric in the pairs of arguments $r_1, \omega_1; \ldots; r_m, \omega_m$, since the pairs can be permuted while the solution (subject to all the imposed conditions) is unique. This can also be seen by noting that $R(z)$ is the unique solution of the homogeneous linear differential equation

$$(D - \omega_1)^{r_1} \cdots (D - \omega_m)^{r_m} y = 0$$

for which $R(0) = \cdots = R^{(r-2)}(0) = 0$ and $R^{(r-1)}(0) = 1$, and the factors in the differential operator may be permuted at will.

We now obtain explicit expressions for $R(z)$ and the $A_k(z)$. Clearly

$$R(z; r_1; \omega_1) = \frac{z^{r_1-1}}{(r_1 - 1)!} e^{\omega_1 z}, \tag{14}$$

since this function has all the requisite properties and there is only one such function. Suppose that $R(z; r_1, \ldots, r_{\mu-1}; \omega_1, \ldots, \omega_{\mu-1})$ has already been determined. Then if J is the operator

$$J = \int_0^z \ldots dz,$$

we have by (13) that

$$(D-\omega_1)^{r_1} \cdots (D-\omega_\mu)^{r_\mu} \{ e^{\omega_\mu z} J^{r_\mu} (e^{-\omega_\mu z} R(z; r_1, \ldots, r_{\mu-1}; \omega_1, \ldots, \omega_{\mu-1})) \}$$

$$= (D-\omega_1)^{r_1} \cdots (D-\omega_{\mu-1})^{r_{\mu-1}}$$

$$\times \{ e^{\omega_\mu z} D^{r_\mu} J^{r_\mu} e^{-\omega_\mu z} R(z; r_1, \ldots, r_{\mu-1}; \omega_1, \ldots, \omega_{\mu-1}) \}$$

$$= (D-\omega_1)^{r_1} \cdots (D-\omega_{\mu-1})^{r_{\mu-1}} R(z; r_1, \ldots, r_{\mu-1}; \omega_1, \ldots, \omega_{\mu-1}) = 0,$$

and since $R(z; r_1, \ldots, r_{\mu-1}; \omega_1, \ldots, \omega_{\mu-1})$ has a zero of order $r_1 + \cdots + r_{\mu-1} - 1$ at 0, the function

$$e^{\omega_\mu z} J^{r_\mu}\big(e^{-\omega_\mu z} R(z; r_1, \ldots, r_{\mu-1}; \omega_1, \ldots, \omega_{\mu-1})\big)$$

has a zero of order $r_1 + \cdots + r_\mu - 1$ at 0, and it clearly has leading coefficient $((r_1 + \cdots + r_\mu - 1)!)^{-1}$. Hence

$$R(z; r_1, \ldots, r_\mu; \omega_1, \ldots, \omega_\mu)$$
$$= e^{\omega_\mu z} J^{r_\mu}\big(e^{-\omega_\mu z} R(z; r_1, \ldots, r_{\mu-1}; \omega_1, \ldots, \omega_{\mu-1})\big), \quad (15)$$

and consequently

$$R(z) = (e^{\omega_m z} J^{r_m})(e^{(\omega_{m-1}-\omega_m)z} J^{r_{m-1}}) \cdots (e^{(\omega_2-\omega_3)z} J^{r_2}) e^{(\omega_1-\omega_2)z} \frac{z^{r_1-1}}{(r_1-1)!}.$$

We now use the standard formula

$$J^\alpha f(z) = \int_0^z \frac{(z-t)^{\alpha-1}}{(\alpha-1)!} f(t)\, dt,$$

which is easily verified by integration by parts. We have

$$e^{(\omega_2-\omega_3)z} J^{r_2}\left(e^{(\omega_1-\omega_2)z} \frac{z^{r_1-1}}{(r_1-1)!}\right)$$

$$= e^{(\omega_2-\omega_3)z} \int_0^z \frac{t_1^{r_1-1}}{(r_1-1)!} \frac{(z-t_1)^{r_2-1}}{(r_2-1)!} e^{(\omega_1-\omega_2)t_1}\, dt_1$$

$$= \int_0^z \frac{t_1^{r_1-1}}{(r_1-1)!} \frac{(z-t_1)^{r_2-1}}{(r_2-1)!} e^{\omega_1 t_1 + \omega_2(z-t_1) - \omega_3 z}\, dt_1,$$

and by induction we see that

$$R(z) = \int_0^z dt_{m-1} \int_0^{t_{m-1}} dt_{m-2} \cdots$$

$$\cdots \int_0^{t_2}\left\{ \frac{t_1^{r_1-1}(t_2-t_1)^{r_2-1} \cdots (t_{m-1}-t_{m-2})^{r_{m-1}-1}(z-t_{m-1})^{r_m-1}}{(r_1-1)!(r_2-1)! \cdots (r_{m-1}-1)!(r_m-1)!} \right.$$

$$\left. \times e^{\omega_1 t_1 + \omega_2(t_2-t_1) + \cdots + \omega_m(z-t_{m-1})} \right\} dt_1. \quad (16)$$

Before deducing an explicit formula for $A_k(z)$, we recall certain properties of inverse operators. The operator D^{-1}, as applied to an integral combination $f(z)$ of polynomials and exponential functions, yields that antiderivative which contains no constant of integration.

Hence

$$D^{-\rho}f(z) = J^{\rho}f(z) + \varphi(z),$$

where φ is a function annihilated by D^{ρ}, that is, a polynomial of degree $\rho - 1$ at most.. More generally, if $\omega \neq 0$ we define, by analogy to (13),

$$(D - \omega)^{-\rho}f(z) = e^{\omega z}D^{-\rho}\big(e^{-\omega z}f(z)\big),$$

so that

$$(D - \omega)^{-\rho}f(z) = e^{\omega z}J^{\rho}\big(e^{-\omega z}f(z)\big) + \psi(z), \tag{17}$$

where $\psi(z)$ is annihilated by $(D - \omega)^{\rho}$; that is, it is $e^{\omega z}$ times a polynomial of degree $\rho - 1$ at most. Since

$$(D - \omega)\left(z^n + \frac{nz^{n-1}}{\omega} + \frac{n(n-1)z^{n-2}}{\omega^2} + \cdots + \frac{n!}{\omega^n}\right) = -\omega z^n,$$

and since no term of the operand is annihilated by $D - \omega$, we can write

$$(D - \omega)^{-1}z^n = -\sum_{r=0}^{n} \frac{n!}{r!\,\omega^{n-r+1}}z^r = -\frac{1}{\omega}\left(1 + \frac{D}{\omega} + \frac{D^2}{\omega^2} + \cdots\right)z^n.$$

More generally, it can be shown that if F is any polynomial of degree n for which $F(0) \neq 0$, then $(F(D))^{-1}z^n$ can be written as

$$(a_0 + a_1 D + \cdots + a_n D^n)z^n, \tag{18}$$

where $a_0 + \cdots + a_n u^n$ is the Maclaurin expansion of $(F(u))^{-1}$ to $n + 1$ terms.*

We can now prove that for $k = 1, \ldots, m$,

$$A_k(z) = \left\{\prod_{\substack{h=1 \\ h \neq k}}^{m} (D + \omega_k - \omega_h)^{-r_h}\right\}\frac{z^{r_k-1}}{(r_k - 1)!}. \tag{19}$$

For $m = 1$, the empty product is interpreted as the identity operator, of course, and in this case the correctness of (19) follows from equation (14). Suppose that it is correct for all polynomials

$$A_k(z; r_1, \ldots, r_{\mu-1}; \omega_1, \ldots, \omega_{\mu-1})$$

---

*A more complete discussion of inverse operators is given in E. L. Ince, *Ordinary Differential Equations*, New York: Longmans, Green & Co., Inc., 1926; reprinted by Dover Publications, New York, 1944; pp. 138–140.

with $\mu - 1$ pairs r_k, ω_k. Then, by (15) and (17),

$$R(z;r_1,\ldots,r_\mu;\omega_1,\ldots,\omega_\mu)$$

$$=e^{\omega_\mu z}J^{r_\mu}\sum_{k=1}^{\mu-1}A_k(z;r_1,\ldots,r_{\mu-1};\omega_1,\ldots,\omega_{\mu-1})e^{(\omega_k-\omega_\mu)z}$$

$$=e^{\omega_\mu z}\sum_{k=1}^{\mu-1}\{e^{-(\omega_\mu-\omega_k)z}(D-\omega_\mu+\omega_k)^{-r_\mu}$$

$$\times A_k(z;r_1,\ldots,r_{\mu-1};\omega_1,\ldots,\omega_{\mu-1})+p_k(z)\}$$

$$=\sum_{k=1}^{\mu-1}e^{\omega_k z}(D-\omega_\mu+\omega_k)^{-r_\mu}A_k(z;r_1,\ldots,r_{\mu-1};\omega_1,\ldots,\omega_{\mu-1})+P(z)e^{\omega_\mu z},$$

where $p_k(z)$ and $P(z)$ are polynomials of degree $r_\mu - 1$ at most. It follows that (19) is correct for $k = 1$, for arbitrary m, and its truth for $k = 2$ follows from the previously noted symmetry of $R(z)$ in the pairs ω_k, r_k.

For fixed complex numbers $\omega_1, \ldots, \omega_m$, our considerations up to this point are valid for all the functions $R(z;r_1,\ldots,r_m;\omega_1,\ldots,\omega_m)$ corresponding to arbitrary sets $r_1,\ldots,r_m$ of positive integers. We now specialize the parameters so as to obtain a collection of functions depending on a single parameter ρ.

For h and k in the sequence $1, 2, \ldots, m$, define

$$\delta_{hk} = \begin{cases} 1 & \text{if } h = k, \\ 0 & \text{if } h \neq k, \end{cases}$$

and put

$$R_h(z)=R_h(z;\rho;\omega_1,\ldots,\omega_m)=R(z;\rho+\delta_{1h},\ldots,\rho+\delta_{mh};\omega_1,\ldots,\omega_m),$$

$$A_{hk}(z)=A_{hk}(z;\rho;\omega_1,\ldots,\omega_m)=A_k(z;\rho+\delta_{1h},\ldots,\rho+\delta_{mh};\omega_1,\ldots,\omega_m).$$

Here ρ is a fixed but arbitrary positive integer. We form the square matrix

$$A(z) = (A_{hk}(z)), \qquad h, k = 1, \ldots, m,$$

having determinant $D(z)$. Let the minor determinant of $A_{hk}(z)$ in $D(z)$ be $D_{hk}(z)$.

Now $A_{hk}(z)$ is a polynomial in z of degree $\rho + \delta_{hk} - 1$, and the coefficient of the highest power of z is, by (19),

$$\frac{1}{(\rho + \delta_{hk} - 1)!}\prod_{\substack{l=1 \\ l \neq k}}^{m}(\omega_k - \omega_l)^{-\rho-\delta_{hl}}.$$

Hence in the expansion of $D(z)$, the term formed from the elements of the main diagonal will be of higher degree than any other term, and $D(z)$ is therefore a polynomial of degree $m\rho$ with the coefficient of the highest power of z equal to

$$\frac{1}{(\rho!)^m} \prod_{k=1}^{m} \prod_{\substack{l=1 \\ l \neq k}}^{m} (\omega_k - \omega_l)^{-\rho}.$$

If, on the other hand, we solve the system of equations

$$\sum_{k=1}^{m} A_{hk}(z)e^{\omega_k z} = R_h(z), \qquad h = 1, \ldots, m,$$

for $e^{\omega_k z}$, we obtain the identity

$$D(z)e^{\omega_k z} = \sum_{h=1}^{m} (-1)^{h+k} D_{hk}(z) R_h(z).$$

Since the expansion of $R_h(z)$ begins with the term $z^{m\rho}/(m\rho)!$, the polynomial $D(z)$ is divisible by $z^{m\rho}$. Hence

$$D(z) = \frac{z^{m\rho}}{(\rho!)^m} \prod_{k=1}^{m} \prod_{\substack{l=1 \\ l \neq k}}^{m} (\omega_k - \omega_l)^{-\rho}, \qquad (20)$$

and $D(z)$ vanishes only at $z = 0$.

Let $c_1, c_2, \ldots$ be positive constants depending only on $m, \omega_1, \ldots, \omega_m$. (In particular, they must not depend on ρ, which will eventually be large.)

Examination of equation (16) shows that for $1 \leq h \leq m$,

$$R_h(1) = O\left(\frac{c_1^\rho}{(\rho!)^m}\right).$$

From (19), we obtain

$$A_{hk}(z) = \left\{ \prod_{\substack{l=1 \\ l \neq k}}^{m} \sum_{\lambda_l=0}^{\infty} \binom{-\rho - \delta_{hl}}{\lambda_l} (\omega_k - \omega_l)^{-\rho - \delta_{hl} - \lambda_l} D^{\lambda_l} \right\} \frac{z^{\rho + \delta_{hk} - 1}}{(\rho + \delta_{hk} - 1)!},$$

where the sums need not be extended past the index ρ. Let

$$\Omega = \prod_{\substack{h,k=1 \\ h < k}}^{m} (\omega_k - \omega_h),$$

so that Ω can be regarded as a polynomial of total degree $m(m-1)/2$

in $\omega_1, \ldots, \omega_m$, with coefficients in Z not exceeding 2^{m^2} in absolute value. Since no exponent $\rho + \delta_{hl} + \lambda_l$ in the above sums exceeds $2\rho + 1$, the expression

$$a_{hk} = \Omega^{2\rho+1}\rho!A_{hk}(1)$$

is a polynomial in $\omega_1, \ldots, \omega_m$, of total degree

$$\frac{m(m-1)}{2}(2\rho+1)$$

at most, whose coefficients are rational integers of the order of magnitude $O(c_2{}^\rho \rho!)$. Finally, we put

$$r_h = \sum_{k=1}^{m} a_{hk}e^{\omega_k}, \qquad h = 1, \ldots, m,$$

so that

$$r_h = \Omega^{2\rho+1}\rho!R_h(1) = O\left(\frac{c_3{}^\rho}{\rho!^{m-1}}\right).$$

The quantities $r_1, \ldots, r_m$ are linear forms in the numbers e^{ω_k}, and they are linearly independent, in the sense that no linear combination of the vectors $(a_{h1}, \ldots, a_{hm})$, for $1 \le h \le m$, is the zero vector. This is equivalent to the assertion that $D(1) \neq 0$, which follows from (20).

THEOREM 5-7. *Suppose that $\omega_1, \ldots, \omega_m$ all lie in an algebraic number field K of degree g, and let*

$$L_h = \sum_{k=1}^{m} b_{hk}e^{\omega_k}, \qquad h = 1, \ldots, \mu,$$

be μ independent linear forms in $e^{\omega_1}, \ldots, e^{\omega_m}$ with coefficients b_{hk} in Z. Suppose that

$$m\left(1 - \frac{1}{g}\right) < \mu < m, \tag{21}$$

and put

$$b = \max_{\substack{1 \le h \le \mu \\ 1 \le k \le m}} (|b_{hk}|), \qquad s = \max_{1 \le h \le \mu} (|L_h|).$$

Then to each $\epsilon > 0$ there corresponds a $b_0(\epsilon)$ such that $s \ge b^{-\tau-\epsilon}$ if $b > b_0(\epsilon)$, where

$$\tau = \frac{m\mu g}{\mu g - m(g-1)} - 1.$$

Proof: By a well-known theorem on independence,* the μ forms $L_1, \ldots, L_\mu$, together with $m - \mu$ of the forms r_h, which we may designate by $r_{h_1}, \ldots, r_{h_{m-\mu}}$, are independent. Hence the determinant

$$
\Delta = \begin{vmatrix}
a_{h_1 1} & \cdots & a_{h_1 m} \\
\vdots & & \vdots \\
a_{h_{m-\mu} 1} & \cdots & a_{h_{m-\mu} m} \\
b_{11} & \cdots & b_{1m} \\
\vdots & & \vdots \\
b_{\mu 1} & \cdots & b_{\mu m}
\end{vmatrix}
$$

is not zero; it is obviously a polynomial in $\omega_1, \ldots, \omega_m$ of degree at most

$$
\sigma = \frac{m(m - 1)(2\rho + 1)(m - \mu)}{2},
$$

with coefficients in Z of the order of magnitude $O(c_4{}^\rho \rho!^{m-\mu} b^\mu)$. It follows, first, that there is a rational integer c_5 such that $c_5{}^\rho \Delta$ is an integer of K, and, second, that

$$
\lceil \Delta \rceil = O((\sigma + 1)^m c_4{}^\rho \rho!^{m-\mu} b^\mu c_6{}^\sigma)
$$
$$
= O(c_7{}^\rho \rho!^{m-\mu} b^\mu),
$$

where c_6 is an upper bound for the various numbers $\lceil \omega_k \rceil$. Hence if $\Delta, \Delta'', \ldots, \Delta^{(g)}$ are the field conjugates of Δ, we have

$$
\left| \frac{1}{\Delta} \right| = \left| \frac{c_5{}^\rho (c_5{}^\rho \Delta'') \cdots (c_5{}^\rho \Delta^{(g)})}{\mathbf{N}(c_5{}^\rho \Delta)} \right|
$$
$$
= O(c_8{}^\rho \rho!^{(g-1)(m-\mu)} b^{(g-1)\mu}).
$$

Moreover, using subscripts on Δ to indicate minors, we have

$$
\Delta_{lk} = O(c_9{}^\rho \rho!^{m-\mu-1} b^\mu), \qquad \Delta_{lk} r_{h_l} = O(c_{10}{}^\rho \rho!^{-\mu} b^\mu),
$$
$$
\text{for } 1 \le l \le m - \mu,
$$

$$
\Delta_{lk} = O(c_{11}{}^\rho \rho!^{m-\mu} b^{\mu-1}), \qquad \Delta_{lk} L_{l-m+\mu} = O(c_{12}{}^\rho \rho!^{m-\mu} b^{\mu-1} s),
$$
$$
\text{for } m - \mu + 1 \le l \le m.
$$

---

* Cf. B. L. van der Waerden, *Modern Algebra* (English edition, translated by Fred Blum from the second revised German edition), New York: Frederick Ungar Publishing Co., 1949, Vol. 1, p. 101.

Using the identity

$$\Delta e^{\omega_k} = \sum_{l=1}^{m-\mu} (-1)^{l+k} \Delta_{lk} r_{h_l} + \sum_{l=m-\mu+1}^{m} (-1)^{l+k} \Delta_{lk} L_{l-m+\mu},$$

it follows that

$$1 = O(c_{13}{}^\rho \rho!^{m(g-1)-\mu g} b^{\mu g}) + O(c_{14}{}^\rho \rho!^{(m-\mu)g} b^{\mu g-1} s),$$

or

$$c_{14}{}^\rho \rho!^{(m-\mu)g} b^{\mu g-1} s \geq c_{15} - c_{16} \cdot c_{13}{}^\rho \rho!^{m(g-1)-\mu g} b^{\mu g}. \qquad (22)$$

From the inequality (21), the exponent $m(g-1) - \mu g$ is negative. Hence the quantity

$$\frac{c_{13}{}^\rho}{\rho!^{\mu g-m(g-1)}}$$

may increase for small values of ρ, but it tends to zero as ρ increases indefinitely.   At any rate, we can say that for b larger than some c_{17}, the smallest value of ρ for which

$$\frac{c_{15}}{2} \geq c_{16} c_{13}{}^\rho \rho!^{m(g-1)-\mu g} b^{\mu g}$$

is so large that $c_{13}{}^\rho$ is negligible as compared to the factorial, and for such ρ we have the asymptotic relation

$$\log \rho! \smallsmile \frac{\mu g}{\mu g - m(g-1)} \log b.$$

By (22), for $b > b_0(\epsilon)$,

$$s > b^{-\tau-\epsilon},$$

where

$$\tau = \mu g - 1 + \frac{(m-\mu)\mu g^2}{\mu g - m(g-1)} = \frac{m\mu g}{\mu g - m(g-1)} - 1.$$

This proves the theorem.

THEOREM 5–8.   *Suppose that $\vartheta_1, \ldots, \vartheta_N$ are elements of an algebraic number field K of degree g, and that they are linearly independent over the rationals, so that no relation of the form*

$$d_1 \vartheta_1 + \cdots + d_N \vartheta_N = 0$$

holds with $d_1, \ldots, d_N$ rational and not all zero.   Then if the coeffi-

cients $b_{\lambda_1\cdots\lambda_N}$ *in the linear form*

$$L = \sum_{\lambda_1=0}^{M_1} \cdots \sum_{\lambda_N=0}^{M_N} b_{\lambda_1\cdots\lambda_N} e^{\lambda_1\vartheta_1+\cdots+\lambda_N\vartheta_N} \tag{23}$$

in the quantities $e^{\lambda_1\vartheta_1+\cdots+\lambda_N\vartheta_N}$ *are rational integers with*

$$b = \max \left(|b_{\lambda_1\cdots\lambda_N}|\right),$$

there is a constant T, *depending only on* g *and* N, *such that for sufficiently large* b,

$$|L| \geq b^{-TM_1\cdots M_N}.$$

Proof: Let $\mu_1, \ldots, \mu_N$ be positive integers, and consider the quantities

$$L_{l_1\cdots l_N} = e^{l_1\vartheta_1+\cdots+l_N\vartheta_N}L,$$

$$l_1 = 0, 1, \ldots, \mu_1; \quad \ldots; \quad l_N = 0, 1, \ldots, \mu_N, \tag{24}$$

their number being

$$\mu = (\mu_1 + 1) \cdots (\mu_N + 1).$$

If we introduce the exponential factor in (24) inside the summation in (23); we see that the various $L_{l_1\cdots l_N}$ may be regarded as linear forms in the quantities

$$e^{\omega_{\lambda_1\cdots\lambda_N}},$$

where

$$\omega_{\lambda_1\cdots\lambda_N} = \lambda_1\vartheta_1 + \cdots + \lambda_N\vartheta_N,$$

$$\lambda_1 = 0, 1, \ldots, M_1 + \mu_1; \quad \ldots; \quad \lambda_N = 0, 1, \ldots, M_N + \mu_N,$$

the number of ω's being

$$m = (M_1 + \mu_1 + 1) \cdots (M_N + \mu_N + 1).$$

The numbers $\omega_{\lambda_1\cdots\lambda_N}$ are distinct on account of the independence of $\vartheta_1, \ldots, \vartheta_N$ over the rationals, so that we can speak of the independence of the forms $L_{l_1\cdots l_N}$. To see that as a matter of fact they are independent, order the subscript sets $\lambda_1 \ldots \lambda_N$ and $l_1 \ldots l_N$ by interpreting the λ's and l's as digits in the base q, for some sufficiently large q. Then there cannot be a linear relation among the coefficient vectors of any set of forms, since the $\omega_{\lambda_1\cdots\lambda_N}$ with largest subscript occurs only in the form $L_{l_1\cdots l_N}$ with largest subscript. Finally, there are positive constants α and β which are independent of the coefficients $b_{\lambda_1 \dots \lambda_N}$, for which

$$\alpha < \left| \frac{L}{L_{l_1 \dots l_N}} \right| < \beta.$$

It follows from Theorem 5–7 that if

$$\frac{m}{\mu} = \prod_{i=1}^{N} \frac{M_i + \mu_i + 1}{\mu_i + 1} < \frac{g}{g - 1}, \tag{25}$$

then for $b > b(\epsilon)$,

$$|L| \geq b^{-\tau - \epsilon},$$

where

$$\tau = \frac{g(M_1 + \mu_1 + 1) \cdots (M_N + \mu_N + 1)(\mu_1 + 1) \cdots (\mu_N + 1)}{g(\mu_1 + 1) \cdots (\mu_N + 1) - (g - 1)(M_1 + \mu_1 + 1) \cdots (M_N + \mu_N + 1)} - 1.$$

Condition (25) is satisfied if

$$\mu_i = \left[\frac{M_i}{\left(\dfrac{2g}{2g - 1} \right)^{1/N} - 1} \right],$$

since then

$$1 + \frac{M_i}{\mu_i + 1} \leq \left(\frac{2g}{2g - 1} \right)^{1/N},$$

$$\prod_{i=1}^{N} \left(1 + \frac{M_i}{\mu_i + 1} \right) \leq \frac{2g}{2g - 1} < \frac{g}{g - 1}.$$

With this choice of μ_i we have

$$\tau = \frac{\mu \displaystyle\prod_{i=1}^{N} \left(1 + \frac{M_i}{\mu_i + 1} \right)}{1 - \dfrac{g - 1}{g} \displaystyle\prod_{i=1}^{N} \left(1 + \frac{M_i}{\mu_i + 1} \right)} - 1 \leq \mu \frac{\dfrac{2g}{2g - 1}}{1 - \dfrac{g - 1}{g} \dfrac{2g}{2g - 1}} - 1$$

$$= 2g\mu - 1. \tag{26}$$

Since $g \geq 1$ and $N \geq 1$, we have $\mu_i \geq M_i \geq 1$. Since $[x] + 1 \leq 2x$ for $x \geq 1$, we have

$$\mu \leq \prod_{i=1}^{N} \frac{2M_i}{\left(\dfrac{2g}{2g - 1} \right)^{1/N} - 1},$$

and we have the theorem with

$$T = \frac{2^{N+1}g}{\prod\limits_{i=1}^{N} \left\{\left(\dfrac{2g}{2g-1}\right)^{1/N} - 1\right\}}.$$

Taking $N = 1$, we have

COROLLARY 1. *If $\vartheta \neq 0$ is algebraic, e^{ϑ} is an S-number, and in particular is transcendental.*

For $\vartheta = \pi i$, $e^{\vartheta} = -1$ is not transcendental. Hence

COROLLARY 2. *π is transcendental.*

We also have the following result, first proved by F. Lindemann in 1882.

COROLLARY 3. *If $\vartheta_1, \ldots, \vartheta_N$ are algebraic and are linearly independent over the rationals, then $e^{\vartheta_1}, \ldots, e^{\vartheta_N}$ are algebraically independent over the field of algebraic numbers, that is, there is no polynomial $P(z_1, \ldots, z_N)$ with algebraic coefficients not all zero for which*

$$P(e^{\vartheta_1}, \ldots, e^{\vartheta_N}) = 0.$$

Finally, for $N = 1$ the brackets can be omitted in the definition of μ_1; then $\mu_1 = (2g - 1)M_1$ and

$$\tau \leq 2g\mu - 1 \leq 2g\left((2g-1)M_1 + 1\right) - 1 = 2g(2g-1)M_1 + 2g - 1.$$

COROLLARY 4. *If $\vartheta \neq 0$ is algebraic of degree g, then the function*

$$\varphi(n, t) = t^{-2g(2g-1)n-2g+1}$$

is a transcendence measure for e^{ϑ}.

5–6 A theorem of Schneider. In addition to the Liouville numbers and values of the exponential function, many other specific numbers are known to be transcendental. To indicate the type of results known, we mention the following:

(a) The Bessel functions $J_0(x)$ and $J_0'(x)$ are transcendental for algebraic $x \neq 0$.

(b) If α and β are algebraic, $\alpha \neq 0$ or 1, and β is irrational, then α^{β} is transcendental. (In particular, $e^{\pi} = (-1)^{-i}$ is included.)

(c) At least one of the numbers $g_2, g_3, \omega_1, \omega_2$ associated with a Weierstrass $\wp$-function is transcendental, and if g_2 and g_3 are algebraic, at least one of z and $\wp(z)$ is transcendental.

(d) If $f(x)$ is a polynomial whose value is in Z for argument in Z, and $f(x) > 0$ for $x > 0$, then the number

$$0.\,f(1)f(2)f(3)\ldots,$$

formed by juxtaposing the decimal representations of the values $f(x)$, is transcendental. (An example is the number $0.1361015\ldots$, generated by $f(x) = (x^2 + x)/2$.)

(e) If ω is a positive quadratic irrationality, then the number

$$\sum_{n=0}^{\infty} [n\omega] z^n$$

is transcendental for algebraic $z \neq 0$.

On the other hand, it is not known whether the following numbers are transcendental:

(a) $\gamma = \lim\limits_{n \to \infty} \left(1 + \dfrac{1}{2} + \cdots + \dfrac{1}{n} - \log n\right),$

(b) $\zeta(2n + 1) = \sum\limits_{k=1}^{\infty} \dfrac{1}{k^{2n+1}},$

(c) $\Gamma(x)$ for algebraic x not in Z,

(d) $e + \pi, \quad e\pi.$

The methods used to prove what little is known about specific transcendental numbers show considerable variety, both in technique and conception. T. Schneider has recently shown, however, that several results which earlier required separate proofs can all be obtained from a single theorem. This theorem says nothing directly about transcendental numbers; rather, its sense is that if several transcendental functions assume algebraic values at a large number of points, then they must either have large rates of growth or be algebraically dependent (as functions). The prototype of Schneider's result, proved by G. Pólya in 1920, asserts that if f is an integral transcendental function which assumes values in Z for $z = 0, 1, 2, \ldots$, then

$$\limsup_{r \to \infty} \frac{M(r)}{2^r} \geq 1,$$

where, as usual,

$$M(r) = \max_{|z|=r} (|f(z)|).$$

There have been many refinements and extensions of Polya's work, of course; we mention only that by A. Gelfond in 1929, where this kind of theorem was first used for transcendence investigations. (His result was that α^β is transcendental for algebraic $\alpha \neq 0, 1$, if β is an imaginary quadratic irrationality.)

In this section we shall prove Schneider's theorem, and in the next we shall apply it to the numbers α^β. (The facts mentioned above concerning the $\wp$-function can also be deduced, but the requisite preliminaries preclude doing so here.) Since the statement of the theorem is complicated, we first introduce some notation.

By the *order* of an entire function $f(z)$ we mean, as usual, the quantity

$$\limsup_{R \to \infty} \frac{\log \log M(R)}{\log R} \, ;$$

if $f(z)$ is of order μ, then

$$f(z) = O(e^{R^{\mu+\epsilon}})$$

as $|z| = R \to \infty$, for every fixed $\epsilon > 0$. Let $\zeta_1, \zeta_2, \ldots$ be an infinite sequence of complex numbers. Designate by

$$z_0(m) = z_0, \quad \ldots, \quad z_k(m) = z_k$$

the distinct numbers among $\zeta_1, \ldots, \zeta_m$, and by $l_\varkappa(m) + 1 = l_\varkappa + 1$ the multiplicity of occurrence of $z_\varkappa$ among $\zeta_1, \ldots, \zeta_m$. Thus

$$\sum_{\varkappa=0}^{k} (l_\varkappa + 1) = m. \tag{27}$$

Let $r(m) = r$ be the radius of the smallest circle about the origin which contains $z_1, \ldots, z_k$, and put

$$\alpha = \liminf_{m \to \infty} \frac{\log m}{\log r} \, , \tag{28}$$

so that $\alpha \leq \infty$. Let

$$l = \max (l_0, \ldots, l_k).$$

Finally, let K be a fixed algebraic number field of degree g, and, as always, let $\boxed{a}$ be the maximum of the absolute values of the conjugates of a, for a in K.

THEOREM 5–9.   *Let $f_1(z), \ldots, f_n(z)$ be meromorphic functions with the property that for each m, the numbers*

$$f_\nu^{(\lambda)}(z_\varkappa), \qquad \lambda = 0, \ldots, l_\varkappa, \quad \varkappa = 0, \ldots, k, \quad \nu = 1, \ldots, n,$$

are in K. Let $H_\nu(z_\varkappa)$ be positive rational integers such that all the numbers

$$H_\nu(z_\varkappa)f_\nu^{(\lambda)}(z_\varkappa), \qquad \lambda = 0, \ldots, l_\varkappa, \quad \varkappa = 0, \ldots, k, \quad \nu = 1, \ldots, n,$$

are integers in K. Suppose that

$$l \leq \frac{m}{\log m} \cdot \tag{29}$$

For each ν, if $f_\nu(z)$ is entire let it be of order μ_ν, and otherwise suppose that there is an entire function $G_\nu(z)$ of order μ_ν such that $G_\nu(z)f_\nu(z)$ is entire and also of order μ_ν. Suppose that

$$\frac{\mu_1 + \cdots + \mu_n}{n - 1} < \alpha, \tag{30}$$

and put

$$\eta_\nu = \frac{\mu_\nu}{\alpha}, \qquad \nu = 1, \ldots, n.$$

Suppose finally that

$$\limsup_{m \to \infty} \frac{\log \log \max_{0 \leq \varkappa \leq k} \left(|G_\nu(z_\varkappa)|^{-1} \right)}{\log m} \leq \eta_\nu, \qquad \nu = 1, \ldots, n, \tag{31}$$

and

$$\limsup_{m \to \infty} \frac{\log \log \max_{\substack{0 \leq \varkappa \leq k \\ 0 \leq \lambda \leq l_\varkappa}} \left(|f_\nu^{(\lambda)}(z_\varkappa)|, \ H_\nu(z_\varkappa) \right)}{\log m} \leq \eta_\nu, \qquad \nu = 1, \ldots, n. \tag{32}$$

Then $f_1, \ldots, f_n$ are algebraically dependent over K.

Proof: We form a polynomial

$$\Phi(z) = \sum_{\tau_1 = 0}^{t_1} \cdots \sum_{\tau_n = 0}^{t_n} C_{\tau_1 \cdots \tau_n} f_1^{\tau_1}(z) \cdots f_n^{\tau_n}(z),$$

and seek to determine the coefficients $C_{\tau_1 \cdots \tau_n}$ so that Φ has a zero of order $l_\varkappa + 1$ at $z_\varkappa$, for $\varkappa = 1, \ldots, k$. Here the numbers k, $z_\varkappa$, and $l_\varkappa$ are all defined in terms of the sequence $\zeta_1, \zeta_2, \ldots$ and an index m, as explained earlier; m is fixed, and will be specified more exactly later. The conditions imposed on Φ require that all the numbers

$$\Phi^{(\lambda)}(z_\varkappa), \qquad \lambda = 0, \ldots, l_\varkappa; \quad \varkappa = 0, \ldots, k,$$

shall vanish, and this in turn yields a set of m homogeneous linear equations of the form

$$\sum_\tau \omega_\mu C_{\tau_1 \cdots \tau_n} = 0, \qquad \mu = 1, \ldots, m, \tag{33}$$

in the $t = (t_1 + 1) \cdots (t_n + 1)$ unknowns $C_{\tau_1 \cdots \tau_n}$. (Of course, the numbers ω_μ also depend on $\tau_1, \ldots, \tau_n$.) We put

$$t_\nu = [(2m^{1+m+\cdots+\eta_n-n\eta_\nu})^{1/n}], \qquad \nu = 1, \ldots, n, \tag{34}$$

so that

$$t = (t_1 + 1) \cdots (t_n + 1) \geq \left\{ \prod_{\nu=1}^{n} 2m^{1+m+\cdots+\eta_n-n\eta_\nu} \right\}^{1/n} = 2m. \tag{35}$$

The coefficients ω_μ in equations (33) are by assumption numbers in K, and after multiplication by the rational integral factor

$$\prod_{\nu=1}^{n} (H_\nu(z_\varkappa))^{t_\nu}$$

they become integers, say Ω_μ, of K. The size of the coefficients Ω_μ is determined in part by this numerical factor, in part by the values of the f_ν and their derivatives at the various points $z_\varkappa$, and in part by the numerical coefficients introduced by differentiation. In the estimate (36) below, the second of these is accounted for by (32). The third depends only on the set of exponents $t_1, \ldots, t_n$ and the order of the derivative considered, and so can be computed from the fact that the sum of all the coefficients in the expansion of

$$\frac{d^\lambda}{dz^\lambda} (z^{\tau_1} \cdots z^{\tau_n})$$

by the product formula is

$$\sum_{\nu=1}^{n} \tau_\nu \left(\sum_{\nu=1}^{n} \tau_\nu - 1 \right) \cdots \left(\sum_{\nu=1}^{n} \tau_\nu - \lambda + 1 \right) \leq \left(\sum_{\nu=1}^{n} \tau_\nu \right)^\lambda.$$

We thus obtain the bound

$$\lceil \Omega_\mu \rceil \leq \prod_{\nu=1}^{n} (H_\nu(z_x))^{t_\nu} \cdot \prod_{\nu=1}^{n} \exp (t_\nu m^{\eta_\nu + \epsilon_\nu}) \cdot \left(\sum_{\nu=1}^{n} t_\nu \right)^{l}, \qquad (36)$$

where $\epsilon_\nu > 0$ and $\epsilon_\nu \to 0$ as $m \to \infty$. (Hereafter we designate any quantity with the latter properties by ϵ, and any positive integer independent of m, l, and r by γ.)

It follows from the inequality (30) and the definition of η_ν that

$$\eta_1 + \cdots + \eta_n < n - 1, \qquad (37)$$

so that, by (34),

$$\left(\sum_{\nu=1}^{n} t_\nu \right)^{l} < (\gamma m)^l.$$

Using this, together with (32) and (36), we obtain

$$\lceil \Omega_\mu \rceil < (\gamma m)^l \exp \left\{ 2 \sum_{\nu=1}^{n} 2^{1/n} m^{(1+\eta_1+\cdots+\eta_n-n\eta_\nu)/n+\eta_\nu+\epsilon} \right\}$$

$$< \gamma^m m^l \gamma^{m^{(1+\eta_1+\cdots+\eta_n)/n+\epsilon}}.$$

By (29), $m^l < \gamma^m$. By (37), $\eta_1 + \cdots + \eta_n = n - 1 - \delta$ with δ a positive constant; hence

$$\frac{1}{n} (1 + \eta_1 + \cdots + \eta_n) + \epsilon = \frac{n - \delta}{n} + \epsilon = 1 - \frac{\delta}{n} + \epsilon.$$

We henceforth require m to be so large that

$$\epsilon < \frac{\delta}{n}. \qquad (38)$$

Then

$$\lceil \Omega_\mu \rceil < \gamma^m. \qquad (39)$$

Using this and (35), we shall now show that *there are coefficients $C_{\tau_1 \cdots \tau_n}$ satisfying (33) which are integers in K, are not all zero, and are such that*

$$\lceil C_{\tau_1 \cdots \tau_n} \rceil < \gamma^m. \qquad (40)$$

To simplify the notation, arrange the $C_{\tau_1 \cdots \tau_n}$ in some fixed linear order, and rewrite (33) in the form

$$\sum_{\tau=1}^{t} \Omega_{\mu\tau} C_\tau = 0, \qquad \mu = 1, \ldots, m.$$

Let $\rho_1, \ldots, \rho_g$ be an integral basis for K, let $h \in Z$ be positive, and let B be the set of integers in K of the form

$$b_1\rho_1 + \cdots + b_g\rho_g,$$

where the b's range independently over the rational integers such that $|b| \leq h$. For each set of elements $X_1, \ldots, X_t$ of B, put

$$y_\mu = \sum_{\tau=1}^{t} \Omega_{\mu\tau} X_\tau, \qquad \mu = 1, \ldots, m.$$

This defines $(2h + 1)gt$ m-tuples $y_1, \ldots, y_m$, not necessarily different from one another. Also, since

$$\lceil X_\tau \rceil < \gamma h, \tag{41}$$

we have from (39) that

$$\lceil y_\mu \rceil < \gamma^m h. \tag{42}$$

Each number y_μ has a basis representation $c_1\rho_1 + \cdots + c_g\rho_g$; similar representations, with the same c_i and with the ρ_i replaced by their conjugates, hold for the conjugates of y_μ. The determinant formed from the ρ_i and their conjugates is not zero, so that it is possible to solve the g equations defining y_μ and its conjugates for the numbers c_i, giving each c_i as a linear expression in the conjugates of y_μ, with coefficients depending only on K. From (42) it follows that for $i = 1, \ldots, g$,

$$|c_i| < \gamma^m h. \tag{43}$$

There are, however, exactly $(2\gamma^m h + 1)^g$ different integers of K whose basis representation satisfies (43); therefore there are at most $(2\gamma^m h + 1)^{gm}$ different systems $y_1, \ldots, y_m$. If

$$(2\gamma^m h + 1)^{gm} < (2h + 1)^{gt}, \tag{44}$$

then two systems $y_1, \ldots, y_m$ corresponding to two different sets $X_1, \ldots, X_t$ coincide, and the respective differences $X_1 - X_1{}', \ldots, X_t - X_t{}'$ constitute a solution of (33). These differences, which we call $C_1, \ldots, C_t$, are not all zero, and by (41), they satisfy the condition

$$\lceil C_\tau \rceil < \gamma h.$$

By (35), $t \geq 2m$, so (44) holds if

$$(2\gamma^m h + 1)^m < (2h + 1)^{2m},$$

which is clearly true if $h = \gamma^m$. But then (40) holds.

We now designate by m_0 a fixed value of m such that (38) holds, and by k_0 and $l_x^{(0)}$ the corresponding values of k and l_x, and define Φ to be a fixed function corresponding to m_0 and having all the properties described up to this point. We are now able to perform an induction.

We know that Φ possesses m_0 zeros, if each is counted with its proper multiplicity. It is asserted that *if m_0 is sufficiently large, then $\Phi(z)$ vanishes at all the points $\zeta_1, \zeta_2, \ldots$.* This is proved inductively by showing that if $\Phi(z) = 0$ for $z = \zeta_1, \ldots, \zeta_m$, with $m \geq m_0$, then also $\Phi(\zeta_{m+1}) = 0$, if m_0 is sufficiently large. More precisely, we assume that Φ has a zero at $z_x(x = 0, \ldots, k)$ of order $l_x + 1$, with

$$\sum_{x=0}^{k} (l_x + 1) = m \geq m_0,$$

and shall deduce that Φ has a zero at $\zeta_{m+1} = \zeta$ of order $\Lambda + 1$, where

$$\Lambda = \begin{cases} 0 & \text{if } \zeta = z_{k+1} \neq z_x \text{ for } x = 0, \ldots, k, \\ l_x + 1 & \text{if } \zeta = z_\sigma \text{ and } 0 \leq \sigma \leq k. \end{cases}$$

Here $k = k(m)$ and $l_x = l_x(m)$.

Put

$$G(z) = \prod_{\nu=1}^{n} G_\nu^{t_\nu}(z);$$

then $G(z)\Phi(z)$ is an entire function which vanishes at the same points z_x as $\Phi(z)$, and to the same order, by (31). We also put

$$Q(z) = \prod_{\substack{x=0 \\ z_x \neq \zeta}}^{k} (z - z_x)^{l_x+1}.$$

By Cauchy's theorem,

$$\frac{d^\Lambda(G(z)\Phi(z)/Q(z))}{dz^\Lambda}\bigg|_{z=\zeta} = \frac{\Lambda!}{2\pi i} \int_\Gamma \frac{G(z)\Phi(z)}{Q(z)} \frac{dz}{(z - \zeta)^{\Lambda+1}}. \quad (45)$$

Here Γ is the circle

$$|z| = R_1 = R^\vartheta, \quad \vartheta > 1,$$

where $R = r(m + 1)$ if $\alpha < \infty$ (we recall that $r(m)$ was defined earlier as the radius of the smallest circle about O containing $z_0, \ldots, z_k$), while if $\alpha = \infty$, R is so chosen that

$$R \geq r(m + 1), \quad \lim_{m \to \infty} \frac{\log m + 1}{\log R} = \infty, \quad \lim_{m \to \infty} R = \infty.$$

Since Φ has a zero of order Λ at ζ, the left side of (45) is simply

$$\left. \frac{G(z)}{Q(z)} \Phi^{(\lambda)}(z) \right|_{z=\zeta},$$

so that

$$|\Phi^{(\lambda)}(\zeta)| = \frac{Q(\zeta)}{G(\zeta)} \frac{\Lambda!}{2\pi} \left| \int_{\Gamma} \frac{G(z)\Phi(z)}{Q(z)} \frac{dz}{(z-\zeta)^{\Lambda+1}} \right|. \tag{46}$$

We shall use this representation to estimate $|\Phi^{(\lambda)}(\zeta)|$. By the inequality (40), we have

$$\max_{|z|=R_1} |G(z)\Phi(z)| < t\gamma^{m_0} \exp\left(\sum_{\nu=1}^{n} t_\nu R_1^{\mu_\nu + \epsilon} \right),$$

where $\epsilon \to 0$ as $R_1 \to \infty$, or equivalently as $m \to \infty$. By the definition (28) of α,

$$R_1 = R^\vartheta < m^{\vartheta/(\alpha-\epsilon)}$$

or

$$R_1 < m^{\vartheta/\alpha + \epsilon},$$

even for $\alpha = \infty$. Hence

$$\max_{|z|=R_1} |G(z)\Phi(z)| < \gamma^{m_0} \exp\left(\sum_{\nu=1}^{n} t_\nu m^{\vartheta \eta_\nu + \epsilon} \right),$$

since it is easily seen from the definition (34) that $t < \gamma^{m_0}$. From (34) we also have that

$$\sum_{\nu=1}^{n} t_\nu m^{\vartheta \eta_\nu + \epsilon} = \sum_{\nu=1}^{n} (2^{1/n} m_0^{(1+\eta_1+\cdots+\eta_n)/n - \eta_\nu}) m^{\vartheta \eta_\nu + \epsilon}$$

$$\leq \sum_{\nu=1}^{n} 2^{1/n} m^{(1+\eta_1+\cdots+\eta_n)/n - \eta_\nu + \vartheta \eta_\nu + \epsilon}$$

$$\leq \sum_{\nu=1}^{n} 2 m^{1 - (\delta/n) + (\vartheta-1)\eta_\nu + \epsilon}.$$

We may suppose, with no loss in generality, that each $\eta_\nu \leq 1$. For suppose that η_n, say, is larger than 1. Then in analogy with (30),

$$\frac{\mu_1 + \cdots + \mu_{n-1}}{n-2} < \alpha,$$

and all hypotheses of the theorem are satisfied by the $n-1$ functions $f_1(z), \ldots, f_{n-1}(z)$. But if $f_1(z), \ldots, f_{n-1}(z)$ can be shown to be algebraically dependent, then $f_1(z), \ldots, f_n(z)$ must also be dependent.

Consequently, if we put $\vartheta = 1 + \delta/2n$, we have

$$(\vartheta - 1) \max_{1 \leq \nu \leq n} (\eta_\nu) \leq \frac{\delta}{2n}.$$

If m_0, and hence also $m \geq m_0$, is so large that

$$\epsilon < \frac{\delta}{2n}, \tag{47}$$

then

$$\sum_{\nu=1}^{n} m^{1 - \delta/n + (\vartheta - 1)\eta_\nu + \epsilon} < nm,$$

and

$$\max_{|z| = R_1} |G(z)\Phi(z)| < \gamma^{m_0} e^{2nm} < \gamma^m. \tag{48}$$

Continuing the estimation of the right side of (46), we notice that since $\vartheta > 1$, and since R grows indefinitely with m, it is possible to choose m_0 so large that

$$\min_{|z| = R_1} |z - z_\varkappa| > \frac{R_1}{2}, \tag{49}$$

for $\varkappa = 0, \ldots, k$, and also

$$\min_{|z| = R_1} |z - \zeta| > \frac{R_1}{2}. \tag{50}$$

In that case,

$$\min_{|z| = R_1} |Q(z)(z - \zeta)^{\Lambda+1}| \geq \left(\frac{R_1}{2}\right)^{m+1} \tag{51}$$

Since ζ and all the $z_\varkappa$ lie in the disk $|z| \leq R$, we have

$$|\zeta - z_\varkappa| < 2R,$$

so that

$$|Q(\zeta)| < (2R)^m. \tag{52}$$

Finally, we see from (31) that

$$|G(\zeta)| > \exp\left(-\sum_{\nu=1}^{n} t_\nu m^{\eta_\nu + \epsilon}\right) > \gamma^{-m}. \tag{53}$$

Combining the relations (48), (51), (52), and (53), we have

$$|\Phi^{(\Lambda)}(\zeta)| < (2R)^m \cdot \gamma^m \Lambda^\Lambda \cdot \gamma^m \cdot \left(\frac{R_1}{2}\right)^{-m-1} \cdot R_1$$

$$< \gamma^m \Lambda^\Lambda \left(\frac{R}{R_1}\right)^m$$

$$< \gamma^m \Lambda^\Lambda R^{-(\vartheta - 1)m}.$$

Since by hypothesis

$$l = \max_{0 \le \varkappa \le k} (l_\varkappa, \Lambda) \le \frac{m+1}{\log (m+1)},$$

the inequality

$$\Lambda^\Lambda < \gamma^m$$

holds; hence

$$|\Phi^{(\Lambda)}(\zeta)| < \gamma^m R^{-\delta m/2n}. \tag{54}$$

Recalling that $\Phi(z)$ is a polynomial in $f_1(z), \ldots, f_n(z)$ with coefficients C_τ which are integers in K, and that all the derivatives of $f_1(z), \ldots, f_n(z)$ up to order Λ have values in K for $z = \zeta$, we see that also $\Phi^{(\Lambda)}(\zeta)$ is a number in K, and that the product

$$\Phi^{(\Lambda)}(\zeta) \prod_{\nu=1}^{n} H_\nu{}^{t_\nu}(\zeta)$$

is an integer in K. By the same reasoning as was used in producing the estimate (36), we have

$$\prod_{\nu=1}^{n} H_\nu{}^{t_\nu}(\zeta) \cdot \overline{|\Phi^{(\Lambda)}(\zeta)|} < \gamma^{m_0} \cdot \prod_{\nu=1}^{n} H_\nu{}^{t_\nu}(\zeta) \cdot \left(\sum_{\nu=1}^{n} t_\nu \right)^\Lambda$$

$$\cdot \prod_{\nu=1}^{n} \exp (t_\nu m^{\eta_\nu + \epsilon}) \cdot \prod_{\nu=1}^{n} (t_\nu + 1).$$

The factor γ^{m_0} comes from the estimate (40) for $\overline{|C_\tau|}$, while the last product is the total number of terms in $\Phi(z)$ itself, which was an unnecessary factor in (36), where we were estimating the terms in a derivative arising from a single term in $\Phi(z)$. By arguments used previously, it follows from the last inequality that

$$\prod_{\nu=1}^{n} H_\nu{}^{t_\nu}(\zeta) \overline{|\Phi^{(\Lambda)}(\zeta)|} < \gamma^m.$$

Combining this with (54), we have

$$\left| \mathbf{N} \left(\Phi^{(\Lambda)}(\zeta) \prod_{\nu=1}^{n} H_\nu{}^{t_\nu}(\zeta) \right) \right| < \gamma^{gm} R^{-\delta m/2n}, \tag{55}$$

and the upper bound here is smaller than 1 for m sufficiently large, say $m \ge m_1$. Hence if m_0 is so large that $m_0 > m_1$, and the inequalities (47), (49), and (50) hold, then it follows from (55) that

$$\Phi^{(\Lambda)}(\zeta) = 0,$$

as asserted.

To complete the proof of Theorem 5–9, we shall make use of the following general considerations. Let $\varphi(z)$ be analytic in the disk bounded by a circle Γ, and let $x_1, \ldots, x_p$ be interior points of this disk. Then $\varphi(z)$ has an expansion

$$\varphi(z) = a_0 + a_1(z-x_1) + a_2(z-x_1)(z-x_2) + \cdots$$
$$+ a_{p-1}(z-x_1)\cdots(z-x_{p-1}) + (z-x_1)\cdots(z-x_p)R_p(z) \qquad (56)$$

with constants $a_0, \ldots, a_{p-1}$ and a function $R_p(z)$ regular in the disk. In fact, if we put

$$\frac{1}{2\pi i} \int_\Gamma \frac{\varphi(t)}{(t-x_1)\cdots(t-x_{q+1})}\, dt = a_q, \qquad (57)$$

for $q = 0, \ldots, p-1$, and

$$\frac{1}{2\pi i} \int_\Gamma \frac{\varphi(t)}{(t-z)(t-x_1)\cdots(t-x_p)}\, dt = R_p(z), \qquad (58)$$

then

$$\varphi(z) - \big(a_0 + a_1(z-x_1) + \cdots + a_{p-1}(z-x_1)\cdots(z-x_{p-1})\big)$$

$$= \frac{1}{2\pi i} \int_\Gamma \left(\frac{1}{t-z} - \frac{1}{t-x_1} - \frac{z-x_1}{(t-x_1)(t-x_2)} - \cdots \right.$$

$$\left. - \frac{(z-x_1)\cdots(z-x_{p-1})}{(t-x_1)\cdots(t-x_p)} \right) \varphi(t)\,dt$$

$$= \frac{1}{2\pi i} \int_\Gamma \frac{(z-x_1)\cdots(z-x_p)}{(t-z)(t-x_1)\cdots(t-x_p)}\, \varphi(t)\,dt$$

$$= (z-x_1)\cdots(z-x_p)R_p(z),$$

and it is clear from (58) that $R_p(z)$ is regular inside Γ.

We apply this with $\varphi(z) = \Phi(z)G(z)$, the "interpolation points" $x_1, \ldots, x_p$ being $\zeta_1, \ldots, \zeta_m$, in this order. For Γ we choose the circle $|z| = R_1 = R^\vartheta$ with $\vartheta > 1$, where $R \geq r(m)$, $\lim_{m\to\infty} R = \infty$, and

$$\liminf_{m\to\infty} \frac{\log m}{\log R} = \alpha.$$

Since $\Phi(z)$ vanishes at all the points $\zeta_1, \zeta_2, \ldots$, the integrand in the expression (57) for a_q is regular in Γ, so that

$$a_q = 0 \qquad \text{for } q = 0, \ldots, m-1.$$

Hence for fixed z with $z \neq \zeta_1, \zeta_2, \ldots$,

$$G(z)\Phi(z) = \lim_{m \to \infty} \left(Q_m(z) \cdot \frac{1}{2\pi i} \int_\Gamma \frac{G(t)\Phi(t)}{Q_m(t)} \frac{dt}{t - z} \right),$$

where

$$Q_m(z) = \prod_{\varkappa=0}^{k} (z - z_\varkappa)^{l_\varkappa+1}, \qquad \sum_{\varkappa=0}^{k} (l_\varkappa + 1) = m.$$

As in the derivation of (54), we have

$$|G(z)\Phi(z)| \leq \gamma^{m_0} \prod_{\nu=1}^{n} (t_\nu + 1) \exp\left(\sum_{\nu=1}^{n} t_\nu R_1^{\mu_\nu+\epsilon} \right) (2r)^m \left(\frac{R_1}{2} \right)^{-m}$$

$$< \gamma^m \left(\frac{R}{R_1} \right)^m = \gamma^m R^{-(\vartheta-1)m}.$$

Since this inequality holds for arbitrarily large m, and since R increases indefinitely with m while γ does not, it must be that

$$G(z)\Phi(z) = 0.$$

Hence $\Phi(z)$ vanishes for all z, which is the assertion of the theorem.

5–7 The Hilbert-Gelfond-Schneider theorem.   As an application of Schneider's theorem of the preceding section, we now prove

THEOREM 5–10.   *If a and b are algebraic numbers, b is irrational, and a is neither 0 nor 1, then a^b is transcendental.*

This theorem settles a question raised by Euler concerning the arithmetic nature of the logarithm of a rational number to a rational base, and repeated in more general form in the seventh of Hilbert's famous list of 23 outstanding problems which seemed to him to be both difficult and important.   The list appeared in 1900, but it was not until 1929 that Gelfond made the first contribution to the solution of this problem.   Further partial results were obtained by Kusmin, Siegel, and Boehle, and in 1934 complete proofs were given almost simultaneously by Gelfond and Schneider.   As mentioned earlier, the proof to be given now is most nearly in the spirit of Gelfond's 1929 paper; it should be instructive to the reader also to examine the original complete proofs by Gelfond and Schneider.

We apply Theorem 5–9 with $n = 2, f_1(z) = a^z, f_2(z) = z$, and $\zeta_m = u + vb$, where u and v range over the positive integers.   On account of the irrationality of b, the numbers ζ_m are distinct; they are to be ordered by the size of $u + v$, and otherwise arbitrarily. Suppose that in the sequence $\zeta_1, \ldots, \zeta_m$, all the numbers occur for which $u + v \leq d$, and possibly some (but not all) of those for which $u + v = d + 1$. Then clearly

$$\frac{d(d - 1)}{2} \leq m < \frac{d(d + 1)}{2},$$

while

$$r = \max \left(|u + vb| \right) \leq (d + 1)(1 + |b|),$$

and (taking $u = d - 1, v = \pm 1$),

$$r \geq d - 1 - |b|.$$

These inequalities show that $\gamma_1 r^2 < m < \gamma_2 r^2$, for some positive γ_1 and γ_2. Thus

$$\alpha = \lim_{m \to \infty} \frac{\log m}{\log r} = 2.$$

By the choice of $f_1(z)$ and $f_2(z)$, $\mu_1 = 1$ and $\mu_2 = 0$, and the inequality (30) holds. Since a^z and z are entire, (31) is without force. If we suppose that z and a^z are elements of an algebraic number field K for $z = b$, then $f_2(\zeta) = u + vb$ and $f_1(\zeta) = a^u(a^b)^v$ are also in K for positive integral u and v. (We need not examine the derivatives, since $\zeta_1, \zeta_2, \ldots$ are distinct.) Moreover, if c is a positive rational integer such that ca, cb, and ca^b are integers of K, then we can choose

$$H_1(z_x) = c^{u+v} \quad \text{and} \quad H_2(z_x) = c$$

for $z_x = u + vb$. It follows from this and the definitions of $f_1(z)$ and $f_2(z)$ that the inequality (32) holds. Thus, under the assumption that a, b, and a^b are all algebraic, all hypotheses of Theorem 5–9 are satisfied, and it follows that z and a^z are algebraically dependent. This being palpably false, the above assumption cannot be maintained, and the theorem is proved.

PROBLEM

Show that e^ϑ is transcendental for algebraic $\vartheta \neq 0$. [*Hint:* Choose $f_1(z) = e^z$, $f_2(z) = z$, and

$$\zeta_{(n-1)^2+1} = \cdots = \zeta_{n^2} = n\vartheta,$$

for $n = 1, 2, \ldots$.]

REFERENCES

Section 5-1

I. Niven, *Bulletin of the American Mathematical Society* **53**, 509 (1947).

Section 5-2

J. Liouville, *Comptes Rendus Hebdomadaires des Séances de l'Académie des Sciences* (Paris) **18**, 883–885, 910–911 (1844); *Journal des Mathematiques Pures et Appliquées* (Paris) **16**, 133–142 (1851).

Sections 5-3, 5-4, 5-5

Most of this material is adapted from Mahler, *Journal für die Reine und Angewandte Mathematik* (Berlin) **66**, 117–150 (1932). For the existence of U-numbers of each degree, see LeVeque, *Journal of the London Mathematical Society* **28**, 220–229 (1953).

Section 5-6

Siegel's work on Bessel functions is to be found in *Abhandlung der Kgl. Preussischen Akademie der Wissenschaften* (Berlin), article no. 1, 70 pp. (1929). The first result stated concerning the $\wp$-function is due to Siegel, *Journal für die Reine und Angewandte Mathematik* **167**, 62–69 (1932); the second to Schneider, *ibid.*, **172**, 70–74 (1934). The transcendence of decimals formed from polynomial values was proved by Mahler, *Proceedings Konink. Nederlandsche Akademie van Wetenschappen* (Amsterdam) **40**, 421–428 (1937), and that of the series $\sum[n\omega]z^n$ by Mahler, *Mathematische Annalen* (Leipzig) **101**, 342–366 (1929) and **103**, 532 (1930), and *Mathematische Zeitschrift* (Berlin) **32**, 545–585 (1930).

Schneider's Theorem 5–9 appeared in *Mathematische Annalen* **121**, 131–140 (1949); he includes a bibliography of work on integral-valued functions. Pólya's work appeared in *Nachrichten von der Gesellschaft der Wissenschaften zu Göttingen*, pp. 1–10 (1920), and Gelfond's in *Tôhoku Mathematical Journal* (Sendai, Japan) **30**, 280–285 (1929).

Section 5-7

Hilbert's problems appeared in *Nachrichten von der Gesellschaft der Wissenschaften zu Göttingen*, pp. 253–297 (1900). The complete solution of the seventh was given by Gelfond, *Comptes Rendus de l'Académie des Sciences de l' U.R.S.S.* (Moscow) **2**, 1–3 (in Russian), 4–6 (in French) (1934), and *Bulletin de l'Académie des Sciences de l' U.R.S.S.* (Leningrad) **7**, 623–640 (1934); and by Schneider, *Journal für die Reine und Angewandte Mathematik* **172**, 65–69 (1934). There is an excellent exposition of Gelfond's method by E. Hille, *American Mathematical Monthly* **49**, 654–661 (1942).

CHAPTER 6

DIRICHLET'S THEOREM

In this chapter and the next we shall consider various questions concerning the distribution of the rational primes. This is a large and difficult field, and we shall be able to obtain only a few of the important results. The first of them, to which this chapter is devoted, is Dirichlet's famous theorem that there are infinitely many primes of the form $km + l$, where k and l are fixed integers which are relatively prime.

6–1 Introduction. Although proofs of certain special cases of Dirichlet's theorem are given in elementary texts,* the methods used cannot be generalized to prove the full theorem. To get an idea of the method used by Dirichlet, let us consider the question of the infinitude of the set of primes of the form $4k + 1$. We base the discussion on the *Riemann ζ-function*, defined for $s > 1$ by the equation

$$\zeta(s) = \sum_{n=1}^{\infty} \frac{1}{n^s}.$$

This is perhaps the simplest of all the *Dirichlet series*

$$\sum_{n=1}^{\infty} \frac{a_n}{n^s},$$

which play an important role in prime number theory. One reason for their importance is exhibited in the following theorem, which gives a relation between the set of primes and the set of positive integers.

THEOREM 6–1. For $s > 1$,

$$\zeta(s) = \prod_p \left(1 - \frac{1}{p^s}\right)^{-1}. \tag{1}$$

Proof: In less abbreviated form, the assertion is that

$$\lim_{N \to \infty} \prod_{p \leq N} \left(1 - \frac{1}{p^s}\right)^{-1} = \lim_{N \to \infty} \sum_{n=1}^{N} \frac{1}{n^s}.$$

_____
* See, for example, Volume I, pp. 9, 46, 59.

The relation

$$\frac{1}{1-x} = 1 + x + x^2 + \cdots = \sum_{n=0}^{\infty} x^n$$

holds for $|x| < 1$; since $|p^{-s}| < 1$, we have

$$\prod_{p \leq N} (1 - p^{-s})^{-1} = \prod_{p \leq N} (1 + p^{-s} + p^{-2s} + \cdots).$$

Multiplying out the product on the right, we obtain terms of the form n^{-s}, where n runs over the integers composed exclusively of primes not exceeding N. Moreover, each such n occurs exactly once, by the Unique Factorization Theorem. The multiplication is permissible, since the series involved are absolutely convergent, and the terms can be arranged in any order. Thus

$$\prod_{p \leq N} (1 - p^{-s})^{-1} = \sum{}' n^{-s},$$

where the accent indicates a summation, in the natural order, over all n such that $p|n$ implies $p \leq N$. In particular, the sum contains all terms n^{-s} for which $n \leq N$. Hence

$$\prod_{p \leq N} (1 - p^{-s})^{-1} = \sum_{n=1}^{N} n^{-s} + \sum_{n>N}{}' n^{-s},$$

and

$$0 < \sum_{n>N}{}' n^{-s} \leq \sum_{n=N+1}^{\infty} n^{-s} < \int_{N}^{\infty} x^{-s}\, dx = \frac{1}{(s-1)N^{s-1}},$$

since $s > 1$. Thus

$$\sum_{n>N}{}' n^{-s} = o(1)$$

as $N \to \infty$, and

$$\lim_{N \to \infty} \prod_{p \leq N} (1 - p^{-s})^{-1} = \lim_{N \to \infty} \sum_{n=1}^{N} n^{-s} = \zeta(s).$$

To see exactly how $\zeta(s)$ behaves as $s \to 1^+$, we use the following standard result.*

LEMMA. *Suppose that* $\lambda_1, \lambda_2, \ldots$ *is a nondecreasing sequence tending to infinity, that* $c_1, c_2, \ldots$ *is an arbitrary sequence of real or complex numbers, and that* $f(x)$ *has a continuous derivative for* $x \geq \lambda_1$. *Put*

$$C(x) = \sum_{\substack{n \\ \lambda_n \leq x}} c_n.$$

---

* See, for example, Volume I, Theorem 6–15.

Then for $x \geq \lambda_1$,

$$\sum_{\lambda_n \leq x} c_n f(\lambda_n) = C(x)f(x) - \int_{\lambda_1}^x C(t)f'(t)\, dt.$$

Applying this with $\lambda_n = n,\, c_n = 1,\, f(x) = x^{-s}$, we obtain

$$\sum_{n \leq x} \frac{1}{n^s} = s \int_1^x \frac{[t]}{t^{s+1}}\, dt + \frac{[x]}{x^s}$$

for $x \geq 1$. If we put $(x) = x - \lfloor x \rfloor$, we have for $s > 1$,

$$\sum_{n \leq x} \frac{1}{n^s} = s \int_1^x \frac{t - (t)}{t^{s+1}}\, dt + \frac{x - (x)}{x^s}$$

$$= s \int_1^x \frac{dt}{t^s} - s \int_1^x \frac{(t)}{t^{s+1}}\, dt + \frac{1}{x^{s-1}} - \frac{(x)}{x^s}$$

$$= \frac{s}{s-1} - \frac{s}{(s-1)x^{s-1}} - s \int_1^x \frac{(t)}{t^{s+1}}\, dt + \frac{1}{x^{s-1}} - \frac{(x)}{x^s}.$$

Letting x increase without bound and noting that $0 \leq (x) < 1$, we have

$$\zeta(s) = \frac{s}{s-1} - s \int_1^\infty \frac{(t)}{t^{s+1}}\, dt. \tag{2}$$

This expression for $\zeta(s)$ agrees with the earlier definition for $s > 1$, but it is also meaningful for $0 < s < 1$, since the integral converges for all $s > 0$. It may therefore be thought of as defining $\zeta(s)$ for $s > 0,\, s \neq 1$. At any rate, (2) shows that

$$\lim_{s \to 1^+} \zeta(s)(s - 1) = 1, \tag{3}$$

and *a fortiori* that

$$\lim_{s \to 1^+} \zeta(s) = \infty. \tag{4}$$

For the remainder of this section, let q and r designate primes of the forms $4k + 1$ and $4k - 1$ respectively. Define the function $\chi(n)$ by the equations

$$\chi(1) = 1, \qquad \chi(q) = 1, \qquad \chi(r) = -1, \qquad \chi(2) = 0,$$

$$\chi(mn) = \chi(m)\chi(n) \text{ for every pair of integers } m,\, n.$$

(A function which satisfies the last of these conditions is said to be *completely multiplicative;* it is entirely determined when its values for all prime arguments are known, since $\chi(p^\alpha) = (\chi(p))^\alpha$.) Inasmuch as $n \equiv 1 \pmod 4$ if and only if $2 \nmid n$ and the total number of r's dividing n is even, we have

$$\chi(n) = \begin{cases} 0 & \text{if } 2 \mid n, \\ (-1)^{\frac{1}{2}(n-1)} & \text{if } 2 \nmid n. \end{cases}$$

We now investigate the function

$$L(s) = \sum_{n=1}^{\infty} \frac{\chi(n)}{n^s}.$$

If we write $\sum a_n \ll \sum b_n$ to mean that $|a_n| \leq b_n$ for $n = 1, 2, \dots,$ then

$$\sum_{n=1}^{\infty} \frac{\chi(n)}{n^s} \ll \sum_{n=1}^{\infty} \frac{1}{n^s}$$

for $s > 1$, so that the series for $L(s)$ is absolutely convergent for $s > 1$. More than this is true, however: *the series for $L(s)$ converges for $s > 0$.* For we note that for any $n > 0$,

$$\chi(n) + \chi(n+1) + \chi(n+2) + \chi(n+3) = 0,$$

so that we have

$$\sum_{n=1}^{N} \chi(n)$$

$$= \sum_{n=1}^{4} \chi(n) + \sum_{n=5}^{8} \chi(n) + \cdots + \sum_{n=4[\frac{1}{4}N]-3}^{4[\frac{1}{4}N]} \chi(n) + \sum_{n=4[\frac{1}{4}N]+1}^{N} \chi(n)$$

$$= 0 + 0 + \cdots + 0 + \sum_{n=4[\frac{1}{4}N]+1}^{N} \chi(n),$$

and hence

$$\left| \sum_{n=1}^{N} \chi(n) \right| \leq 1.$$

The truth of the assertion is therefore a weak consequence of the following theorem, which is due to Abel.

THEOREM 6–2. *If $\{a_n\}$ is a sequence of constants for which*

$$\sum_{n=1}^{N} a_n = O(1)$$

as $N \to \infty$, and if $\{b_n(s)\}$ is a sequence of positive-valued functions which converges monotonically and uniformly to zero for s in some *interval J, then the series*

$$\sum_{n=1}^{\infty} a_n b_n(s)$$

converges uniformly for s in J.

Proof: Put

$$A_n = \sum_{k=1}^{n} a_k,$$

so that $|A_n| < A$ for some A and all n. Using the monotonicity of $b_n(s)$, we have

$$\left| \sum_{n=j}^{k} a_n b_n(s) \right| = \left| \sum_{n=j}^{k} (A_n - A_{n-1}) b_n(s) \right|$$

$$= \left| \sum_{n=j}^{k-1} A_n \big(b_n(s) - b_{n+1}(s) \big) + A_k b_k(s) - A_{j-1} b_j(s) \right|$$

$$\leq A \big(b_j(s) - b_k(s) \big) + A b_k(s) + A b_j(s) = 2 A b_j(s).$$

By hypothesis, this upper bound can be made uniformly small, for s in J, by taking j sufficiently large. This proves the theorem.

Here we have a situation which does not arise in the case of power series. For while a power series converges absolutely at every interior point of its interval of convergence, the Dirichlet series for $L(s)$ converges for $s > 0$, but converges absolutely only for $s > 1$, since the series

$$\sum_{n=1}^{\infty} \left| \frac{\chi(n)}{n} \right| = \sum_{n=0}^{\infty} \frac{1}{2n+1}$$

diverges.

On account of the complete multiplicativity of χ, we have

$$\left(1 - \frac{\chi(p)}{p^s} \right)^{-1} = 1 + \frac{\chi(p)}{p^s} + \frac{(\chi(p))^2}{p^{2s}} + \cdots$$

$$= 1 + \frac{\chi(p)}{p^s} + \frac{\chi(p^2)}{p^{2s}} + \cdots.$$

Using this idea, the proof of Theorem 6–1 can easily be modified to yield

THEOREM 6–3. *If f is completely multiplicative, and the series*

$$\sum_{n=1}^{\infty} \frac{f(n)}{n^s}$$

converges absolutely for $s > s_0$, then

$$\sum_{n=1}^{\infty} \frac{f(n)}{n^s} = \prod_p \left(1 - \frac{f(p)}{p^s}\right)^{-1}$$

for $s > s_0$.

COROLLARY. *For $s > 1$,*

$$L(s) = \prod_p \left(1 - \frac{\chi(p)}{p^s}\right)^{-1}.$$

We are finally in a position to prove Dirichlet's theorem for primes of the form $4k + 1$. Let s be greater than 1. We have

$$\zeta(s) = \prod_p (1 - p^{-s})^{-1} = (1 - 2^{-s})^{-1} \prod_q (1 - q^{-s})^{-1} \prod_r (1 - r^{-s})^{-1},$$

and, from the corollary to Theorem 6–3,

$$L(s) = \prod_q (1 - q^{-s})^{-1} \prod_r (1 + r^{-s})^{-1}.$$

Hence

$$\zeta(s)L(s) = (1 - 2^{-s})^{-1} \prod_q (1 - q^{-s})^{-2} \prod_r (1 - r^{-2s})^{-1}. \qquad (5)$$

Now, for $s \geq 1$,

$$L(s) = 1 - \frac{1}{3^s} + \frac{1}{5^s} - \frac{1}{7^s} + \cdots > 1 - \frac{1}{3^s} \geq \frac{2}{3},$$

and so

$$\lim_{s \to 1^+} \zeta(s)L(s) = \infty,$$

by (4). If there were only finitely many primes q, the expression on the right side of (6) would remain bounded as $s \to 1^+$, since for $s \geq 1$,

$$\prod_r (1 - r^{-2s})^{-1} \leq \prod_r (1 - r^{-2})^{-1} \leq \prod_p (1 - p^{-2})^{-1} = \zeta(2).$$

This contradiction shows not only that there are infinitely many primes q, but also that they occur sufficiently frequently that

$$\lim_{s \to 1^+} \prod_q (1 - q^{-s})^{-1} = \infty.$$

The proof which has just been given contains most of the essential features of the general proof. The major formal difference which will arise in the general case is that we shall have to consider a number of functions like χ above, and each will have an associated Dirichlet series, some aspects of whose behavior must be investigated. The most difficult part of the proof lies in showing that these series do not vanish at $s = 1$, a point which caused no trouble in the present case.

PROBLEM

Let $b_n - 1$ or 0, according as the equation $n - x^2 + y^2$ has or does not have a solution in integers x, y. It is known* that $b_n = 1$ if and only if every prime $r \equiv -1 \pmod 4$ which divides n occurs to an even power in the canonical factorization of n. Show that the series

$$\sum_{n=1}^{\infty} \frac{b_n}{n^s}$$

converges for $s > 1$, and diverges for $s \leq 1$. [*Hint:* Establish a relation among $\zeta(s)$, $L(s)$, and the square of the given series.]

6–2 Characters. We recall that the elements of a reduced residue system (mod k) form an abelian group under multiplication (mod k), which we designate by $M(k)$. The number of elements of $M(k)$, called its *order*, is $\varphi(k)$; hereafter we shall use h as an abbreviation for $\varphi(k)$.

One of the fundamental theorems on finite abelian groups is that every such group has a *basis:* if it is a multiplicative group, this means that there is a set of elements $A_1, \ldots, A_r$ such that every element of the group can be written uniquely in the form

$$A_1{}^{x_1} \cdots A_r{}^{x_r},$$

where each x_i is one of the integers $0, 1, \ldots,$ ord $A_i - 1$, and ord A_i is the order of the cyclic subgroup generated by A_i. Moreover, the product of all the numbers ord A_i is the order of the group. The following theorem, for which we give a proof based on the theory of primitive roots† is a special case.

THEOREM 6–4.   (a) *Let* $k = p_1{}^{\alpha_1} \cdots p_r{}^{\alpha_r}$, *where* $p_i \neq p_j$ *and each of the prime powers* $p_i{}^{\alpha_i}$ *has a primitive root, say* g_i. *Then the numbers*

---

* See, for example, Volume I, Theorem 7–3.
† See, for example, Volume I, Chapter 4.

$A_1, \ldots, A_r$ *form a basis for* $M(k)$ *if, for each* i,

$$A_i \equiv \begin{cases} g_i \pmod{p_i^{\alpha_i}} \\ 1 \pmod{p_j^{\alpha_j}} \end{cases} \quad \text{if } j \neq i, \quad 1 \leq j \leq r.$$

(b) *Let* $k = 2^\alpha p_2^{\alpha_2} \cdots p_r^{\alpha_r}$, *where* $\alpha \geq 3$, *and let* g_i *be a primitive root of* $p_i^{\alpha_i}$ *for* $2 \leq i \leq r$. *Then the numbers* $A_0, A_1, \ldots, A_r$ *constitute a basis for* $M(k)$, *where*

$$A_0 \equiv \begin{cases} -1 \pmod{2^\alpha} \\ 1 \pmod{p_i^{\alpha_i}} \end{cases} \quad \text{for } 2 \leq i \leq r,$$

$$A_1 \equiv \begin{cases} 5 \pmod{2^\alpha} \\ 1 \pmod{p_i^{\alpha_i}} \end{cases} \quad \text{for } 2 \leq i \leq r,$$

and for $2 \leq i \leq r$,

$$A_i \equiv \begin{cases} g_i \pmod{p_i^{\alpha_i}} \\ 1 \pmod{p_j^{\alpha_j}} \end{cases} \quad \text{for } j \neq i, \quad 1 \leq j \leq r.$$

Proof: Let a be relatively prime to k. Then it is also prime to every divisor of k, so that there are unique elements $a_1, \ldots, a_r$ of $M(p_1^{\alpha_1}), \ldots, M(p_r^{\alpha_r})$, respectively, such that

$$a \equiv a_1 \pmod{p_1^{\alpha_1}},$$
$$\vdots \tag{6}$$
$$a \equiv a_r \pmod{p_r^{\alpha_r}}.$$

Conversely, for any choice of $a_1, \ldots, a_r$ in $M(p_1^{\alpha_1}), \ldots, M(p_r^{\alpha_r})$, respectively, the system (6) has a solution a which is unique modulo k, by the Chinese Remainder Theorem, and a is prime to k. Moreover, if a is the solution of (6), and if, for $1 \leq i \leq r$, b_i is the solution of the system

$$b_i \equiv \begin{cases} a_i \pmod{p_i^{\alpha_i}} \\ 1 \pmod{p_j^{\alpha_j}} \end{cases} \quad \text{for } j \neq i, \quad 1 \leq j \leq r, \tag{7}$$

then

$$b_1 \cdots b_r \equiv 1 \cdots 1 \cdot a_i \cdot 1 \cdots 1 \equiv a_i \pmod{p_i^{\alpha_i}}, \quad \text{for } 1 \leq i \leq r,$$

so that

$$a \equiv b_1 \cdots b_r \pmod{k}. \tag{8}$$

(Thus, in the language of group theory, $M(k)$ is the direct product of $M(p_1^{\alpha_1}), \ldots, M(p_r^{\alpha_r})$.)

Now if $p_i{}^{\alpha_i}$ has a primitive root g_i, and

$$A_i \equiv \begin{cases} g_i \pmod{p_i{}^{\alpha_i}} \\ 1 \pmod{p_j{}^{\alpha_j}} \end{cases} \quad \text{for } j \neq i, \quad 1 \leq j \leq r,$$

then, since

$$a_i \equiv g_i{}^{\text{ind } a}{}_i \pmod{p_i{}^{\alpha_i}},$$

we have that

$$b_i \equiv A_i{}^{\text{ind } a_i} \pmod{p_j{}^{\alpha_j}}, \quad \text{for } 1 \leq j \leq r,$$

and hence

$$b_i \equiv A_i{}^{\text{ind } a_i} \pmod{k}.$$

Thus by the congruence (8), if all $p_i{}^{\alpha_i}$ have primitive roots,

$$a \equiv A_1{}^{\text{ind } a_1} \cdots A_r{}^{\text{ind } a_r} \pmod{k},$$

and this representation is unique if each index is given its smallest non-negative value, so that $0 \leq \text{ind } a_i < \varphi(p_i{}^{\alpha_i})$.

On the other hand, if $p_1{}^{\alpha_1} = 2^\alpha$ with $\alpha \geq 3$, then -1 and 5 constitute a basis for $M(p_1{}^{\alpha_1})$. For* 5 is a primitive λ-root of 2^α, so that the $2^{\alpha-2}$ numbers

$$5, 5^2, \ldots, 5^{2^{\alpha-2}}$$

are distinct $\pmod{2^\alpha}$; since they are all congruent to $1 \pmod{4}$, and since there are exactly $2^{\alpha-2}$ numbers in a reduced residue system $\pmod{2^\alpha}$ which are congruent to $1 \pmod{4}$, these must be the numbers. Likewise, their negatives are all the numbers congruent to $-1 \pmod{4}$ in a reduced residue system $\pmod{2^\alpha}$.† Hence, if a is in $M(2^\alpha)$, then, for some choice of x_0 and x_1,

$$a \equiv (-1)^{x_0} 5^{x_1} \pmod{2^\alpha}.$$

Thus if $A_0, \ldots, A_r$ are defined as in part (b) of the theorem, we have

$$a \equiv A_0{}^{x_0} A_1{}^{x_1} A_2{}^{\text{ind } a_2} \cdots A_r{}^{\text{ind } a_r} \pmod{k},$$

and the representation is again unique if we require that

$$0 \leq x_0 < \text{ord } A_0 = 2,$$

$$0 \leq x_1 < \text{ord } A_1 = 2^{\alpha-2},$$

$$0 \leq \text{ind } a_i < \text{ord } A_i = \varphi(p_i{}^{\alpha_i}).$$

---

* See Volume I, Theorem 4–9.

† A similar argument is used in Volume I, in the proof of Theorem 5–1.

Notice that in the two cases we have

$$\operatorname{ord} A_1 \cdots \operatorname{ord} A_r = \varphi(p_1{}^{\alpha_1}) \cdots \varphi(p_r{}^{\alpha_r}) = h,$$

$$\operatorname{ord} A_0 \cdot \operatorname{ord} A_1 \cdots \operatorname{ord} A_r = 2 \cdot 2^{\alpha-2} \cdot \varphi(p_2{}^{\alpha_2}) \cdots \varphi(p_r{}^{\alpha_r}) = h.$$

To obviate the distinction between cases (a) and (b), we rename the basis elements $B_1, \ldots, B_m$, and put ord $B_i = h_i$ for $i = 1, \ldots, m$.

A complex-valued function χ, defined over the group $M(k)$ (more generally, over any finite abelian group), is called a *character* (mod k) (or a *character of the group*) if it is completely multiplicative and not identically zero, that is, if

$$\chi(ab) = \chi(a)\chi(b), \qquad \text{for } a \text{ and } b \text{ in } M(k),$$

$$\chi(a) \neq 0, \qquad \text{for some } a \text{ in } M(k).$$

Since in the group $M(k)$ we identify integers which are congruent (mod k), we have

$$\chi(a) = \chi(a'), \qquad \text{if } a \equiv a' \pmod{k} \quad \text{and} \quad (a, k) = (a', k) = 1,$$

so that one could also think of characters as being defined over the residue classes themselves. Notice that necessarily $\chi(1) = 1$, since for any a for which $\chi(a) \neq 0$, we have $\chi(a) = \chi(a \cdot 1) = \chi(a)\chi(1)$. Moreover, if a is in $M(k)$ and ord $a = t$, then

$$\big(\chi(a)\big)^t = \chi(a^t) = \chi(1) = 1.$$

Since $t|h$, it follows that every value of every character is an hth root of unity.

On account of its complete multiplicativity, any character is totally determined when its value is specified for each basis element B_j. Thus the characters are contained in the set of all completely multiplicative functions over $M(k)$ for which

$$\chi(B_j) = e^{2\pi i \beta_j / h_j}, \qquad 0 \leq \beta_j < h_j, \tag{9}$$

for $j = 1, \ldots, m$. But conversely, every such function is obviously a character, and different choices of the β's lead to different characters. Thus there are h different characters, corresponding to the $h_1 \cdots h_m$ different m-tuples $(\beta_1, \ldots, \beta_m)$.

Two groups G and G', with elements $a, b, \ldots$ and $a', b', \ldots$, are said to be *isomorphic* if it is possible to find a pairing of elements of G with elements of G', such that each element of G corresponds to precisely one element of G', and conversely, and such that if $a \leftrightarrow a'$ and $b \leftrightarrow b'$, then $ab \leftrightarrow a'b'$. In this case the groups are abstractly identical, and any theorem concerning one group has an immediate analog for the other group. To construct such an isomorphism between two finite abelian groups, it suffices to find a one-to-one correspondence of basis elements such that corresponding elements have equal orders. For let the bases be $C_1, \ldots, C_s$ and $C_1', \ldots, C_s'$, so named that ord $C_i = $ ord C_i', for $i = 1, \ldots, s$. Then we can make a and a' correspond if

$$ a = C_1^{x_1} \cdots C_s^{x_s} \qquad \text{and} \qquad a' = C_1'^{x_1} \cdots C_s'^{x_s}, $$

$$ 0 \leq x_i < \text{ord } C_i. $$

For if also

$$ b = C_1^{y_1} \cdots C_s^{y_s} \qquad \text{and} \qquad b' = C_1'^{y_1} \cdots C_s'^{y_s}, $$

then

$$ ab = C_1^{x_1+y_1} \cdots C_s^{x_s+y_s}, \qquad a'b' = C_1'^{x_1+y_1} \cdots C_s'^{x_s+y_s}, $$

and

$$ (ab)' = C_1'^{x_1+y_1} \cdots C_s'^{x_s+y_s} = a'b'. $$

Moreover, this is a one-to-one correspondence, since the representations by basis elements are unique for the ranges $0 \leq x_i < $ ord $C_i = $ ord C_i', $1 \leq i \leq s$.

For the basis $B_1, \ldots, B_m$ of $M(k)$, define characters $\chi_1, \ldots, \chi_m$ as follows:

$$ \chi_\mu(B_j) = \begin{cases} e^{2\pi i / h_\mu} & \text{if } j = \mu, \\ 1 & \text{if } j \neq \mu, \quad 1 \leq j \leq m. \end{cases} \tag{10} $$

Then from the sentence containing equation (9), we see that every character can be represented uniquely in the form

$$ \chi = \chi_1^{\beta_1} \cdots \chi_m^{\beta_m}, \qquad 0 \leq \beta_i < h_i \quad \text{for } i = 1, \ldots, m, $$

since this gives $\chi(B_j)$ as in (9). (We say that two characters are equal if they have the same value for every element of the group, and define the product of two characters as the function whose values are the products of the component values; this function is also a character, by the sentence following (9).) Under multiplication, the characters form a group $X(k)$, having basis $\chi_1, \ldots, \chi_m$; since ord $\chi_i = $ ord B_i, the groups $X(k)$ and $M(k)$ are isomorphic. The unit element of $X(k)$ is the character χ_0, the *principal character*, such that $\chi_0(a) = 1$ for every a in $M(k)$.

We summarize the chief results obtained so far.

THEOREM 6–5.   *There are h distinct characters* (mod k), *and these form a group* $X(k)$ *which is isomorphic to* $M(k)$. *Every value* $\chi(a)$ *is an hth root of unity. The characters* $\chi_1, \ldots, \chi_m$ *defined in* (10) *form a basis for* $X(k)$.

We shall also need the following result.

THEOREM 6–6.   *If* χ *is in* $X(k)$, *then*

$$\sum_{a \in M(k)} \chi(a) = \begin{cases} h & \text{if } \chi = \chi_0, \\ 0 & \text{if } \chi \neq \chi_0, \end{cases}$$

while if a is in $M(k)$, *then*

$$\sum_{\chi \in X(k)} \chi(a) = \begin{cases} h & \text{if } a \equiv 1 \ (\text{mod } k), \\ 0 & \text{if } a \not\equiv 1 \ (\text{mod } k). \end{cases}$$

Proof: We have

$$\sum_{a \in M(k)} \chi_0(a) = \sum_{a \in M(k)} 1 = h.$$

If $\chi \neq \chi_0$, then for some $\bar{a}$ in $M(k)$, $\chi(\bar{a}) \neq 1$. For this $\bar{a}$,

$$\chi(\bar{a}) \sum_a \chi(a) = \sum_a \chi(a)\chi(\bar{a}) = \sum_a \chi(a\bar{a}),$$

and, as a runs over a reduced residue system, so does $a\bar{a}$, so that

$$\chi(\bar{a}) \sum_a \chi(a) = \sum_a \chi(a),$$

$$\sum_a \chi(a) = 0.$$

If $a \not\equiv 1 \ (\text{mod } k)$, and $a = B_1^{x_1} \cdots B_m^{x_m}$, then some $x_i \neq 0$. For this i, $\chi_i(a) \neq 1$, and

$$\chi_i(a) \sum_\chi \chi(a) = \sum_\chi \chi_i(a)\chi(a) = \sum \chi_i'(a),$$

where $\chi_i' = \chi_i \chi$. As χ runs over $X(k)$, so also does $\chi_i \chi = \chi_i'$, and

$$\chi_i(a) \sum_\chi \chi(a) = \sum_\chi \chi(a),$$

$$\sum_\chi \chi(a) = 0.$$

$\chi(a)$ has so far been defined only for arguments relatively prime to k. For simplicity in later formulas, we define

$$\chi(a) = 0, \qquad \text{if } (a, k) > 1.$$

This does not affect the validity of Theorem 6–6.

The duality of the relations of Theorem 6–6 is a reflection of the isomorphism of $X(k)$ and $M(k)$. In a sense, the reason for the importance of characters in the investigation of primes in progressions lies in the second relation, since it singles out the elements of a particular residue class $(\bmod k)$, so that by use of the relation

$$\sum_{\substack{u \leq a < v \\ a \equiv 1 \,(\bmod k)}} g(a) = \frac{1}{h} \sum_{u \leq a < v} g(a) \sum_{\chi} \chi(a),$$

sums can be extended over an entire interval instead of a finite or infinite arithmetic progression $kt + 1$. Moreover, by a slight modification, any other residue class can be distinguished in the same way.

Theorem 6–7. *If $(a, k) = (b, k) = 1$, then*

$$\sum_{\chi \in X(k)} \frac{\chi(a)}{\chi(b)} = \begin{cases} h & \text{if } a \equiv b \,(\bmod k), \\ 0 & \text{otherwise.} \end{cases}$$

Proof: Choose c so that $bc \equiv 1 \,(\bmod k)$. Then

$$\sum_{\chi \in X(k)} \frac{\chi(a)}{\chi(b)} = \sum_{\chi \in X(k)} \chi(ac),$$

and, by Theorem 6–6, the last sum is h or 0 according as ac is or is not congruent to 1 $(\bmod k)$, that is, according as a is or is not congruent to b $(\bmod k)$.

It should be noticed that the function

$$\chi(n) = \begin{cases} (-1)^{\frac{1}{2}(n-1)} & \text{for } n \text{ odd}, \\ 0 & \text{for } n \text{ even}, \end{cases}$$

introduced in Section 6–1 is a character modulo 4. It and the principal character

$$\chi_0(n) = \begin{cases} 1 & \text{for } n \text{ odd}, \\ 0 & \text{for } n \text{ even}, \end{cases}$$

constitute the group $X(4)$ of order $\varphi(4) = 2$. The correspondence

$$\chi_0 \leftrightarrow 1, \qquad \chi \leftrightarrow 3,$$

describes the isomorphism between $X(4)$ and $M(4)$; each is the cyclic group of two elements.

6–3 The L-functions. For each character χ, we define a function $L(s, \chi)$ for $s > 1$ by the equation

$$L(s, \chi) = \sum_{n=1}^{\infty} \frac{\chi(n)}{n^s},$$

or equivalently (according to Theorem 6–3) by the equation

$$L(s, \chi) = \prod_p \left(1 - \frac{\chi(p)}{p^s}\right)^{-1}. \tag{11}$$

In particular,

$$L(s, \chi_0) = \prod_{p \nmid k} (1 - p^{-s})^{-1} = \prod_{p \mid k} (1 - p^{-s}) \zeta(s),$$

so that, by equation (2),

$$L(s, \chi_0) = \prod_{p \mid k} (1 - p^{-s}) \left(\frac{s}{s-1} - s \int_1^{\infty} \frac{(t)}{t^{s+1}} \, dt\right). \tag{12}$$

This latter representation for $L(s, \chi_0)$ is consistent with the series definition for $s > 1$, and may be taken as the definition for $0 < s < 1$.

For the proof of Dirichlet's theorem, it is necessary to know some of the properties of these L-functions. All the relevant properties can be proved by elementary arguments, but the proofs frequently can be simplified considerably if use is made of the theory of functions of a complex variable. In these cases alternative proofs will be given.

THEOREM 6–8. $L(s, \chi_0)$ *is continuous for $s > 1$, and*

$$\lim_{s \to 1^+} (s - 1)L(s, \chi_0) = \frac{h}{k}.$$

Proof: For $s \geq s_0 > 1$,

$$L(s, \chi_0) = \sum_{(n,k)=1} \frac{1}{n^s} \ll \sum_{n=1}^{\infty} \frac{1}{n^{s_0}} = \zeta(s_0),$$

so that the series for $L(s, \chi_0)$ converges uniformly in any interval to the right of $s = 1$. Since the separate terms are continuous, the sum is also continuous. Moreover, by (12),

$$\lim_{s \to 1^+} (s - 1)L(s, \chi_0) = \prod_{p \mid k} (1 - p^{-1}) = \frac{\varphi(k)}{k} = \frac{h}{k}.$$

For $\chi \neq \chi_0$, Theorem 6–6 shows that for arbitrary n_0,

$$\sum_{n=n_0}^{n_0+k} \chi(n) = 0,$$

so that by grouping the terms of

$$\sum_{n=u}^{v} \chi(n)$$

in blocks of k, with perhaps part of a block left over, we see that

$$\left| \sum_{n=u}^{v} \chi(n) \right| \leq h.$$

It follows from Theorem 6–2 that the Dirichlet series for $L(s, \chi)$ is convergent for $s > 0$. We need a slightly stronger result, which is proved in the following theorem.

THEOREM 6–9. *If* $\chi \neq \chi_0$, *then* $L(s, \chi)$ *has a continuous derivative (and is therefore itself continuous) for* $s > 0$.

Elementary proof: We use the standard theorem from analysis, that if the series resulting from termwise differentiation of a given series converges uniformly over an interval, then its sum is the derivative of the original series. The termwise derivative of

$$\sum_{n=1}^{\infty} \frac{\chi(n)}{n^s}$$

is

$$-\sum_{n=1}^{\infty} \frac{\chi(n) \log n}{n^s}, \tag{13}$$

and for $0 < s_0 \leq s \leq s_1$, the result follows from Theorem 6–2 by taking $a_n = \chi(n)$, $b_n(s) = n^{-s} \log n$. But s_0 may be arbitrarily small, and s_1 arbitrarily large, so that every $s > 0$ can be included in an interval in which $L(s, \chi)$ is continuously differentiable.

Alternative proof: Applying Theorem 6–2 and the fact that

$$\sum_{n=1}^{\infty} \frac{\chi(n)}{n^s} \ll \sum_{n=1}^{\infty} \frac{1}{n^\sigma},$$

where $\sigma = \operatorname{Re} s$, we see that the series for $L(s, \chi)$ is uniformly convergent for $\operatorname{Re} s \geq \sigma_0 > 0$. Since each term of the series is an analytic function of s, the sum is also analytic, and is therefore differentiable.

6–4 Nonelementary proof of Dirichlet's theorem. There is a proof of Dirichlet's theorem which is remarkably simple and illuminating, and which fails to be elementary only in the sense that logarithms of complex numbers are used. If the student who is not familiar with this extension will assume that the usual properties of logarithms of positive numbers (including the form of the Maclaurin expansion of $\log(1+x)$ for $|x| < 1$) carry over to logarithms of nonzero complex numbers, he will find this proof much more straightforward than the elementary proof given in the following section, where use is made of the relation

$$\frac{d}{dx}\log f(x) = \frac{f'(x)}{f(x)}$$

to avoid logarithms entirely.

For $s > 1$, $|\chi(p)/p^s| < 1$, so that for such s we can describe a branch of the function $\log(1 - \chi(p)/p^s)$ by the equation

$$\log\left(1 - \frac{\chi(p)}{p^s}\right) = -\sum_{m=1}^{\infty} \frac{1}{m}\left(\frac{\chi(p)}{p^s}\right)^m = -\sum_{m=1}^{\infty} \frac{\chi(p^m)}{mp^{ms}}.$$

By (11), this induces the choice

$$\log L(s,\chi) = \sum_p \sum_{m=1}^{\infty} \frac{\chi(p^m)}{mp^{ms}}, \qquad \text{for } s > 1. \tag{14}$$

THEOREM 6–10. *For each χ, the function*

$$F(s,\chi) = \log L(s,\chi) - \sum_p \frac{\chi(p)}{p^s} \tag{15}$$

is bounded in absolute value for $s \geqq 1$.

Proof: We rewrite (14) in the form

$$\log L(s,\chi) = \sum_p \frac{\chi(p)}{p^s} + \sum_p \sum_{m=2}^{\infty} \frac{\chi(p^m)}{mp^{ms}}.$$

Here,

$$\sum_p \sum_{m=2}^{\infty} \frac{\chi(p^m)}{mp^{ms}} \ll \sum_p \sum_{m=2}^{\infty} \frac{1}{2p^{ms}} = \frac{1}{2}\sum_p \frac{1}{p^{2s}(1-p^{-s})}$$

$$\ll \frac{1}{2}\sum_p \frac{1}{p^{2s}(1-2^{-s})} = \frac{(1-2^{-s})^{-1}}{2}\sum_p \frac{1}{p^{2s}}$$

$$\ll \frac{(1-2^{-s})^{-1}}{2}\sum_{n=1}^{\infty} \frac{1}{n^{2s}} = \frac{(1-2^{-s})^{-1}}{2}\zeta(2s),$$

and since $\zeta(2s)$ is bounded for $2s \geq 1 + \epsilon$, the theorem follows.

We can now complete the proof of Dirichlet's theorem, except for one gap which will be considered later.

THEOREM 6–11 (*Dirichlet's theorem*).　If $(k, l) = 1$, *then there are infinitely many primes of the form* $kt + l$.

Proof: Multiply equation (15) by $1/\chi(l)$ and sum over all χ in $X(k)$. This gives

$$\sum_\chi \frac{\log L(s, \chi)}{\chi(l)} = \sum_\chi \sum_p \frac{\chi(p)}{\chi(l)p^s} + \sum_\chi \frac{F'(s, \chi)}{\chi(l)}$$

$$= \sum_p \frac{1}{p^s} \sum_\chi \frac{\chi(p)}{\chi(l)} + \sum_\chi \frac{F(s, \chi)}{\chi(l)},$$

and, by Theorem 6–7,

$$\sum_\chi \frac{\log L(s, \chi)}{\chi(l)} = h \sum_{p \equiv l \,(\mathrm{mod}\, k)} \frac{1}{p^s} + \sum_\chi \frac{F(s, \chi)}{\chi(l)}. \qquad (16)$$

Let $s \to 1^+$ in (16). The second term on the right remains bounded, by Theorem 6–10. We know that

$$\lim_{s \to 1^+} L(s, \chi_0) = \infty,$$

so that

$$\lim_{s \to 1^+} \frac{\log L(s, \chi_0)}{\chi_0(l)} = \infty.$$

Suppose for the moment that it had been shown that the remaining functions $L(s, \chi)$ (which we know to be continuous at $s = 1$) have nonzero values $L(1, \chi)$ at $s = 1$. It would follow that

$$\left| \lim_{s \to 1^+} \sum_{\chi \neq \chi_0} \frac{\log L(s, \chi)}{\chi(l)} \right| < \infty,$$

and (16) would then imply that

$$\lim_{s \to 1^+} \sum_{p \equiv l \,(\mathrm{mod}\, k)} \frac{1}{p^s} = \infty,$$

an equation which is possible only if the sum has infinitely many terms. Thus when we show that $L(1, \chi) \neq 0$ if $\chi \neq \chi_0$, we shall have proved not only Dirichlet's theorem but the stronger result that the series

$$\sum_{p \equiv l \,(\mathrm{mod}\, k)} \frac{1}{p}$$

diverges.

6–5 Elementary proof of Dirichlet's theorem. It is possible to avoid the complex logarithm $\log L(s, \chi)$ by using its derivative instead:

$$\frac{d}{dx} \log L(s, \chi) = \frac{L'(s, \chi)}{L(s, \chi)} = \frac{L'}{L}(s, \chi).$$

If we could use the relation (14), we could immediately deduce that

$$\frac{L'(s, \chi)}{L(s, \chi)} = -\sum_p \sum_{m=1}^{\infty} \frac{\chi(p^m) \log p}{p^{ms}};$$

since we cannot, we arrive at the same result by the rather more awkward method of dividing $L'(s, \chi)$ by $L(s, \chi)$. In the process, we shall have occasion to use some properties of the Möbius μ-function, which is defined by the following relations:

$$\mu(n) = \begin{cases} 1 & \text{if } n = 1, \\ 0 & \text{if } n \text{ is divisible by a square larger than 1,} \\ (-1)^r & \text{if } n \text{ is the product of } r \text{ distinct primes.} \end{cases}$$

Alternatively, μ is the multiplicative function (that is, $\mu(mn) = \mu(m)\mu(n)$ whenever $(m, n) = 1$) such that

$$\mu(n) = \begin{cases} 1 & \text{if } n = 1, \\ -1 & \text{if } n = p, \text{ a prime,} \\ 0 & \text{if } n = p^\alpha, \quad \alpha > 1. \end{cases}$$

The properties we shall need are these.*

(a) $\sum_{d|n} \mu(d) = \begin{cases} 1 & \text{if } n = 1, \\ 0 & \text{if } n > 1. \end{cases}$

(b) If f is any number-theoretic function and

$$F(n) = \sum_{d|n} f(d),$$

then

$$f(n) = \sum_{d|n} \mu(d) F\left(\frac{n}{d}\right).$$

THEOREM 6–12.   *If f is a completely multiplicative function, and the series*

$$\sum_{n=1}^{\infty} \frac{f(n)}{n^s}$$

---

* See, for example, Volume I, Theorems 6–5 and 6–6.

converges absolutely for $s > s_0$, then

$$\left(\sum_{n=1}^{\infty} \frac{f(n)}{n^s} \right)^{-1} = \sum_{n=1}^{\infty} \frac{f(n)\mu(n)}{n^s}$$

for $s > s_0$.

Proof: We have

$$\sum_{m=1}^{\infty} \frac{f(m)}{m^s} \sum_{n=1}^{\infty} \frac{f(n)\mu(n)}{n^s} = \sum_{m,n=1}^{\infty} \frac{f(mn)\mu(n)}{(mn)^s}$$

$$= \sum_{j=1}^{\infty} \frac{\sum_{d|j} \mu(d)}{j^s} f(j) = 1.$$

THEOREM 6–13. *For each χ, the relation*

$$\frac{L'}{L}(s, \chi) = - \sum_{n=1}^{\infty} \frac{\chi(n)\Lambda(n)}{n^s} \tag{17}$$

holds for $s > 1$, where

$$\Lambda(n) = \begin{cases} \log p & \text{if } n = p^\alpha \text{ for some } \alpha > 0 \text{ and prime } p, \\ 0 & \text{otherwise.} \end{cases}$$

Proof: By the preceding theorem and the expression (13) for $L'(s, \chi)$, we have, for $s > 1$,

$$\frac{L'}{L}(s, \chi) = - \sum_{n=1}^{\infty} \frac{\chi(m) \log m}{m^s} \cdot \sum_{j=1}^{\infty} \frac{\chi(j)\mu(j)}{j^s}$$

$$= - \sum_{m,j=1}^{\infty} \frac{\chi(mj)\mu(j) \log m}{(mj)^s}$$

$$= - \sum_{n=1}^{\infty} \frac{\chi(n) \sum_{d|n} \mu(d) \log \dfrac{n}{d}}{n^s}.$$

But from the obvious relation

$$\log n = \sum_{d|n} \Lambda(d)$$

and the Möbius inversion formula quoted above, we have

$$\Lambda(n) = \sum_{d|n} \mu(d) \log \frac{n}{d},$$

and the theorem follows.

THEOREM 6-14. *For each χ, the function*

$$G(s, \chi) = \frac{L'}{L}(s, \chi) + \sum_p \chi(p) \frac{\log p}{p^s} \tag{18}$$

is bounded in absolute value for $s \geq 1$.

Proof: Equation (17) may be rewritten in the form

$$\frac{L'}{L}(s, \chi) = -\sum_p \frac{\chi(p) \log p}{p^s} - \sum_p \sum_{m=2}^{\infty} \frac{\chi(p^m) \log p}{p^{ms}},$$

and

$$\sum_p \sum_{m=2}^{\infty} \frac{\chi(p^m) \log p}{p^{ms}} \ll \sum_p \sum_{m=2}^{\infty} \frac{\log p}{p^{ms}} = \sum_p \frac{\log p}{p^{2s}(1 - p^{-s})}$$

$$\ll \sum_p \frac{\log p}{p^{2s}(1 - 2^{-s})},$$

and the last series clearly converges for $s > \frac{1}{2}$.

We can now complete the proof of Dirichlet's theorem in much the same way as before. Multiplying both sides of (18) by $1/\chi(l)$, and summing over all χ in $X(k)$, we obtain

$$\sum_{\chi} \frac{1}{\chi(l)} \frac{L'}{L}(s, \chi) = -\sum_{\chi} \sum_p \frac{\chi(p)}{\chi(l)} \frac{\log p}{p^s} + \sum_{\chi} \frac{1}{\chi(l)} G(s, \chi)$$

$$= -h \sum_{p \equiv l \,(\mathrm{mod}\, k)} \frac{\log p}{p^s} + \sum_{\chi} \frac{1}{\chi(l)} G(s, \chi).$$

Now let $s \to 1^+$. The second term on the right remains bounded. Assuming again that $L(1, \chi) \neq 0$ for $\chi \neq \chi_0$, the quantity $1/L(s, \chi)$ is also bounded for s sufficiently close to 1, since L is continuous at $s = 1$. For $\chi \neq \chi_0$, $L'(s, \chi)$ remains bounded, by Theorem 6-9. On the other hand,

$$\left| \frac{L'}{L}(s, \chi_0) \right| = \sum_{(n,k)=1} \frac{\Lambda(n)}{n^s} = \sum_{p \nmid k} \frac{\log p}{p^s} > \sum_{p \nmid k} \frac{1}{p^s}$$

$$= \log L(s, \chi_0) + F(s, \chi_0),$$

by Theorem 6-10, and the quantity $\log L(s, \chi_0) + F(s, \chi_0)$ increases without bound as $s \to 1^+$. It follows that

$$\lim_{s \to 1^+} \sum_{p \equiv l \,(\mathrm{mod}\, k)} \frac{\log p}{p^s} = \infty,$$

and the theorem is again proved, except for the verification of the fact that $L(1, \chi) \neq 0$ for $\chi \neq \chi_0$.

6–6 Proof that $L(1, \chi) \neq 0$

THEOREM 6–15. *If χ assumes a nonreal value for some n, then $L(1, \chi) \neq 0$.*

Proof: Let χ be such a character, and let $\bar{\chi}$ be the function whose value for each a is the complex conjugate of that of χ. Clearly $\bar{\chi}$ is also a character, and $\bar{\chi} \neq \chi$. But if $L(1, \chi) = 0$, then also

$$L(1, \bar{\chi}) = \overline{L(1, \chi)} = 0,$$

so at least two L-functions must vanish in this case. Since $L(s, \chi)$ is differentiable at $s = 1$, the quantities

$$L'(1, \chi) = \lim_{s \to 1} \frac{L(s, \chi)}{s - 1} \quad \text{and} \quad L'(1, \bar{\chi}) = \lim_{s \to 1} \frac{L(s, \bar{\chi})}{s - 1}$$

exist, so that there is a number A such that

$$\lim_{s \to 1^+} \frac{\prod\limits_{\chi \neq \chi_0} L(s, \chi)}{(s - 1)^2} = A.$$

Since

$$\lim_{s \to 1^+} (s - 1)L(s, \chi_0) = \frac{h}{k},$$

we deduce that

$$\lim_{s \to 1^+} \prod_{\chi} L(s, \chi) = \lim_{s \to 1^+} \left\{ (s - 1)\big((s - 1)L(s, \chi_0)\big) \frac{\prod\limits_{\chi \neq \chi_0} L(s, \chi)}{(s - 1)^2} \right\}$$

$$= 0 \cdot \frac{h}{k} \cdot A = 0.$$

But by (14),

$$\sum_{\chi} \log L(s, \chi) = \sum_{\chi} \sum_{p} \sum_{m=1}^{\infty} \frac{\chi(p^m)}{mp^{ms}}$$

$$= \sum_{p} \sum_{m=1}^{\infty} \frac{\sum\limits_{\chi} \chi(p^m)}{mp^{ms}}$$

$$= h \sum_{\substack{p,m \\ p^m \equiv 1 \,(\text{mod } k)}} \frac{1}{mp^{ms}} > 0$$

for $s > 1$, so that

$$\lim_{s \to 1^+} \prod_\chi L(s, \chi) > e^0 = 1.$$

This contradiction establishes the theorem.

It would not be easy to avoid the use of the complex logarithm in this proof, since the Dirichlet series for $\prod_\chi L(s, \chi)$ has very complicated coefficients. To obtain an elementary proof, it is simpler to use a different combination of L-functions. Unfortunately, the choice we make can hardly be motivated by an elementary argument, but must remain a *deus ex machina* until Section 7–3. It is the left side of the inequality

$$L^3(s, \chi_0)|L(s, \chi)|^4|L(s, \chi^2)|^2 \geq 1; \tag{19}$$

this inequality we now show to be valid for $s > 1$.

Note first that for $z = r (\cos \theta + i \sin \theta)$,

$$|1 - z|^2 = |1 - r \cos \theta - ir \sin \theta|^2 = 1 - 2r \cos \theta + r^2,$$

and that for arbitrary real θ,

$$2 \cos \theta + \cos 2\theta = 2 \cos \theta + 2 \cos^2 \theta - 1 = 2(\cos \theta + \tfrac{1}{2})^2 - \tfrac{3}{2} \geq -\tfrac{3}{2}.$$

Using the fact that the geometric mean of three positive numbers is at most equal to their arithmetic mean, we see that, if $p \nmid k$ and

$$\chi(p) = \cos \theta_p + i \sin \theta_p,$$

then

$$\left(\left| 1 - \frac{\chi(p)}{p^s} \right|^2 \right)^2 \left| 1 - \frac{\chi^2(p)}{p^s} \right|^2$$

$$= (1 - 2p^{-s} \cos \theta_p + p^{-2s})^2 (1 - 2p^{-s} \cos 2\theta_p + p^{-2s})$$

$$\leq (1 - \tfrac{2}{3} p^{-s}(2 \cos \theta_p + \cos 2\theta_p) + p^{-2s})^3$$

$$\leq (1 + p^{-s} + p^{-2s})^3 \leq \left(\frac{1}{1 - p^{-s}} \right)^3,$$

or

$$(1 - \chi_0(p)p^{-s})^3|1 - \chi(p)p^{-s}|^4|1 - \chi^2(p)p^{-s}| \leq 1.$$

This inequality also holds if $p|k$, and, multiplying over all p, we obtain (19).

It is now simple to prove that $L(1, \chi) \neq 0$ if $\chi^2 \neq \chi_0$, that is, if χ is nonreal. Supposing the opposite, and using the fact that $L'(s, \chi)$ is continuous at $s = 1$, we have that for $1 \leq s \leq s_1$,

$$|L(s, \chi)| = |L(s, \chi) - L(1, \chi)| = \left| \int_1^s L'(u, \chi)\, du \right| \leq A_1 (s - 1),$$

where
$$A_1 = \max_{1 \leq s \leq s_1} |L'(s, \chi)|.$$

But now (19) can be recast in the form

$$(s - 1)((s - 1)L(s, \chi_0))^3 \left| \frac{L(s, \chi)}{s - 1} \right|^4 |L(s, \chi^2)|^2 \geq 1,$$

in which the first factor tends to zero and the others remain bounded, as $s \to 1^+$. This inequality is false for some $s > 1$, and the contradiction shows that $L(1, \chi) \neq 0$.

No device of this sort has been found for the case that $\chi(n)$ is real for all n. Showing that $L(1, \chi) \neq 0$ for a real character is the most difficult point in the entire proof. Dirichlet effected it by showing that $L(1, \chi)$ is a factor in the class number of a certain quadratic field. This and other algebraic proofs require a considerable amount of background; we shall content ourselves with an elementary and a function-theoretic proof. We first sketch the idea.

If $s > 1$, then

$$\zeta(s)L(s, \chi) = \sum_{m=1}^{\infty} \frac{1}{m^s} \sum_{n=1}^{\infty} \frac{\chi(n)}{n^s} = \sum_{m,n=1}^{\infty} \frac{\chi(n)}{(mn)^s} = \sum_{t=1}^{\infty} \frac{\sum_{n|t} \chi(n)}{t^s},$$

so that if we put

$$f(n) = \sum_{d|n} \chi(d), \tag{20}$$

then
$$\zeta(s)L(s, \chi) = \sum_{n=1}^{\infty} \frac{f(n)}{n^s} \tag{21}$$

for $s > 1$.

By Theorem 6–17, below,

$$\sum_{n=1}^{\infty} \frac{f(n)}{n^s} \geq \sum_{m=1}^{\infty} \frac{1}{(m^2)^s} = \zeta(2s),$$

so that even if the series $\sum f(n)n^{-s}$ converges to the right of $s = \frac{1}{2}$, it is certainly not bounded near $s = \frac{1}{2}$. In the analytic proof, we show that (21) is correct for $s \geq \frac{1}{2}$ if $L(1, \chi) = 0$, and obtain the contradiction

$$\lim_{s \to \frac{1}{2}^+} L(s, \chi)\zeta(s) = L(\tfrac{1}{2}, \chi)\zeta(\tfrac{1}{2}) = \infty.$$

In the elementary proof, questions of convergence are avoided by considering partial sums for $s = \frac{1}{2}$ rather than the full series for s near $\frac{1}{2}$. It will be shown that

$$\sum_{n=1}^{x} \frac{f(n)}{\sqrt{n}} = 2\sqrt{x}\, L(1, \chi) + O(1),$$

and also that the sum on the left tends to infinity with x, so that the relation $L(1, \chi) = 0$ is impossible.

THEOREM 6–16.   *With f as in (20),*

$$f(n) \geq \begin{cases} 0 & \text{for all } n, \\ 1 & \text{for square } n. \end{cases}$$

Proof: Being the arithmetic sum function of a multiplicative function,* f is itself multiplicative,* so that

$$f(p_1^{\alpha_1} \cdots p_r^{\alpha_r}) = f(p_1^{\alpha_1}) \cdots f(p_r^{\alpha_r}).$$

Since χ is a real character, $\chi(p) = 0$ or ± 1 for each prime p, and

$$f(p^\alpha) = \sum_{\beta=0}^{\alpha} \chi(p^\beta)$$

$$= \sum_{\beta=0}^{\alpha} (\chi(p))^\beta = \begin{cases} 1 + 0 + \cdots + 0 & \text{if } \chi(p) = 0, \\ 1 + 1 + \cdots + 1 & \text{if } \chi(p) = 1, \\ 1 - 1 + \cdots + (-1)^\alpha & \text{if } \chi(p) = -1. \end{cases}$$

Hence

$$f(p^\alpha) = \begin{cases} 1 & \text{if } \chi(p) = 0, \\ \alpha + 1 & \text{if } \chi(p) = 1, \\ 1 & \text{if } \chi(p) = -1, \alpha \text{ even}, \\ 0 & \text{if } \chi(p) = -1, \alpha \text{ odd}, \end{cases}$$

and the theorem follows.

THEOREM 6–17.   *The relations*

$$\sum_{n=1}^{x} \chi(n) = O(1) \tag{22}$$

and

$$\sum_{n=x}^{\infty} \frac{\chi(n)}{n^s} = O\left(\frac{1}{x^s}\right) \tag{23}$$

hold as $x \to \infty$, for $s > 0$.

---

* Cf. Volume I, Theorem 6–3.

Proof: We have already noticed that if

$$S(x) = \sum_{n=1}^{x} \chi(n),$$

then $|S(x)| \leq h$, which implies (22). Using this, we have

$$\left| \sum_{n=x}^{\infty} \frac{\chi(n)}{n^s} \right| = \left| \sum_{n=x}^{\infty} \frac{S(n) - S(n-1)}{n^s} \right|$$

$$= \left| \sum_{n=x}^{\infty} S(n) \left(\frac{1}{n^s} - \frac{1}{(n+1)^s} \right) - \frac{S(x-1)}{x^s} \right|$$

$$\leq h \sum_{n=x}^{\infty} \left(\frac{1}{n^s} - \frac{1}{(n+1)^s} \right) + \frac{h}{x^s} = \frac{2h}{x^s}$$

which implies (23).

THEOREM 6–18. *There is a constant C such that*

$$\sum_{n=1}^{x} \frac{1}{\sqrt{n}} = 2\sqrt{x} + C + O\left(\frac{1}{\sqrt{x}} \right).$$

Proof: Put

$$t_n = 2\sqrt{n} - 2\sqrt{n-1} - \frac{1}{\sqrt{n}} = \int_{n-1}^{n} \frac{dx}{\sqrt{x}} - \frac{1}{\sqrt{n}},$$

so that

$$\sum_{n=2}^{x} t_n = 2\sqrt{x} - 2 - \sum_{n=2}^{x} \frac{1}{\sqrt{n}}.$$

Now t_n, being the area of the triangular region bounded by the curve $y = x^{-\frac{1}{2}}$ and the lines $x = n - 1$ and $y = n^{-\frac{1}{2}}$, is positive and smaller than $(n-1)^{-\frac{1}{2}} - n^{-\frac{1}{2}}$, so that the series

$$\sum_{n=1}^{\infty} t_n$$

converges, and

$$\sum_{n=x+1}^{\infty} t_n < \sum_{n=x+1}^{\infty} \left(\frac{1}{\sqrt{n-1}} - \frac{1}{\sqrt{n}} \right) = \frac{1}{\sqrt{x}}.$$

Hence

$$2\sqrt{x} - \sum_{n=1}^{x} \frac{1}{\sqrt{n}} = 1 + \sum_{n=2}^{\infty} t_n - \sum_{n=x+1}^{\infty} t_n = 1 + \sum_{n=2}^{\infty} t_n + O\left(\frac{1}{\sqrt{x}} \right).$$

This proves the theorem for integral x; its extension to real x is immediate.

THEOREM 6–19. *If $\chi \neq \chi_0$ is a real character, then $L(1, \chi) \neq 0$.*

Proof: Put

$$G(x) = \sum_{n=1}^{x} \frac{f(n)}{\sqrt{n}}.$$

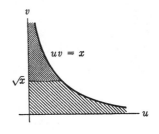

FIGURE 6–1

By Theorem 6–16,

$$G(x) \geq \sum_{m=1}^{\sqrt{x}} \frac{1}{\sqrt{m^2}} = \sum_{m=1}^{\sqrt{x}} \frac{1}{m},$$

so that $G(x) \to \infty$ with x.

On the other hand,

$$G(x) = \sum_{j=1}^{x} \frac{1}{\sqrt{j}} \sum_{d|j} \chi(d) = \sum_{uv \leq x} \frac{\chi(v)}{\sqrt{uv}}.$$

This sum, extended over the lattice points u, v for which $u \geq 1$, $v \geq 1$, $uv \leq x$, we split into two parts, as indicated in Fig. 6–1:

$$G(x) = \sum_{u=1}^{\sqrt{x}} \sum_{v=\sqrt{x}+1}^{x/u} \frac{\chi(v)}{\sqrt{uv}} + \sum_{v=1}^{\sqrt{x}} \sum_{u=1}^{x/v} \frac{\chi(v)}{\sqrt{uv}}$$

$$= \sum_{u=1}^{\sqrt{x}} \frac{1}{\sqrt{u}} \sum_{v=\sqrt{x}+1}^{x/u} \frac{\chi(v)}{\sqrt{v}} + \sum_{v=1}^{\sqrt{x}} \frac{\chi(v)}{\sqrt{v}} \sum_{u=1}^{x/v} \frac{1}{\sqrt{u}}$$

$$= \sum_{u=1}^{\sqrt{x}} \frac{1}{\sqrt{u}} O(x^{-\frac{1}{4}}) + \sum_{v=1}^{\sqrt{x}} \frac{\chi(v)}{\sqrt{v}} \left(2\sqrt{\frac{x}{v}} + C + O\left(\sqrt{\frac{v}{x}}\right) \right)$$

$$= O(x^{-\frac{1}{4}}) \cdot O(x^{\frac{1}{4}}) + 2\sqrt{x} \sum_{v=1}^{\sqrt{x}} \frac{\chi(v)}{v} + C \cdot O(1) + \frac{1}{\sqrt{x}} O(1)$$

$$= 2\sqrt{x} \left(L(1, \chi) + O\left(\frac{1}{\sqrt{x}}\right) \right) + O(1),$$

so that

$$\sum_{n=1}^{x} \frac{f(n)}{\sqrt{n}} = 2\sqrt{x}\, L(1, \chi) + O(1). \tag{24}$$

Thus, if $L(1, \chi)$ were zero, $G(x)$ would remain bounded as $x \to \infty$, which is not the case.

A rather more straightforward (function-theoretic) proof can be obtained by extending (21), which we know to be valid for $s > 1$, to the range $s > \frac{1}{2}$, under the assumption that $L(1, \chi) = 0$. By an argument quite similar to, but slightly simpler than that which yielded (24), it can be shown that if $L(1, \chi) = 0$, then

$$\sum_{n=1}^{x} f(n) = O(\sqrt{x}).$$

Theorem 6–2 implies that the series

$$\sum_{n=1}^{\infty} \frac{f(n)}{n^s} = \sum_{n=1}^{\infty} \frac{f(n)/n^{\frac{1}{2}}}{n^{s-\frac{1}{2}}} = \sum_{n=1}^{\infty} \frac{O(1)}{n^{s-\frac{1}{2}}}$$

converges for $s > \frac{1}{2}$. Now let σ_0 be a real number greater than $\frac{1}{2}$, and let s be a complex number with $\operatorname{Re} s = \sigma > \sigma_0$. Then for $v > u > 1$, we have

$$\sum_{n=u}^{v} \frac{f(n)}{n^s} = \sum_{n=u}^{v} \frac{f(n)/n^{\sigma_0}}{n^{s-\sigma_0}}$$

$$= \sum_{n=u}^{v} \frac{\displaystyle\sum_{m=1}^{n} \frac{f(m)}{m^{\sigma_0}} - \sum_{m=1}^{n-1} \frac{f(m)}{m^{\sigma_0}}}{n^{s-\sigma_0}}$$

$$= \sum_{n=u}^{v-1} \sum_{m=1}^{n} \frac{f(m)}{m^{\sigma_0}} \left(\frac{1}{n^{s-\sigma_0}} - \frac{1}{(n+1)^{s-\sigma_0}} \right)$$

$$+ v^{-(s-\sigma_0)} \sum_{m=1}^{v} \frac{f(m)}{m^{\sigma_0}} - u^{-(s-\sigma_0)} \sum_{m=1}^{u-1} \frac{f(m)}{m^{\sigma_0}},$$

so that

$$\left| \sum_{n=u}^{v} \frac{f(n)}{n^s} \right|$$

$$\leq A \sum_{n=u}^{v-1} \left| \frac{1}{n^{s-\sigma_0}} - \frac{1}{(n+1)^{s-\sigma_0}} \right| + A|v^{-(s-\sigma_0)}| + A|u^{-(s-\sigma_0)}|$$

$$= A \sum_{n=u}^{v-1} \left| (s - \sigma_0) \int_{n}^{n+1} \frac{dz}{z^{s-\sigma_0+1}} \right| + Av^{-(\sigma-\sigma_0)} + Au^{-(\sigma-\sigma_0)}$$

$$\leq A \sum_{n=u}^{v-1} \frac{1}{n^{\sigma-\sigma_0+1}} + Av^{-(\sigma-\sigma_0)} + Au^{-(\sigma-\sigma_0)},$$

where A is such that

$$\left| \sum_{n=1}^{N} \frac{f(n)}{n^{\sigma_0}} \right| < A, \quad \text{for all } N \geq 1.$$

It follows that the series

$$\sum_{n=1}^{\infty} \frac{f(n)}{n^s}$$

converges to an analytic function in the half-plane $\sigma > \frac{1}{2}$. Since it coincides with $L(s, \chi)\zeta(s)$ for s real and larger than 1, it represents an analytic continuation of $L(s, \chi)\zeta(s)$ for $\sigma > \frac{1}{2}$. But it is unbounded near $s = \frac{1}{2}$, while $L(s, \chi)\zeta(s)$ is not.

CHAPTER 7

THE PRIME NUMBER THEOREM

7-1 Introduction. It is shown in elementary number theory texts* that if

$$\lim_{x \to \infty} \frac{\pi(x)}{x/\log x}$$

exists, it must have the value 1, and that there are positive constants c and c' such that for $x \geq 2$,

$$c < \frac{\pi(x)}{x/\log x} < c'.$$

Neither of these results implies the other, of course; together they show that

$$0 < \liminf_{x \to \infty} \frac{\pi(x)}{x/\log x} \leq 1 \leq \limsup_{x \to \infty} \frac{\pi(x)}{x/\log x} < \infty.$$

(Here, as always, $\pi(x)$ denotes the number of primes less than or equal to x.) Both results were obtained by Chebyshev in 1851 and 1852 (in rather more precise form), but it was not until some forty-five years later that the final link was supplied by Hadamard and de la Vallée Poussin, who showed independently that the limit actually exists, and thus proved the Prime Number Theorem. Both proofs made essential use of the theory of functions of a complex variable, and despite much effort it seemed for many years impossible to give a proof entirely free of considerations as sophisticated as this theory. In 1948, however, P. Erdös and A. Selberg gave a completely elementary proof. More precisely, Selberg proved the fundamental relation

$$\sum_{p \leq x} \log^2 p + \sum_{pq \leq x} \log p \log q = 2x \log x + O(x),$$

and he and Erdös deduced the Prime Number Theorem from it.†

---

* See, for example, Volume I, Sections 6–6 and 6–7.

† Excellent expositions of this proof are given in T. Nagell, *Introduction to Number Theory* (New York: John Wiley & Sons, 1951) and in G. H. Hardy and E. M. Wright, *An Introduction to the Theory of Numbers* (3rd edition, New York: Oxford University Press, 1954).

We present a proof based on the behavior of the ζ-function for complex s. Throughout this chapter, familiarity with the contents of a standard course in the theory of functions of a complex variable is presupposed.

Before going into detail, we outline the proof. Our object is to get an estimate for

$$\pi(x) = \sum_{p \le x} 1 = \sum_{n=1}^{x} P(n),$$

where P is the characteristic function of the primes:

$$P(n) = \begin{cases} 1 & \text{if } n \text{ is prime,} \\ 0 & \text{otherwise.} \end{cases}$$

While P itself does not arise in a natural way, the function P^* such that

$$P^*(n) = \begin{cases} \dfrac{1}{m} & \text{if } n = p^m \text{ for some } m, p, \\ 0 & \text{otherwise,} \end{cases}$$

occurs in the Dirichlet series for $\log \zeta(s)$:

$$\log \zeta(s) = \sum_{m,p} \frac{1}{mp^{ms}} = \sum_{n=1}^{\infty} \frac{P^*(n)}{n^s}. \tag{1}$$

For fixed m, the number of mth powers of primes which do not exceed x is equal to the number of primes which do not exceed $\sqrt[m]{x}$, so that

$$\sum_{n=1}^{x} P^*(n) = \sum_{n=1}^{x} P(n) + \frac{1}{2} \sum_{n=1}^{\sqrt{x}} P(n) + \frac{1}{3} \sum_{n=1}^{\sqrt[3]{x}} P(n) + \cdots$$

$$= \pi(x) + \frac{\pi(\sqrt{x})}{2} + \frac{\pi(\sqrt[3]{x})}{3} + \cdots,$$

and since, for $m \ge 2$,

$$\pi(x^{1/m}) < cx^{1/m} \le c\sqrt{x} = o\left(\frac{x}{\log x}\right),$$

it is to be expected that

$$\sum_{n=1}^{x} P^*(n) \sim \pi(x).$$

In light of (1), the present case is a specialization of the following problem: given a function

$$f(s) = \sum_{n=1}^{\infty} \frac{a_n}{n^s},$$

to estimate

$$\sum_{n=1}^{x} a_n. \tag{2}$$

It will be shown that

$$J(w) = \frac{1}{2\pi i} \int_{2-\infty i}^{2+\infty i} \frac{e^{ws}}{s^2} \, ds = \begin{cases} w & \text{if } w \geq 0, \\ 0 & \text{if } w \leq 0, \end{cases}$$

so that $J(w)$ is closely related to the characteristic function of the positive real numbers. If we put

$$w = \log \frac{x}{n},$$

this gives

$$\frac{1}{2\pi i} \int_{2-\infty i}^{2+\infty i} \frac{(x/n)^s}{s^2} \, ds = \begin{cases} \log (x/n) & \text{if } n \leq x, \\ 0 & \text{if } n \geq x, \end{cases}$$

so that

$$\frac{1}{2\pi i} \int_{2-\infty i}^{2+\infty i} \frac{x^s}{s^2} f(s) \, ds = \sum_{n \leq x} a_n \log \frac{x}{n}.$$

If $\delta = \delta(x)$ tends monotonically to zero as $x \to \infty$, then

$$\sum_{n \leq x(1+\delta)} a_n \log \frac{x(1+\delta)}{n} - \sum_{n \leq x} a_n \log \frac{x}{n}$$

$$= \log (1+\delta) \sum_{n \leq x} a_n + \sum_{x < n \leq x + \delta x} a_n \log \frac{x(1+\delta)}{n}$$

$$= \log (1+\delta) \sum_{n \leq x} a_n + O\left(\log (1+\delta) \sum_{x < n \leq x+\delta x} a_n \right).$$

If the remainder term here is of smaller order of magnitude than the first term for suitable choice of δ, then

$$\sum_{n \leq x} a_n \sim \frac{1}{2\pi i \delta} \int_{2-\infty i}^{2+\infty i} \frac{x^s((1+\delta)^s - 1)}{s^2} f(s) \, ds,$$

and the problem reduces to that of obtaining an adequate estimate of this integral. To do this, we replace the line of integration by a suitable large closed contour, inside and on which we have sufficient information about $f(s)$ to apply standard contour-integral techniques.

In the case at hand, the estimation of the integral in the last relation requires some knowledge of the zeros, poles, and size of $\zeta(s)$.

7-2 Preliminary results. Following the odd but harmless tradition in analytic number theory, we designate by σ and t the real and imaginary parts of the complex variable s. For $x > 0$, x^s means $e^{s \log x}$, where $\log x$ indicates the real logarithm.

When we have proved the Prime Number Theorem, we shall consider some other rather similar problems, and for one of these it will be necessary to use not the Riemann ζ-function but the so-called *Hurwitz ζ-function*, defined for $0 < w \leq 1$ and $\sigma > 1$ by the equation

$$\zeta(s, w) = \sum_{n=0}^{\infty} \frac{1}{(n + w)^s}.$$

Since $\zeta(s, 1) = \zeta(s)$, and since the requisite properties are no more difficult to prove for $\zeta(s, w)$ than for $\zeta(s)$, we consider the more general function.

THEOREM 7–1. *For any $\sigma_0 > 1$, the series*

$$\sum_{n=0}^{\infty} (n + w)^{-s}$$

converges uniformly for $\sigma \geq \sigma_0$, so that $\zeta(s, w)$ is regular (or analytic) for $\sigma > 1$.

Proof: We have

$$|(n + w)^{-s}| = |e^{-(\sigma + it) \log (n+w)}| = e^{-\sigma \log (n+w)} = (n + w)^{-\sigma},$$

so that for $\sigma \geq \sigma_0$,

$$\sum_{n=0}^{\infty} (n + w)^{-s} \ll \sum_{n=0}^{\infty} (n + w)^{-\sigma_0}.$$

Thus we have a series of analytic functions which is dominated throughout the region $\sigma \geq \sigma_0$ by a convergent series of positive constants, and which is therefore uniformly convergent, and the result follows from Weierstrass' theorem.

THEOREM 7–2. *If a and b are integers with $b > a \geq 0$, and if f has a continuous derivative over $a \leq x \leq b$, then*

$$\sum_{n=a+1}^{b} f(n) = \int_{a}^{b} f(u) \, du + \int_{a}^{b} (u - [u]) f'(u) \, du.$$

Proof: We have

$$\int_{n-1}^{n} uf'(u)\, du = nf(n) - (n-1)f(n-1) - \int_{n-1}^{n} f(u)\, du$$

$$= f(n) + (n-1)\int_{n-1}^{n} f'(u)\, du - \int_{n-1}^{n} f(u)\, du$$

$$= f(n) + \int_{n-1}^{n} [u]f'(u)\, du - \int_{n-1}^{n} f(u)\, du,$$

from which the result follows by summing on n from $a+1$ to b.

THEOREM 7-3. *If m is a non-negative integer, and $\sigma > 1$, then*

$$\zeta(s, w) - \frac{1}{(s-1)(m+w)^{s-1}} = \sum_{n=0}^{m} \frac{1}{(n+w)^s} - s\int_{m}^{\infty} \frac{u-[u]}{(u+w)^{s+1}}\, du. \quad (3)$$

It follows that $\zeta(s, w) - 1/(s-1)$ is regular for $\sigma > 0$, and that (3) holds for $\sigma > 0$.

Proof: If $\sigma > 1$ and

$$f(u) = \frac{1}{(u+w)^s},$$

then the equation of Theorem 7-2 continues to hold if $b \to \infty$, and, replacing a by m, we have

$$\sum_{n=m+1}^{\infty} \frac{1}{(n+w)^s} = \frac{1}{(s-1)(m+w)^{s-1}} - s\int_{m}^{\infty} \frac{u-[u]}{(u+w)^{s+1}}\, du,$$

from which (3) follows. Since

$$\left| \frac{u-[u]}{(u+w)^{s+1}} \right| < \frac{1}{(u+w)^{\sigma+1}} < \frac{1}{u^{\sigma+1}},$$

the integral on the right side of (3) converges absolutely for $\sigma > 0$, and uniformly for $\sigma \geq \sigma_0 > 0$. For arbitrary $n \geq 0$, the quantity

$$\int_{n}^{n+1} \frac{u-[u]}{(u+w)^{s+1}}\, du = \int_{n}^{n+1} \frac{u-n}{(u+w)^{s+1}}\, du$$

is a regular function of s for $\sigma > 0$; the same is true of

$$\sum_{n=m}^{\infty} \int_{n}^{n+1} \frac{u-[u]}{(u+w)^{s+1}}\, du = \int_{m}^{\infty} \frac{u-[u]}{(u+w)^{s+1}}\, du,$$

for $m \geq 0$.   Finally, taking $m = 0$ in (3), we have

$$\zeta(s, w) - \frac{1}{s-1} = \frac{1}{w^s} + \frac{w^{1-s} - 1}{s-1} - s \int_0^\infty \frac{u - [u]}{(u+w)^{s+1}} \, du,$$

and the right side is regular for $\sigma > 0$.

Equation (3) thus provides an analytic continuation of $\zeta(s, w)$ over the half-plane $\sigma > 0$.   The function is actually analytic over the entire plane, except for the pole at $s = 1$, but this fact is not needed.

Hereafter c will denote a positive constant which depends only on the arguments indicated; it need not have the same value in different occurrences, unless it has a subscript.

THEOREM 7–4.   *For $\frac{1}{2} \leq \sigma \leq 2$ and $t > c(w)$,*

$$|\zeta(s, w)| < t^{\frac{3}{4}}.$$

For $t \geq 8$ and $1 - (\log t)^{-1} \leq \sigma \leq 2$,

$$|\zeta(s, w)| < c(w) \log t.$$

Proof: For $\frac{1}{2} \leq \sigma \leq 2$ and $t \geq 3$ we have $|s| < 2 + t < 2t$ and $|s - 1| \geq t > 1$.   Hence if we take $m = [t] + 1$ in (3), we have

$$|\zeta(s, w)| < \frac{1}{([t] + 1 + w)^{\sigma - 1}} + \frac{1}{w^\sigma} + \sum_{n=1}^{[t]+1} \frac{1}{n^\sigma} + 2t \int_t^\infty \frac{du}{u^{\sigma+1}}$$

$$\leq \frac{1}{([t] + 1 + w)^{\sigma - 1}} + c(w) + \sum_{n=1}^{[t]} \frac{1}{n^\sigma} + \frac{2t}{\sigma t^\sigma},$$

or

$$|\zeta(s, w)| < \frac{1}{([t] + 1 + w)^{\sigma - 1}} + c(w) + \sum_{n=1}^{[t]} \frac{1}{n^\sigma} + 4t^{1-\sigma}. \qquad (4)$$

Thus, for this same range of σ and t,

$$|\zeta(s, w)| < \frac{1}{([t] + 1 + w)^{-\frac{1}{2}}} + c(w) + \sum_{n=1}^{[t]} \frac{1}{\sqrt{n}} + 4\sqrt{t}$$

$$< 2\sqrt{t} + c(w) + \int_0^t \frac{du}{\sqrt{u}} + 4\sqrt{t} \leq 8\sqrt{t} + c(w),$$

and this is smaller than $t^{\frac{3}{4}}$ for $t > c(w)$.

Now take $t \geq 8 > e^2$.   Then $1 - (\log t)^{-1} \geq \frac{1}{2}$, so   that   if $1 - (\log t)^{-1} \leq \sigma \leq 2$, the inequality (4) gives

$$|\zeta(s, w)| < (2t)^{1/\log t} + c(w) + \sum_{n=1}^{[t]} \frac{n^{1/\log t}}{n} + 4i^{1/\log t}$$

$$< 2^{\frac{1}{2}}e + c(w) + e \sum_{n=1}^{[t]} \frac{1}{n} + 4e < c(w) \log t.$$

THEOREM 7–5. *If, for* $|x| \leq 1$,

$$f(x) = \sum_{n=1}^{\infty} a_n x^n$$

is regular and Re $f(x) \leq \frac{1}{2}$, *then* $|a_n| \leq 1$ *for* $n \geq 1$.

Proof: Since $|f(x)| \leq |1 - f(x)|$ for $|x| \leq 1$, the function

$$\frac{f(x)}{1 - f(x)} = \frac{a_1 x + \cdots}{1 - a_1 x - \cdots} = a_1 x + b_2 x^2 + \cdots$$

is regular and has modulus at most 1 for $|x| \leq 1$. But the function

$$f_1(x) = \frac{f(x)}{x(1 - f(x))}$$

is also regular for $|x| \leq 1$, and its value at $x = 0$ is a_1; by the maximum-modulus principle, its absolute value is at least as large at some point on $|x| = 1$. Since for $|x| = 1$,

$$|f_1(x)| = \left| \frac{f(x)}{1 - f(x)} \right|,$$

it follows that

$$|a_1| \leq 1. \tag{5}$$

The theorem will therefore be proved if we show that each of the functions

$$F_n(x) = a_n x + a_{2n} x^2 + \cdots$$

fulfills the same hypotheses as $f(x)$ itself. This depends on the fact that if $\eta = e^{2\pi i/n}$, then

$$\sum_{l=0}^{n-1} \eta^{lk} = \begin{cases} n & \text{if } n|k, \\ (\eta^{kn} - 1)/(\eta^k - 1) = 0 & \text{if } n \nmid k. \end{cases}$$

We have

$$\sum_{l=0}^{n-1} f(\eta^l x) = \sum_{l=0}^{n-1} \sum_{k=1}^{\infty} a_k \eta^{kl} x^k = \sum_{k=1}^{\infty} a_k x^k \sum_{l=0}^{n-1} \eta^{kl}$$

$$= n \sum_{n|k} a_k x^k = nF_n(x^n),$$

so that $F_n(x)$ is regular for $|x| \leq 1$, and for such x,

$$\operatorname{Re} F_n(x^n) = \frac{1}{n} \sum_{l=0}^{n-1} \operatorname{Re} f(\eta^l x) \leq \frac{1}{n} \sum_{l=0}^{n-1} \frac{1}{2} = \frac{1}{2}.$$

THEOREM 7–6.   *Let R be positive, and suppose that*

$$f(x) = \sum_{n=0}^{\infty} a_n (x - x_0)^n$$

is regular and $\operatorname{Re} f(x) \leq M$ for $|x - x_0| \leq R$.   Then, for $n \geq 1$,

$$|a_n| \leq \frac{2}{R^n} (M - \operatorname{Re} a_0).$$

Proof: If $\operatorname{Re} a_0 = M$, then $a_n = 0$ for $n \geq 1$, by the maximum-modulus principle.

If $\operatorname{Re} a_0 < M$, put

$$g(x) = \frac{f(x_0 + Rx) - a_0}{2(M - \operatorname{Re} a_0)}.$$

Then g is regular for $|x| \leq 1$, $g(0) = 0$, and

$$\operatorname{Re} g(x) = \frac{\operatorname{Re} f(x_0 + Rx) - \operatorname{Re} a_0}{2(M - \operatorname{Re} a_0)} \leq \frac{M - \operatorname{Re} a_0}{2(M - \operatorname{Re} a_0)} = \frac{1}{2}.$$

Hence g satisfies the hypotheses of Theorem 7–5, so that

$$\left| \frac{a_n R^n}{2(M - \operatorname{Re} a_0)} \right| \leq 1,$$

and the theorem follows.

THEOREM 7–7.   *If f satisfies the hypotheses of Theorem 7–6, and $0 < r < R$, then for $|x - x_0| \leq r$,*

$$|f(x)| \leq |a_0| + \frac{2r}{R - r} (|M| + |a_0|)$$

and $$|f'(x)| \leq \frac{2R}{(R - r)^2} (|M| + |a_0|).$$

Proof: We have

$$|f(x)| \leq |a_0| + \sum_{n=1}^{\infty} |a_n| r^n \leq |a_0| + 2(|M| + |a_0|) \sum_{n=1}^{\infty} \left(\frac{r}{R} \right)^n$$

$$= |a_0| + \frac{2r}{R - r} (|M| + |a_0|),$$

and

$$|f'(x)| \leq \sum_{n=1}^{\infty} |a_n| n r^{n-1} \leq \frac{2(|M| + |a_0|)}{R} \sum_{n=1}^{\infty} n \left(\frac{r}{R}\right)^{n-1}$$

$$= \frac{2R}{(R-r)^2} (|M| + |a_0|).$$

THEOREM 7-8. *Let r be positive and M real, and suppose that $f(s_0) \neq 0$ and that, for $|s - s_0| \leq r$, $f(s)$ is regular and*

$$\left| \frac{f(s)}{f(s_0)} \right| < e^M.$$

Suppose also that $f(s) \neq 0$ in the semicircular region $|s - s_0| \leq r$, $\operatorname{Re} s > \operatorname{Re} s_0$. Then

$$- \operatorname{Re} \frac{f'}{f} (s_0) \leq \frac{4M}{r},$$

and if there is a zero ρ of f on the open line segment between $s_0 - r/2$ and s_0, then

$$- \operatorname{Re} \frac{f'}{f} (s_0) \leq \frac{4M}{r} - \frac{1}{s_0 - \rho}.$$

Proof: There is clearly no loss in generality in supposing that $f(s_0) = 1$ and $s_0 = 0$. In this case, the hypotheses can be listed as follows.

(1) For $|s| \leq r$, $f(s)$ is regular and $|f(s)| < e^M$, where $M > 0$.

(2) $f(0) = 1$.

(3) $f(s) \neq 0$ for $|s| \leq r$, $\sigma > 0$.

We look for an upper bound for $- \operatorname{Re} f'(0)$.

If ρ runs through the zeros of f in the circle $|s| \leq r/2$, then the function

$$g(s) = \frac{f(s)}{\prod_{\rho} (1 - s/\rho)}$$

is regular for $|s| \leq r$. On the circle $|s| = r$, we have

$$\left| 1 - \frac{s}{\rho} \right| \geq \left| \frac{s}{\rho} \right| - 1 \geq 1,$$

so that here $|g(s)| \leq |f(s)| < e^M$. By the maximum-modulus principle,

$$|g(s)| < e^M, \qquad \text{for } |s| \leq \frac{r}{2}.$$

Since $g(s) \neq 0$ for $|s| \leq r/2$, and $g(0) = 1$, we can write

$$g(s) = e^{G(s)}, \qquad \text{for } |s| \leq \frac{r}{2},$$

where G is regular and $\operatorname{Re} G(s) < M$, $G(0) = 0$. By Theorem 7–6, with $r/2$ instead of R,

$$|G'(0)| < \frac{2}{r/2} M = \frac{4M}{r}.$$

But

$$\frac{g'}{g}(s) = \frac{f'}{f}(s) - \sum_\rho \frac{-1/\rho}{1 - s/\rho} = \frac{f'}{f}(s) + \sum_\rho \frac{1}{\rho - s},$$

so that

$$\left| f'(0) + \sum_\rho \frac{1}{\rho} \right| = \left| \frac{g'}{g}(0) \right| = |G'(0)| \leq \frac{4M}{r},$$

$$-\operatorname{Re} f'(0) - \sum_\rho \frac{1}{\operatorname{Re}\rho} \leq \frac{4M}{r},$$

$$-\operatorname{Re} f'(0) \leq \frac{4M}{r} + \sum_\rho \operatorname{Re} \frac{1}{\rho}.$$

Since we have supposed that all zeros ρ have nonpositive real parts, the theorem follows.

If f is regular on the vertical line $\sigma_0 + ti$, and if

$$\lim_{\substack{a \to \infty \\ b \to \infty}} \int_{\sigma_0 - ai}^{\sigma_0 + bi} f(s)\, ds = \lim_{\substack{a \to \infty \\ b \to \infty}} \int_{-a}^b f(\sigma_0 + ti) i\, dt$$

exists, then we abbreviate this limit to

$$\int_{(\sigma_0)} f(s)\, ds.$$

THEOREM 7–9. *We have*

$$\frac{1}{2\pi i} \int_{(2)} \frac{y^s}{s^2}\, ds = \begin{cases} 0 & \text{for } 0 < y < 1, \\ \log y & \text{for } y \geq 1. \end{cases}$$

Proof: The integral converges, because

$$\left|\frac{y^s}{s^2}\right| = \frac{y^2}{4 + t^2}.$$

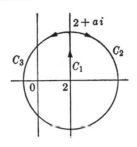

FIGURE 7–1

First suppose that $0 < y < 1$. Then in the region bounded by C_1 and C_2 (see Fig. 7–1), the integrand is regular, so that by Cauchy's theorem,

$$\int_{C_1} \frac{y^s}{s^2}\,ds + \int_{C_2} \frac{y^s}{s^2}\,ds = 0.$$

But along C_2, which is of length πa, we have

$$\left|\frac{y^s}{s^2}\right| < \frac{1}{a^2},$$

so that

$$\left|\int_{C_2} \frac{y^s}{s^2}\,ds\right| < \frac{\pi a}{a^2} = \frac{\pi}{a}.$$

Hence, as $a \to \infty$,

$$\int_{C_2} \frac{y^s}{s^2}\,ds \longrightarrow 0, \qquad \int_{C_1} \frac{y^s}{s^2}\,ds \longrightarrow \int_{(2)} \frac{y^s}{s^2}\,ds,$$

and the result follows.

Now suppose that $y \geq 1$, and that $a > 2$. Then the pole $s = 0$ of the integrand lies in the region bounded by C_1 and C_3, and since

$$\frac{y^s}{s^2} = \frac{1 + s \log y + (s^2 \log^2 y)/2 + \cdots}{s^2} = \frac{1}{s^2} + \frac{\log y}{s} + \cdots,$$

we have by the residue theorem that

$$\frac{1}{2\pi i} \int_{C_1} \frac{y^s}{s^2}\,ds + \frac{1}{2\pi i} \int_{C_3} \frac{y^s}{s^2}\,ds = \log y.$$

But along C_3, which is of length πa, we have

$$\left|\frac{y^s}{s^2}\right| = \frac{y^\sigma}{|s|^2} \leq \frac{y^2}{(a - 2)^2},$$

so that, for $a > 4$,

$$\left| \int_{C_3} \frac{y^s}{s^2} ds \right| < \frac{\pi a y^2}{(a-2)^2} < \frac{4\pi y^2}{a}.$$

Hence, as $a \to \infty$,

$$\int_{C_3} \frac{y^s}{s^2} ds \longrightarrow 0, \qquad \int_{C_1} \frac{y^s}{s^2} ds \longrightarrow \int_{(2)} \frac{y^s}{s^2} ds,$$

and the result follows.

7–3 The Prime Number Theorem. It will be necessary in what follows to know something about the location of the zeros of the ζ-function. For $|t|$ large, this information is supplied by Theorem 7–13; for $|t|$ small, we use only the fact that $\zeta(s)$ *does not vanish for* $\sigma \geq 1$. Historically, this was the first nontrivial result obtained concerning the zeros of the ζ-function. (A trivial fact is that $\zeta(s) \neq 0$ for $\sigma > 1$, which follows immediately from the product representation

$$\zeta(s) = \prod_p (1 - p^{-s})^{-1},$$

valid for $\sigma > 1$.) The proof below that $\zeta(1 + ti) \neq 0$ is due to de la Vallée Poussin; it may have been suggested by the following considerations.

For $\sigma > 1$ we have

$$\log \zeta(s) = \sum_{m,p} \frac{1}{mp^{ms}} = \sum_p \frac{1}{p^s} + f(s),$$

and f is easily seen to be regular for $\sigma > \frac{1}{2}$. Since ζ has a pole at $s = 1$, with residue 1, it follows that as $\sigma \to 1^+$,

$$\sum_p \frac{1}{p^\sigma} \sim \log \frac{1}{\sigma - 1}. \tag{6}$$

We now reason heuristically. If $1 + t_0 i$ is a zero of ζ, and we put $s = \sigma + t_0 i$, then as $\sigma \to 1^+$,

$$\log |\zeta(s)| \sim \log (\sigma - 1)$$

and

$$\mathrm{Re} \log \zeta(s) - \mathrm{Re}\, f(s) = \log |\zeta(s)| - \mathrm{Re}\, f(s)$$

$$= \sum_p \frac{\cos (t_0 \log p)}{p^\sigma} \sim \log (\sigma - 1).$$

Comparing this with (6) we see that for most p, $\cos (t_0 \log p)$ must be close to -1. But then $\cos (2t_0 \log p)$ must usually be nearly 1, and

$$\sum_p \frac{\cos (2t_0 \log p)}{p^\sigma} \sim \log \frac{1}{\sigma - 1}.$$

But this requires that ζ have a pole at $1 + 2t_0 i$, which is not the case.

To make this argument rigorous, note that for all real θ,

$$3 + 4 \cos \theta + \cos 2\theta = 2(1 + \cos \theta)^2 \geq 0.$$

Hence for $\sigma > 1$,

$$\log |\zeta^3(\sigma)\zeta^4(\sigma + t_0 i)\zeta(\sigma + 2t_0 i)|$$

$$= 3 \log |\zeta(\sigma)| + 4 \log |\zeta(\sigma + t_0 i)| + \log |\zeta(\sigma + 2t_0 i)|$$

$$= 3 \sum_{n,p} \frac{1}{np^{n\sigma}} + 4 \sum_{n,p} \frac{\cos (t_0 n \log p)}{np^{n\sigma}} + \sum_{n,p} \frac{\cos (2t_0 n \log p)}{np^{n\sigma}}$$

$$= \sum_{n,p} \frac{3 + 4 \cos (t_0 n \log p) + \cos (2t_0 n \log p)}{np^{n\sigma}}$$

$$\geq 0.$$

Thus

$$((\sigma - 1)\zeta(\sigma))^3 \left| \frac{\zeta(\sigma + t_0 i)}{\sigma - 1} \right|^4 |\zeta(\sigma + 2t_0 i)| \geq \frac{1}{\sigma - 1},$$

and if $1 + t_0 i$ were a zero of ζ, the left side in this inequality would remain bounded as $\sigma \to 1^+$, while the right side increases without limit.

We now use this technique, together with Theorem 7–8, to show that $\zeta(s)$ does not vanish at any point too close to the line $\sigma = 1$ and sufficiently far from the real axis.

THEOREM 7–10. *For $\sigma > 1$,*

$$\text{Re} \left(-3 \frac{\zeta'}{\zeta} (\sigma) - 4 \frac{\zeta'}{\zeta} (\sigma + ti) - \frac{\zeta'}{\zeta} (\sigma + 2ti) \right) \geq 0.$$

Proof: Differentiating the relation

$$\log \zeta(s) = \sum_{m,p} \frac{1}{mp^{ms}},$$

we obtain

$$\frac{\zeta'}{\zeta} (s) = - \sum_{m,p} \frac{\log p}{p^{ms}} = - \sum_{n=1}^{\infty} \frac{\Lambda(n)}{n^s}, \qquad (7)$$

where

$$\Lambda(n) = \begin{cases} \log p & \text{if } n = p^m, \text{ for any } m > 0 \text{ and prime } p, \\ 0 & \text{otherwise.} \end{cases}$$

The termwise differentiation is justified because the series for $\log \zeta(s)$ converges uniformly in any region to the right of $\sigma = 1$. Hence

$$\text{Re}\left(-3\frac{\zeta'}{\zeta}(\sigma) - 4\frac{\zeta'}{\zeta}(\sigma+ti) - \frac{\zeta'}{\zeta}(\sigma+2ti)\right)$$

$$= \text{Re} \sum_{n=1}^{\infty} \frac{(3 + 4n^{-ti} + n^{-2ti})\Lambda(n)}{n^\sigma}$$

$$= \sum_{n=1}^{\infty} \frac{(3 + 4\cos(t\log n) + \cos(2t\log n))\Lambda(n)}{n^\sigma}$$

$$\geq 0.$$

THEOREM 7–11. (a) *For $\sigma \geq \frac{1}{2}$ and $t > c$, we have $|\zeta(s)| < t$.*
 (b) *For $t \geq 8$ and $\sigma \geq 1 - (\log t)^{-1}$, we have $|\zeta(s)| < c \log t$.*

Proof: For $\sigma > 2$ and $t \geq 8$,

$$|\zeta(s)| < \sum_{n=1}^{\infty} \frac{1}{n^2} < 2 < \begin{cases} t, \\ \log t. \end{cases}$$

For $\sigma \leq 2$, both inequalities of the theorem follow from Theorem 7–4.

THEOREM 7–12. For $\sigma > 1$,

$$\frac{1}{\zeta(s)} = \sum_{n=1}^{\infty} \frac{\mu(n)}{n^s},$$

where μ is the Möbius function.

Proof: This follows immediately from Theorem 6–12 for s real and greater than 1; by analytic continuation, it is correct for $\sigma > 1$.

THEOREM 7–13. *There are constants $c_1 > 8$ and $c_2 > 0$ such that $\zeta(s) \neq 0$ for*

$$t > c_1 \quad \text{and} \quad \sigma > 1 - \frac{c_2}{\log t}.$$

Proof: In accordance with Theorem 7–11 (a), choose $c_3 > 8$ such that

$$|\zeta(s)| < t, \quad \text{for } \sigma \geq \tfrac{1}{2}, \quad t > c_3. \tag{8}$$

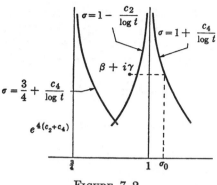

FIGURE 7-2

Inasmuch as

$$\frac{3}{4} + \frac{c_4}{\log x} < 1 - \frac{c_2}{\log x}, \qquad \text{for } x > e^{4(c_2+c_4)},$$

it suffices to show that any zero $\beta + \gamma i$ of ζ with γ sufficiently large (in particular, larger than 8) and for which

$$\beta > \frac{3}{4} + \frac{c_4}{\log \gamma},$$

is such that

$$\beta < 1 - \frac{c_2}{\log \gamma}.$$

Put

$$\sigma_0 = \sigma_0(\gamma) = 1 + \frac{c_4}{\log \gamma},$$

and suppose that $\beta + \gamma i$ is a zero of ζ for which $\gamma > e^{4(1+c_2+c_4)}$ and $\beta > \sigma_0 - \frac{1}{4}$. We shall apply Theorem 7–8, once with $s_0 = \sigma_0 + \gamma i$ and once with $s_0 = \sigma_0 + 2\gamma i$. In either case, since $\sigma_0 > 1$, we have that for $\gamma \geq c_3 + \frac{1}{2}$, the circle $|s - s_0| \leq \frac{1}{2}$ lies in the quadrant $\sigma \geq \frac{1}{2}, t \geq c_3$. Since $\gamma > e^{c_4}$, we have $\sigma_0 < 2$, and, by Theorem 7–12,

$$\left| \frac{1}{\zeta(s_0)} \right| \leq \sum_{n=1}^{\infty} \frac{1}{n^{\sigma_0}} < 1 + \int_1^{\infty} \frac{du}{u^{\sigma_0}} = 1 + \frac{1}{\sigma_0 - 1} < \frac{2}{\sigma_0 - 1} = \frac{2}{c_4} \log \gamma.$$

Thus for each $\epsilon_1 > 0$ there is a c_5 such that for $\gamma > c_5 > c_3 + \frac{1}{2}$, the inequality

$$\left| \frac{\zeta(s)}{\zeta(s_0)} \right| \leq \frac{2}{c_4} \left(2\gamma + \frac{1}{2} \right) \log \gamma < \gamma^{1+\epsilon_1}$$

holds at every point s of the circular disk $|s - s_0| \leq \frac{1}{2}$, since at every such point, $c_3 < t \leq 2\gamma + \frac{1}{2}$. If $\gamma \geq c_5$, we can now apply Theorem 7–8, with $r = \frac{1}{2}$, $f(s) = \zeta(s)$, $M = (1 + \epsilon_1) \log \gamma$. Using the first inequality of that theorem with $s_0 = \sigma_0 + 2\gamma i$, we obtain

$$-\operatorname{Re} \frac{\zeta'}{\zeta} (\sigma_0 + 2\gamma i) < 8(1 + \epsilon_1) \log \gamma; \tag{9}$$

using the second with $s_0 = \sigma_0 + \gamma i$ we have

$$-\operatorname{Re} \frac{\zeta'}{\zeta} (\sigma_0 + \gamma i) < 8(1 + \epsilon_1) \log \gamma - \frac{1}{\sigma_0 - \beta}, \tag{10}$$

since

$$\sigma_0 - r/2 = \sigma_0 - \tfrac{1}{4} < \beta \leq 1 < \sigma_0.$$

Finally, since $\sigma_0 \to 1^+$ as $t \to \infty$, we have from (6) that for $\epsilon_2 > 0$,

$$-\frac{\zeta'}{\zeta} (\sigma_0) < \frac{1 + \epsilon_2}{\sigma_0 - 1} = \frac{1 + \epsilon_2}{c_4} \log \gamma \tag{11}$$

for $\gamma > c_6$. Using the estimates (9), (10), and (11) in Theorem 7–10 gives

$$\frac{3(1 + \epsilon_2)}{c_4} \log \gamma + 4 \cdot 8(1 + \epsilon_1) \log \gamma - \frac{4}{\sigma_0 - \beta} + 8(1 + \epsilon_1) \log \gamma \geq 0.$$

This inequality can easily be simplified to

$$\sigma_0 - \beta > \frac{c_7}{\log \gamma},$$

where

$$c_7 = \frac{4c_4}{3(1 + \epsilon_2) + 40(1 + \epsilon_1)c_4},$$

and this gives

$$\beta < 1 - \frac{c_7 - c_4}{\log \gamma}.$$

It is clear that $c_7 > c_4$ if $\epsilon_1 < \frac{1}{3}$ and c_4 is sufficiently small, and we can then take $c_2 = c_7 - c_4$ and $c_1 = \max(c_5, c_6)$.

THEOREM 7–14. *If* $0 < c_8 < c_2$, *then*

$$|\log \zeta(s)| < \log^2 t \qquad for\ t > c_9\ and\ \sigma \geq 1 - \frac{c_8}{\log t}.$$

Proof: We use Theorem 7–7, with $s_0 = 2 + t_0 i$, for some $t_0 > 8$ to be determined. For t sufficiently large, the circular region

$$|s - s_0| \leq 1 + \frac{\frac{1}{2}(c_2 + c_8)}{\log t_0} \qquad (12)$$

lies entirely in the region described in the preceding theorem, in which ζ has no zeros. Hence the function $\log \zeta(s)$ is regular in this disk, and by Theorem 7–11(b),

$$\text{Re} \log \zeta(s) = \log |\zeta(s)| < \log (c \log t)$$
$$< \log (c \log (t_0 + 2)) < c_{10} \log \log t_0.$$

Hence, by Theorem 7–7, we have that for s in the region (12),

$$|\log \zeta(s)| \leq |\zeta(s_0)| + \frac{2 \cdot 2(c_{10} \log \log t_0 + |\zeta(s_0)|)}{\dfrac{c_8 - c_2}{2} \cdot \dfrac{1}{\log t_0}}$$

$$\leq c + c \log t_0 \log \log t_0 < \log^2 t_0,$$

if t_0 is sufficiently large. This inequality holds on the radius extending toward the left from s_0, for every large t_0, and hence throughout a region $t \geq c_9$, $1 - c_8 (\log t)^{-1} \leq \sigma \leq 2$. Finally, $|\zeta(s)|$ and $|1/\zeta(s)|$ are bounded in the half-plane $\sigma > 2$, and $|\log \zeta(s)|$ is consequently smaller than $\log^2 t$ for t large and $\sigma > 2$.

THEOREM 7–15. *There is a constant $\alpha > 0$ such that as $x \to \infty$,*

$$\sum_{p \leq x} \log \frac{x}{p} = \int_c^1 \frac{x^s}{s^2} \, ds + O(x e^{-\alpha \sqrt{\log x}}),$$

for some c with $0 < c < 1$.

Proof: Using Theorem 7–9, we have

$$\frac{1}{2\pi i} \int_{(2)} \frac{x^s}{s^2} \log \zeta(s) \, ds = \frac{1}{2\pi i} \int_{(2)} \frac{1}{s^2} \sum_{n=1}^{\infty} \frac{\Lambda(n)}{\log n} \left(\frac{x}{n}\right)^s \, ds$$

$$= \frac{1}{2\pi i} \sum_{n=1}^{\infty} \frac{\Lambda(n)}{\log n} \int_{(2)} \frac{(x/n)^s}{s^2} \, ds = \sum_{n \leq x} \frac{\Lambda(n)}{\log n} \log \frac{x}{n} = \sum_{\substack{m,p \\ p^m \leq x}} \frac{1}{m} \log \frac{x}{p^m}$$

$$= \sum_{p \leq x} \log \frac{x}{p} + \sum_{\substack{m,p \\ m \geq 2 \\ p^m \leq x}} \frac{1}{m} \log \frac{x}{p^m} \cdot$$

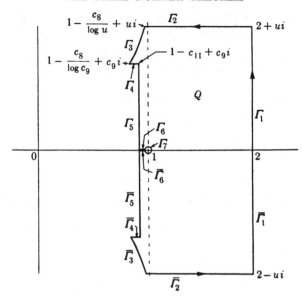

FIGURE 7–3

As noted earlier, the number of terms in the last sum is

$$\pi(x^{\frac{1}{2}}) + \pi(x^{\frac{1}{3}}) + \cdots < x^{\frac{1}{2}} + x^{\frac{1}{3}} + \cdots + x^{1/u} < ux^{\frac{1}{2}},$$

where u is the smallest integer such that $x^{1/u} < 2$. Thus

$$\pi(x^{\frac{1}{2}}) + \pi(x^{\frac{1}{3}}) + \cdots = O(x^{\frac{1}{2}} \log x),$$

so that

$$2\pi i \sum_{p \leq x} \log \frac{x}{p} = \int_{(2)} \frac{x^s}{s^2} \log \zeta(s) \, ds + O(xe^{-\sqrt{\log x}}),$$

since

$$\sum_{\substack{m \geq 2 \\ p^m \leq x}} \frac{1}{m} \log \frac{x}{p^m} \leq \sum_{\substack{m \geq 2 \\ p^m \leq x}} \log x = O(\sqrt{x} \log^2 x) = O(xe^{-\sqrt{\log x}}).$$

We now cut the complex plane along the real axis, extending the cut from $s = 1$ to the left, and examine the function $\log \zeta(s)$ in the cut plane. If $\bar{z}$ is the complex conjugate of z, then $\zeta(\bar{s}) = \overline{\zeta(s)}$ and $\log \bar{z} = \overline{\log z}$, so that $\log \zeta(\bar{s}) = \overline{\log \zeta(s)}$. Hence, by Theorem 7–13, $\zeta(s) \neq 0$ for $|t| > c_9 > c_1$ and $\sigma \geq 1 - c_8 (\log |t|)^{-1}$. Moreover, since $\zeta(s)$ does not vanish on the line $\sigma = 1$, and since its zeros have no finite limit point in any half-plane $\sigma \geq \sigma_0 > 0$ (since $\zeta(s)$ is regular there), there is a constant $c_{11} > 0$ such that $\zeta(s) \neq 0$ in the rectangle

$$1 - c_{11} \leq \sigma \leq 1, \qquad |t| \leq c_9.$$

Finally, $\zeta(s) \neq 0$ for $1 \leq \sigma \leq 2$, and the only singularity of the function in the half-plane $\sigma > 0$ is at $s = 1$. Consequently, for arbitrary $u > c_9$, $\log \zeta(s)$ is a single-valued analytic function in the region Q shown in Fig. 7–3, bounded by the arcs $\Gamma_1, \Gamma_2, \ldots, \Gamma_6, \Gamma_7,$ $\bar{\Gamma}_6, \ldots, \bar{\Gamma}_2, \bar{\Gamma}_1$. Hence if we denote by Γ the complete boundary of this region (so that we might write symbolically $\Gamma = \Gamma_1 + \Gamma_2 + \cdots + \bar{\Gamma}_1$), we have by Cauchy's theorem that

$$\int_\Gamma \frac{x^s}{s^2} \log \zeta(s) \, ds = 0.$$

It follows that, if the integrals are taken in the positive direction,

$$\int_{(2)} \frac{x^s}{s^2} \log \zeta(s) \, ds$$

$$= \left(\int_{2-\infty i}^{2-ui} + \int_{\bar{\Gamma}_1} + \int_{\Gamma_1} + \int_{2+ui}^{2+\infty i} \right) \frac{x^s}{s^2} \log \zeta(s) \, ds$$

$$= \left(\int_{2-\infty i}^{2-ui} - \int_{\Gamma_2 + \cdots + \Gamma_6 + \Gamma_7 + \bar{\Gamma}_6 + \cdots + \bar{\Gamma}_2} + \int_{2+ui}^{2+\infty i} \right) \frac{x^s}{s^2} \log \zeta(s) \, ds.$$

We shall show that all the other integrals are small in comparison with those along Γ_6 and $\bar{\Gamma}_6$, if u is sufficiently large. For brevity, put

$$\psi(x, s) = \frac{x^s}{s^2} \log \zeta(s).$$

By Theorem 7–14, we have that for $u > u_0(\epsilon)$,

$$\left| \int_{2+ui}^{2+\infty i} \psi(x, s) \, ds \right| \leq \int_{2+ui}^{2+\infty i} \frac{x^2}{|s|^2} |\log \zeta(s)||ds|$$

$$\leq x^2 \int_u^\infty \frac{\log^2 t}{t^2} \, dt \leq x^2 \int_u^\infty \frac{dt}{t^{2-\epsilon}} < \frac{cx^2}{u^{1-\epsilon}},$$

so that

$$\lim_{u \to \infty} \int_{2+ui}^{2+\infty i} \psi(x, s) \, ds = 0.$$

The same estimate applies to the integral from $2 - \infty i$ to $2 - ui$.

Since the length of Γ_2 is less than 2, and the integrand is again smaller than $x^2 \log^2 u/u^2$ for u large, we have

$$\lim_{u \to \infty} \int_{\Gamma_2} \psi(x, s)\, ds = 0,$$

and similar considerations give

$$\lim_{u \to \infty} \int_{\bar{\Gamma}_2} \psi(x, s)\, ds = 0.$$

Along Γ_3 we have $s = 1 - c_8 (\log t)^{-1} + ti$, so that

$$\left| \int_{\Gamma_3} \psi(x, s)\, ds \right| \leq \int_{c_9}^{u} \frac{x^{1 - c_8 (\log t)^{-1}}}{t^2} \log^2 t \left| \frac{c_8}{t \log^2 t} + i \right| dt.$$

Now suppose that x, and then u, are chosen large enough that

$$c_9 < e^{\sqrt{2 c_8 \log x}} < u.$$

Then

$$\int_{\Gamma_3} \psi(x, s)\, ds$$

$$= O\left(\int_{c_9}^{e^{\sqrt{2 c_8 \log x}}} x \cdot x^{-c_8 (2 c_8 \log x)^{-\frac{1}{2}}} \frac{\log^2 t}{t^2}\, dt + x \int_{e^{\sqrt{2 c_8 \log x}}}^{u} \frac{\log^2 t}{t^{\frac{1}{2}} \cdot t^{\frac{3}{2}}}\, dt \right)$$

$$= O\left(x e^{-\sqrt{\frac{1}{2} c_8 \log x}} \int_{c_9}^{\infty} \frac{\log^2 t}{t^2}\, dt + \frac{x}{e^{\sqrt{\frac{1}{2} c_8 \log x}}} \cdot \int_{e^{\sqrt{2 c_8 \log x}}}^{u} \frac{\log^2 t\, dt}{t^{\frac{3}{2}}} \right)$$

$$= O(x e^{-\alpha \sqrt{\log x}}),$$

where $\alpha = \sqrt{c_8/2}$.

By symmetry,

$$\int_{\bar{\Gamma}_3} \psi(x, s)\, ds = O(x e^{-\alpha \sqrt{\log x}}).$$

The paths Γ_4, Γ_5, $\bar{\Gamma}_4$, and $\bar{\Gamma}_5$ are of fixed lengths, and on them

$$\psi(x, s) = O(x^{1 - c_{11}}) = o(x e^{-\alpha \sqrt{\log x}}),$$

so that the same estimate holds for the integrals themselves.

Γ_7 is described by the relations $s = 1 + \delta e^{i\theta}$, $|\theta| \leq \pi$, where $\delta > 0$. Since $(s - 1)\zeta(s) \to 1$ as $s \to 1$, we have

$$\text{Re log } \zeta(s) = \log |\zeta(s)| \sim - \log |s - 1| = - \log \delta,$$

$$\text{Im log } \zeta(s) = \arg \zeta(s) = O(1)$$

as $\delta \to 0^+$. Hence

$$\int_{\Gamma_7} \psi(x, s) \, ds = O\left(2\pi\delta \frac{x^{1+\delta}}{(1 - \delta)^2} \log \delta\right) = o(1).$$

Combining all these results, we can take the limit as $u \to \infty$ and $\delta \to 0^+$ and obtain

$$2\pi i \sum_{p \leq x} \log \frac{x}{p} = \int_{1-c_{11}}^{1} \psi(x, s) \, ds + \int_{1}^{1-c_{11}} \psi(x, s) \, ds + o(xe^{-\alpha\sqrt{\log x}}),$$

where the first integral is along the upper edge, and the second along the lower edge, of the cut. We know that $(1 - s)\zeta(s) = R(s)$ is regular in the region $\sigma > 0$, and that it has no zeros in the region $\sigma > 1 - c_{11}, |t| < c_9$. Hence the function

$$\log ((s - 1)\zeta(s)) = \log (s - 1) + \log \zeta(s)$$

is single-valued in this region; since $\log (s - 1)$ has, on the upper and lower edges of the cut, values which differ by $2\pi i$, the same is true of $\log \zeta(s)$, if the difference is taken in reverse order. Hence if s^+ indicates the upper edge of the cut, and s^- the lower edge, then

$$\int_{1-c_{11}}^{1} \psi(x, s^+) \, ds^+ + \int_{1}^{1-c_{11}} \psi(x, s^-) \, ds^-$$

$$= \int_{1-c_{11}}^{1} \frac{x^{s^+}}{(s^+)^2} \log \zeta(s^+) \, ds^+ - \int_{1-c_{11}}^{1} \frac{x^{s^+}}{(s^+)^2} (\log \zeta(s^+) - 2\pi i) \, ds^+$$

$$= 2\pi i \int_{1-c_{11}}^{1} \frac{x^s}{s^2} \, ds,$$

and

$$\sum_{p \leq x} \log \frac{x}{p} = \int_{1-c_{11}}^{1} \frac{x^s}{s^2} \, ds + O(xe^{-\alpha\sqrt{\log x}}). \tag{13}$$

The theorem is proved.

THEOREM 7–16.   *As* $x \to \infty$,

$$\pi(x) = \int_2^x \frac{du}{\log u} + O(xe^{-\frac{1}{2}\alpha\sqrt{\log x}}).$$

Proof: Replace $1 - c_{11}$ by C in (13), and put

$$\delta = \delta(x) = e^{-\frac{1}{2}\alpha\sqrt{\log x}}.$$

Then since $\log (1 + \delta) \sim \delta$ as $x \to \infty$, we have

$$\sum_{p \le x(1+\delta)} \log \frac{x(1+\delta)}{p} - \sum_{p \le x} \log \frac{x}{p}$$

$$= \sum_{p \le x} \log (1 + \delta) + \sum_{x < p \le x(1+\delta)} \log \frac{x(1+\delta)}{p}$$

$$= \log (1 + \delta)\pi(x) + O(\log (1 + \delta) \cdot \delta x)$$

$$= \int_C^1 \frac{x^s}{s^2} \left((1+\delta)^s - 1\right) ds + O(xe^{-\alpha\sqrt{\log x}}),$$

so that

$$\pi(x) = \frac{1}{\log (1 + \delta)} \int_C^1 \frac{(1+\delta)^s - 1}{s^2} x^s\, ds + O(\delta x) + O\left(\frac{xe^{-\alpha\sqrt{\log x}}}{\delta}\right).$$

Now

$$(1 + \delta)^s - 1 = s\delta + \frac{s(s-1)}{2!} (1 + \vartheta\delta)^{s-2}\delta^2,$$

where $0 < \vartheta < 1$, so that for $0 < s < 1$,

$$|(1 + \delta)^s - 1 - s\delta| \le \frac{s|s-1|}{2} \delta^2 < \delta^2.$$

Thus, making the change of variable $x^s = u$, we obtain

$$\int_C^1 \frac{(1+\delta)^s - 1}{s^2} x^s\, ds = \delta \int_C^1 \frac{x^s}{s}\, ds + O\left(\delta^2 \int_C^1 \frac{x^s}{s^2}\, ds\right)$$

$$= \delta \int_{x^C}^x \frac{du}{\log u} + O\left(\delta^2 \int_C^1 x^s\, ds\right)$$

$$= \delta \int_2^x \frac{du}{\log u} + O(\delta^2 x).$$

Finally,

$$\pi(x) = \frac{\delta}{\log(1+\delta)} \int_2^x \frac{du}{\log u} + O(\delta x) + O\left(\frac{xe^{-\alpha\sqrt{\log x}}}{\delta}\right)$$

$$= (1 + O(\delta)) \int_2^x \frac{du}{\log u} + O(\delta x) + O(xe^{-\frac{1}{2}\alpha\sqrt{\log x}})$$

$$= \int_2^x \frac{du}{\log u} + O(\delta x) = \int_2^x \frac{du}{\log u} + O(xe^{-\frac{1}{2}\alpha\sqrt{\log x}}).$$

The Prime Number Theorem is a very weak consequence of Theorem 7–16, since

$$\int_2^x \frac{du}{\log u} = \frac{u}{\log u}\Big]_2^x + \int_2^x \frac{du}{\log^2 u} \sim \frac{x}{\log x}$$

and

$$xe^{-c\sqrt{\log x}} = o\left(\frac{x}{\log x}\right)$$

for every $c > 0$. In fact, we see by repeated integration by parts that the relation

$$\pi(x) = \frac{x}{\log x} + \frac{2!x}{\log^2 x} + \frac{3!x}{\log^3 x} + \cdots + \frac{m!x}{\log^m x} + o\left(\frac{x}{\log^m x}\right)$$

holds for every positive integer m.

The coefficient α occurring in the remainder term in Theorem 7–16 can easily be bounded explicitly; it can be shown for example that $\alpha = \frac{1}{45}$ is an allowable value, by choosing $c_2 = \frac{1}{1000}$, $c_4 = \frac{1}{200}$, $c_8 = \frac{1}{1002}$, $\epsilon_1 = \frac{1}{10}$, $\epsilon_2 = \frac{3}{100}$. However, no result of this type is as good as the known result that

$$\pi(x) = \int_2^x \frac{du}{\log u} + O(xe^{-c\sqrt{\log x \log\log x}}).$$

In a variant of the proof given here, the factor $\log \zeta(s)$ in the integrand is replaced by $\zeta'(s)/\zeta(s)$. The logarithmic singularity at $s = 1$ is then replaced by a simple pole, which makes the analysis somewhat less complicated. On the other hand this gives an estimate not of

$$\pi(x) = \sum_{p \le x} 1,$$

but of

$$\psi(x) = \sum_{p \le x} \log p \left\lfloor \frac{\log x}{\log p} \right\rfloor,$$

and an additional step is needed to obtain the final result.

7–4 Extension to primes in an arithmetic progression. For relatively prime integers k and l, let $\pi(x; k, l)$ be the number of primes $p \equiv l \pmod{k}$ which do not exceed x. For given k, there are $\varphi(k) = h$ choices of l which are distinct modulo k, so that if the primes are more or less evenly dispersed among the various progressions, it is to be expected that

$$\pi(x; k, l) \sim \frac{1}{h} \frac{x}{\log x} .$$

It is the object of this section to show that this is the case, and in fact to obtain an estimate for $\pi(x; k, l)$ similar to that given in Theorem 7–16 for $\pi(x) = \pi(x; 1, 0)$. Several proofs will not be given in full detail since they are similar to those of the preceding sections. As in the preceding chapter, we isolate the primes lying in a given arithmetic progression by use of characters and L-functions. The L-functions are in turn simple combinations of Hurwitz ζ-functions, as the following theorem shows.

THEOREM 7–17.   *For $\sigma > 1$,*

$$L(s, \chi) = \frac{1}{k^s} \sum_{a=1}^{k} \chi(a) \zeta\left(s, \frac{a}{k}\right) . \tag{14}$$

Proof: Since χ is periodic of period k,

$$L(s, \chi) = \sum_{n=1}^{\infty} \frac{\chi(n)}{n^s}$$

$$= \sum_{a=1}^{k} \chi(a) \sum_{m=0}^{\infty} \frac{1}{(km + a)^s}$$

$$= \frac{1}{k^s} \sum_{a=1}^{k} \chi(a) \zeta\left(s, \frac{a}{k}\right) .$$

The domain of validity of (14) can be extended somewhat. If we put

$$E(\chi) = \begin{cases} 1 & \text{if } \chi = \chi_0, \\ 0 & \text{if } \chi \neq \chi_0, \end{cases}$$

then the first equation of Theorem 6–6 becomes

$$\sum_{a=1}^{k} \chi(a) = E(\chi)h.$$

Hence, for $\sigma > 1$,

$$L(s, \chi) - \frac{E(\chi)h}{k} \cdot \frac{1}{s-1}$$

$$= \frac{E(\chi)h}{s-1}\left(\frac{1}{k^s} - \frac{1}{k}\right) + \frac{1}{k^s}\sum_{a=1}^{k}\chi(a)\left\{\zeta\left(s, \frac{a}{k}\right) - \frac{1}{s-1}\right\}.$$

By Theorem 7–3, each summand on the right is regular for $\sigma > 0$, and

$$\frac{1}{s-1}\left(\frac{1}{k^s} - \frac{1}{k}\right) = \frac{k^{1-s} - 1}{k(s-1)}$$

is an integral function. By analytic continuation, we have

THEOREM 7–18. *The relation* (14) *holds for* $\sigma > 0$, *except at* $s = 1$. *Moreover*,

$$\lim_{s \to 1} (s-1)L(s, \chi) = \frac{hE(\chi)}{k}. \tag{15}$$

Hence $L(s, \chi)$ *is regular for* $\sigma > 0$, *except that* $L(s, \chi_0)$ *has a simple pole at* $s = 1$.

For $\sigma > 2$ and $t \geq 8$,

$$|L(s, \chi)| < \sum_{n=1}^{\infty}\frac{1}{n^2} < 2 < \begin{cases} t, \\ \log t, \end{cases}$$

while for $\sigma > 0$ and $t > 0$,

$$|L(s, \chi)| \leq \sum_{a=1}^{k}\left|\zeta\left(s, \frac{a}{k}\right)\right|,$$

so that Theorem 7–4 yields

THEOREM 7–19.   (a) *For* $\sigma \geq \frac{1}{2}$ *and* $t > c_{12}(k)$, *we have* $|L(s, \chi)| < t$.
   (b) *For* $t \geq 8$ *and* $\sigma > 1 - (\log t)^{-1}$, *we have* $|L(s, \chi)| < c_{13}(k) \log t$.

The proof of the nonvanishing of $\zeta(s)$ on $\sigma = 1$ can be generalized in a simple way.

THEOREM 7–20. $L(s, \chi)$ *does not vanish on the line* $\sigma = 1$.

Proof: For $\sigma > 1$,

$$L(s, \chi) = \prod_{p}(1 - \chi(p)p^{-s})^{-1},$$

so that we can choose

$$\log L(s, \chi) = \sum_{m,p} \frac{\chi(p^m)}{mp^{ms}}.$$

Hence

$$\log |L^3(\sigma, \chi_0)L^4(\sigma + ti, \chi)L(\sigma + 2ti, \chi^2)|$$
$$= 3 \log |L(\sigma, \chi_0)| + 4 \log |L(\sigma + ti, \chi)| + \log |L(\sigma + 2ti, \chi^2)|$$
$$= 3 \log L(\sigma, \chi_0) + 4 \operatorname{Re} \log L(\sigma + ti, \chi) + \operatorname{Re} \log L(\sigma + 2ti, \chi^2)$$
$$= \sum_{m,p} \left(\frac{3\chi_0(p^m)}{mp^{m\sigma}} + \operatorname{Re} \frac{4\chi(p^m)}{mp^{m(\sigma+ti)}} + \operatorname{Re} \frac{\chi^2(p^m)}{mp^{m(\sigma+2ti)}} \right)$$
$$= \sum_{\substack{m,p \\ p \nmid k}} \frac{3 + 4 \cos \left(\eta(p^m) - t \log p^m \right) + \cos 2\left(\eta(p^m) - t \log p^m \right)}{mp^{m\sigma}}$$
$$\geq 0,$$

where $\chi(p^m) = e^{i\eta(p^m)}$. Thus

$$((\sigma - 1)L(\sigma, \chi_0))^3 \left| \frac{L(\sigma + ti, \chi)}{\sigma - 1} \right|^4 |L(\sigma + 2ti, \chi^2)| \geq \frac{1}{\sigma - 1},$$

and the falsity of the theorem would contradict Theorem 7–18.

By now the proof of the following analog of Theorem 7–10 should be a simple exercise for the reader.

THEOREM 7–21. *For $\sigma > 1$,*

$$-3 \frac{L'}{L} (\sigma, \chi_0) - 4 \operatorname{Re} \frac{L'}{L} (\sigma + ti, \chi) - \operatorname{Re} \frac{L'}{L} (\sigma + 2ti, \chi^2) \geq 0.$$

Theorem 7–13 becomes

THEOREM 7–22. *There is a $c_1(k) \geq 8$ such that $L(s, \chi) \neq 0$ for $t > c_1(k)$ and $\sigma \geq 1 - c_2/\log t$.*

The only difference in the proofs is that now the first inequality of Theorem 7–8 is applied with $f(s) = L(s, \chi^2)$ and $s_0 = \sigma + 2ti$, while the second is applied with $f(s) = L(s, \chi)$ and $s_0 = \sigma + ti$. Also, the constants now depend on k. After these trivial modifications, the proofs are identical.

Similarly, replacing $\zeta(s)$ by $L(s, \chi)$ throughout, Theorem 7–14 becomes

THEOREM 7-23. *For* $t \geq c_9(k) > 8$ *and* $\sigma \geq 1 - c_8 (\log t)^{-1}$, $|\log L(s, \chi)| < \log^2 t$.

The constant $c_9(k)$ may be different from the c_9 of Theorem 7–14; the subscript is retained to facilitate reference to Fig. 7–3. In the same way, c_{11} becomes $c_{11}(k)$.

Instead of proceeding directly to the analog of Theorem 7–15, it is convenient to break the argument into two steps.

THEOREM 7-24. *For* $(k, l) = 1$, *we have*

$$\sum_{\substack{p \leq x \\ p \equiv l \,(\text{mod } k)}} \log \frac{x}{p} = \frac{1}{2\pi i h} \sum_{\chi} \frac{1}{\chi(l)} \int_{(2)} \frac{x^s}{s^2} \log L(s, \chi) \, ds + O(\sqrt{x} \log^2 x).$$

Proof: Using Theorem 7–9 and the series expansion for $\log L(s, \chi)$, we obtain

$$\frac{1}{2\pi i} \int_{(2)} \frac{x^s}{s^2} \log L(s, \chi) \, ds = \frac{1}{2\pi i} \int_{(2)} \frac{x^s}{s^2} \sum_{m,p} \frac{\chi(p^m)}{mp^{ms}} \, ds$$

$$= \frac{1}{2\pi i} \sum_{m,p} \frac{\chi(p^m)}{m} \int_{(2)} \frac{(x/p^m)^s}{s^2} \, ds$$

$$= \sum_{\substack{m,p \\ p^m \leq x}} \frac{\chi(p^m) \log (x/p^m)}{m}$$

$$= \sum_{\substack{p \leq x \\ p \nmid k}} \chi(p) \log \frac{x}{p} + \sum_{\substack{p^m \leq x \\ m \geq 2}} \frac{\chi(p^m) \log (x/p^m)}{m}$$

$$= \sum_{\substack{p \leq x \\ p \nmid k}} \chi(p) \log \frac{x}{p} + O(\sqrt{x} \log^2 x).$$

Multiplying by $1/\chi(l)$ and summing over all characters modulo k, we deduce with the help of Theorem 6–7 that

$$\sum_{\chi} \frac{1}{\chi(l)} \sum_{\substack{p \leq x \\ p \nmid k}} \chi(p) \log \frac{x}{p} = h \sum_{\substack{p \leq x \\ p \equiv l \,(\text{mod } k)}} \log \frac{x}{p}$$

$$= \frac{1}{2\pi i} \sum_{\chi} \frac{1}{\chi(l)} \int_{(2)} \frac{x^s}{s^2} \log L(s, \chi) \, ds + O(\sqrt{x} \log^2 x),$$

which is the theorem. (Here and throughout the remainder of this section, the implied constant in the O-symbol may depend on k.)

To estimate the integrals appearing in Theorem 7–24, we must distinguish two cases. First consider the case $\chi = \chi_0$. Every property of the integrand which was used in estimating

$$\int_{(2)} \frac{x^s}{s^2} \log \zeta(s) \, ds$$

carries over to the integrand of

$$\int_{(2)} \frac{x^s}{s^2} \log L(s, \chi_0) \, ds.$$

It follows that for suitable c with $0 < c < 1$,

$$\int_{(2)} \frac{x^s}{s^2} \log L(s, \chi_0) \, ds = 2\pi i \int_c^1 \frac{x^s}{s^2} \, ds + O(xe^{-\alpha\sqrt{\log x}}).$$

On the other hand, if $\chi \neq \chi_0$, then $L(s, \chi)$ has no pole at $s = 1$, but the other properties used earlier still obtain. Hence, if we do not cut the plane, but consider the line segments Γ_5 and $\bar{\Gamma}_5$ in Fig. 7–3 as a single segment Γ_8, and omit Γ_6, Γ_7, and $\bar{\Gamma}_6$, then the function

$$\psi(s, \chi) = \frac{x^s}{s^2} \log L(s, \chi)$$

is regular in the region bounded by Γ_1, Γ_2, Γ_3, Γ_4, Γ_8, $\bar{\Gamma}_4$, $\bar{\Gamma}_3$, $\bar{\Gamma}_2$, $\bar{\Gamma}_1$, so that

$$\int_{(2)} \psi(s, \chi) \, ds = \left(\int_{2-\infty i}^{2-ui} - \int_{\bar{\Gamma}_2 + \bar{\Gamma}_3 + \bar{\Gamma}_4 + \Gamma_8 + \Gamma_4 + \Gamma_3 + \Gamma_2} + \int_{2+ui}^{2+\infty i} \right) \psi(s, \chi) \, ds.$$

Moreover, the integral along each of these new arcs either tends to zero or is

$$O(xe^{-\alpha\sqrt{\log x}}).$$

It follows that

$$\sum_{\substack{p \leq x \\ p \equiv l \pmod k}} \log \frac{x}{p} = \frac{1}{h} \int_c^1 \frac{x^s \, ds}{s^2} + O(xe^{-\alpha\sqrt{\log x}}), \qquad (16)$$

which is the analog of Theorem 7–15. In exactly the same way as Theorem 7–16 was deduced from Theorem 7–15, equation (16) leads to the desired result:

THEOREM 7–25. *If k is a fixed integer and $(k, l) = 1$, then, as $x \to \infty$,*

$$\pi(x; k, l) = \frac{1}{\varphi(k)} \int_2^x \frac{du}{\log u} + O(xe^{-\frac{1}{2}\alpha\sqrt{\log x}}).$$

As consequences of Theorem 7–25, we have that

$$\pi(x; k, l) \sim \frac{1}{\varphi(k)} \frac{x}{\log x},$$

and that, if $(k, l_1) = (k, l_2) = 1$, then

$$\lim_{x \to \infty} \frac{\pi(x; k, l_1)}{\pi(x; k, l_2)} = 1,$$

so that asymptotically there are equally many primes in the progressions $kt + l_1$ and $kt + l_2$.

A serious drawback of Theorem 7–25 is that the error term is not uniform in k. This precludes applying this version of the theorem to problems in which k increases with x, and these unfortunately are among the most important applications of this kind of theorem. It is known that the error term in Theorem 7–25 is uniform in k for $k < \log^m x$ for some $m > 0$, in other words, that the relation displayed in the theorem can be used if k increases sufficiently slowly with x. The proof of the more general theorem, while similar to that given here, is more complicated. The chief difficulty is this: when dealing with fixed k, it is enough to prove that $L(s, \chi) \neq 0$ for $s = 1 + ti$ in order to deduce that for some $c_{11}(k)$, $L(s, \chi) \neq 0$ for $1 - c_{11} < \sigma \leq 1$, $|t| < c_9(k)$. When k increases, however, c_{11} might tend to zero quite rapidly as a function of k, in which case the integral along Γ_8 would not be negligible. It is therefore necessary to investigate further the zeros of the L-functions near the line $\sigma = 1$ for small $|t|$.

7–5 The integers representable as a sum of two squares. As a final illustration of the methods of this chapter, we shall obtain an asymptotic estimate for $B(x)$, the number of integers not exceeding x which can be written as a sum of two squares. The integers counted are exactly those in whose prime-power factorization the primes $r \equiv 3 \pmod 4$ occur only to even powers.* The following heuristic argument indicates that it is to be expected that $B(x)$ is of the order of magnitude of $x/\sqrt{\log x}$, which is in agreement with the result to be obtained.

Take x very large. Since one out of every p integers is divisible by p, the number of integers up to x not divisible by p is about

---

* Cf. Volume I, Theorem 7–3.

$x(1 - 1/p)$. Hence the number not divisible by any $p \leq \sqrt{x}$ is roughly

$$x \prod_{p \leq \sqrt{x}} \left(1 - \frac{1}{p}\right),$$

so that, by the Prime Number Theorem,

$$x \prod_{p \leq \sqrt{x}} \left(1 - \frac{1}{p}\right) \approx \frac{x}{\log x},$$

where the symbol "$\approx$" means "is probably of the order of magnitude of." To count the integers contributing to $B(x)$, we do not want to eliminate all composite numbers, but only those divisible by an odd power of any of the various primes $r \equiv 3 \pmod 4$. As in the cross-classification principle,* we can omit all those divisible by r, then reintroduce those divisible by r^2, then take out those divisible by r^3, etc., giving

$$x \left(1 - \frac{1}{r}\right)\left(1 + \frac{1}{r^2}\right)\left(1 - \frac{1}{r^3}\right) \cdots$$

as the number left after accounting for the one prime r. (The product has only finitely many factors.) Hence

$$B(x) \approx x \prod_{r \leq \sqrt{x}} \left(1 - \frac{1}{r}\right) \prod_{r^2 \leq \sqrt{x}} \left(1 + \frac{1}{r^2}\right) \cdots,$$

and since each product after the first converges as $x \to \infty$, we can write simply

$$B(x) \approx x \prod_{r \leq \sqrt{x}} \left(1 - \frac{1}{r}\right).$$

Now

$$\log \prod_{p \leq \sqrt{x}} \left(1 - \frac{1}{p}\right) = \sum_{p \leq \sqrt{x}} \log \left(1 - \frac{1}{p}\right) \approx -\log \log x,$$

and since, by the results of the preceding section, about half the p's are r's, we have

$$\log \prod_{r \leq \sqrt{x}} \left(1 - \frac{1}{r}\right) \approx -\frac{1}{2} \log \log x = -\log \sqrt{\log x},$$

---

* Cf. Volume I, Theorem 6–4.

so that
$$B(x) \approx \frac{x}{\sqrt{\log x}} \cdot$$

Probably the most that can be said for this argument is that after seeing it, the reader should not be very surprised to learn that, for some $b > 0$,

$$B(x) \sim \frac{bx}{\sqrt{\log x}} \cdot \qquad (17)$$

Nevertheless, it is just this type of reasoning which underlies the proof of (17) which will now be developed.

If we put
$$b_n = \begin{cases} 1 & \text{if } n = x^2 + y^2 \text{ for some integers } x, y, \\ 0 & \text{otherwise,} \end{cases}$$
then
$$B(x) = \sum_{n \le x} b_n.$$

For $\sigma > 1$ let
$$f(s) = \sum_{n=1}^{\infty} \frac{b_n}{n^s} ;$$

the series converges absolutely in this domain, and uniformly in any closed bounded region to the right of the line $\sigma = 1$. Using q and r to denote primes congruent to 1 and 3 (mod 4) respectively, we deduce from the definition of b_n and Theorem 6–3 that, for $\sigma > 1$,

$$f(s) = \left(1 + \frac{1}{2^s} + \frac{1}{2^{2s}} + \cdots\right) \prod_q \left(1 + \frac{1}{q^s} + \frac{1}{q^{2s}} + \cdots\right)$$

$$\times \prod_r \left(1 + \frac{1}{r^{2s}} + \frac{1}{r^{4s}} + \cdots\right)$$

$$= (1 - 2^{-s})^{-1} \prod_q (1 - q^{-s})^{-1} \prod_r (1 - r^{-2s})^{-1}.$$

As was pointed out in formula (5) of the preceding chapter,

$$\zeta(s)L(s) = (1 - 2^{-s})^{-1} \prod_q (1 - q^{-s})^{-2} \prod_r (1 - r^{-2s})^{-1}, \quad (18)$$

where $L(s) = L(s, \chi)$ is the L-function for the nonprincipal character (mod 4),

$$\chi(n) = \begin{cases} 0 & \text{if } 2|n, \\ (-1)^{\frac{1}{2}(n-1)} & \text{if } 2 \nmid n. \end{cases}$$

(The relation (18) was proved earlier only for $s > 1$ and real; the extension to the half-plane $\sigma > 1$ is immediate.) Hence, for $\sigma > 1$,

$$f^2(s) = (1 - 2^{-s})^{-1} \prod_r (1 - r^{-2s})^{-1} \zeta(s) L(s). \qquad (19)$$

Since L is regular for $\sigma > 0$ and

$$L(1) = 1 - \frac{1}{3} + \frac{1}{5} - \cdots = \frac{\pi}{4},$$

the function ζL has a simple pole at $s = 1$, with residue $\pi/4$, but is otherwise regular for $\sigma > 0$. Moreover, neither $\zeta(s)$ nor $L(s)$ is zero for s in the region Q of Fig. 7–3, for suitable positive c_{11} and c_9. Since the functions

$$(1 - 2^{-s})^{-1} \quad \text{and} \quad \prod_r (1 - r^{-2s})^{-1} \qquad (20)$$

are regular and different from zero for $\sigma > \frac{1}{2}$, and bounded in absolute value for $\sigma \geq \sigma_0 > \frac{1}{2}$, we deduce the following properties of f from known properties of ζ and L.

THEOREM 7–26.   (a) $f^2(s)$ *is regular and different from zero in the region Q of Fig. 7–3, for suitable c_{11} and c_9, and it has a simple pole at $s = 1$, with residue*

$$\frac{\pi}{2} \prod_r (1 - r^{-2})^{-1} = b^2. \qquad (21)$$

Hence f is also regular in Q, and $f^2(s) \cdot (s - 1)$ is regular in the uncut region Q' formed from Q by omitting Γ_6, Γ_7, and $\bar{\Gamma}_6$, and joining Γ_5 and $\bar{\Gamma}_5$.

(b) *For $|t| \geq 8$ and s in Q, the inequality $|f(s)| < c_{14} \log |t|$ holds (cf. Theorem 7–11 (b)).*

From this follows the usual consequence.

THEOREM 7–27.   *For suitable $c < 1$,*

$$\sum_{n \leq x} b_n \log \frac{x}{n} = \frac{1}{\pi i} \int_c^1 \frac{x^s}{s^2} f(s) \, ds + O(xe^{-\alpha\sqrt{\log x}}).$$

Proof: The proof follows the lines of that of Theorem 7–15 as regards changing the path of integration in the relation

$$\sum_{n \leq x} b_n \log \frac{x}{n} = \frac{1}{2\pi i} \int_{(2)} \frac{x^s}{s^2} f(s) \, ds$$

and estimating the new integrals along those paths which are bounded away from $s = 1$; the only change is that the estimate $|f(s)| < c_{14} \log |t|$ is used rather than $|\log \zeta(s)| < \log^2 |t|$. Omitting the tiresome details, we arrive at the relation

$$\sum_{n=1}^{x} b_n \log \frac{x}{n} = \frac{1}{2\pi i} \left(-\int_{\bar{\Gamma}_6} - \int_{\Gamma_7} - \int_{\Gamma_6} \right) \frac{x^s}{s^2} f(s) \, ds + O(xe^{-\alpha\sqrt{\log x}}).$$

In the neighborhood of $s = 1, f(s)$ has the expansion

$$f(s) = \frac{b}{\sqrt{s-1}} + \cdots,$$

with b as in (21). Here $\sqrt{s-1} > 0$ for $s > 1$. Putting $s = 1 + \delta e^{i\theta}$, we have

$$\int_{\Gamma_7} \frac{x^s}{s^2} f(s) \, ds = O\left(\frac{x^{1+\delta}}{(1-\delta)^2} \cdot \frac{1}{\sqrt{\delta}} \cdot 2\pi\delta \right) = o(1)$$

as $\delta \to 0$.

Since $f^2(s)(s-1)$ is single-valued in Q', the quantity $2 \arg f(s) + \arg(s-1)$ is unchanged by traversing a path in Q from Γ_6 to $\bar{\Gamma}_6$. Since $\arg(s-1)$ increases by 2π, $\arg f(s)$ decreases by π, so that $f(s)$ has opposite signs on the two edges of the cut. Hence

$$\int_{\Gamma_6} \frac{x^s}{s^2} f(s) \, ds + \int_{\bar{\Gamma}_6} \frac{x^s}{s^2} f(s) \, ds = 2\int_{\bar{\Gamma}_6} \frac{x^s}{s^2} f(s) \, ds,$$

and $$\sum_{n=1}^{x} b_n \log \frac{x}{n} = \frac{1}{\pi i} \int_{1-c_{11}}^{1} \frac{x^s}{s^2} f(s) \, ds + O(xe^{-\alpha\sqrt{\log x}}).$$

The proof is complete.

THEOREM 7–28.   *As* $x \to \infty$,

$$B(x) = \frac{Bx}{\sqrt{\log x}} + O\left(\frac{x}{(\log x)^{\frac{3}{4}}} \right),$$

where $$B = \frac{1}{\sqrt{2}} \prod_r \left(1 - \frac{1}{r^2} \right)^{-\frac{1}{2}}.$$

Proof: On $\bar{\Gamma}_6$ we have

$$\frac{f(s)}{s^2} = \frac{bi}{\sqrt{1-s}\,(1-(1-s))^2} + O(\sqrt{1-s})$$

$$= \frac{bi}{\sqrt{1-s}} + O(\sqrt{1-s})$$

as $s \to 1^-$, so that

$$\sum_{n=1}^{x} b_n \log \frac{x}{n}$$

$$= \frac{1}{\pi i} \int_{1-c_{11}}^{1} \frac{x^s b i}{\sqrt{1-s}}\, ds + O\left(\int_{1-c_{11}}^{1} x^s \sqrt{1-s}\, ds\right) + O\left(\frac{x}{\log^2 x}\right)$$

$$= \frac{b}{\pi} \int_{0}^{c_{11}} x^{1-u} u^{-\frac{1}{2}}\, du + O\left(\int_{0}^{c_{11}} x^{1-u} u^{\frac{1}{2}}\, du\right) + O\left(\frac{x}{\log^2 x}\right)$$

$$= \frac{bx}{\pi} \int_{0}^{c_{11}} e^{-u \log x} u^{-\frac{1}{2}}\, du + O\left(x \int_{0}^{c_{11}} e^{-u \log x} u^{\frac{1}{2}}\, du\right) + O\left(\frac{x}{\log^2 x}\right)$$

$$= \frac{bx}{\pi} \int_{0}^{c_{11} \log x} e^{-v} \left(\frac{v}{\log x}\right)^{-\frac{1}{2}} \frac{dv}{\log x} + O\left(x \int_{0}^{c_{11} \log x} e^{-v} \left(\frac{v}{\log x}\right)^{\frac{1}{2}} \frac{dv}{\log x}\right)$$
$$+ O\left(\frac{x}{\log^2 x}\right)$$

$$= \frac{bx}{\pi \sqrt{\log x}} \int_{0}^{c_{11} \log x} e^{-v} v^{-\frac{1}{2}}\, dv + O\left(\frac{x}{\log^{\frac{3}{2}} x} \int_{0}^{\infty} e^{-v} v^{\frac{1}{2}}\, dv\right) + O\left(\frac{x}{\log^2 x}\right)$$

$$= \frac{bx}{\pi \sqrt{\log x}} \left(\Gamma\left(\frac{1}{2}\right) - \int_{c_{11} \log x}^{\infty} e^{-v} v^{-\frac{1}{2}}\, dv\right) + O\left(\frac{x}{\log^{\frac{3}{2}} x}\right)$$

$$= \frac{bx}{\pi \sqrt{\log x}} \left\{\sqrt{\pi} + O\left(\int_{c_{11} \log x}^{\infty} e^{-v}\, dv\right)\right\} + O\left(\frac{x}{\log^{\frac{3}{2}} x}\right).$$

Thus
$$\sum_{n=1}^{x} b_n \log \frac{x}{n} = \frac{Bx}{\sqrt{\log x}} + O\left(\frac{x}{\log^{\frac{3}{2}} x}\right),$$

where
$$B = \frac{b}{\sqrt{\pi}} = \frac{1}{\sqrt{2}} \prod_{r} (1 - r^2)^{-\frac{1}{2}}.$$

Now let $\delta = \delta(x)$ be positive. Then

$$\sum_{n=1}^{x+\delta x} b_n \log \frac{x + \delta x}{n} - \sum_{n=1}^{x} b_n \log \frac{x}{n}$$

$$= \log (1 + \delta) \sum_{n=1}^{x} b_n + \sum_{n=x}^{x+\delta x} b_n \log \frac{x + \delta x}{n}$$

$$= \log (1 + \delta) B(x) + O\left(\log (1 + \delta) \cdot \delta x\right),$$

while

$$\frac{Bx\,(1+\delta)}{\sqrt{\log\,(x+\delta x)}} - \frac{Bx}{\sqrt{\log x}} = \frac{Bx}{\sqrt{\log x}}\left(\frac{1+\delta}{\sqrt{\log\,(x+\delta x)/\log x}} - 1\right)$$

$$= \frac{Bx}{\sqrt{\log x}}\left(\frac{1+\delta}{\sqrt{1+\log\,(1+\delta)/\log x}} - 1\right)$$

$$= \frac{Bx}{\sqrt{\log x}}\left(\frac{1+\delta}{1+O(\delta/\log x)} - 1\right)$$

$$= \frac{Bx}{\sqrt{\log x}}\left(\frac{\delta + O(\delta/\log x)}{1+O(\delta/\log x)}\right)$$

$$= \frac{Bx}{\sqrt{\log x}}\,(\delta + O(\delta/\log x)).$$

Hence, since $\log\,(1+\delta) = \delta + O(\delta^2)$ as $\delta \to 0$,

$$B(x) = \frac{Bx}{\sqrt{\log x}}\left\{\frac{\delta}{\log\,(1+\delta)} + O\left(\frac{1}{\log x}\right)\right\} + O(\delta x) + O\left(\frac{x}{\delta \log^{\frac{3}{2}} x}\right)$$

$$= \frac{Bx}{\sqrt{\log x}}\,(1+O(\delta)) + O\left(\frac{x}{\log^{\frac{3}{2}} x}\right) + O(\delta x) + O\left(\frac{x}{\delta \log^{\frac{3}{2}} x}\right).$$

Choosing $\delta(x) = \log^{-\frac{3}{4}} x$, we obtain

$$B(x) = \frac{Bx}{\sqrt{\log x}} + O\left(\frac{x}{\log^{\frac{5}{4}} x}\right) + O\left(\frac{x}{\log^{\frac{3}{2}} x}\right)$$

$$+ O\left(\frac{x}{\log^{\frac{3}{4}} x}\right) + O\left(\frac{x}{\log^{\frac{3}{4}} x}\right)$$

$$= \frac{Bx}{\sqrt{\log x}} + O\left(\frac{x}{\log^{\frac{3}{4}} x}\right),$$

and the proof is complete.

SUPPLEMENTARY READING

Chapter 1

DICKSON, L. E., *Introduction to the Theory of Numbers*, Chicago: University of Chicago Press, 1929.

DICKSON, L. E., *Modern Elementary Theory of Numbers*, Chicago, University of Chicago Press, 1939.

FORD, L. R., *An Introduction to the Theory of Automorphic Functions*, London: George Bell & Sons, Ltd., 1915. Reprinted, Chelsea Publishing Company, New York, 1951.

JONES, B. W., *The Arithmetic Theory of Quadratic Forms*, Carus Mathematical Monograph #10, Buffalo, N.Y.: Mathematical Association of America, 1950. Distributed by John Wiley & Sons, Inc., New York.

KLEIN, F., *Vorlesungen über die Theorie der Elliptischen Modulfunktionen*, Leipzig: Teubner Verlagsgesellschaft, 1890–1892.

Chapter 2

HECKE, E., *Vorlesungen über die Theorie der Algebraischen Zahlen*, Leipzig: Akademische Verlagsgesellschaft m.b.H., 1923. Reprinted, Chelsea Publishing Company, New York, 1948.

LANDAU, E., *Vorlesungen über Zahlentheorie*, vol. 3, Leipzig: S. Hirzel Verlag, 1927. Reprinted, Chelsea Publishing Company, New York, 1947.

POLLARD, H., *The Theory of Algebraic Numbers*, Carus Mathematical Monograph #9, Buffalo, N.Y.: Mathematical Association of America, 1950. Distributed by John Wiley & Sons, Inc., New York.

REID, L. W., *Elements of the Theory of Algebraic Numbers*, New York: The Macmillan Company, 1910.

WEYL, H., *Algebraic Theory of Numbers*, Annals of Mathematics Studies, #1, Princeton: Princeton University Press, 1940.

Chapter 3

LANDAU, E., *Vorlesungen über Zahlentheorie*, vol. 3.

MORDELL, L. J., *Three Lectures on Fermat's Last Theorem*, New York: Cambridge University Press, 1921.

Chapter 5

GELFOND, A. O., *The Approximation of Algebraic Numbers by Algebraic Numbers and the Theory of Transcendental Numbers*, American Mathematical Society Translation #65, Providence: American Mathematical Society, 1952. Translated from *Uspekhi Matematicheskikh Nauk* (Moscow) **4**, no. 4 (32), 19–49 (1949).

KOKSMA, J. F., *Diophantische Approximationen*, Berlin: Springer-Verlag OHG, 1936. (Ergebnisse der Mathematik, vol. 4, no. 4.) Reprinted, Chelsea Publishing Company, New York, 1951.

PERRON, O., *Irrationalzahlen*, 2nd edition, Berlin: Walter De Gruyter & Co., 1929.

SIEGEL, C. L., *Transcendental Numbers*, Annals of Mathematics Studies, #16, Princeton: Princeton University Press, 1949.

Chapter 6

HASSE, H., *Vorlesungen über Zahlentheorie*, Berlin: Springer-Verlag OHG, 1950.

LANDAU, E., *Vorlesungen über Zahlentheorie*, vol. 1.

Chapter 7

ESTERMANN, T., *Introduction to Modern Prime Number Theory*, Cambridge Tracts, #41, New York: Cambridge University Press, 1952.

INGHAM, A. E., *The Distribution of Prime Numbers*, Cambridge Tracts, #30, New York: Cambridge University Press, 1932.

LANDAU, E., *Vorlesungen über Zahlentheorie*, vol. 2.

LANDAU, E., *Handbuch der Lehre von der Verteilung der Primzahlen*, Leipzig: Teubner Verlagsgesellschaft, 1909. Reprinted, Chelsea Publishing Company, New York, 1953.

LANDAU, E., *Einführung in die Elementare und Analytische Theorie der Algebraischen Zahlen und der Ideale*, Leipzig: Teubner Verlagsgesellschaft, 1918. Reprinted, Chelsea Publishing Company, New York, 1949.

LIST OF SYMBOLS

Γ, unimodular group, 8

$[a,\ b,\ c]$, quadratic form, 15

$\Gamma_A(f)$, group of automorphs, 25

R, rational field, 38

$R[x]$, polynomials over R, 38

deg p, degree of a polynomial, 38

$R(\vartheta)$, algebraic number field, 42

Z, rational integers, 48

$R[\vartheta]$, integral domain, 48

$\mathbf{N}$, norm, 48, 68

$|$, divides, 53, 65

$\mathbf{S}$, trace, 73

$\sqcap$, 75, 124

$\sim$, equivalent ideals, 82

K_p, cyclotomic field, 85

π, prime in K_p, 85

$\|$, exactly divides, 111

$\|\ \|$, 124

H, height of polynomial, 124

$M(\xi)$, Markov's constant, 166

$\zeta(s)$, Riemann's function, 201

$M(k)$, group of residues prime to k, 207

h, $\varphi(k)$, 207

$\chi(a)$, character, 210

$X(k)$, group of characters, 211

$L(s,\ \chi)$, Dirichlet's function, 214

$\zeta(s,\ w)$, Hurwitz' function, 232

Q, region of integration, 247

$\pi(x;\ k,\ l)$ number of primes $p \equiv l \pmod{k}$ with $p \leq x$, 252.

INDEX

Errata in **Topics in Number Theory**, Vol. II

page	line	replace	with
vii	-11	y^2	y^3
11	4	$\lvert \gamma z - \alpha \rvert$	$\lvert \gamma z' - \alpha \rvert$
16	-3	$-b/a$	$-b/2a$
17	-4	$\begin{pmatrix} 1 & 0 \\ n_1 & 1 \end{pmatrix}$	$\begin{pmatrix} 1 & 0 \\ -n_1 & 1 \end{pmatrix}$
23	2	$\dfrac{a}{t}$	$\dfrac{a}{2t}$
23	2	at	$\dfrac{at}{2}$
25	-3	$\frac{1}{2}(t_0 + u_0\sqrt{D})^n$	$\left(\dfrac{t_0 + u_0\sqrt{D}}{2}\right)^n$
27	-5	$z = x + iy$	$z' = x + iy$
33	$-1, -2$	(cf. Problem 3, Section 1-10).	*delete this*
35	6	x_n is	x_n, and of degree g in x_1, is
35	9	$P(x_1, \ldots, x_n)$. In	$P(x_1, \ldots, x_n)$. It is of total degree g in the former. In
35	12	$n,$	n, and leading coefficient 1,
36	-3	every polynomial	every nonconstant polynomial
44	8,9,11,13	φ	ψ
45	17	ζi	ζ_i
54	-14	$2b \equiv 1 \pmod 4$	$2b \equiv 1 \pmod 2$

271

69	-10	are distinct ideals	are ideals
83	-7	a field basis	an integral basis
87	2	$1 \le r \le p-1$	$1 \le r \le p-1,\ r \ne s$
125	6	a $(t+1)$th	a primitive $(t+1)$th
129	$-11, -14$	dependent	dependent over K
151	-1	$mr_1(1+\delta)$	mr_1
165	7	3–19	4–1
168	11,15	$\displaystyle\prod_{j=2}^{n}$	$\displaystyle\prod_{j=2}^{g}$
169	7	$[2\omega + 2]$	$[2\omega + 1]$
206	-10	(6)	(5)
223	-7	6–17	6–16
228	3	converges to	converges uniformly for $\sigma > \sigma_1 > \sigma_0$, and hence to
236	8	principle.	principle applied to $e^{f(z)}$.
238	-5	*the line*	$\displaystyle\lim_{a\to\infty}\int_{\sigma_0-ai}^{\sigma_0+ai} f(s)\,ds =$ $\displaystyle\lim_{a\to 0}\int_{-a}^{a} f(\sigma_0 + ti)i\,dt$
242	-1	*add new line*	Clearly we may suppose $c_2 < 1$. Suppose that $c_4 > 0$.
243	2	*add at end of line*	and hence for $x > e^{4(1+c_4)}$.
243	-9	$e^{4(1+c_2+c_4)}$	$e^{4(1+c_4)}$

243	-4	$1 + \dfrac{1}{\sigma_0^{-1}} < \dfrac{2}{\sigma_0^{-1}}$	$1 + \dfrac{1}{\sigma_0 - 1} < \dfrac{2}{\sigma_0 - 1}$
244	-3	$\max(c_5, c_6).$	$\max\left(c_5, c_6, e^{4(1+c_4)}\right).$
245	9	the region (12),	the region $|s - s_0| < 1 + \dfrac{c_8}{\log t_0},$
245	10	$c_8 - c_2$	$c_2 - c_8$
250	-1	$+ O(\delta^2 x).$	$+ O(x^C) + O(\delta^2 x).$

A CATALOG OF SELECTED
DOVER BOOKS
IN SCIENCE AND MATHEMATICS

A CATALOG OF SELECTED

DOVER BOOKS

IN SCIENCE AND MATHEMATICS

Astronomy

BURNHAM'S CELESTIAL HANDBOOK, Robert Burnham, Jr. Thorough guide to the stars beyond our solar system. Exhaustive treatment. Alphabetical by constellation: Andromeda to Cetus in Vol. 1; Chamaeleon to Orion in Vol. 2; and Pavo to Vulpecula in Vol. 3. Hundreds of illustrations. Index in Vol. 3. 2,000pp. 6⅛ x 9¼.
23567-X, 23568-8, 23673-0 Three-vol. set

THE EXTRATERRESTRIAL LIFE DEBATE, 1750–1900, Michael J. Crowe. First detailed, scholarly study in English of the many ideas that developed from 1750 to 1900 regarding the existence of intelligent extraterrestrial life. Examines ideas of Kant, Herschel, Voltaire, Percival Lowell, many other scientists and thinkers. 16 illustrations. 704pp. 5⅜ x 8½. 40675-X

A HISTORY OF ASTRONOMY, A. Pannekoek. Well-balanced, carefully reasoned study covers such topics as Ptolemaic theory, work of Copernicus, Kepler, Newton, Eddington's work on stars, much more. Illustrated. References. 521pp. 5⅜ x 8½.
65994-1

AMATEUR ASTRONOMER'S HANDBOOK, J. B. Sidgwick. Timeless, comprehensive coverage of telescopes, mirrors, lenses, mountings, telescope drives, micrometers, spectroscopes, more. 189 illustrations. 576pp. 5⅝ x 8¼. (Available in U.S. only.)
24034-7

STARS AND RELATIVITY, Ya. B. Zel'dovich and I. D. Novikov. Vol. 1 of *Relativistic Astrophysics* by famed Russian scientists. General relativity, properties of matter under astrophysical conditions, stars, and stellar systems. Deep physical insights, clear presentation. 1971 edition. References. 544pp. 5⅝ x 8¼. 69424-0

Chemistry

CHEMICAL MAGIC, Leonard A. Ford. Second Edition, Revised by E. Winston Grundmeier. Over 100 unusual stunts demonstrating cold fire, dust explosions, much more. Text explains scientific principles and stresses safety precautions. 128pp. 5⅜ x 8½. 67628-5

THE DEVELOPMENT OF MODERN CHEMISTRY, Aaron J. Ihde. Authoritative history of chemistry from ancient Greek theory to 20th-century innovation. Covers major chemists and their discoveries. 209 illustrations. 14 tables. Bibliographies. Indices. Appendices. 851pp. 5⅜ x 8½. 64235-6

CATALYSIS IN CHEMISTRY AND ENZYMOLOGY, William P. Jencks. Exceptionally clear coverage of mechanisms for catalysis, forces in aqueous solution, carbonyl- and acyl-group reactions, practical kinetics, more. 864pp. 5⅜ x 8½.
65460-5

THE HISTORICAL BACKGROUND OF CHEMISTRY, Henry M. Leicester. Evolution of ideas, not individual biography. Concentrates on formulation of a coherent set of chemical laws. 260pp. 5⅜ x 8½. 61053-5

A SHORT HISTORY OF CHEMISTRY, J. R. Partington. Classic exposition explores origins of chemistry, alchemy, early medical chemistry, nature of atmosphere, theory of valency, laws and structure of atomic theory, much more. 428pp. 5⅜ x 8½. (Available in U.S. only.) 65977-1

GENERAL CHEMISTRY, Linus Pauling. Revised 3rd edition of classic first-year text by Nobel laureate. Atomic and molecular structure, quantum mechanics, statistical mechanics, thermodynamics correlated with descriptive chemistry. Problems. 992pp. 5⅜ x 8½. 65622-5

Engineering

DE RE METALLICA, Georgius Agricola. The famous Hoover translation of greatest treatise on technological chemistry, engineering, geology, mining of early modern times (1556). All 289 original woodcuts. 638pp. 6¾ x 11. 60006-8

FUNDAMENTALS OF ASTRODYNAMICS, Roger Bate et al. Modern approach developed by U.S. Air Force Academy. Designed as a first course. Problems, exercises. Numerous illustrations. 455pp. 5⅜ x 8½. 60061-0

DYNAMICS OF FLUIDS IN POROUS MEDIA, Jacob Bear. For advanced students of ground water hydrology, soil mechanics and physics, drainage and irrigation engineering and more. 335 illustrations. Exercises, with answers. 784pp. 6⅛ x 9¼. 65675-6

ANALYTICAL MECHANICS OF GEARS, Earle Buckingham. Indispensable reference for modern gear manufacture covers conjugate gear-tooth action, gear-tooth profiles of various gears, many other topics. 263 figures. 102 tables. 546pp. 5⅜ x 8½. 65712-4

MECHANICS, J. P. Den Hartog. A classic introductory text or refresher. Hundreds of applications and design problems illuminate fundamentals of trusses, loaded beams and cables, etc. 334 answered problems. 462pp. 5⅜ x 8½. 60754-2

MECHANICAL VIBRATIONS, J. P. Den Hartog. Classic textbook offers lucid explanations and illustrative models, applying theories of vibrations to a variety of practical industrial engineering problems. Numerous figures. 233 problems, solutions. Appendix. Index. Preface. 436pp. 5⅜ x 8½. 64785-4

STRENGTH OF MATERIALS, J. P. Den Hartog. Full, clear treatment of basic material (tension, torsion, bending, etc.) plus advanced material on engineering methods, applications. 350 answered problems. 323pp. 5⅜ x 8½. 60755-0

A HISTORY OF MECHANICS, René Dugas. Monumental study of mechanical principles from antiquity to quantum mechanics. Contributions of ancient Greeks, Galileo, Leonardo, Kepler, Lagrange, many others. 671pp. 5⅜ x 8½. 65632-2

METAL FATIGUE, N. E. Frost, K. J. Marsh, and L. P. Pook. Definitive, clearly written, and well-illustrated volume addresses all aspects of the subject, from the historical development of understanding metal fatigue to vital concepts of the cyclic stress that causes a crack to grow. Includes 7 appendixes. 544pp. 5⅜ x 8½. 40927-9

STATISTICAL MECHANICS: Principles and Applications, Terrell L. Hill. Standard text covers fundamentals of statistical mechanics, applications to fluctuation theory, imperfect gases, distribution functions, more. 448pp. 5⅜ x 8½. 65390-0

THE VARIATIONAL PRINCIPLES OF MECHANICS, Cornelius Lanczos. Graduate level coverage of calculus of variations, equations of motion, relativistic mechanics, more. First inexpensive paperbound edition of classic treatise. Index. Bibliography. 418pp. 5⅜ x 8½. 65067-7

THE VARIOUS AND INGENIOUS MACHINES OF AGOSTINO RAMELLI: A Classic Sixteenth-Century Illustrated Treatise on Technology, Agostino Ramelli. One of the most widely known and copied works on machinery in the 16th century. 194 detailed plates of water pumps, grain mills, cranes, more. 608pp. 9 x 12. 28180-9

ORDINARY DIFFERENTIAL EQUATIONS AND STABILITY THEORY: An Introduction, David A. Sánchez. Brief, modern treatment. Linear equation, stability theory for autonomous and nonautonomous systems, etc. 164pp. 5⅜ x 8¼. 63828-6

ROTARY WING AERODYNAMICS, W. Z. Stepniewski. Clear, concise text covers aerodynamic phenomena of the rotor and offers guidelines for helicopter performance evaluation. Originally prepared for NASA. 537 figures. 640pp. 6⅛ x 9¼. 64647-5

INTRODUCTION TO SPACE DYNAMICS, William Tyrrell Thomson. Comprehensive, classic introduction to space-flight engineering for advanced undergraduate and graduate students. Includes vector algebra, kinematics, transformation of coordinates. Bibliography. Index. 352pp. 5⅜ x 8½. 65113-4

HISTORY OF STRENGTH OF MATERIALS, Stephen P. Timoshenko. Excellent historical survey of the strength of materials with many references to the theories of elasticity and structure. 245 figures. 452pp. 5⅜ x 8½. 61187-6

ANALYTICAL FRACTURE MECHANICS, David J. Unger. Self-contained text supplements standard fracture mechanics texts by focusing on analytical methods for determining crack-tip stress and strain fields. 336pp. 6⅛ x 9¼. 41737-9

Mathematics

HANDBOOK OF MATHEMATICAL FUNCTIONS WITH FORMULAS, GRAPHS, AND MATHEMATICAL TABLES, edited by Milton Abramowitz and Irene A. Stegun. Vast compendium: 29 sets of tables, some to as high as 20 places. 1,046pp. 8 x 10½. 61272-4

FUNCTIONAL ANALYSIS (Second Corrected Edition), George Bachman and Lawrence Narici. Excellent treatment of subject geared toward students with background in linear algebra, advanced calculus, physics and engineering. Text covers introduction to inner-product spaces, normed, metric spaces, and topological spaces; complete orthonormal sets, the Hahn-Banach Theorem and its consequences, and many other related subjects. 1966 ed. 544pp. 6⅛ x 9¼. 40251-7

ASYMPTOTIC EXPANSIONS OF INTEGRALS, Norman Bleistein & Richard A. Handelsman. Best introduction to important field with applications in a variety of scientific disciplines. New preface. Problems. Diagrams. Tables. Bibliography. Index. 448pp. 5⅜ x 8½. 65082-0

FAMOUS PROBLEMS OF GEOMETRY AND HOW TO SOLVE THEM, Benjamin Bold. Squaring the circle, trisecting the angle, duplicating the cube: learn their history, why they are impossible to solve, then solve them yourself. 128pp. 5⅜ x 8½. 24297-8

VECTOR AND TENSOR ANALYSIS WITH APPLICATIONS, A. I. Borisenko and I. E. Tarapov. Concise introduction. Worked-out problems, solutions, exercises. 257pp. 5⅝ x 8¼. 63833-2

THE ABSOLUTE DIFFERENTIAL CALCULUS (CALCULUS OF TENSORS), Tullio Levi-Civita. Great 20th-century mathematician's classic work on material necessary for mathematical grasp of theory of relativity. 452pp. 5⅝ x 8¼. 63401-9

AN INTRODUCTION TO ORDINARY DIFFERENTIAL EQUATIONS, Earl A. Coddington. A thorough and systematic first course in elementary differential equations for undergraduates in mathematics and science, with many exercises and problems (with answers). Index. 304pp. 5⅜ x 8½. 65942-9

FOURIER SERIES AND ORTHOGONAL FUNCTIONS, Harry F. Davis. An incisive text combining theory and practical example to introduce Fourier series, orthogonal functions and applications of the Fourier method to boundary-value problems. 570 exercises. Answers and notes. 416pp. 5⅜ x 8½. 65973-9

COMPUTABILITY AND UNSOLVABILITY, Martin Davis. Classic graduate-level introduction to theory of computability, usually referred to as theory of recurrent functions. New preface and appendix. 288pp. 5⅜ x 8½. 61471-9

ASYMPTOTIC METHODS IN ANALYSIS, N. G. de Bruijn. An inexpensive, comprehensive guide to asymptotic methods–the pioneering work that teaches by explaining worked examples in detail. Index. 224pp. 5⅜ x 8½ 64221-6

ESSAYS ON THE THEORY OF NUMBERS, Richard Dedekind. Two classic essays by great German mathematician: on the theory of irrational numbers; and on transfinite numbers and properties of natural numbers. 115pp. 5⅜ x 8½. 21010-3

APPLIED COMPLEX VARIABLES, John W. Dettman. Step-by-step coverage of fundamentals of analytic function theory–plus lucid exposition of five important applications: Potential Theory; Ordinary Differential Equations; Fourier Transforms; Laplace Transforms; Asymptotic Expansions. 66 figures. Exercises at chapter ends. 512pp. 5⅜ x 8½. 64670-X

INTRODUCTION TO LINEAR ALGEBRA AND DIFFERENTIAL EQUATIONS, John W. Dettman. Excellent text covers complex numbers, determinants, orthonormal bases, Laplace transforms, much more. Exercises with solutions. Undergraduate level. 416pp. 5⅜ x 8½. 65191-6

MATHEMATICAL METHODS IN PHYSICS AND ENGINEERING, John W. Dettman. Algebraically based approach to vectors, mapping, diffraction, other topics in applied math. Also generalized functions, analytic function theory, more. Exercises. 448pp. 5⅝ x 8¼. 65649-7

CALCULUS OF VARIATIONS WITH APPLICATIONS, George M. Ewing. Applications-oriented introduction to variational theory develops insight and promotes understanding of specialized books, research papers. Suitable for advanced undergraduate/graduate students as primary, supplementary text. 352pp. 5⅜ x 8½. 64856-7

COMPLEX VARIABLES, Francis J. Flanigan. Unusual approach, delaying complex algebra till harmonic functions have been analyzed from real variable viewpoint. Includes problems with answers. 364pp. 5⅜ x 8½. 61388-7

AN INTRODUCTION TO THE CALCULUS OF VARIATIONS, Charles Fox. Graduate-level text covers variations of an integral, isoperimetrical problems, least action, special relativity, approximations, more. References. 279pp. 5⅜ x 8½. 65499-0

CATASTROPHE THEORY FOR SCIENTISTS AND ENGINEERS, Robert Gilmore. Advanced-level treatment describes mathematics of theory grounded in the work of Poincaré, R. Thom, other mathematicians. Also important applications to problems in mathematics, physics, chemistry and engineering. 1981 edition. References. 28 tables. 397 black-and-white illustrations. xvii + 666pp. 6⅛ x 9¼. 67539-4

INTRODUCTION TO DIFFERENCE EQUATIONS, Samuel Goldberg. Exceptionally clear exposition of important discipline with applications to sociology, psychology, economics. Many illustrative examples; over 250 problems. 260pp. 5⅜ x 8½. 65084-7

NUMERICAL METHODS FOR SCIENTISTS AND ENGINEERS, Richard Hamming. Classic text stresses frequency approach in coverage of algorithms, polynomial approximation, Fourier approximation, exponential approximation, other topics. Revised and enlarged 2nd edition. 721pp. 5⅜ x 8½. 65241-6

INTRODUCTION TO NUMERICAL ANALYSIS (2nd Edition), F. B. Hildebrand. Classic, fundamental treatment covers computation, approximation, interpolation, numerical differentiation and integration, other topics. 150 new problems. 669pp. 5⅜ x 8½. 65363-3

THE FUNCTIONS OF MATHEMATICAL PHYSICS, Harry Hochstadt. Comprehensive treatment of orthogonal polynomials, hypergeometric functions, Hill's equation, much more. Bibliography. Index. 322pp. 5⅜ x 8½. 65214-9

THREE PEARLS OF NUMBER THEORY, A. Y. Khinchin. Three compelling puzzles require proof of a basic law governing the world of numbers. Challenges concern van der Waerden's theorem, the Landau-Schnirelmann hypothesis and Mann's theorem, and a solution to Waring's problem. Solutions included. 64pp. 5¾ x 8½. 40026-3

CALCULUS REFRESHER FOR TECHNICAL PEOPLE, A. Albert Klaf. Covers important aspects of integral and differential calculus via 756 questions. 566 problems, most answered. 431pp. 5⅜ x 8½. 20370-0

THE PHILOSOPHY OF MATHEMATICS: An Introductory Essay, Stephan Körner. Surveys the views of Plato, Aristotle, Leibniz & Kant concerning propositions and theories of applied and pure mathematics. Introduction. Two appendices. Index. 198pp. 5⅜ x 8½. 25048-2

INTRODUCTORY REAL ANALYSIS, A.N. Kolmogorov, S. V. Fomin. Translated by Richard A. Silverman. Self-contained, evenly paced introduction to real and functional analysis. Some 350 problems. 403pp. 5⅜ x 8½. 61226-0

APPLIED ANALYSIS, Cornelius Lanczos. Classic work on analysis and design of finite processes for approximating solution of analytical problems. Algebraic equations, matrices, harmonic analysis, quadrature methods, much more. 559pp. 5⅜ x 8½. 65656-X

AN INTRODUCTION TO ALGEBRAIC STRUCTURES, Joseph Landin. Superb self-contained text covers "abstract algebra": sets and numbers, theory of groups, theory of rings, much more. Numerous well-chosen examples, exercises. 247pp. 5⅜ x 8½. 65940-2

SPECIAL FUNCTIONS, N. N. Lebedev. Translated by Richard Silverman. Famous Russian work treating more important special functions, with applications to specific problems of physics and engineering. 38 figures. 308pp. 5⅜ x 8½. 60624-4

QUALITATIVE THEORY OF DIFFERENTIAL EQUATIONS, V. V. Nemytskii and V.V. Stepanov. Classic graduate-level text by two prominent Soviet mathematicians covers classical differential equations as well as topological dynamics and ergodic theory. Bibliographies. 523pp. 5⅜ x 8½. 65954-2

NUMBER THEORY AND ITS HISTORY, Oystein Ore. Unusually clear, accessible introduction covers counting, properties of numbers, prime numbers, much more. Bibliography. 380pp. 5⅜ x 8½. 65620-9

THEORY OF MATRICES, Sam Perlis. Outstanding text covering rank, nonsingularity and inverses in connection with the development of canonical matrices under the relation of equivalence, and without the intervention of determinants. Includes exercises. 237pp. 5⅜ x 8½. 66810-X

CATALOG OF DOVER BOOKS

INTRODUCTION TO ANALYSIS, Maxwell Rosenlicht. Unusually clear, accessible coverage of set theory, real number system, metric spaces, continuous functions, Riemann integration, multiple integrals, more. Wide range of problems. Undergraduate level. Bibliography. 254pp. 5⅜ x 8½. 65038-3

MODERN NONLINEAR EQUATIONS, Thomas L. Saaty. Emphasizes practical solution of problems; covers seven types of equations. ". . . a welcome contribution to the existing literature...."–*Math Reviews*. 490pp. 5⅜ x 8½. 64232-1

MATRICES AND LINEAR ALGEBRA, Hans Schneider and George Phillip Barker. Basic textbook covers theory of matrices and its applications to systems of linear equations and related topics such as determinants, eigenvalues and differential equations. Numerous exercises. 432pp. 5⅜ x 8½. 66014-1

MATHEMATICS APPLIED TO CONTINUUM MECHANICS, Lee A. Segel. Analyzes models of fluid flow and solid deformation. For upper-level math, science and engineering students. 608pp. 5⅜ x 8½. 65369-2

ELEMENTS OF REAL ANALYSIS, David A. Sprecher. Classic text covers fundamental concepts, real number system, point sets, functions of a real variable, Fourier series, much more. Over 500 exercises. 352pp. 5⅜ x 8½. 65385-4

AN INTRODUCTION TO MATRICES, SETS AND GROUPS FOR SCIENCE STUDENTS, G. Stephenson. Concise, readable text introduces sets, groups, and most importantly, matrices to undergraduate students of physics, chemistry, and engineering. Problems. 164pp. 5⅜ x 8½. 65077-4

SET THEORY AND LOGIC, Robert R. Stoll. Lucid introduction to unified theory of mathematical concepts. Set theory and logic seen as tools for conceptual understanding of real number system. 496pp. 5⅝ x 8¼. 63829-4

TENSOR CALCULUS, J.L. Synge and A. Schild. Widely used introductory text covers spaces and tensors, basic operations in Riemannian space, non-Riemannian spaces, etc. 324pp. 5⅝ x 8¼. 63612-7

ORDINARY DIFFERENTIAL EQUATIONS, Morris Tenenbaum and Harry Pollard. Exhaustive survey of ordinary differential equations for undergraduates in mathematics, engineering, science. Thorough analysis of theorems. Diagrams. Bibliography. Index. 818pp. 5⅜ x 8½. 64940-7

INTEGRAL EQUATIONS, F. G. Tricomi. Authoritative, well-written treatment of extremely useful mathematical tool with wide applications. Volterra Equations, Fredholm Equations, much more. Advanced undergraduate to graduate level. Exercises. Bibliography. 238pp. 5⅜ x 8½. 64828-1

FOURIER SERIES, Georgi P. Tolstov. Translated by Richard A. Silverman. A valuable addition to the literature on the subject, moving clearly from subject to subject and theorem to theorem. 107 problems, answers. 336pp. 5⅜ x 8½. 63317-9

POPULAR LECTURES ON MATHEMATICAL LOGIC, Hao Wang. Noted logi-
cian's lucid treatment of historical developments, set theory, model theory, recursion
theory and constructivism, proof theory, more. 3 appendixes. Bibliography. 1981 edition. ix + 283pp. 5⅜ x 8½.   67632-3

CALCULUS OF VARIATIONS, Robert Weinstock. Basic introduction covering
isoperimetric problems, theory of elasticity, quantum mechanics, electrostatics, etc.
Exercises throughout. 326pp. 5⅜ x 8½.   63069-2

THE CONTINUUM: A Critical Examination of the Foundation of Analysis,
Hermann Weyl. Classic of 20th-century foundational research deals with the conceptual problem posed by the continuum. 156pp. 5⅜ x 8½.   67982-9

CHALLENGING MATHEMATICAL PROBLEMS WITH ELEMENTARY
SOLUTIONS, A. M. Yaglom and I. M. Yaglom. Over 170 challenging problems on
probability theory, combinatorial analysis, points and lines, topology, convex poly-
gons, many other topics. Solutions. Total of 445pp. 5⅜ x 8½. Two-vol. set.
                                        Vol. I: 65536-9   Vol. II: 65537-7

A SURVEY OF NUMERICAL MATHEMATICS, David M. Young and Robert
Todd Gregory. Broad self-contained coverage of computer-oriented numerical algorithms for solving various types of mathematical problems in linear algebra, ordinary
and partial, differential equations, much more. Exercises. Total of 1,248pp. 5⅜ x 8½.
Two volumes.   Vol. I: 65691-8   Vol. II: 65692-6

INTRODUCTION TO PARTIAL DIFFERENTIAL EQUATIONS WITH
APPLICATIONS, E. C. Zachmanoglou and Dale W. Thoe. Essentials of partial differential equations applied to common problems in engineering and the physical sciences. Problems and answers. 416pp. 5⅜ x 8½.   65251-3

THE THEORY OF GROUPS, Hans J. Zassenhaus. Well-written graduate-level text
acquaints reader with group-theoretic methods and demonstrates their usefulness in
mathematics. Axioms, the calculus of complexes, homomorphic mapping, *p*-group
theory, more. Many proofs shorter and more transparent than older ones. 276pp.
5⅜ x 8½.   40922-8

DISTRIBUTION THEORY AND TRANSFORM ANALYSIS: An Introduction to
Generalized Functions, with Applications, A. H. Zemanian. Provides basics of distribution theory, describes generalized Fourier and Laplace transformations. Numerous
problems. 384pp. 5⅜ x 8½.   65479-6

Math–Decision Theory, Statistics, Probability

ELEMENTARY DECISION THEORY, Herman Chernoff and Lincoln E.
Moses. Clear introduction to statistics and statistical theory covers data process-
ing, probability and random variables, testing hypotheses, much more. Exercises.
364pp. 5⅜ x 8½.   65218-1

CATALOG OF DOVER BOOKS

STATISTICS MANUAL, Edwin L. Crow et al. Comprehensive, practical collection of classical and modern methods prepared by U.S. Naval Ordnance Test Station. Stress on use. Basics of statistics assumed. 288pp. 5⅜ x 8½. 60599-X

SOME THEORY OF SAMPLING, William Edwards Deming. Analysis of the problems, theory and design of sampling techniques for social scientists, industrial managers and others who find statistics important at work. 61 tables. 90 figures. xvii +602pp. 5⅜ x 8½. 64684-X

STATISTICAL ADJUSTMENT OF DATA, W. Edwards Deming. Introduction to basic concepts of statistics, curve fitting, least squares solution, conditions without parameter, conditions containing parameters. 26 exercises worked out. 271pp. 5⅜ x 8½. 64685-8

LINEAR PROGRAMMING AND ECONOMIC ANALYSIS, Robert Dorfman, Paul A. Samuelson and Robert M. Solow. First comprehensive treatment of linear programming in standard economic analysis. Game theory, modern welfare economics, Leontief input-output, more. 525pp. 5⅜ x 8½. 65491-5

DICTIONARY/OUTLINE OF BASIC STATISTICS, John E. Freund and Frank J. Williams. A clear concise dictionary of over 1,000 statistical terms and an outline of statistical formulas covering probability, nonparametric tests, much more. 208pp. 5⅜ x 8½. 66796-0

PROBABILITY: An Introduction, Samuel Goldberg. Excellent basic text covers set theory, probability theory for finite sample spaces, binomial theorem, much more. 360 problems. Bibliographies. 322pp. 5⅜ x 8½. 65252-1

GAMES AND DECISIONS: Introduction and Critical Survey, R. Duncan Luce and Howard Raiffa. Superb nontechnical introduction to game theory, primarily applied to social sciences. Utility theory, zero-sum games, n-person games, decision-making, much more. Bibliography. 509pp. 5⅜ x 8½. 65943-7

FIFTY CHALLENGING PROBLEMS IN PROBABILITY WITH SOLUTIONS, Frederick Mosteller. Remarkable puzzlers, graded in difficulty, illustrate elementary and advanced aspects of probability. Detailed solutions. 88pp. 5⅜ x 8½. 65355-2

PROBABILITY THEORY: A Concise Course, Y. A. Rozanov. Highly readable, self-contained introduction covers combination of events, dependent events, Bernoulli trials, etc. 148pp. 5⅜ x 8¼. 63544-9

STATISTICAL METHOD FROM THE VIEWPOINT OF QUALITY CONTROL, Walter A. Shewhart. Important text explains regulation of variables, uses of statistical control to achieve quality control in industry, agriculture, other areas. 192pp. 5⅜ x 8½. 65232-7

THE COMPLEAT STRATEGYST: Being a Primer on the Theory of Games of Strategy, J. D. Williams. Highly entertaining classic describes, with many illustrated examples, how to select best strategies in conflict situations. Prefaces. Appendices. 268pp. 5⅜ x 8½. 25101-2

Math–Geometry and Topology

ELEMENTARY CONCEPTS OF TOPOLOGY, Paul Alexandroff. Elegant, intuitive approach to topology from set-theoretic topology to Betti groups; how concepts of topology are useful in math and physics. 25 figures. 57pp. 5⅜ x 8½. 60747-X

COMBINATORIAL TOPOLOGY, P. S. Alexandrov. Clearly written, well-organized, three-part text begins by dealing with certain classic problems without using the formal techniques of homology theory and advances to the central concept, the Betti groups. Numerous detailed examples. 654pp. 5⅜ x 8½. 40179-0

EXPERIMENTS IN TOPOLOGY, Stephen Barr. Classic, lively explanation of one of the byways of mathematics. Klein bottles, Moebius strips, projective planes, map coloring, problem of the Koenigsberg bridges, much more, described with clarity and wit. 43 figures. 210pp. 5⅜ x 8½. 25933-1

CONFORMAL MAPPING ON RIEMANN SURFACES, Harvey Cohn. Lucid, insightful book presents ideal coverage of subject. 334 exercises make book perfect for self-study. 55 figures. 352pp. 5⅝ x 8¼. 64025-6

THE GEOMETRY OF RENÉ DESCARTES, René Descartes. The great work founded analytical geometry. Original French text, Descartes's own diagrams, together with definitive Smith-Latham translation. 244pp. 5⅜ x 8½. 60068-8

THE THIRTEEN BOOKS OF EUCLID'S ELEMENTS, translated with introduction and commentary by Sir Thomas L. Heath. Definitive edition. Textual and linguistic notes, mathematical analysis. 2,500 years of critical commentary. Unabridged. 1,414pp. 5⅜ x 8½. Three-vol. set.
Vol. I: 60088-2 Vol. II: 60089-0 Vol. III: 60090-4

GEOMETRY OF COMPLEX NUMBERS, Hans Schwerdtfeger. Illuminating, widely praised book on analytic geometry of circles, the Moebius transformation, and two-dimensional non-Euclidean geometries. 200pp. 5⅝ x 8¼. 63830-8

DIFFERENTIAL GEOMETRY, Heinrich W. Guggenheimer. Local differential geometry as an application of advanced calculus and linear algebra. Curvature, transformation groups, surfaces, more. Exercises. 62 figures. 378pp. 5⅜ x 8½. 63433-7

CURVATURE AND HOMOLOGY: Enlarged Edition, Samuel I. Goldberg. Revised edition examines topology of differentiable manifolds; curvature, homology of Riemannian manifolds; compact Lie groups; complex manifolds; curvature, homology of Kaehler manifolds. New Preface. Four new appendixes. 416pp. 5⅜ x 8½. 40207-X

TOPOLOGY, John G. Hocking and Gail S. Young. Superb one-year course in classical topology. Topological spaces and functions, point-set topology, much more. Examples and problems. Bibliography. Index. 384pp. 5⅝ x 8¼. 65676-4

LECTURES ON CLASSICAL DIFFERENTIAL GEOMETRY, Second Edition, Dirk J. Struik. Excellent brief introduction covers curves, theory of surfaces, fundamental equations, geometry on a surface, conformal mapping, other topics. Problems. 240pp. 5⅜ x 8½. 65609-8

Math–History of

A SHORT ACCOUNT OF THE HISTORY OF MATHEMATICS, W. W. Rouse Ball. One of clearest, most authoritative surveys from the Egyptians and Phoenicians through 19th-century figures such as Grassman, Galois, Riemann. Fourth edition. 522pp. 5⅜ x 8½. 20630-0

THE HISTORY OF THE CALCULUS AND ITS CONCEPTUAL DEVELOP-MENT, Carl B. Boyer. Origins in antiquity, medieval contributions, work of Newton, Leibniz, rigorous formulation. Treatment is verbal. 346pp. 5⅜ x 8½. 60509-4

THE HISTORICAL ROOTS OF ELEMENTARY MATHEMATICS, Lucas N. H. Bunt, Phillip S. Jones, and Jack D. Bedient. Fundamental underpinnings of modern arithmetic, algebra, geometry and number systems derived from ancient civilizations. 320pp. 5⅜ x 8½. 25563-8

A HISTORY OF MATHEMATICAL NOTATIONS, Florian Cajori. This classic study notes the first appearance of a mathematical symbol and its origin, the competition it encountered, its spread among writers in different countries, its rise to popularity, its eventual decline or ultimate survival. Original 1929 two-volume edition presented here in one volume. xxviii+820pp. 5⅜ x 8½. 67766-4

GAMES, GODS & GAMBLING: A History of Probability and Statistical Ideas, F. N. David. Episodes from the lives of Galileo, Fermat, Pascal, and others illustrate this fascinating account of the roots of mathematics. Features thought-provoking references to classics, archaeology, biography, poetry. 1962 edition. 304pp. 5⅜ x 8½. (Available in U.S. only.) 40023-9

OF MEN AND NUMBERS: The Story of the Great Mathematicians, Jane Muir. Fascinating accounts of the lives and accomplishments of history's greatest mathematical minds–Pythagoras, Descartes, Euler, Pascal, Cantor, many more. Anecdotal, illuminating. 30 diagrams. Bibliography. 256pp. 5⅜ x 8½. 28973-7

HISTORY OF MATHEMATICS, David E. Smith. Nontechnical survey from ancient Greece and Orient to late 19th century; evolution of arithmetic, geometry, trigonometry, calculating devices, algebra, the calculus. 362 illustrations. 1,355pp. 5⅜ x 8½. Two-vol. set. Vol. I: 20429-4 Vol. II: 20430-8

A CONCISE HISTORY OF MATHEMATICS, Dirk J. Struik. The best brief history of mathematics. Stresses origins and covers every major figure from ancient Near East to 19th century. 41 illustrations. 195pp. 5⅜ x 8½. 60255-9

Physics

OPTICAL RESONANCE AND TWO-LEVEL ATOMS, L. Allen and J. H. Eberly. Clear, comprehensive introduction to basic principles behind all quantum optical resonance phenomena. 53 illustrations. Preface. Index. 256pp. 5⅜ x 8½. 65533-4

ULTRASONIC ABSORPTION: An Introduction to the Theory of Sound Absorption and Dispersion in Gases, Liquids and Solids, A. B. Bhatia. Standard reference in the field provides a clear, systematically organized introductory review of fundamental concepts for advanced graduate students, research workers. Numerous diagrams. Bibliography. 440pp. 5⅜ x 8½. 64917-2

QUANTUM THEORY, David Bohm. This advanced undergraduate-level text presents the quantum theory in terms of qualitative and imaginative concepts, followed by specific applications worked out in mathematical detail. Preface. Index. 655pp. 5⅜ x 8½. 65969-0

ATOMIC PHYSICS (8th edition), Max Born. Nobel laureate's lucid treatment of kinetic theory of gases, elementary particles, nuclear atom, wave-corpuscles, atomic structure and spectral lines, much more. Over 40 appendices, bibliography. 495pp. 5⅜ x 8½. 65984-4

AN INTRODUCTION TO HAMILTONIAN OPTICS, H. A. Buchdahl. Detailed account of the Hamiltonian treatment of aberration theory in geometrical optics. Many classes of optical systems defined in terms of the symmetries they possess. Problems with detailed solutions. 1970 edition. xv + 360pp. 5⅜ x 8½. 67597-1

THIRTY YEARS THAT SHOOK PHYSICS: The Story of Quantum Theory, George Gamow. Lucid, accessible introduction to influential theory of energy and matter. Careful explanations of Dirac's anti-particles, Bohr's model of the atom, much more. 12 plates. Numerous drawings. 240pp. 5⅜ x 8½. 24895-X

ELECTRONIC STRUCTURE AND THE PROPERTIES OF SOLIDS: The Physics of the Chemical Bond, Walter A. Harrison. Innovative text offers basic understanding of the electronic structure of covalent and ionic solids, simple metals, transition metals and their compounds. Problems. 1980 edition. 582pp. 6⅛ x 9¼. 66021-4

HYDRODYNAMIC AND HYDROMAGNETIC STABILITY, S. Chandrasekhar. Lucid examination of the Rayleigh-Benard problem; clear coverage of the theory of instabilities causing convection. 704pp. 5⅜ x 8¼. 64071-X

INVESTIGATIONS ON THE THEORY OF THE BROWNIAN MOVEMENT, Albert Einstein. Five papers (1905–8) investigating dynamics of Brownian motion and evolving elementary theory. Notes by R. Fürth. 122pp. 5⅜ x 8½. 60304-0

THE PHYSICS OF WAVES, William C. Elmore and Mark A. Heald. Unique overview of classical wave theory. Acoustics, optics, electromagnetic radiation, more. Ideal as classroom text or for self-study. Problems. 477pp. 5⅜ x 8½. 64926-1

CATALOG OF DOVER BOOKS

PHYSICAL PRINCIPLES OF THE QUANTUM THEORY, Werner Heisenberg. Nobel Laureate discusses quantum theory, uncertainty, wave mechanics, work of Dirac, Schroedinger, Compton, Wilson, Einstein, etc. 184pp. 5⅜ x 8½. 60113-7

ATOMIC SPECTRA AND ATOMIC STRUCTURE, Gerhard Herzberg. One of best introductions; especially for specialist in other fields. Treatment is physical rather than mathematical. 80 illustrations. 257pp. 5⅜ x 8½. 60115-3

AN INTRODUCTION TO STATISTICAL THERMODYNAMICS, Terrell L. Hill. Excellent basic text offers wide-ranging coverage of quantum statistical mechanics, systems of interacting molecules, quantum statistics, more. 523pp. 5⅜ x 8½. 65242-4

THEORETICAL PHYSICS, Georg Joos, with Ira M. Freeman. Classic overview covers essential math, mechanics, electromagnetic theory, thermodynamics, quantum mechanics, nuclear physics, other topics. First paperback edition. xxiii + 885pp. 5⅜ x 8½. 65227-0

PROBLEMS AND SOLUTIONS IN QUANTUM CHEMISTRY AND PHYSICS, Charles S. Johnson, Jr. and Lee G. Pedersen. Unusually varied problems, detailed solutions in coverage of quantum mechanics, wave mechanics, angular momentum, molecular spectroscopy, more. 280 problems plus 139 supplementary exercises. 430pp. 6½ x 9¼. 65236-X

THEORETICAL SOLID STATE PHYSICS, Vol. 1: Perfect Lattices in Equilibrium; Vol. II: Non-Equilibrium and Disorder, William Jones and Norman H. March. Monumental reference work covers fundamental theory of equilibrium properties of perfect crystalline solids, non-equilibrium properties, defects and disordered systems. Appendices. Problems. Preface. Diagrams. Index. Bibliography. Total of 1,301pp. 5⅜ x 8½. Two volumes. Vol. I: 65015-4 Vol. II: 65016-2

A TREATISE ON ELECTRICITY AND MAGNETISM, James Clerk Maxwell. Important foundation work of modern physics. Brings to final form Maxwell's theory of electromagnetism and rigorously derives his general equations of field theory. 1,084pp. 5⅜ x 8½. Two-vol. set. Vol. I: 60636-8 Vol. II: 60637-6

OPTICKS, Sir Isaac Newton. Newton's own experiments with spectroscopy, colors, lenses, reflection, refraction, etc., in language the layman can follow. Foreword by Albert Einstein. 532pp. 5⅜ x 8½. 60205-2

THEORY OF ELECTROMAGNETIC WAVE PROPAGATION, Charles Herach Papas. Graduate-level study discusses the Maxwell field equations, radiation from wire antennas, the Doppler effect and more. xiii + 244pp. 5⅜ x 8½. 65678-5

INTRODUCTION TO QUANTUM MECHANICS With Applications to Chemistry, Linus Pauling & E. Bright Wilson, Jr. Classic undergraduate text by Nobel Prize winner applies quantum mechanics to chemical and physical problems. Numerous tables and figures enhance the text. Chapter bibliographies. Appendices. Index. 468pp. 5⅜ x 8½. 64871-0

CATALOG OF DOVER BOOKS

METHODS OF THERMODYNAMICS, Howard Reiss. Outstanding text focuses on physical technique of thermodynamics, typical problem areas of understanding, and significance and use of thermodynamic potential. 1965 edition. 238pp. 5⅜ x 8½. 69445-3

TENSOR ANALYSIS FOR PHYSICISTS, J. A. Schouten. Concise exposition of the mathematical basis of tensor analysis, integrated with well-chosen physical examples of the theory. Exercises. Index. Bibliography. 289pp. 5⅜ x 8½. 65582-2

RELATIVITY IN ILLUSTRATIONS, Jacob T. Schwartz. Clear nontechnical treatment makes relativity more accessible than ever before. Over 60 drawings illustrate concepts more clearly than text alone. Only high school geometry needed. Bibliography. 128pp. 6⅛ x 9¼. 25965-X

THE ELECTROMAGNETIC FIELD, Albert Shadowitz. Comprehensive undergraduate text covers basics of electric and magnetic fields, builds up to electromagnetic theory. Also related topics, including relativity. Over 900 problems. 768pp. 5⅝ x 8¼. 65660-8

GREAT EXPERIMENTS IN PHYSICS: Firsthand Accounts from Galileo to Einstein, edited by Morris H. Shamos. 25 crucial discoveries: Newton's laws of motion, Chadwick's study of the neutron, Hertz on electromagnetic waves, more. Original accounts clearly annotated. 370pp. 5⅜ x 8½. 25346-5

RELATIVITY, THERMODYNAMICS AND COSMOLOGY, Richard C. Tolman. Landmark study extends thermodynamics to special, general relativity; also applications of relativistic mechanics, thermodynamics to cosmological models. 501pp. 5⅜ x 8½. 65383-8

LIGHT SCATTERING BY SMALL PARTICLES, H. C. van de Hulst. Comprehensive treatment including full range of useful approximation methods for researchers in chemistry, meteorology and astronomy. 44 illustrations. 470pp. 5⅜ x 8½. 64228-3

STATISTICAL PHYSICS, Gregory H. Wannier. Classic text combines thermodynamics, statistical mechanics and kinetic theory in one unified presentation of thermal physics. Problems with solutions. Bibliography. 532pp. 5⅜ x 8½. 65401-X

Paperbound unless otherwise indicated. Available at your book dealer, online at **www.doverpublications.com**, or by writing to Dept. GI, Dover Publications, Inc., 31 East 2nd Street, Mineola, NY 11501. For current price information or for free catalogues (please indicate field of interest), write to Dover Publications or log on to **www.doverpublications.com** and see every Dover book in print. Dover publishes more than 500 books each year on science, elementary and advanced mathematics, biology, music, art, literary history, social sciences, and other areas.